Special Purpose Nozzles

Jingwei Shi • Li Zhou • Xiaobo Zhang
Zhanxue Wang

Special Purpose Nozzles

Design Methods and Applications

SCIENCE PRESS

Springer

Jingwei Shi
School of Power and Energy
Northwestern Polytechnical University
Xi'an, Shaanxi, China

Li Zhou
School of Power and Energy
Northwestern Polytechnical University
Xi'an, Shaanxi, China

Xiaobo Zhang
School of Power and Energy
Northwestern Polytechnical University
Xi'an, Shaanxi, China

Zhanxue Wang
School of Power and Energy
Northwestern Polytechnical University
Xi'an, Shaanxi, China

ISBN 978-981-95-2938-4 ISBN 978-981-95-2939-1 (eBook)
https://doi.org/10.1007/978-981-95-2939-1

Jointly published with Science Press

The print edition is not for sale in China mainland. Customers from China mainland please order the print book from: Science Press.

This Springer imprint is published by the registered company Springer Nature Singapore Pte Ltd.
The registered company address is: 152 Beach Road, #21-01/04 Gateway East, Singapore 189721, Singapore

Preface

This book focuses on special-configuration nozzles urgently needed for future fighter aircraft. It is compiled based on the research achievements of the Institute of Jet Propulsion Theory and Engineering at Northwestern Polytechnical University, while integrating the recent research progress and state-of-the-art technological capabilities domestically and internationally.

With the continuous improvement of cycle parameters in aero-engines and the growing demands for mission adaptability, conventional convergent, and convergent-divergent nozzles can no longer satisfy the requirements of mission adaptability or high cycle parameters. For example, to meet specific requirements such as super-stealth and high maneuverability in fourth-generation fighter aircraft, aero-engines equipped with conventional nozzles will be unable to fulfill mission demands, and nozzles such as serpentine nozzles and thrust vectoring nozzles must be adopted. For short take-off and landing (STOL) fighter aircraft requiring vertical take-off and vertical landing capabilities, three-bearing swivel nozzles with 90-degree thrust vectoring have become indispensable. For high-altitude and high-speed fighter aircraft, the nozzle pressure ratio increases significantly. From the perspective of reducing the structural weight of nozzles, single expansion ramp nozzles are needed. It is precisely due to the evolving requirements such as multi-mission adaptability in aero-engines that special-purpose nozzles are receiving increasing attention.

To provide a theoretical foundation and technical support for the application of special-purpose nozzles in aero-engines under various mission requirements in China, the authors have summarized foundational research achievements in the fields of serpentine nozzles, three-bearing swivel nozzles, and single expansion ramp nozzles both domestically and internationally, resulting in the book Design Methods and Applications of Special-Purpose Nozzles. This book aims to offer a valuable reference for researchers and engineers in relevant technical fields, helping readers gain a deeper understanding of the fundamental knowledge and theories related to special-purpose nozzles, and supporting the development of design and application studies.

This book consists of six chapters. Chapter 1, starting from the urgent demands of future fighter aircraft for special-purpose nozzles, analyzes the configurations of three types of nozzles required by different fighter aircraft—serpentine stealth nozzles, three-bearing swivel nozzles, and single expansion ramp nozzles—and provides geometric feature parameters and performance evaluation metrics applicable to these special-purpose nozzles. Chapters 2 and 3 introduce the design methods of serpentine stealth nozzles and the effects of key parameters on their flow and infrared radiation characteristics. Chapter 4 focuses on three-bearing swivel nozzles, introducing their design methods, flow characteristics, ground interference effects, and cold/hot model tests. Chapter 5 presents the design methods of single expansion ramp nozzles, as well as active/passive flow control methods to improve over-expansion performance of tandem and parallel single expansion ramp nozzles. Chapter 6 introduces the overall coupling analysis between serpentine stealth nozzles, three-bearing swivel nozzles, and single expansion ramp nozzles with the aero-engine.

This book is organized and strongly supported by Professor Wang Zhanxue of Northwestern Polytechnical University. Chapter 1 was authored by Wang Zhanxue; Chaps. 2 and 3 were authored by Zhou Li; Chaps. 4 and 5 were authored by Shi Jingwei and Chap. 6 by Zhang Xiaobo. The entire book was compiled by Shi Jingwei.

During the writing process, Dr. Cheng Wen, Dr. Liu Shuai, Dr. Sun Xiaolin, Master Xiao Hua, Master Qi Shaobo, and Master Liu Aihua of Northwestern Polytechnical University provided part of the writing materials. Dr. Sun Peng, Dr. Shi Jie, Dr. Meng Yubo, and Dr. Yang Yuming participated in the drafting of certain sections and the creation of illustrations and charts. Their contributions are gratefully acknowledged. In addition, the compilation of this book received strong support from the School of Power and Energy of Northwestern Polytechnical University and colleagues, to whom sincere thanks are also extended.

Due to the limited capabilities of the authors, flaws and deficiencies are inevitable in this book. Readers' criticism and corrections are sincerely welcomed. It is hoped that the publication of this book will contribute to advancing the application of special-purpose nozzles in meeting the multi-mission adaptability demands of aero-engines in China.

The Authors

Xi'an, Shaanxi, China

Jingwei Shi
Li Zhou
Xiaobo Zhang
Zhanxue Wang

About this Book

With the continuous improvement of cycle parameters in aero-engines and the increasing demands for mission adaptability, conventional convergent, and convergent-divergent nozzles can no longer meet the requirements of mission adaptability or high cycle parameters. From the perspective of future fighter technology development or multi-mission adaptability of aero-engines, this book systematically and comprehensively discusses three types of special-purpose nozzles urgently needed for future fighter aircraft—serpentine nozzles, three-bearing swivel nozzles, and single expansion ramp nozzles—focusing on the aerodynamic design, flow characteristics, infrared radiation characteristics, cold/hot state experiments, performance enhancement methods, and the overall coupling effects between special-purpose nozzles and the aero-engine. It provides relatively comprehensive fundamental theories and data support for researchers engaged in related studies.

This book can serve as a reference for engineers and technical personnel, teachers, and graduate students engaged in the design of special-configuration exhaust systems for aero-engines.

Contents

Chapter 1
Concept and Parameter Definition of Nozzles

1.1 Functions and Layout Forms of Nozzles

1.1.1 Functions of Nozzles

As a critical component of an aircraft engine, the nozzle's primary function is to expand and accelerate the high-temperature, high-pressure exhaust gases from the turbine and expel them from the airframe, thereby generating engine thrust. By adjusting the nozzle throat area, the distribution of gas expansion ratios between the turbine and the nozzle can be altered, changing the operating point of the compressor and turbine to control the engine's working state. This adjustment modifies the engine's thrust and fuel consumption rate, improves startup performance, and minimizes the impact on engine operation when engaging or disengaging afterburners (Lian 2005). With advancements in aviation technology and evolving air combat requirements, the functions of nozzles have expanded. For example, they now provide thrust vectoring, supplementing or replacing aerodynamic control surfaces during low-speed and high-angle-of-attack flight to enable post-stall maneuvers. This reduces the weight, drag, and radar cross-section of aerodynamic surfaces while shortening takeoff and landing distances. Additionally, nozzles can control infrared radiation signatures, radar cross-sections, and noise levels, enhancing the aircraft's infrared stealth, radar stealth, and acoustic stealth capabilities, thereby improving survivability (Gao et al. 1995). Therefore, the fundamental requirements for aircraft engine nozzles are: minimal internal losses and external drag; low infrared radiation levels and small effective scattering surfaces; thrust vector control during takeoff, landing, and combat maneuvers (for agile aircraft); and low noise levels (within permissible limits) (Cui and Xu 2018; Liu 2002).

Nozzle performance significantly impacts engine thrust and fuel consumption rate. Whether in supersonic flight or subsonic cruise conditions, a 1% decrease in nozzle efficiency results in a net engine thrust loss exceeding 1% (Liu 2002). Therefore, nozzle design should strive for maximum performance while

J. Shi et al., *Special Purpose Nozzles*,
https://doi.org/10.1007/978-981-95-2939-1_1

comprehensively considering factors such as weight, structural complexity, reliability, maintainability, and cost. An optimal nozzle must account for its design shape, installed aircraft type, position on the engine, and performance parameters—minimizing external engine drag while providing axial thrust. Key design objectives include: minimal flow losses, near-complete expansion, exhaust direction alignment with desired orientation, adjustable cross-sectional geometry, and low noise. Additionally, future nozzle development should enhance: post-stall maneuverability at high angles of attack (breaking the stall barrier), aircraft performance/ agility improvements, reduced takeoff/landing distances, and stealth capabilities.

1.1.2 Nozzle Configuration Types

Aircraft exhaust nozzles vary widely, with selection depending on comprehensive analysis of the aircraft's mission profile, primary flight regimes, engine cycle parameters, and installation position. Classified by flow path characteristics: convergent nozzles and convergent-divergent nozzles; by exit area adjustability: fixed or variable; by cross-section shape: axisymmetric or non-axisymmetric; and by thrust direction: conventional, thrust-vectoring, thrust-reversing, or thrust-vectoring types. Below introduces common nozzle types (Cui and Xu 2018; Liu 2002).

(1) Fixed convergent nozzle (subsonic exhaust nozzle with un-adjustable exit area): The flow path area gradually decreases along the streamwise direction, limiting maximum exit flow velocity to Mach 1. This nozzle type features simple structure, light weight, high reliability, easy adjustment, and disassembly convenience. It is typically used in subsonic aircraft, short-duration supersonic/ low-supersonic aircraft with non-afterburning turbojet engines, and turboprop/ turbofan engines with small post-turbine enthalpy drop.
(2) Adjustable convergent nozzle: Enables the engine to achieve good performance under various operating conditions. Afterburner-equipped engines must use adjustable nozzles to ensure the nozzle is appropriately enlarged during afterburner operation. Some engines change operating conditions by adjusting the nozzle area. The flow path of an adjustable convergent nozzle consists of a fixed duct, adjustable fish-scale flaps, and an actuation system that regulates the flap opening. It is suitable for engines with afterburners requiring significant changes in nozzle exit area. Main types include: multi-flap, dual-flap, movable tail cone, and pneumatic adjustment systems.
(3) Adjustable convergent-divergent nozzle: Also known as a convergent-divergent (CD) nozzle, it is a type of supersonic nozzle where the flow path area first contracts and then expands. The smallest cross-sectional area, where the nozzle transitions from convergent to divergent, is called the 'throat.' Supersonic aircraft equipped with afterburner gas turbine engines typically use convergent-divergent nozzles, where both the throat and exit areas are adjustable.

1.2 Typical Special-Purpose Nozzles

With continuous improvement in aeroengine cycle parameters and increasing demands for mission adaptability, conventional convergent and convergent-divergent nozzles can no longer meet requirements for either mission flexibility or high cycle parameters. For instance, fourth-generation fighters' requirements of super-stealth and high maneuverability cannot be satisfied by conventional nozzles, necessitating specialized designs like serpentine nozzles (Ai et al. 2017) and thrust vectoring nozzles (Cui and Xu 2018). For short/vertical takeoff and landing (S/VTOL) fighters requiring vertical flight capability, 90-degree thrust-vectoring three-bearing swivel nozzles become essential (Wang et al. 2014). For high-altitude high-speed fighters with significantly increased nozzle pressure ratios, single-expansion ramp nozzles are adopted to reduce structural weight (Liu 2007). As aeroengine multi-mission adaptability requirements evolve, special-purpose nozzles are gaining increasing attention. Below are introductions to several typical special-purpose nozzles.

1.2.1 S-Shaped Nozzle

Low observability is one of the most essential characteristics that future military aircraft should possess (Gan and Zhang 2016; Song et al. 2012). Among the detectable signal features of an aircraft, the engine exhaust nozzle is the strongest infrared radiation source and a major radar reflection source. Therefore, reducing the infrared radiation intensity and radar reflection signals of the exhaust nozzle can significantly decrease the detectable signal features of the aircraft, enhancing its stealth capabilities (Ai et al. 2017).

The infrared radiation from the nozzle originates from the detectable hot surfaces (including the turbine exit surface, afterburner cylinder, and nozzle walls) and the hot exhaust flow expelled through the nozzle. The reflection of radar waves by the nozzle depends on the cavity reflection of the nozzle and the frontal echo of the turbine blades. Regarding the radiation characteristics of the nozzle, the S-shaped nozzle has distinct low-observability advantages and has garnered widespread attention: (1) The S-shaped curved profile of the nozzle effectively shields the high-temperature internal components of the engine, significantly reducing the infrared radiation intensity of the hot nozzle surfaces; (2) The non-axisymmetric nozzle exit enhances the mixing of the hot exhaust flow with the surrounding air, shortening the high-temperature core length of the exhaust flow and substantially lowering its infrared radiation intensity; (3) The S-shaped flow path also prevents rotating turbine blades and afterburner flame stabilizers from being directly exposed to electromagnetic waves, while increasing the number of reflections of incident electromagnetic waves within the S-shaped channel, thereby weakening the energy of reflected waves and markedly reducing the radar signal features of the nozzle. As a result, S-shaped nozzle technology has been identified as one of the key technologies for stealth aircraft (Gridley and Walker 1996; Sun 2018).

The low-observability advantages of the S-shaped nozzle have drawn significant attention from research institutions worldwide, particularly in aeronautically advanced countries like the United States, where it has been adopted as a critical component for reducing the detectable signals of stealth aircraft. Examples include the U.S. F-117A "Nighthawk" fighter, B-2 "Spirit" stealth bomber, X-47B unmanned combat aircraft, Sweden's "Eikon" drone, France's "Neuron" drone, and the UK's "Taranis" drone, all of which are equipped with S-shaped nozzles, as shown in Figs. 1.1, 1.2, 1.3 and 1.4 (Chali, 2002; Wen 2001; Perry 2011; Johansson 2006; Xing and Dong 2014; Xu et al. 2012). Given the U.S. sixth-generation aircraft's demand for ultra-stealth capabilities and the mature application of S-shaped nozzle technology in stealth bombers and drones, the U.S. has listed S-shaped nozzle technology as one of the key technologies for its next-generation aircraft propulsion systems. Figure 1.5 illustrates the dual-external-flow variable cycle engine structure proposed under the ADVENT program, which includes not only conventional internal and external flows but also a third cooling airflow to enhance power output, improve thermal management, reduce installation drag, and increase inlet total pressure recovery. Additionally, the cooling air can mix with the exhaust flow to lower the temperature and infrared signature of the exhaust system. The rear nozzle adopts a three-flow S-shaped configuration to shield turbine components, reducing radar reflection and infrared radiation, thereby achieving ultra-stealth capabilities for the aircraft (Simmons 2009).

1.2.2 Three-Bearing Swivel Nozzle

Short Takeoff/Vertical Landing (S/VTOL) aircraft combine the advantages of helicopters and traditional fighter jets. These aircraft achieve short takeoff/vertical landing through lift generated by their propulsion systems, making S/VTOL propulsion

Fig. 1.1 Simplified Diagram of B-2 Stealth Strategic Bomber Engine Installation (Wen 2001)

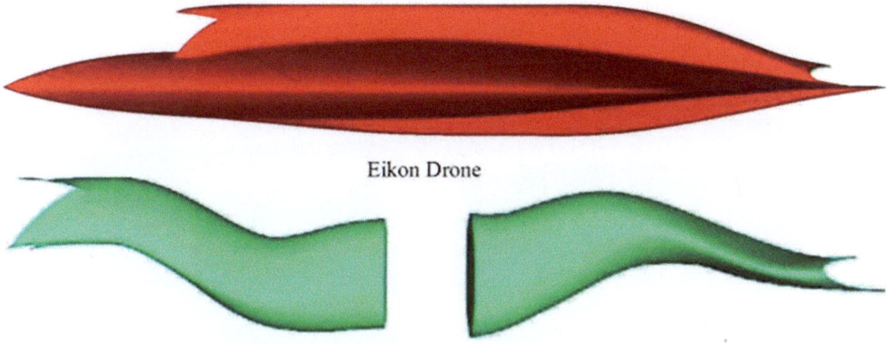

Fig. 1.2 Eikon UAV with S-shaped Inlet and S-bend Nozzle (Johansson 2006)

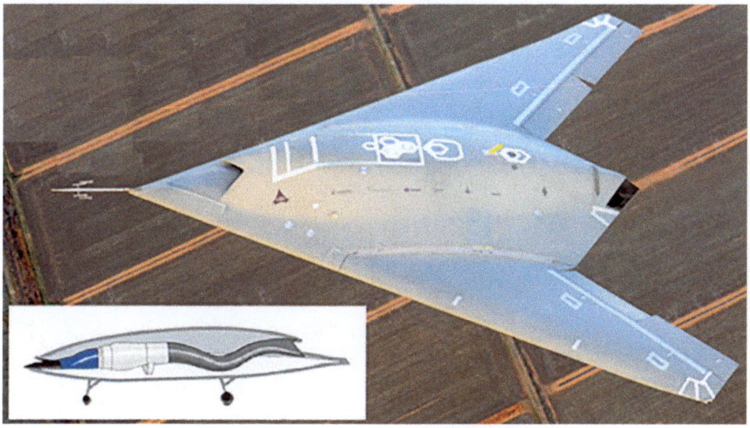

Fig. 1.3 'Neuron' unmanned combat aerial vehicle (Xing and Dong 2014)

systems a key research focus for aviation-developed nations and a primary combat asset for future light aircraft carriers and amphibious assault ships (Suo et al. 2014; Liu 2016). Modern S/VTOL aircraft propulsion system configurations are all based on the thrust vectoring concept, which can be categorized as: integrated propulsion systems, combined propulsion systems, and hybrid propulsion systems. Representative aircraft for these configurations are the Harrier Jet, Yak-141 fighter, and F-35B fighter, respectively (Suo et al. 2014; Anderson 1983). The British Harrier was a pioneer in operational S/VTOL fighters. The success of the Harrier series was largely due to its Pegasus thrust-vectoring engine. The Pegasus engine converts between lift and thrust via vectoring nozzles, with both lift and thrust generated by the same engine. The Pegasus engine (Fig. 1.6) is a twin-spool turbofan with a bypass ratio of approximately 1.4. Air enters the engine through intakes on both sides of the aircraft. After being pressurized by the fan, about 58% of the air (at a temperature of around 150 °C) flows into the bypass duct and is expelled through

Fig. 1.4 'Taranis' unmanned combat aerial vehicle (Xu et al. 2012)

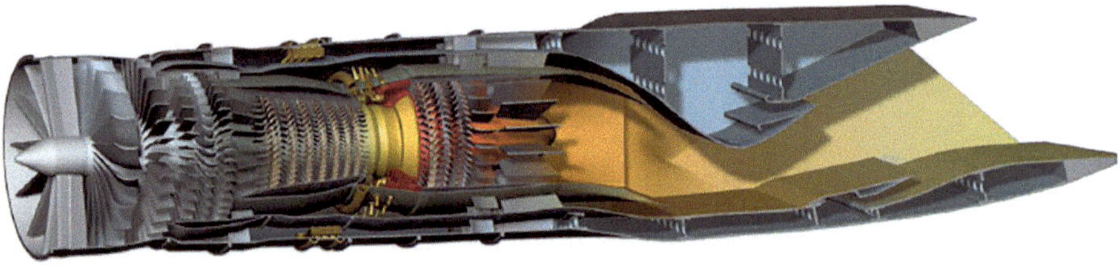

Fig. 1.5 Structure of a dual-bypass variable cycle engine (Simmons 2009)

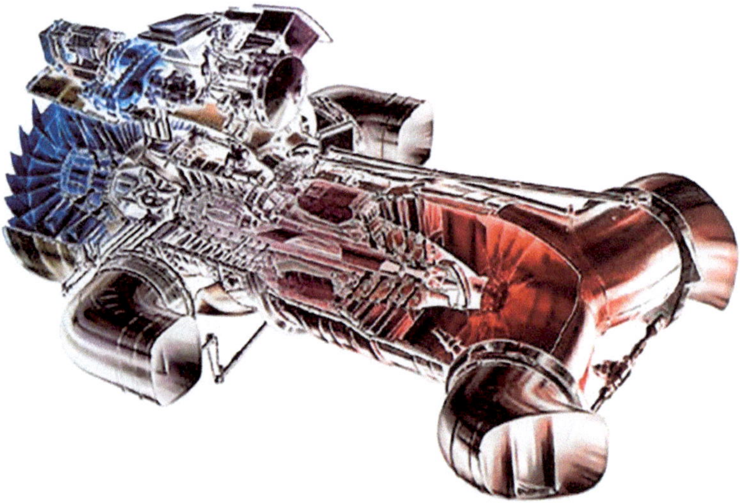

Fig. 1.6 "Pegasus" engine structural diagram (Frost and Bishop 1965; Hooker 1981)

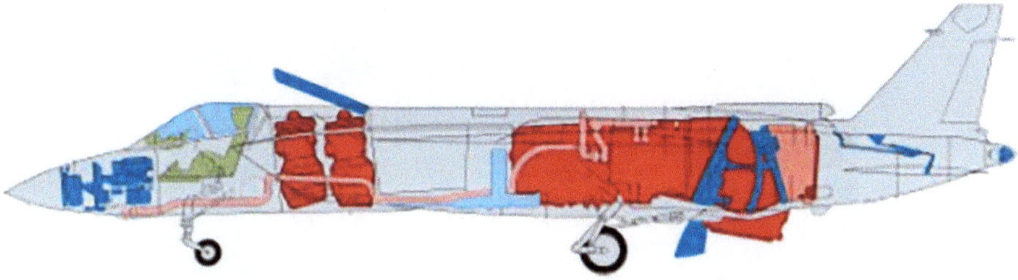

Fig. 1.7 Yak-141 propulsion system configuration (Russian Fedration 2014)

front nozzles on either side of the engine, generating partial lift. The remaining air enters the engine core and is expelled through rear nozzles (at a gas temperature of about 670°C), producing another portion of lift that balances the front nozzles. The front and rear nozzle exits employ louvered guide vanes to reduce flow separation along the nozzle walls, ensuring more uniform exhaust flow and thereby increasing the nozzle's thrust coefficient. During engine mode transitions, the front and rear nozzles rotate synchronously via a transmission system to maintain aircraft balance. Flight tests have shown that this propulsion system configuration increases aircraft weight, reduces flight performance, and imposes higher thrust-to-weight ratio requirements on the lift engine. Additionally, transitioning between lift and cruise engines during mode changes complicates aircraft control (Frost and Bishop 1965; Hooker 1981).

The Yak-141's powerplant consists of an R-79 lift/cruise engine with a vectoring nozzle, producing 103 kN of lift and 152 kN of thrust, along with two RD-41 lift engines mounted in tandem behind the cockpit, each generating 41 kN of lift (Fig. 1.7). The Yak-141 can only perform vertical takeoff and landing (VTOL) and not short takeoff and landing (STOL). During VTOL, the cruise engine's nozzle rotates downward to a vertical position, working together with the lift engines to generate lift. In level flight, the lift engines are inactive. However, during VTOL, the R-79's afterburner exit is too close to the ground, causing severe ground erosion. Additionally, the intake is prone to ingesting the high-temperature, high-pressure exhaust from the lift engines, disrupting the normal operation of the lift/cruise engine. To address this issue, the Yak-141 is equipped with retractable baffles below the intake to prevent exhaust gas ingestion. The Yak-141's three-bearing swivel nozzle (Fig. 1.8) has a slightly curved profile at 0° deflection and can deflect up to 95°. For short takeoff, the nozzle is set at 65°. Although the Yak-141 successfully completed STOL/VTOL test flights, its development was terminated in 1991 due to the dissolution of the Soviet Union and lack of funding (Russian Federation, 2010, 2014).

The lift-fan-equipped composite propulsion system used in the F-35B is currently the best-performing design of its kind (Fig. 1.9). To date, no other S/VTOL fighter's powerplant has surpassed its overall performance (Bevilaqua 1990, 2003, 2009a, b; Lewis 2000). Compared to the integrated propulsion systems of the Harrier and X-32B, the lift-fan composite system offers a more aerodynamically

Fig. 1.8 Yak-141 three-bearing swivel nozzle assembly (Russian Fedration 2010)

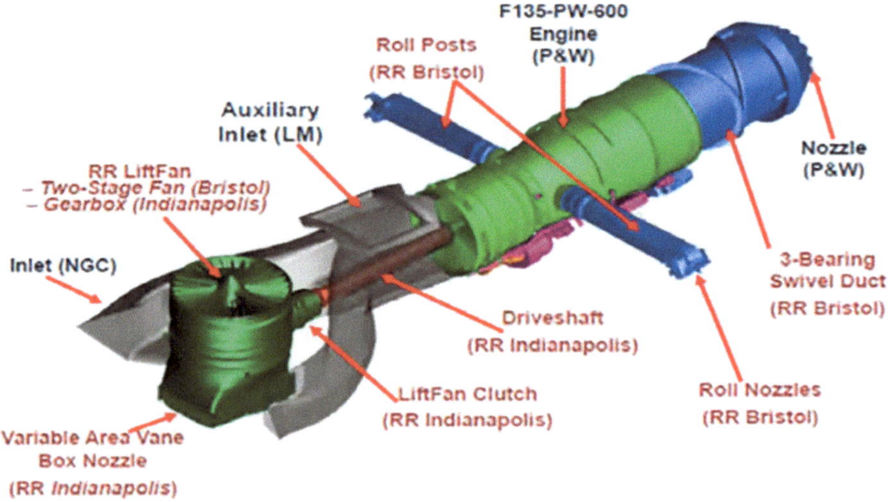

Fig. 1.9 Schematic diagram of the propulsion system for F-35B (Bevilaqua 2009a, b)

favorable layout and simpler power distribution for balance control. Unlike the combined propulsion systems of the Yak series, this system uses only a single engine, reducing fuel consumption during STOL/VTOL operations. The forward-mounted lift fan generates cool airflow, preventing the ingestion of high-temperature exhaust from the lift/cruise engine and eliminating ground erosion effects.

The three-bearing swivel nozzle for the X-35B uses three hydraulic motors to independently drive the rotation of three cylindrical sections, as shown in Fig. 1.10 (Bevilaqua 2009a, b). The drive motors for the second and third nozzle sections are both mounted on the second cylindrical section, with the motors operating at the same speed and direction. This design facilitates nozzle modification and results in a more compact nozzle profile. The three bearings in the nozzle are

Fig. 1.10 Three-bearing swivel nozzle for X-35B (Bevilaqua 2009a, b)

high-temperature-resistant bearings developed by Germany's FAG company. The inner and outer rings of the bearings are made of high-temperature and corrosion-resistant steel, while the bearing balls use ceramic materials that are heat-resistant, have low thermal expansion coefficients, and possess self-lubricating properties. The nozzle employs a double-walled cylindrical structure for cooling, where high-pressure cold air drawn from the compressor flows through the gap between the inner and outer cylinders to cool each section. The inner cylinder is rigidly connected to the outer cylinder via a torque transmission structure (Farah 2007) and rotates along with the outer cylinder.

The capability for short takeoff and vertical landing depends on the technical level and structural choices of the propulsion system. Judging from the structural configurations of propulsion systems used in short/vertical takeoff and landing (S/VTOL) aircraft across various countries, the most feasible approach is the combination of a main engine with a three-bearing swivel nozzle (similar to those in the F-35B and Yak-141) and a front-mounted lift engine/fan. The performance of the propulsion system directly determines whether the aircraft's design can be realized. Therefore, the performance of the vector-capable propulsion system and its core component—the three-bearing swivel nozzle—is the most critical aspect of S/VTOL technology.

The primary function of the three-bearing swivel nozzle is twofold. First, during short/vertical takeoff and landing, it works in conjunction with the lift fan and roll nozzles to provide power output and adjust the aircraft's attitude, ensuring smooth ascent or landing. Second, during maneuvering or combat, the three-bearing swivel nozzle can deliver multi-axis thrust vectoring, providing power in pitch, roll, and yaw directions to enhance the aircraft's agility in dogfights and pursuits. It can be said that the three-bearing swivel nozzle is key to achieving thrust vector control for S/VTOL aircraft during takeoff/landing and maneuvering, making it an essential requirement for modern S/VTOL fighters.

1.2.3 Single Expansion Ramp Nozzle

Hypersonic vehicles refer to aircraft with a flight Mach number greater than 5, powered by air-breathing engines or combined cycle engines, capable of achieving hypersonic flight within and across the atmosphere (Liu 2004). They are hailed as the third revolution in world aviation history following propeller and jet aircraft, representing the development trend in the aerospace field of the 21st century. In recent years, major aerospace nations have launched numerous hypersonic vehicle development programs focusing on hypersonic technology. Among these, air-breathing hypersonic propulsion technology is the key to achieving breakthroughs in hypersonic flight and has become a core technology in hypersonic flight. The Turbine Based Combined Cycle (TBCC) engine, as one of the most ideal power choices for air-breathing hypersonic vehicles, offers advantages such as high specific impulse, a wide flight envelope, and reusability, garnering widespread attention (Watanabe et al. 1993). Major aerospace nations have successively initiated a series of TBCC technology development programs, such as the U.S. RTA program (Bartolotta and McNelis 2003), Japan's HYPR program (Miyagi et al. 1998), and Europe's LAPCAT program (Steelant 2008).

The exhaust nozzle is the primary component for generating engine thrust. Research has found that at a flight Mach number of Ma = 6, the nozzle can contribute up to about 70% of the total engine thrust. Therefore, the quality of nozzle design directly impacts the overall engine performance (Bradford 2002). Generally, to achieve optimal nozzle performance while meeting the wide flight envelope of hypersonic vehicles, the nozzle pressure ratio needs to reach hundreds or even thousands. Due to mechanical limitations and sealing mechanisms of the expansion surface, traditional axisymmetric nozzles cannot meet these requirements. Thus, TBCC exhaust nozzles typically employ a Single Expansion Ramp Nozzle (SERN), as shown in Fig. 1.11 (Grarnland and Berens 1995). The configuration of SERN allows

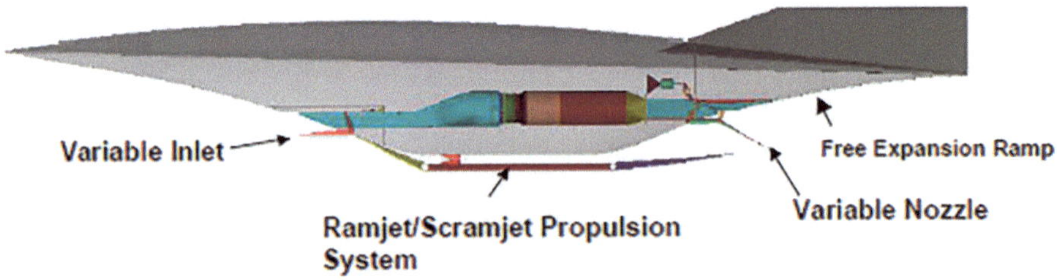

Fig. 1.11 Exhaust nozzle of an X-43-type hypersonic vehicle (Grarnland and Berens 1995)

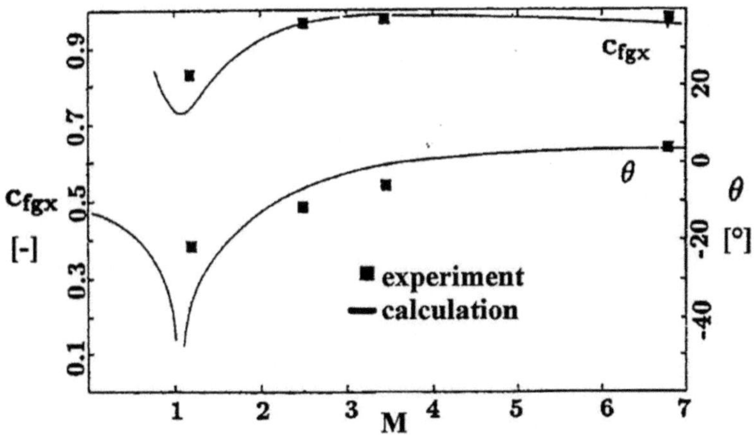

Fig. 1.12 Variation of axial thrust coefficient and thrust vector angle of the SERN nozzle with flight Mach number (Jürgen and Alfred 1998)

its outer surface to integrate well with the lower surface of hypersonic vehicles, facilitating the integrated design of the propulsion system and the vehicle. This not only enables SERN to achieve very high pressure ratios but also ensures good thrust performance across a wide flight envelope. Studies also indicate that the asymmetric configuration of SERN can reduce vehicle weight, lower noise, minimize infrared signature, and, compared to traditional nozzles, more easily achieve thrust vector control (Le et al., 2003; MacLean 1993).

SERN adopts a highly integrated configuration between the rear body of the hypersonic vehicle and the nozzle. Although this design exhibits excellent thrust performance at hypersonic Mach numbers, under lower flight Mach numbers and lower pressure ratios, the airflow inside the SERN becomes over-expanded. Complex structures such as shock waves and boundary layer separation occur within the nozzle, severely affecting its performance. The nozzle performance deteriorates sharply, leading to a significant reduction in the thrust coefficient, as shown in Fig. 1.12 (Jürgen and Alfred 1998). Therefore, improving the thrust and thrust vector performance of large expansion ratio SERN under transonic conditions is one of the key issues in current hypersonic vehicle design.

1.3 Typical Cross-Sections and Geometric Parameters of the Nozzle

1.3.1 S-Shaped Nozzle

The S-shaped nozzle features an S-bend configuration, a circular-to-rectangular cross-section transition, and a two-dimensional exit geometry. The S-bend design effectively shields the high-temperature and rotating components of the engine, reducing the infrared radiation intensity and electromagnetic scattering intensity of the exhaust nozzle. The two-dimensional exit configuration enhances the mixing of the core flow with the surrounding atmosphere, thereby suppressing the infrared radiation intensity of the hot exhaust. The surface structure of the S-shaped nozzle has the following characteristics:

(1) The centerline of the S-bend configuration. The shape of the centerline determines the geometry of the nozzle passage and the deflection of the airflow within the nozzle. Both the inlet and outlet airflow directions are parallel to the engine axis.
(2) The circular-to-rectangular flow cross-section. The nozzle inlet is connected to the circular exit of the mixing chamber. To facilitate integrated design with the aircraft's rear fuselage and enhance the mixing of the core flow with the surrounding atmosphere, the nozzle exit is typically a two-dimensional configuration. Therefore, the flow cross-section of the nozzle must smoothly transition from a circular inlet to a rectangular (two-dimensional) outlet. The two-dimensional exit cross-section generally includes rectangular and elliptical configurations.
(3) Low observability conditions. The S-shaped nozzle achieves effective shielding of the engine's high-temperature and rotating components through its curved configuration. Therefore, during the surface design of the S-shaped nozzle, it is necessary to consider establishing low observability conditions and integrate these conditions into the nozzle design process. To ensure that the high-temperature and rotating components cannot be detected under any flight attitude of the aircraft, a low observability condition that fully shields the high-temperature turbine is typically adopted.

Figure 1.13 presents the main geometric parameters involved in the double S-bend nozzle, including: the exit area A_1 and exit width W_1 (or aspect ratio W_1/H_1) of the first S-bend channel, the axial length L_1 , the vertical offset ΔY_1 , the axial length L_2 and vertical offset ΔY_2 of the second S-shaped channel, the exit aspect ratio W_e/H_e , the straight section length L_3, and the centerline variation pattern of the two S-bend curves. The purpose of the straight section is to ensure uniform airflow at the exit.

1.3.2 Three-Bearing Swivel Nozzle

The three-bearing swivel nozzle has become a research hotspot for vectoring nozzles in S/VTOL aircraft due to its simple structure, lightweight, and high degree of freedom. Mechanical vectoring nozzles generally rely on complex actuation systems to alter the geometric profile of the nozzle, thereby changing the direction of engine thrust. The three-bearing swivel nozzle consists of three rotatable cylindrical segments. By adjusting the rotation angles of these three segments and combining them, the deflection angle of the nozzle can be modified.

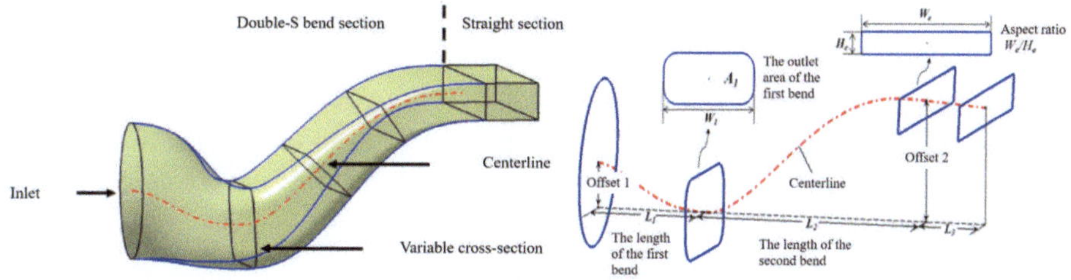

Fig. 1.13 Schematic of the double S-bend nozzle geometry and design parameters. (**a**) Geometric schematic. (**b**) Design parameter schematic

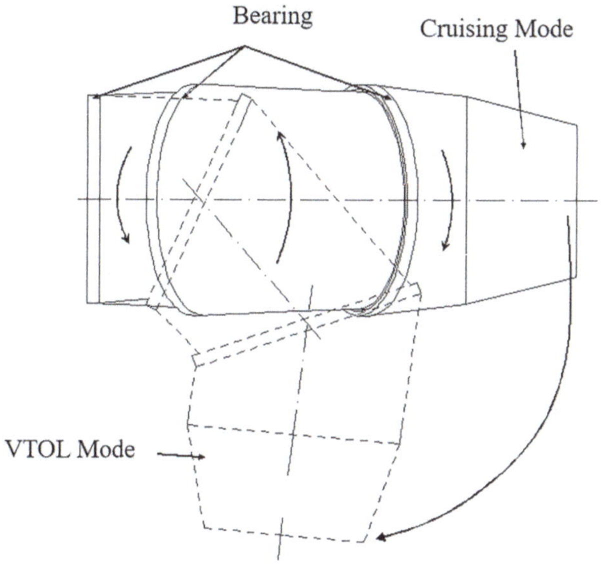

Fig. 1.14 Schematic diagram of the working principle of the three-bearing swivel nozzle

Figure 1.14 illustrates the working principle of the three-bearing swivel nozzle. The solid lines represent the nozzle's installation view in its non-vectoring state, while the dashed lines depict its operational view at a 95° deflection angle. As shown, the nozzle achieves deflection through mutual rotation of its three cylindrical segments, with adjacent segments connected by bearings to reduce friction during rotation. During deflection, the first and third segments rotate in the same direction, while the second segment rotates in the opposite direction relative to the first and third. The axes of all three segments remain in the same pitch plane at all times, meaning the nozzle generates only pitch forces and no lateral forces during deflection. Regarding the advantages of the three-bearing swivel nozzle over other mechanical vectoring nozzles, Curtiss-Wright Corporation has provided the following description:

(1) During deflection, the vector thrust remains within the same pitch plane;
(2) During deflection, the vector angle of the nozzle can be continuously varied;
(3) The deflection mechanism design of the nozzle does not affect the aerodynamic performance of the nozzle in its non-vectoring state.

Figure 1.15 shows the schematic diagram of the key geometric features of the three-bearing swivel nozzle. Figure 1.15a illustrates the profile of the nozzle in its non-vectoring state, where sections 0, 2, and 4 are the installation sections for the nozzle bearings and also the rotational joint surfaces between adjacent cylinder segments. To ensure proper bearing installation and that the geometric boundaries of the two rotating joint surfaces remain aligned during rotation, sections 0, 2, and 4 must be circular. According to geometric principles, the oblique cross-section of an ellipsoid can be circular. Therefore, based on the geometric condition that sections 0, 2, and 4 are circular, the nozzle cylinder between sections 1 and 5 should be an elliptical cylinder, while the segment between sections 0 and 1 is a transition from a circular cylinder to an elliptical cylinder.

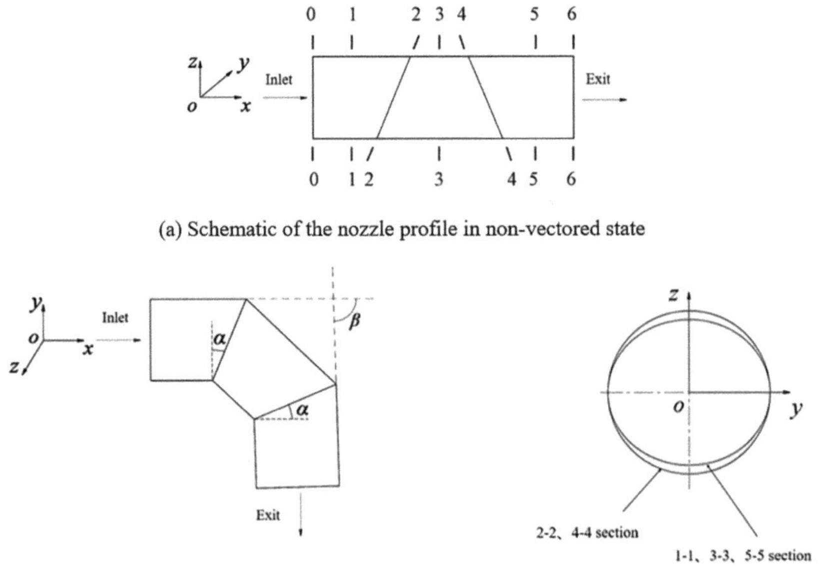

(a) Schematic of the nozzle profile in non-vectored state

(b) Schematic diagram of the nozzle profile at maximum deflection angle

(c) Schematic diagram of the nozzle axial cross-section

Fig. 1.15 Schematic diagram of key geometric features of the three-bearing swivel nozzle. (**a**) Schematic of the nozzle profile in non-vectored state. (**b**) Schematic diagram of the nozzle profile at maximum deflection angle. (**c**) Schematic diagram of the nozzle axial cross-section

Figure 1.15b shows a schematic diagram of the nozzle profile at the maximum deflection angle, where the maximum deflection angle of the nozzle is β. The oblique cut angles of sections 2 and 4 are α, and the relationship between α and β can be seen from the figure as

$$\alpha = \frac{\pi}{2} - \frac{\pi - \beta / 2}{2} = \frac{\beta}{4} \tag{1.1}$$

Figure 1.15c shows the axial schematic diagram of each section of the nozzle. Assuming the radii of sections 2 and 4 are R, it can be deduced that the major axis of the ellipse for sections 1, 3, and 5 is R, and the minor axis is R/cosα. Therefore, the elliptical surface formula for sections 1, 3, and 5 is

$$y^2 + z^2 \cos^2 \alpha = R^2 \tag{1.2}$$

Through the above analysis, the key geometric features of the three-bearing swivel nozzle were obtained and summarized as follows:

(1) The three rotating joint surfaces of the nozzle are circular in cross-section;
(2) The first section of the nozzle includes a transition segment from a circular cross-section to an elliptical cross-section;
(3) The second section of the nozzle is an elliptical cylinder;
If the maximum deflection angle of the nozzle is set to β, then the inclination angle α of the inclined section of each cylinder segment is α = β/4, and the elliptical section equation can be expressed by the inclination angle α.

1.3.3 Single Expansion Ramp Nozzle

The two-dimensional profile of the SERN for a series-connected TBCC engine is shown in Fig. 1.16. Considering the influence of external flow, the internal nozzle profile is supplemented with upper and lower external profiles. The structural parameters of the SERN include the area A_l at the exit of the lower inclined surface of the nozzle; the area A_e at the exit of the upper expansion inclined surface; the area A_r of the extended section of the upper expansion inclined surface; the length L_r of the extended section of the upper expansion inclined surface; the throat height H_t; and the length L_C of the lower expansion surface.

The tail nozzle for a parallel-connected TBCC must meet the requirements of two engines (a turbine engine and a ramjet engine) operating under different design conditions. The geometric structure of the SERN is shown in Fig. 1.17, typically featuring upper and lower channels for the turbine nozzle and ramjet nozzle passages, as well as dual throats. Therefore, determining the optimal positional relationship between the two nozzle channels and ensuring the performance of the TBCC tail nozzle across the entire operational envelope are key concerns in the design process.

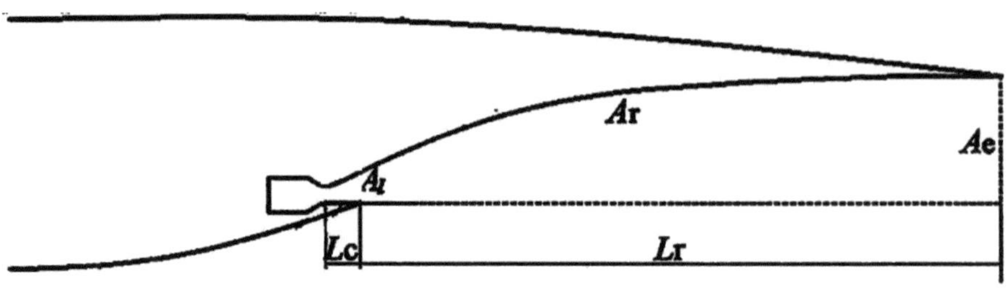

Fig. 1.16 Two-dimensional geometric structure of the SERN for a series-connected TBCC engine

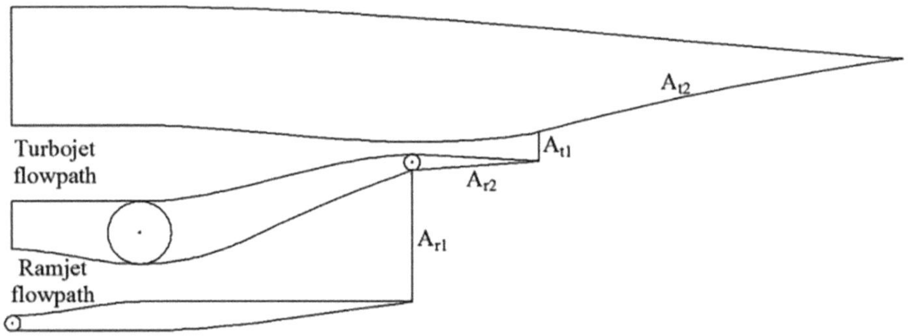

Fig. 1.17 Geometric structure of the SERN for a parallel-connected TBCC engine

1.4 Nozzle Performance Parameters

1.4.1 S-Shaped Nozzle

The aerodynamic performance parameters of the nozzle mainly include: total pressure recovery coefficient σ, flow coefficient C_D, and thrust coefficient C_{fg}. This paper conducts a comparative analysis of the aerodynamic performance of low-observable S-duct nozzles under different working conditions and geometric configurations through these three parameters, aiming to obtain the influence patterns of key parameters on the aerodynamic performance of low-observable S-shape nozzles.

The total pressure recovery coefficient σ is the ratio of the mass-weighted average total pressure at the nozzle outlet to the total pressure at the inlet, expressed as

$$\sigma = \left(\frac{1}{\dot{m}_e} \int P_e^* d\dot{m}_e \right) / \left(\frac{1}{\dot{m}_{\text{in}}} \int P_{\text{in}}^* d\dot{m}_{\text{in}} \right) \tag{1.3}$$

Among them, $\dot{m}_{in}$ and $\dot{m}_e$ represent the flow rates at the inlet and outlet sections, respectively, while P_{in}^* and P_e^* denote the total pressures at each calculation point on the inlet and outlet sections.

The flow coefficient C_D is the ratio of the flow rate obtained through numerical simulation to the ideal flow rate, i.e.,

$$C_{\mathrm{D}} = \dot{m} / \dot{m}_{\mathrm{i}} \tag{1.4}$$

$\dot{m}$ and $\dot{m}_i$ are calculated respectively through Eqs. (1.5) and (1.6).

$$\dot{m} = \int(\rho U)dA_{\mathrm{e}} \tag{1.5}$$

$$\dot{m}_{\mathrm{i}} = \begin{cases} K \cdot P_{\mathrm{in}}^{*} \cdot A_{\mathrm{e}} \cdot q(\lambda) / \sqrt{T_{\mathrm{in}}^{*}}, & NPR < 1.893 \\ K \cdot P_{\mathrm{in}}^{*} \cdot A_{\mathrm{e}} / \sqrt{T_{\mathrm{in}}^{*}}, & NPR \geq 1.893 \end{cases} \tag{1.6}$$

Among them, ρ and U represent the density and velocity at each calculation point on the exit cross-section, respectively, A_e denotes the area of the exit cross-section, T_{in}^{*} represents the inlet total temperature, $q(\lambda) = ((k+1)/2)^{1/(k-1)} \cdot \sqrt{(k+1)/(k-1) \cdot \left[1-(P_b/P_{\mathrm{in}}^{*})^{(k-1)/k}\right]} \cdot (P_b/P_{\mathrm{in}}^{*})^{1/k}$, P_b represents the ambient pressure, and for air, k = 1.4.

The thrust coefficient C_{fg} is obtained by numerically simulating the ratio of the axial thrust F_x to the ideal thrust F_i, i.e.,

$$C_{\mathrm{fg}} = F_{\mathrm{x}} / F_{\mathrm{i}} \tag{1.7}$$

F_x and F_i are calculated using Eqs. (1.6) and (1.7), respectively.

$$F_{\mathrm{x}} = \int\left[\rho \cdot U^2 + (P_{\mathrm{e}} - P_{\mathrm{b}})\right]dA_{\mathrm{e}} \tag{1.8}$$

$$F_{\mathrm{i}} = \begin{cases} \dot{m}_{\mathrm{i}} \cdot \sqrt{2k/(k-1) \cdot R \cdot T_{\mathrm{in}}^{*} \cdot \left[1-(P_{\mathrm{b}}/P_{\mathrm{in}}^{*})^{(k-1)/k}\right]}, & NPR < 1.893 \\ \dot{m}_{\mathrm{i}} \cdot \sqrt{2k/(k+1) \cdot R \cdot T_{\mathrm{in}}^{*}} + (P_{\mathrm{in}}^{*}/1.893 - P_{\mathrm{b}}) \cdot A_{\mathrm{e}}, & NPR \geq 1.893 \end{cases} \tag{1.9}$$

Among them, P_e represents the static pressure at each calculation point on the exit cross-section. For air, R = 287.06J/(kg·K).

1.4.2 Three-Bearing Swivel Nozzle

The thrust vector angle (δ_p) is a key performance parameter of the three-bearing swivel nozzle, which can generally be determined by the ratio of the lateral force to the axial force acting on the nozzle, yielding the tangent of δ_p. The lateral force on the nozzle can be obtained by integrating the wall pressure or derived from the momentum theorem. Due to the non-uniform pressure distribution on the inner wall surface of the nozzle at certain vector angles, with significant local pressure differences, the accuracy of the lateral force obtained through wall integration may be affected. In contrast, the momentum theorem approaches the problem from the overall perspective of the nozzle, avoiding errors caused by solver precision. Therefore, the momentum theorem can be used to solve for the lateral and axial forces of the nozzle.

Through the derivation of the momentum theorem, the force on the nozzle wall A_{out} can be expressed by the pressure difference force and momentum force at the nozzle exit section. By directly summing the pressure differences on each micro-element surface at the nozzle exit section, the pressure difference force at the nozzle exit can be obtained.

$$\mathbf{F}_{\text{p}} = \sum_{i=1}^{n} (p_i - p_0) \cdot \mathbf{A}_{\text{out},i} \tag{1.10}$$

The components of $\mathbf{F}_p$ in the x and y directions are

$$\mathbf{F}_{\text{px}} = \mathbf{F}_{\text{p}} \times \cos\beta \tag{1.11}$$

$$\mathbf{F}_{\text{py}} = \mathbf{F}_{\text{p}} \times \sin\beta \tag{1.12}$$

The momentum force in the x-direction at the nozzle exit $\mathbf{F}_x$ can be obtained by summing the product of the flow rate and the x-direction velocity over the micro-elements of the exit cross-section.

$$\mathbf{F}_x = \sum_{i=1}^{n} (\rho \mathbf{V}_i \cdot \mathbf{A}_i) \cdot \mathbf{V}_{x,i} \tag{1.13}$$

The momentum force in the y-direction at the nozzle exit $\mathbf{F}_y$, can be obtained by summing the product of the flow rate and the y-direction velocity over the micro-elements of the nozzle exit cross-section

$$\mathbf{F}_y = \sum_{i=1}^{n} (\rho \mathbf{V}_i \cdot \mathbf{A}_i) \cdot \mathbf{V}_{y,i} \tag{1.14}$$

The total thrust is

$$F_{\text{noz}} = \sqrt{(\mathbf{F}_{\text{px}} + \mathbf{F}_x)^2 + (\mathbf{F}_{\text{py}} + \mathbf{F}_y)^2} \tag{1.15}$$

From the momentum force $\mathbf{F}_x$ and $\mathbf{F}_y$, the thrust vector angle δ_p can be derived as

$$\delta_{\text{p}} = \tan^{-1}\left(\frac{F_{\text{py}} + F_y}{F_{\text{px}} + F_x}\right) \tag{1.16}$$

The nozzle thrust coefficient C_{fg} is a key parameter for measuring nozzle thrust performance, defined as the ratio of actual thrust to ideal thrust, expressed as follows:

$$C_{\text{fg}} = F_{\text{noz}} / F_{\text{i}} \tag{1.17}$$

In the equation, $\mathbf{F}_i$ represents the total thrust when the nozzle is in a non-vectoring state, with the airflow inside the nozzle undergoing isentropic full expansion and the exit direction parallel to the nozzle axis, which can be calculated using Eq. (1.9).

The definitions of nozzle flow coefficient C_d and nozzle total pressure recovery coefficient σ are consistent with those of the S-shaped nozzle.

1.4.3 Single Expansion Ramp Nozzle

1.4.3.1 Tandem TBCC Engine with SERN

This paper adopts flow coefficient C_d, axial thrust coefficient, thrust vector angle δ, and nozzle base drag X_b as parameters to evaluate SERN performance.

(1) Flow coefficient C_d, defined consistently with the S-shaped nozzle.
(2) Axial thrust coefficient C_{fgx}:

$$C_{fgx} = F_j / F_i \tag{1.18}$$

Where: F_j is the actual axial thrust of the nozzle, F_i is the ideal thrust of the nozzle. The ideal thrust F_i is calculated as per Eq. (1.9), and the actual axial thrust F_j is calculated by the following formula:

$$F_j = \int_{A_l} \rho V_x^2 dA + \int_{A_l} (P - P_\infty) dA + \int_{A_r} (P - P_\infty) dA_x - X_{fx} \tag{1.19}$$

Among them: the integral $\int_{A_l} \rho V_x^2 dA$ on the right side represents the axial thrust generated by the momentum of the airflow discharged from surface A_l, $\int_{A_l} (P - P_\infty) dA$ is the axial thrust produced by the pressure difference at surface A_l, $\int_{A_r} (P - P_\infty) dA_x$ is the axial thrust caused by the pressure difference at surface A_r, X_{fx} is the projection of the friction force at surface A_r in the axial direction, and the definitions of A_l and A_r are shown in Fig. 1.16.

(3) Thrust vector angle δ_p:

$$\delta_p = \tan^{-1}(F_n / F_j) \tag{1.20}$$

Where: F_n is the force generated by the nozzle perpendicular to the axis direction.

$$F_n = \int_{A_l} -\rho V_Y^2 dA + \int_{A_r} (p - p_\infty) dA_Y + X_{fY} \tag{1.21}$$

Where: $\int_{A_l} -\rho V_Y^2 dA$ is the force perpendicular to the axis direction generated by the momentum of the airflow discharged from surface A_l, $\int_{A_r} (p - p_\infty) dA_Y$ is the force perpendicular to the axis direction generated by the pressure difference at surface A_r , and X_{fY} is the projection of the friction force at surface A_r in the direction perpendicular to the axis.

(4) Nozzle tail resistance X b :

The interaction between the exhaust gas ejected from the tail nozzle outlet and the external airflow passing over the rear fuselage can cause flow separation both inside and outside the nozzle, increasing aircraft drag. The effective thrust of the nozzle, which is the actual thrust minus the tail drag. The calculation formula for tail drag is:

$$X_b = -\int_M^9 (p - p_0)dA + X_f \tag{1.22}$$

Where: 9 and M represent the tail nozzle exit cross-section and maximum cross-section, respectively; X_f denotes the frictional drag.

1.4.3.2 Parallel-Type TBCC Engine Tail Nozzle

The parallel TBCC exhaust nozzle consists of two sets of asymmetric nozzles arranged side by side, with the flow coefficient defined as follows:

$$C_d = (m_{TJ} + m_{RJ})/(m_{TI} + m_{RI}) \tag{1.23}$$

Where: m_J is the actual flow rate of the nozzle, m_I is the ideal flow rate of the nozzle, and the subscripts T and R represent the turbine nozzle channel and the ramjet nozzle channel, respectively. The calculation formula for the ideal flow rate is given in Eq. (1.6).

The axial thrust coefficient is defined as:

$$C_{fgx} = (F_{TJ} + F_{RJ})/(F_{TI} + F_{SI}) \tag{1.24}$$

Where: F_J is the actual axial thrust of the nozzle, F_I is the ideal thrust of the nozzle, and the subscripts T and R represent the turbine nozzle channel and the ramjet nozzle channel, respectively.

The ideal thrust can be calculated using Eq. (1.9), and the actual axial thrust is calculated as follows:

$$F_J = \int_{A_1} \rho V_x^2 dA + \int_{A_1} (P - P_\infty)dA + \int_{A_2} (P - P_\infty)dA_x - X_{fx} \tag{1.25}$$

In the equation: The first term on the right-hand side represents the axial momentum force generated by the exhaust flow from surface A_1 , the second term represents the axial pressure difference force caused by the pressure difference at surface A_1 , the third term represents the axial pressure difference force caused by the pressure difference at surface A_2 , X_{fx} is the projection of the friction force at surface A_2 in the axial direction, and the definitions of A_1 and A_2 are shown in Fig. 1.17.

The calculation formula for the vertical force of the nozzle is as follows:

$$F_N = \int_{A_1} \rho V_y V_x dA + \int_{A_2} (P - P_\infty)dA_y - X_{fy} \tag{1.26}$$

In the equation: The first term on the right-hand side represents the vertical momentum force generated by the exhaust flow from surface A_1, the second term represents the vertical pressure difference force caused by the pressure difference at surface A_2, and X_{fy} is the projection of the friction force at surface A_2 in the vertical direction.

The definition of thrust vector angle is:

$$\delta_p = \arctan\left[\left(F_{TN} + F_{RN}\right)/\left(F_{TJ} + F_{RJ}\right)\right] \tag{1.27}$$

Chapter 2
Design Method and Flow Characterization of S-Shaped Nozzle

2.1 Introduction

Compared to conventional axisymmetric nozzles, S-shaped nozzles have a unique S-shaped bending configuration (Brunet et al. 2009) and round-to-square cross-section transition characteristics (Laughrey et al. 1979; Coates and Page 2012), which can significantly reduce the infrared radiation intensity of the engine exhaust system (Liu et al. 2013; Zhou 2017) and electromagnetic scattering intensity (Gao et al. 2015; Li et al. 2014). However, this large curvature anomalous bending structure will lead to complex three-dimensional flow characteristics such as uneven aerodynamic parameters (Wang et al. 2017), localized acceleration, and lateral flow inside the nozzle (Sun et al. 2016). While in the real turbofan engine operating environment, phenomena such as internal/external two-streams based on different drop pressure ratios and temperatures, tail cones, complex wave flap mixer structures (Liu et al. 2015), and nozzle inlet cyclone generated by the turbine stub plates exacerbate the inhomogeneity and complexity of the flow inside the nozzle (Crowe and Martin 2019), in comparison to the S-shaped of the homogeneous inlet condition of the nozzle, the internal flow field of the S-shaped nozzle in the real turbofan engine operating environment appears some new and special complex flow phenomena (Crowe and Martin 2015), which in turn changes the aerodynamic performance of the S-shaped nozzle (Sang et al. 2019), and brings new problems and challenges for the research of the profile design and flow characteristics of the high-performance S-shaped nozzle. Therefore, the research on the design method of S-shaped nozzle for turbofan engine and its flow characteristics in the real turbofan engine working environment, and the in-depth analysis of the real refined flow field characteristics of the nozzle under the influence of the complex intake structure and the influence law of the geometrical parameter can provide a theoretical basis for the subsequent research on the stealthy design of the S-shaped nozzle based on high aerodynamic performance.

J. Shi et al., *Special Purpose Nozzles*,
https://doi.org/10.1007/978-981-95-2939-1_2

This chapter firstly introduces the design process of S-shaped nozzle profile based on multi-parameter coupling; secondly, it carries out the cold model experiment of S-shaped nozzle for turbofan engine, and effectively verifies the accuracy of the numerical simulation method through the experimental results; then, it analyzes the characteristics of the complex flow field and its generating mechanism inside the S-shaped nozzle for turbofan engine in detail; and finally, it studies the influence law of the key geometric parameter on the flow characteristics of the S-shaped nozzle to clarify the change mechanism of aerodynamic performance of the S-shaped nozzle. Finally, the influence of key geometrical parameters on the flow characteristics of the S-shaped nozzle is investigated, and the mechanism of the aerodynamic performance change of the S-shaped nozzle is clarified.

2.2 S-Shaped Nozzle Design Method

The S-shaped nozzle is a large curvature multi-bend shaped tube flow component, which has geometrical features such as multi-dimensional deflection configuration, shaped cross-section transition, and fusion of the binary shaped nozzle with the rear body of the aircraft. Multi-dimensional deflection and complex shaped cross-section can block the high temperature parts of the nozzle inlet, reducing the infrared radiation characteristics of the engine; in the binary shaped cross-section is easy to realize the fusion of the nozzle and the rear body of the aircraft, which reduces the fuselage rear body drag, and improves the stealth capability of the aircraft. Therefore, the design of the S-shaped nozzle focuses on the multi-dimensional deflection configuration and heteromorphic cross-section characteristics under the low detectability constraint criterion. However, the buried installation of the S-shaped nozzle makes the nozzle geometric configuration subject to the spatial constraints of the fuselage profile, which makes it difficult to realize the S-shaped nozzle profile design with high aerodynamic performance under the constraints of the limited layout. Based on this, this section proposes the idea of variable cross-section design based on multi-parameter coupling, and develops the S-shaped nozzle profile design method under the constrained spatial layout conditions, which can effectively extract the key parameters characterizing the geometrical configuration of the S-shaped nozzle and design the high-performance nozzle to meet the requirements of the spatial layout. The following S-shaped nozzle profile design process is described in detail.

2.2.1 Overview of S-Shaped Nozzle Profile Design Methods

When designing the S-shaped nozzle under the restricted space layout, firstly, based on the restricted space layout characteristics of the S-shaped nozzle, the selection range of key parameters affecting the geometric configuration of the nozzle is determined. On this basis, a multi-parameter coupled variable cross-section design method is adopted to establish a low-detectability criterion for completely shielding the high-temperature components, and based on the cross-section area constraints and the low-detectability criterion constraints, the S-shaped nozzle profile meeting the requirements of the constrained spatial layout is finally designed, as shown in Fig. 2.1, and the specific design flow is as follows:

1. Determine the layout of the S-shaped nozzle based on the scope of the space layout and the location of the inlet and outlet;
2. Determine the selection range of key geometric parameters based on the layout form of the S-shaped nozzle;
3. The centerline design of the S-shaped nozzle is carried out according to the values of key parameters and the nozzle layout form;
4. Design of the nozzle along the flow section based on the shape of the nozzle cross-section;
5. Establishment of low detectability guidelines for complete shielding of high temperature components;
6. Completing the generation of the S-shaped nozzle profile, and trimming the nozzle cross-section according to the space layout and the rear body of the airplane.

Based on the S-shaped nozzle design process outlined above, under the constraints of limited spatial layout and low observability requirements, various practical engineering-oriented S-shaped nozzle configurations can be developed as shown in Fig. 2.2. These include unidirectional/multidimensional deflection nozzles, with the number of bends ranging from single, double to triple curves. Cross-sectional

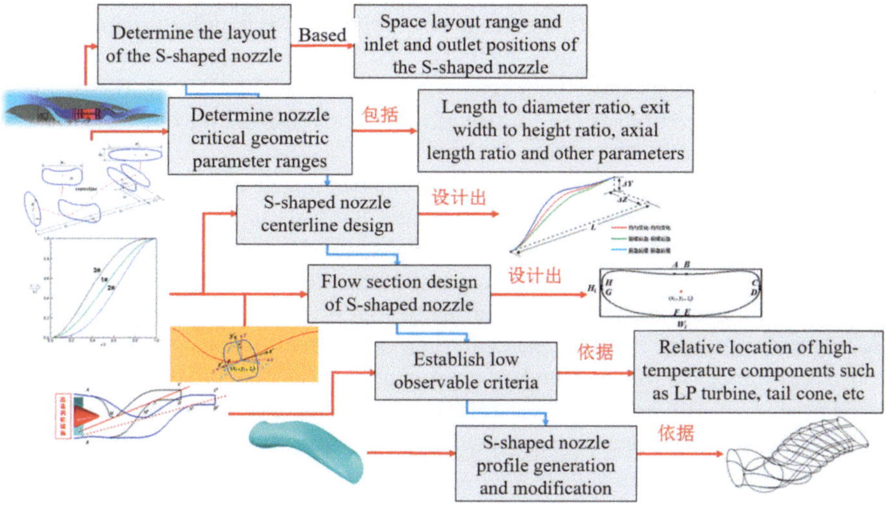

Fig. 2.1 S-shaped nozzle design flow for constrained space layout

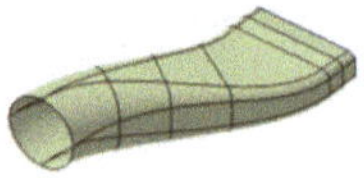

(a) Spreading deflection single-bend rectangular profile

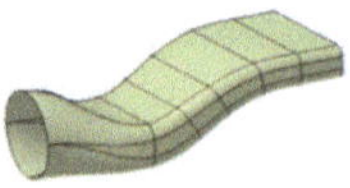

(b) Spreading deflection double-bend rectangular profile

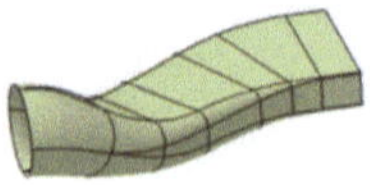

(c) Multi-dimensional deflection multi-bend rectangular profile

(d) Multi-dimensional deflected multi-bending ellipsoid-like surface

(e) Multi-dimensional deflected multi-bending pea-shaped surface

Fig. 2.2 Multiple S-shaped nozzle configurations can be designed. (**a**) Spreading deflection single-bend rectangular profile. (**b**) Spreading deflection double-bend rectangular profile. (**c**) Multi-dimensional deflection multi-bend rectangular profile. (**d**) Multi-dimensional deflected multi-bending ellipsoid-like surface. (**e**) Multi-dimensional deflected multi-bending pea-shaped surface

shapes may adopt rectangular-like, oval-like, or pea-like profiles, where the specific configuration depends on design constraint requirements.

The spatial layout of the S-shaped nozzle range and the location of the import and export of the S-shaped nozzle layout form, including the direction of deflection and bending form, such as unidirectional deflection or multi-dimensional deflection, a single bend or multiple bends, the direction of the bending is upward or downward, and the exit cross-section shape of the class of rectangle or class of ellipse. According to the layout of the S-shaped nozzle to further determine the selection range of key geometric parameters, including the length-to-diameter ratio L/D, the exit width-to-height ratio W_e/H_e, the first bend of the longitudinal deflection $\Delta Y_1/L_1$, the two bends of the axial length of the ratio of L_1/L_2 and so on. The two-step analysis basically completes the construction of the overall profile characteristics of the S-shaped nozzle in the restricted space.

After clarifying the spatial layout of the S-shaped nozzle and the range of values of key geometric parameters, according to the multi-dimensional deflection configuration of the S-shaped nozzle, shaped cross-section transition and binary shaped nozzle and other geometric features, a multi-parameter coupling of variable cross-section design method can be used for the design of S-shaped nozzle profiles, the core content of which includes the centerline design and the design of the circulation cross-section. In view of the Lee curve method is conducive to the advantages of the pipeline parametric study, can refer to the Lee curve method for the centerline of the S-shaped nozzle and flow section design, the design should follow the following requirements:

1. Centerline design of the S-shaped nozzle. The shape of the centerline is determined by the spatial layout form of the nozzle, and when the centerline design is completed, the curved shape of the nozzle channel is determined, and the deflection of the airflow inside the nozzle is also determined. And the length and deflection of the centerline are determined by the key geometric parameters. In addition, the centerline design must be completed to determine the center of the section before the nozzle flow section design. Therefore, the centerline design is the basis of the S-shaped nozzle profile design.
2. Binary shaped flow cross-section design. The shape of the flow section is also determined by the spatial layout of the nozzle, the inlet of the nozzle is connected to the outlet of the circular mixing chamber, and the outlet is usually a binary nozzle, so the flow section of the nozzle needs to be smoothly transitioned from the circular inlet cross-section to the binary heterogeneous outlet cross-section. The dimensions of the flow cross-section are determined by key geometrical parameters, including cross-section width, cross-section height, etc. The cross-section area of the nozzle is constrained by the cross-section area. Under the constraint of the cross sectional area, the geometrical parameters along the cross section it involves need to be corrected. In order to meet the requirements of spatial layout, the local circulation cross section of the S-shaped nozzle needs to be trimmed.

In order to improve the stealth capability of the S-shaped nozzle, it is necessary to realize the complete shielding of the inlet high-temperature parts through its bending configuration during the nozzle design process, i.e., to establish the low-detectability criterion applicable to different S-shaped nozzle configurations, and to make clear the coupling relationship of the geometric parameters of the nozzle brought about by the low-detectability criterion, and to incorporate this condition into the design process of the nozzle. After completing the profile generation of the S-shaped nozzle, it is necessary to carry out various modifications to the nozzle exit shape, such as beveling, adding convergent segments, etc., in order to realize the profile fusion of the nozzle exit and the aft body of the aircraft, and to improve the practical engineering application value of the nozzle.

2.2.2 *Critical Geometric Parameters of S-Shaped Nozzle*

The key geometrical parameters of the S-shaped nozzle are determined by its spatial layout, including the deflection form of the S-shaped nozzle and the number of S-curves, such as unidirectional or multidimensional deflection, single S-shaped or double S-shaped, and whether the outlet cross-section is rectangle-like or ellipse-like. Two S-shaped nozzle configurations are selected here for detailed description. The first is a unidirectionally deflected double S-shaped nozzle, which has a rectangular-like outlet cross-section; the second is a multidimensionally deflected spreading single S-shaped, longitudinal double S-shaped nozzle, which has an elliptical-like outlet cross-section. The key geometrical parameters involved in the unidirectionally deflected S-shaped nozzle include the exit area A_1, exit width W_1, axial length L_1, longitudinal deflection ΔY_1 for the first S-shaped channel, axial length L_2, longitudinal deflection ΔY_2 for the second S-shaped channel, the exit width-to-height ratio W_e/H_e, and the isotropic section length L_3 as shown in the Fig. 2.3a, which are converted into 7 sets of dimensionless parameters, including the aspect ratio L/D, the isotropic section length L_3/L, the ratio of the axial lengths of the two S-shaped lines L_1/L_2, the area of the first bend cross-section A_1/A_{in}, the width of the first bend cross-section W_1/D, the longitudinal deflection of the first bend $\Delta Y_1/L_1$, and the exit aspect ratio W_e/H_e. In contrast, the multidimensional deflected S-shaped nozzle adds the first bend spreading deflection ΔZ_1 in addition to the above geometrical parameters, which corresponds to the dimensionless parameter $\Delta Z_1/(L_1 + L_2)$, as shown in Fig. 2.3b. Determining the range of values of the key parameters provides specific design constraints for the subsequent design of the S-shaped nozzle centerline and flow section.

2.2.3 S-Shaped Nozzle Centerline Design

The shape of the centerline of the S-shaped nozzle is determined by the spatial layout of the nozzle, and the curved configuration of the centerline is designed to achieve effective shielding of the high-temperature engine parts, making them difficult to be directly detected by infrared-guided weapons. The direction and number of centerline deflection determine the name of the S-shaped nozzle. For the nozzle with unidirectional deflection of the centerline, it is named as longitudinal (spreading) S-shaped nozzle according to its deflection direction; the centerline consists of one or two S-shaped lines, named as single (double) S-shaped nozzle, as shown in Fig. 2.4a, b. For the nozzle with multi-dimensional deflection of the centerline, it is named by combining the number of S-shaped curves in each direction of deflection,

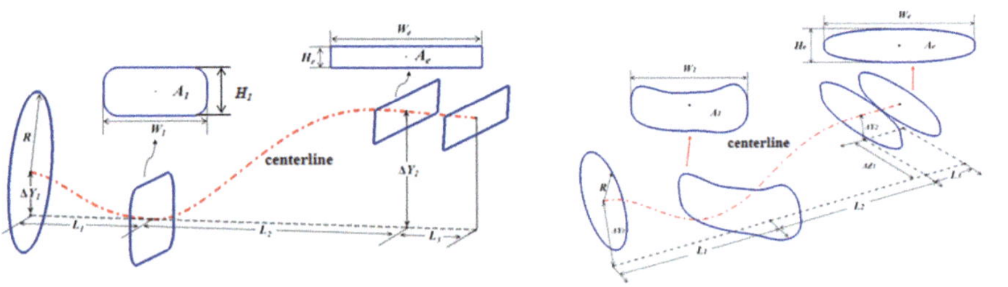

(a) Double S-shaped nozzle with unidirectional deflection

(b) S-shaped nozzle with multi-dimensional deflection

Fig. 2.3 Key Geometric Parameters of the S-Bend Nozzle. (**a**) Double S-shaped nozzle with unidirectional deflection. (**b**) S-shaped nozzle with multi-dimensional deflection

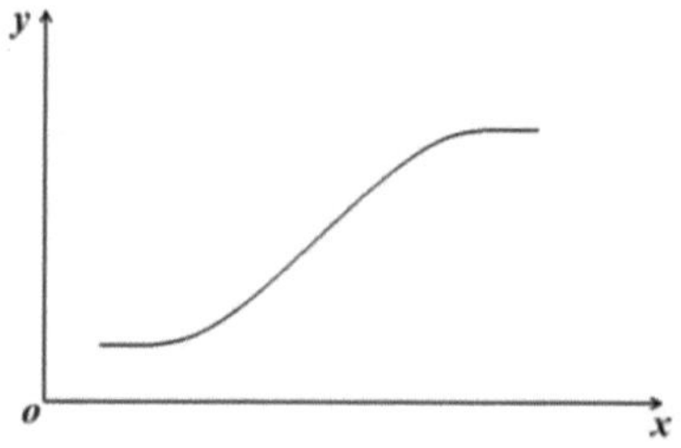

(a) Longitudinal single S-shaped nozzle centerline

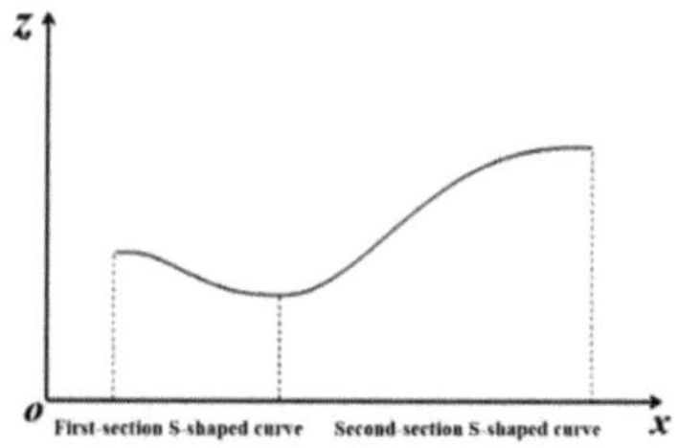

(b) Spreading double S-shaped nozzle centerline

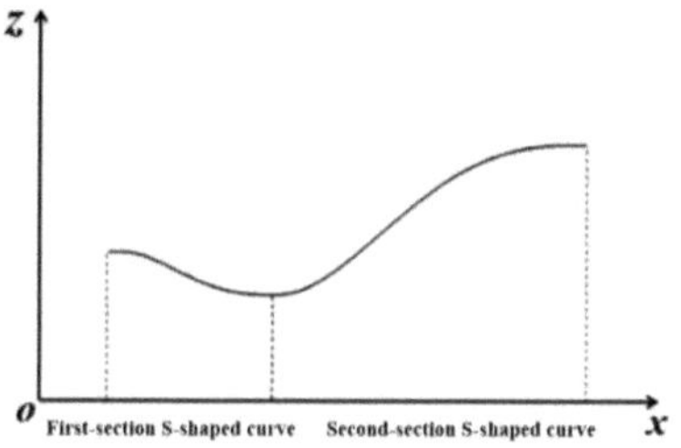

(c) Centerline of a multi-dimensional deflected S-shaped nozzle with two bends in the longitudinal direction and one bend in the spreading direction

Fig. 2.4 Schematic of the centerline of the S-shaped nozzle. (**a**) Longitudinal single S-shaped nozzle centerline. (**b**) Spreading double S-shaped nozzle centerline. (**c**) Centerline of a multi-dimensional deflected S-shaped nozzle with two bends in the longitudinal direction and one bend in the spreading direction

e.g. For example, if the centerline of the S-shaped nozzle consists of two S-shapeds along the longitudinal direction and one S-shaped along the spreading direction, then it is named as a multi-dimensional deflected S-shaped nozzle with two bends in the longitudinal direction and one bend in the spreading direction, as shown in Fig. 2.4c. The derivatives of the centerline at both the start and end points are equal to zero, in order to ensure that the airflow directions of the nozzle inlet and outlet are both parallel to the engine axis.

Many mathematical formulas have been used to construct curved profiles in aero-engine exhaust systems, such as Vickers curves are commonly used to generate convergent nozzle profiles, and CST curves are used to construct nacelle profiles. In view of the Lee curve method (Lee et al. 1985) is mostly used for the centerline design of curved pipes, and is conducive to the advantages of pipe parameterization research, this section mainly uses Lee curve for the centerline design of S-shaped nozzle, Lee curve change rule there are three kinds of changes, such as Fig. 2.5 shown in the mathematical expressions such as formula (2.1) to formula (2.3).

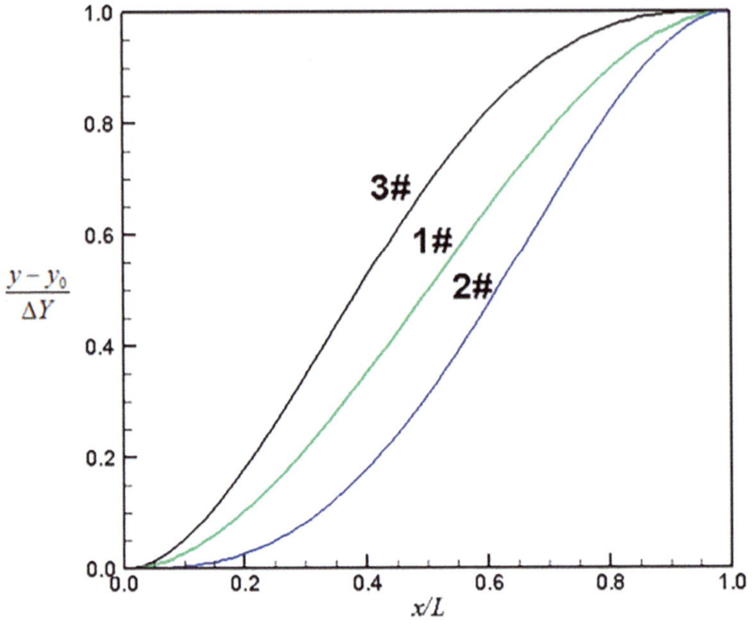

Fig. 2.5 Lee curve change pattern

$$1\#,\text{uniform variation}, y_i = y_0 + \Delta Y \times \left[3 \times \left(\frac{x_i}{L}\right)^2 - 2 \times \left(\frac{x_i}{L}\right)^3\right] \tag{2.1}$$

$$2\#,\text{front slow, back fast}, y_i = y_0 + \Delta Y \times \left[-3 \times \left(\frac{x_i}{L}\right)^4 + 4 \times \left(\frac{x_i}{L}\right)^3\right] \tag{2.2}$$

$$3\#,\text{front rush}, y_i = y_0 + \Delta Y \times \left[3 \times \left(\frac{x_i}{L}\right)^4 - 8 \times \left(\frac{x_i}{L}\right)^3 + 6 \times \left(\frac{x_i}{L}\right)^2\right] \tag{2.3}$$

where y_0 is the longitudinal coordinate of the starting point of each segment of the S-shapeding line, ΔY is the difference between the longitudinal coordinates of the starting point and the termination point of each segment of the S-shapeding line, L is the axial length of the S-shapeding line, and i denotes an arbitrary coordinate position on the S-shapeding line.

The rule of change of the center line determines the S-shaped nozzle inside the flow direction of the pressure gradient distribution, "uniform change", "before the slow after the rapid" and "before the rapid after the slow" rule of change of the curve will lead to different velocity gradient distribution inside the nozzle. S-shaped nozzle centerline change rule and constitute the centerline of the number of S-shaped line, for single S-shaped nozzle, the centerline consists of a Lee curve, the corresponding change rule of the centerline has three kinds. For the unidirectional deflection of the double S-shaped nozzle, the centerline consists of two Lee curves, the corresponding change rule of the centerline of a total of nine kinds of laws, Fig. 2.6 gives "uniform change–uniform change", "before the slow and then the rapid–before the slow and then the rapid," and The three typical patterns of centerline change are "uniform change-uniform change", "slow before slow after slow" and "slow before slow after slow". For the multi-dimensional deflection S-shaped nozzle, take the multi-dimensional deflection S-shaped nozzle with longitudinal double bends and one bend in the spread direction as an example, there are 27 kinds of corresponding centerline variation rules, and so on. Figure 2.7 show the "uniform change" in the spreading direction, "uniform change-uniform change", "slow front and rapid back–slow front and rapid back" and "slow front and rapid back" in the longitudinal direction.

The centerline design of the S-shaped nozzle requires determining the number of curves in the centerline based on the spatial layout and obtaining the design parameters of the centerline based on the key geometric parameters, including: the total axial length of the centerline (including the centerline length and isoclinic length), the ratio of the axial lengths of each segment of the S-shaped line, and the longitudinal and spreading offsets of the centerline. The design process of the centerline of the S-shaped nozzle is as follows:

1. According to the key geometric parameters of the nozzle, determine the axial length L, longitudinal deviation ΔY and spreading deviation ΔZ of the centerline of the S-shaped nozzle, take the nozzle inlet center as the origin, x-axis as the axial direction, y-axis as the longitudinal direction, and z-axis as the spreading direction, and arrange a series of uniformly arranged control points along the axial direction from the inlet to the outlet of the nozzle, and the axial coordinates of the control points are x_i, y_i and z_i (where i = 1, 2,…K, then there are a total of K cross sections along the course). The axial coordinates of each control point are x_i, the longitudinal coordinates are y_i, and the spreading coordinates are z_i (where i = 1, 2,…K, there are a total of K cross sections along the course).
2. According to the spatial layout of the nozzle, determine the number of S-shaped lines M_z and M_y along the centerline in the spreading direction and longitudinal direction, respectively. Based on the proportionality of the axial lengths of each segment of the S-shapeding line, determine the axial lengths of each segment of

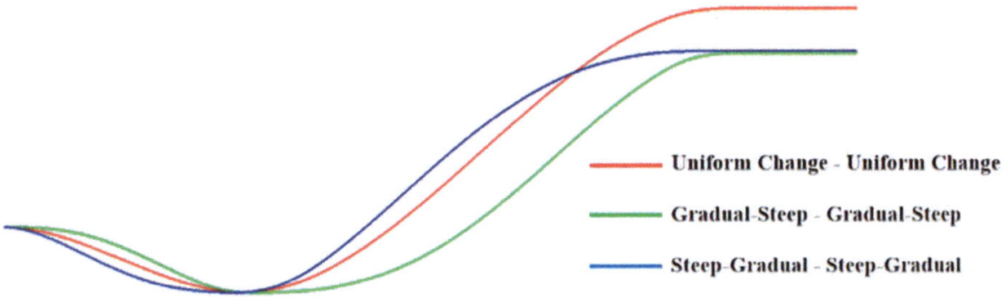

Fig. 2.6 Change rule of centerline of double S-shaped nozzle

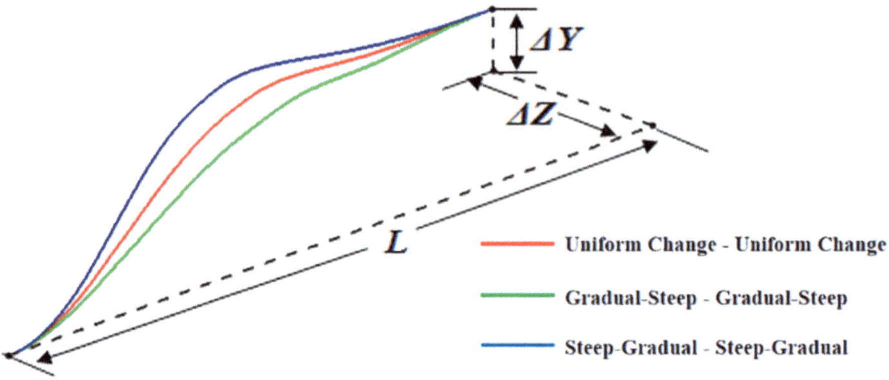

Fig. 2.7 Changing law of centerline of S-shaped nozzle with multi-dimensional deflection

the S-shapeding line in the longitudinal direction L_{jy} (j = 1, 2…M) and the

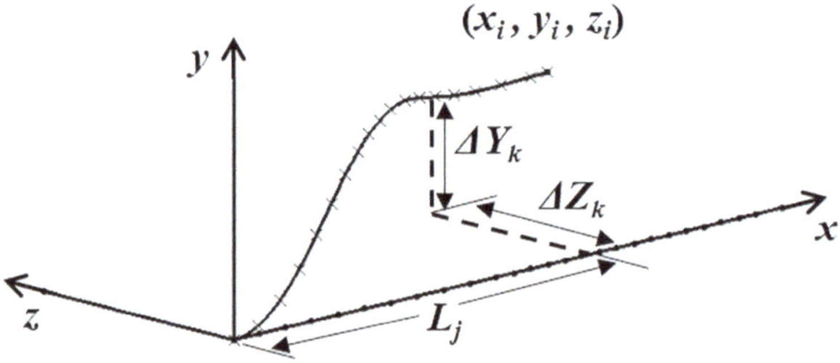

Fig. 2.8 Centerline of a multi-dimensional deflected S-shaped nozzle with one bend in the spread and two bends in the longitudinal direction

spreading direction L_{jz} (j = 1, 2…M), respectively, and obtain the corresponding longitudinal deviation of each segment of the S-shapeding line ΔY_j (j = 1, 2…M) or spreading deviation ΔZ_j (j = 1, 2…M) corresponding to each S-shapeding line.

3. According to the changing law of Lee curve shown in Fig. 2.5, different relations (2.1) to (2.3) are chosen for different S-shapeding lines to calculate the longitudinal coordinates $y_i$and z_i of each control point, so as to obtain a series of coordinates (x_i, y_i, z_i) of the discrete points.
4. Three times natural spline interpolation is used to fit these discrete points into a spatial 3D S-shaped line to complete the centerline design of the S-shaped nozzle. Figure 2.8 show the centerline of the multidimensional deflected S-shaped nozzle with one bend in the spreading direction and two bends in the longitudinal direction.

2.2.4 *S Curved Nozzle Flow Cross-Section Design*

The shape of the circulation cross section of the S-shaped nozzle is determined by the spatial layout of the nozzle, the nozzle inlet is connected with the outlet of the circular mixing chamber, and the outlet is usually a binary nozzle, thus the circulation cross section of the nozzle is gradually transitioned from the circular inlet cross section to the binary heteromorphic outlet cross section, which results in the change of the area, width, height, and cross-section shape of the circulation cross section. Based on the multi-parameter coupling variable cross-section design method, according to the nozzle inlet and outlet cross-section shape and cross-section parameters, the Lee curve change rule is used to obtain the cross-section area, cross-section width, cross-section height and other parameters along the cross-section, and through the cross-section area constraints on the cross-section parameters along the correction, and ultimately obtained along the cross-section geometric parameters, to complete the design of the S-curve nozzle circulation cross-section. The center of the circulation section of the S-shaped nozzle is located at the discrete control points on the centerline, so the design of the circulation section is to calculate the cross-section parameters and cross-section shape design at the discrete points on the centerline.

Binary shaped outlet cross sections are widely used in both rectangular and elliptical configurations. Three different cross-sectional shapes are commonly chosen for the design of S-shaped nozzle flow cross sections as shown in Fig. 2.9 are: "rectangular-like", "elliptical-like", and "pea-shaped". "Pea-shaped".

2.2.4.1 "Rectangular-like" Circulation Cross Sections

The "rectangle-like" circulation cross-section, shown in Fig. 2.10, consists of four straight segments and four 1/4 arcs, which are tangent to the arcs, forming eight characteristic points. The shape of the cross-section is uniquely determined by its area A_i, width W_i, height H_i, and chamfer radius R_i, and satisfies the area eq. (2.4).

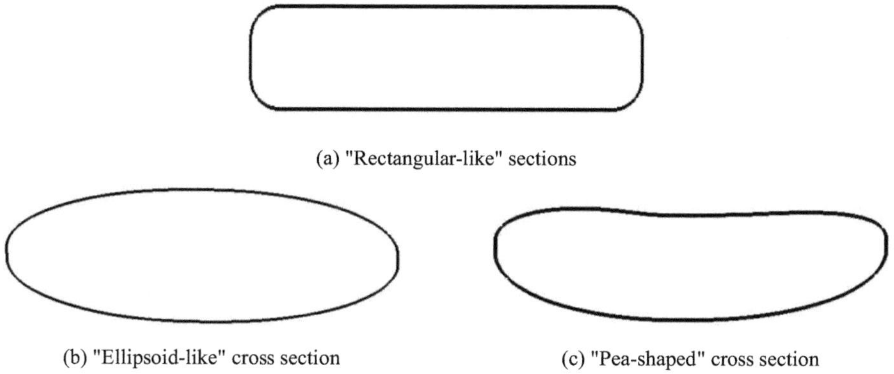

Fig. 2.9 Three different flow cross-section shapes. (**a**) "Rectangular-like" sections. (**b**) "Ellipsoid-like" cross section. (**c**) "Pea-shaped" cross section

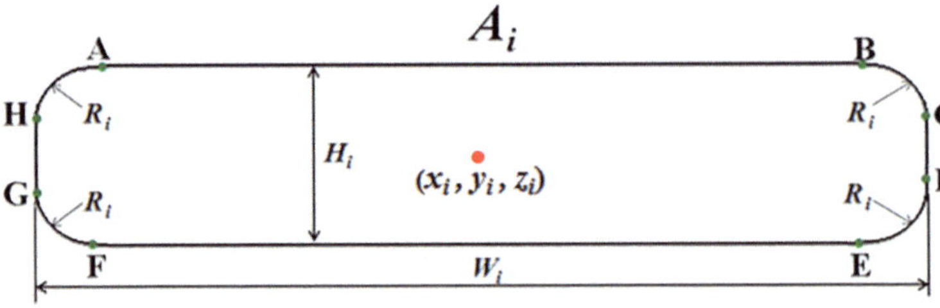

Fig. 2.10 "Rectangle-like" circulation cross-section

$$A_i = W_i \times H_i - (4 - \pi) \times R_i^2 \tag{2.4}$$

In the design of circulation section of "rectangle-like", the area A_i, width W_i, height H_i and chamfer R_i of the cross-section are usually calculated by using the rule of change of Lee curve, and the height of the cross-section is corrected according to the area formula (2.4). Then determine whether the section width W_i and section height H_i satisfy the constraints in Eq. (2.5):

$$W_i \geq 2R_i, H_i \geq 2R_i \tag{2.5}$$

If satisfied, the parametric design of the cross-section shape is completed. According to the determined parameters A_i, W_i, H_i and R_i of the cross-section, the design of the cross-section profile line is started. If they are not satisfied, the design conditions for the cross-section shape are satisfied by correcting the chamfer $R_{i.}$

2.2.4.2 "Ellipsoid" Shaped Circulation Cross Section

The "ellipsoid-like" shape of the circulation cross-section is shown in Fig. 2.11, which consists of 4 straight line segments and 4 F-value curves, which are tangent to the curves, forming 8 characteristic points. The shape of the cross-section is uniquely determined by its area A_i, width W_i, height H_i and quadratic curve F_i and satisfies the area eq. (2.6).

$$A_i = W_i \times H_i - 4 \times A_{Fi}^2 \tag{2.6}$$

where A_{Fi} is the area enclosed by the quadratic curve and the rectangular border.

In the design of circulation cross-section of "elliptic-like", the area A_i, width W_i, height H_i of the cross-section and F value F_i of the quadratic curve are usually calculated by the rule of change of the Lee curve, and then the equations of the four F-value curves are calculated, and A_{Fi} is obtained by integrating, and the height of the cross-section is corrected according to the area formula (2.7), so as to complete the design of the cross-section profile line. The area eq. (2.7) is corrected for the height of the section, so as to complete the design of the profile line of the section.

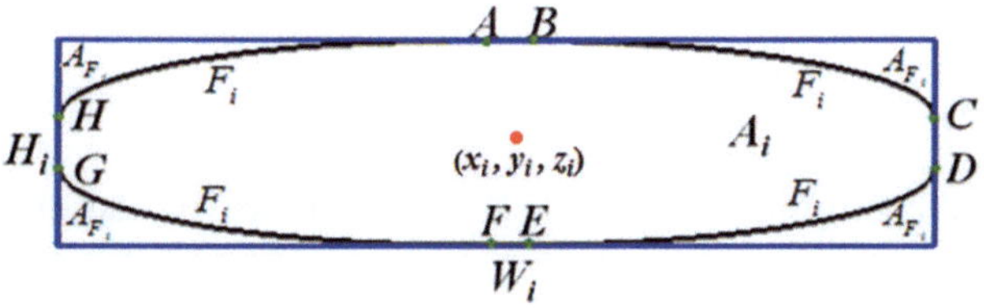

Fig. 2.11 "Ellipse-like" circulation cross-section

Fig. 2.12 Schematic of F-value curve

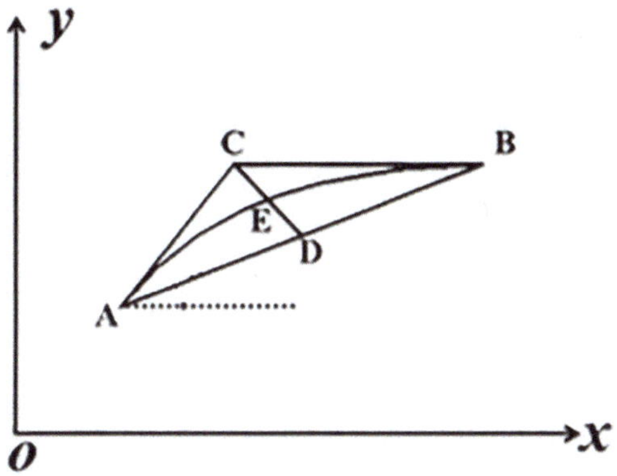

The F-value curve refers to the quadratic curve that determines the bending degree of the curve by the magnitude of the F-value, and the expression of the curve is:

$$ax^2 + bx + cxy + dy^2 + ey + f = 0 \tag{2.7}$$

Solve the F-value curve equation, that is, through the known conditions to solve the unknown variables a ~ f, known conditions include the curve F-value, the coordinates of the two endpoints, and the slope of the tangent line of the curve at the two endpoints. The solution process is as follows:

For the quadratic curve AB shown in Fig. 2.12, the coordinates of the two endpoints A (x_1, y_1) and B (x_2, y_2) and the slopes of the tangents to the curve at the two endpoints are known.

1. Find the intersection C of the tangents at the two endpoints A and B (x_c, y_c).
2. Find the midpoint D (x_m, y_m) of the line segment AB.

$$x_m = \frac{x_1 + x_2}{2} \tag{2.8}$$

$$y_\text{m} = \frac{y_1 + y_2}{2} \tag{2.9}$$

3. The magnitude of the F-value is defined as the ratio of the length of the line segment DE to the length of the line segment CD.

$$F = \frac{|DE|}{|CD|} \tag{2.10}$$

4. Determine the coordinates of point E (x_0, y_0), based on the magnitude of the value of F.

$$x_0 = x_m + (x_c - x_m) \times F \tag{2.11}$$

$$y_0 = y_m + (y_c - y_m) \times F \tag{2.12}$$

5. Provide that the slope of the tangent to the quadratic curve at point E is equal to the slope of the line AB, and thus the values of the coordinates of the points A, B, and E, as well as the slopes of the tangents at that location, are known, and the value of the quadratic equation a ~ f can be solved and the quadratic equation can be found.

2.2.4.3 "Pea-Shaped" Circulation Cross-Section

Compared with other cross-section shapes, the "pea-shaped" circulation cross-section can significantly reduce the longitudinal and transverse deflections of the S-shaped nozzle and minimize the bending characteristics of the nozzle while effectively shielding the high-temperature components. The shape of the cross section is shown in Fig. 2.13. The cross section consists of 4 straight line segments and 8 curves, the tangent points of the straight line segments and the curves form 8 characteristic points, and each curve segment consists of two F-value curves spliced together. The design process of the "pea-shaped" cross-section is the same as that of the "ellipsoid-like" circulation cross-section, and its area A_i, width W_i, height H_i and quadratic curve $F_i$are uniquely determined and satisfy the area eq. (2.6).

The design of the flow section of the S-shaped nozzle requires the determination of the outlet cross-section shape based on the spatial layout and the acquisition of the outlet cross-section parameters of the nozzle based on the key geometric parameters, including: cross-section width, cross-section height, cross-section area, and

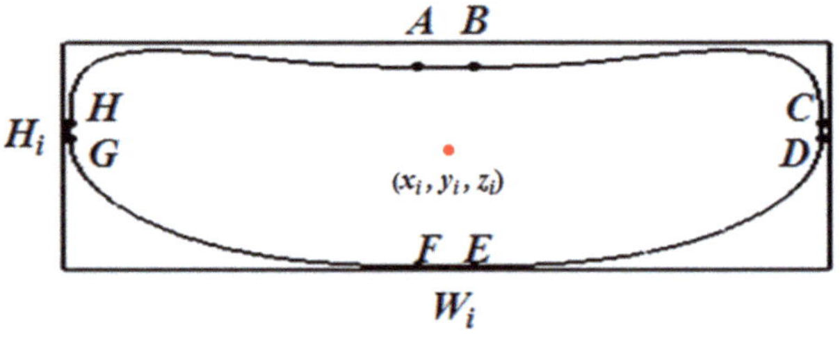

Fig. 2.13 "Pea-shaped" circulation cross-section

chamfering radius or curve F. The process of designing the flow section of the S-shaped nozzle is as follows:

1. Determine whether the outlet cross-section shape is "rectangular-like" or "elliptic-like" according to the constraints of the profile in the outlet region of the S-shaped nozzle, and determine the inlet cross-section diameter *d*, outlet cross-section width W_{ex}, and cross-section height H_{ex} through the range of geometric parameter selection. The geometric parameters are selected to determine the inlet section diameter *d*, outlet section width W_{ex} and section height H_{ex}.
2. Complete the calculation of the cross-sectional parameters of the nozzle inlet and outlet, and the formula for calculating the cross-sectional area of the inlet is as follows:

$$A_{\text{in}} = \frac{\pi}{4} d^2 \tag{2.13}$$

where *d* is the engine diameter, the inlet width W_{in} and inlet height H_{in} of the nozzle are equal to the diameter.

The exit area of the nozzle is calculated according to the engine performance parameters, and is obtained from the gas flow rate at the nozzle inlet $\dot{m}$, the total pressure P^*, the total temperature T^*, and the nozzle total pressure recovery coefficient σ, that is:

$$A_{\text{e}} = \frac{\dot{m} \cdot \sqrt{T^*}}{K \cdot \sigma \cdot P^*} \tag{2.14}$$

3. Adopt the rule of change of Lee curve, through the import and export cross-section parameter calculation to get the circulation cross-section parameter at each control point on the center line. For "rectangle-like" cross-section, the cross-section parameters include cross-section area A_i, cross-section width W_i, cross-section height H_i and chamfering angle R_i, and the computational expressions for solving the cross-section parameters by the law of "uniform change" are as follows:

$$\begin{cases} A_i = A_{in} + \left[3\times\left(\frac{x_i}{L}\right)^2 - 2\times\left(\frac{x_i}{L}\right)^3\right]\times\left(A_{ex} - A_{in}\right) \\ W_i = W_{in} + \left[3\times\left(\frac{x_i}{L}\right)^2 - 2\times\left(\frac{x_i}{L}\right)^3\right]\times\left(W_{ex} - W_{in}\right) \\ H_i = H_{in} + \left[3\times\left(\frac{x_i}{L}\right)^2 - 2\times\left(\frac{x_i}{L}\right)^3\right]\times\left(H_{ex} - H_{in}\right) \\ R_i = R_{in} + \left[3\times\left(\frac{x_i}{L}\right)^2 - 2\times\left(\frac{x_i}{L}\right)^3\right]\times\left(R_{ex} - R_{in}\right) \end{cases} \tag{2.15}$$

4. According to the shape of the circulation cross-section, different ways of cross-section correction are selected from to determine the profile line characteristics of the cross-section. Knowing the coordinates of each characteristic point on the cross-section and the curve equations between each characteristic point, each curve segment is discretized into a finite number of points by the discretization method. The discretization diagram of the cross-section profile line of the "rectangle-like" shape is shown in Fig. 2.14, so as to get the coordinates of each discretized point on the cross-section profile line (x_k, y_k, z_k).
5. The layout of the circulation cross section of the S-shaped nozzle is usually made in two ways, the box cut section and the normal section, as shown in Fig. 2.15, where the box cut section is to lay out different numbers of sections along the direction perpendicular to the axis, and the normal section is to lay out different numbers of sections along the direction perpendicular to the centerline. In the actual flow field inside the curved channel, the tangent direction of the centerline is basically the same as the airflow direction. In order to ensure that the airflow meets the changing characteristics of the nozzle flow path, the normal profile is used to ensure that the flow sections along the nozzle are perpendicular to the tangent direction of the centerline, which can more accurately reflect the flow state of the airflow through the sections along the nozzle. Therefore, it is necessary to rotate the coordinates of the discrete points on the flow section based on the centerline, as shown in Fig. 2.16.

   Assume that the circulation cross-section at the point (x_i, y_i, z_i) on the centerline of the S-shaped nozzle is rotated around the z-axis by an angle α and around the y-axis by an angle β. The coordinates of a point on the circulation cross-section before rotation are (x_k, y_k, z_k), after rotation around the z-axis (x_m, y_m, z_m), and after rotation around the y-axis (x_n, y_n, z_n), then:

$$x_m = (x_k - x_i)\cdot\cos\alpha + (y_k - y_i)\cdot\sin\alpha + x_i \tag{2.16}$$

$$y_m = -(x_k - x_i)\cdot\sin\alpha + (y_k - y_i)\cdot\cos\alpha + y_i \tag{2.17}$$

$$z_m = z_k \tag{2.18}$$

$$x_n = (x_m - x_i)\cdot\cos\beta + (z_m - z_i)\cdot\sin\beta + x_i \tag{2.19}$$

$$z_n = -(x_m - x_i)\cdot\sin\beta + (z_m - z_i)\cdot\cos\beta + z_i \tag{2.20}$$

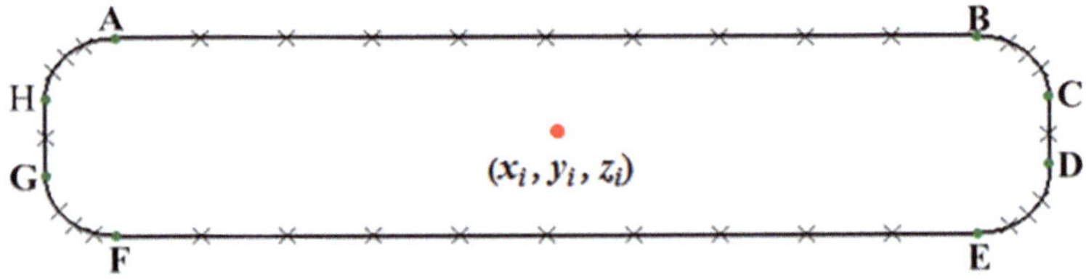

Fig. 2.14 Discretization of the cross-section lines of a "rectangle-like" shape

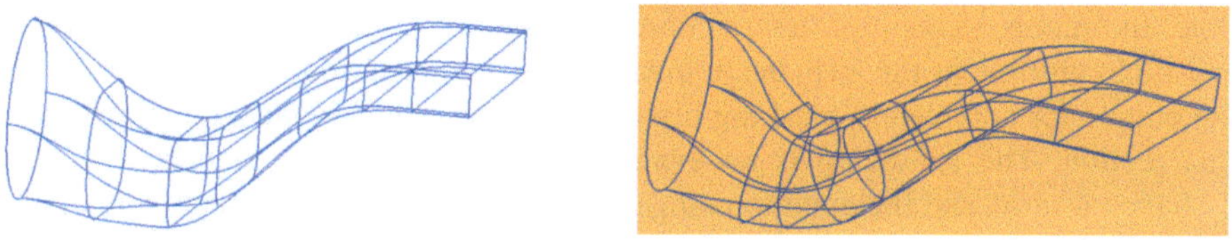

(a) Box section (b) Normal section

Fig. 2.15 S-shaped nozzle circulation cross-section layout method. (**a**) Box section. (**b**) Normal section

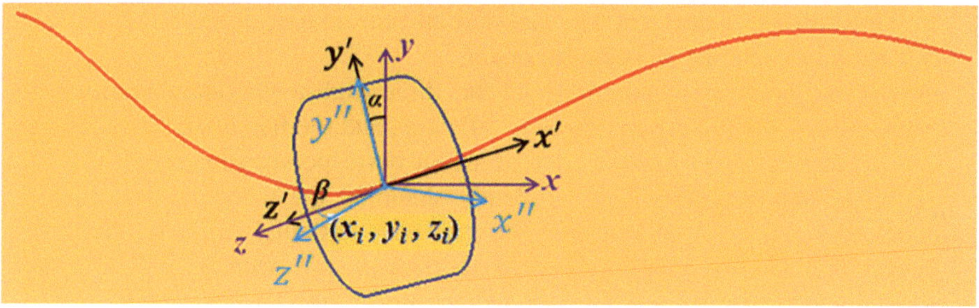

Fig. 2.16 Schematic diagram of rotational transformation of the flow section

$$y_n = y_m \tag{2.21}$$

The coordinates of the discrete points on the rotated circulation section (x_n, y_n, z_n) are obtained from the above equation, and the three times natural spline interpolation is used to fit these discrete points into the profile line, and finally the design of the circulation section of the S-shaped nozzle is completed.

2.2.5 Establishment of Low Detectability Criteria for S-Shaped Nozzles

The high-temperature component masking technology is considered to be a way that can effectively enhance the stealth capability of the exhaust system (Baranwal and Mahulikar 2016). The S-curved profile of the S-curved nozzle can effectively mask the high-temperature components of the engine, which significantly reduces the intensity of infrared radiation and electromagnetic scattering intensity of the engine. Compared with turbojet engines, it is more difficult to realize effective shading of high-temperature parts of turbofan engine exhaust system, which includes high-temperature parts flushed by hot jets, such as tail cone and flap mixer in the exhaust mixer, in addition to low-pressure turbine parts, thus it is necessary to consider the effective shading of multiple high-temperature parts at the same time. In order to make high-temperature components impossible to be detected in any flight attitude of the airplane, a low-detectability criterion based on the turbofan engine for completely masking high-temperature components is established in this section. The following is an example of a unidirectional deflected double S-shaped nozzle, and the common tangent law is chosen to realize complete masking of the high-temperature parts of the S-shaped nozzle. The specific process is as follows:

1. Analyze the type and number of high-temperature components to be shielded by the S-shaped nozzle based on the engine type; in the case of a turbofan engine, for example, the high-temperature components in front of the inlet of the S-shaped nozzle include the low-pressure turbine as well as the tail cone in the exhaust mixer and the wavelength mixer.
2. Determine the relative positions of the three high-temperature components, namely the low-pressure turbine, the tail cone, and the flap mixer, through geometric measurements, and then determine the corresponding longitudinal offsets of the S-shaped nozzles by completely shielding the three high-temperature components, so as to make clear the shielding difficulty of each component. Taking the exhaust mixer configuration given in this section as an example, when the flap mixer is completely blocked, the low-pressure turbine and the tail cone are not visible, which means that the blocking difficulty of the flap mixer is relatively the greatest.
3. The common tangent rule is used so that the common tangent of the lower longitudinal line on the symmetry plane of the double S-shaped nozzle passes through the upper apex C of the nozzle outlet or the lower apex B of the end of the wave mixer, i.e., M, N, and C are co-located or M, N, and B are co-located as shown in Fig. 2.17.

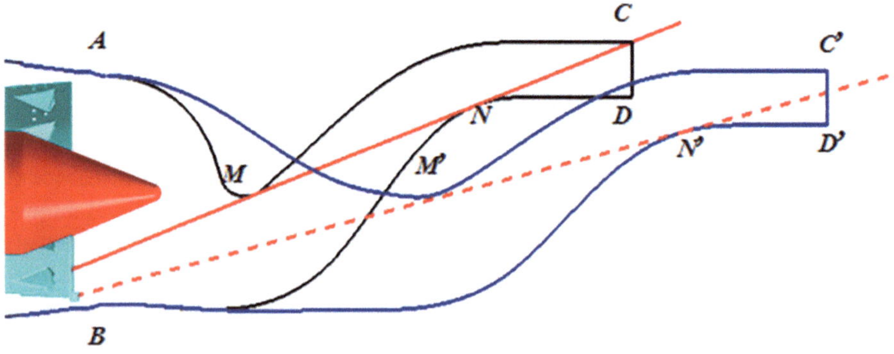

Fig. 2.17 Schematic diagram of a unidirectionally deflected double S-shaped nozzle completely shielding high temperature components

2.2.6 *Profile Generation and Trimming for S-Shaped Nozzles*

After completing the design of the centerline and circulation section of the S-shaped nozzle, the corresponding feature points on each section are fitted into 8 longitudinal lines using three times natural spline interpolation, and the multisection surface is generated according to the section profile line and the 8 longitudinal lines, that is, the S-shaped nozzle's profile, as shown in Fig. 2.18.

Under some extreme space layout constraints, the S-shaped nozzle profile designed by the above method cannot fully meet the space layout requirements, and it is necessary to carry out local cross-section modification according to the space layout of the nozzle, as shown in Fig. 2.19, to modify the symmetric cross-section of the "pea-shape" into an asymmetric cross-section in order to meet the space layout requirements. In the integration design of the S-shaped nozzle outlet and the rear body of the aircraft, it is necessary to trim the nozzle outlet according to the shape of the rear body of the aircraft to realize the integration of the nozzle outlet and the rear body of the fuselage, as shown in Fig. 2.20.

2.3 Flow Characterization of S-Shaped Nozzle for Turbofan Engine

Compared with the turbojet engine, the complex structure of the exhaust mixer before the inlet of the S-shaped nozzle for the turbofan engine will have an impact on the internal flow field characteristics of the nozzle, which in turn will change the aerodynamic performance of the S-shaped nozzle (Sun et al. 2018). In this section, a numerical simulation study of the flow characteristics of the S-shaped nozzle under double culvert conditions is carried out. Firstly, experimental measurements of the flow field distribution characteristics and aerodynamic performance parameters of the S-shaped nozzle under double culvert conditions are carried out to obtain

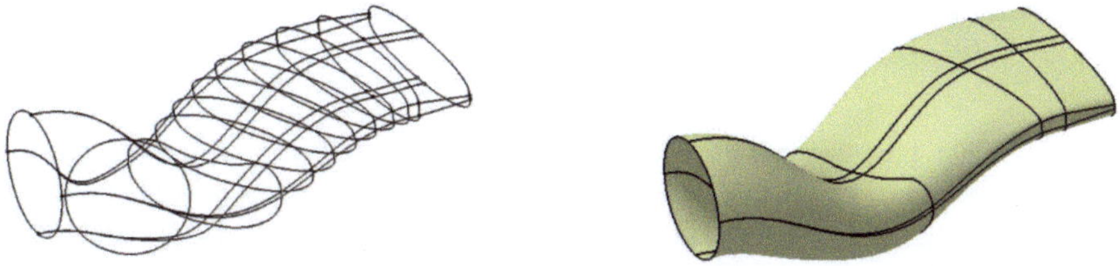

(a) Section lines and longitudinal lines of the cross section

(b) Profile of the S-shaped nozzle

Fig. 2.18 Profile generation for S-shaped nozzle. (**a**) Section lines and longitudinal lines of the cross section. (**b**) Profile of the S-shaped nozzle

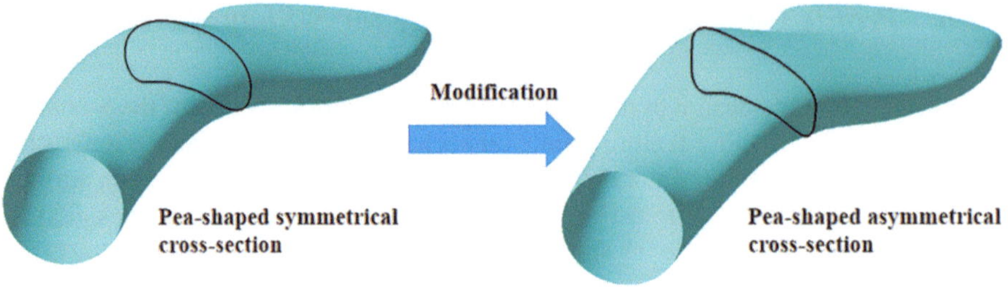

Fig. 2.19 Local cross-section modification of S-shaped nozzle based on spatial layout

the basic characteristics of the S-shaped nozzle under double culvert conditions

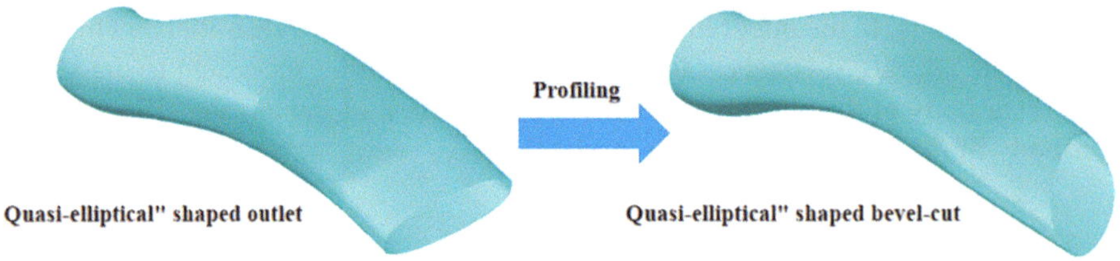

Fig. 2.20 S-curved nozzle outlet cross-section modifications

under different operating conditions. Secondly, the accuracy of the numerical simulation method is verified by comparing the numerical calculation and experimental measurement results of the S-shaped nozzle under double culvert conditions; and then, the typical flow mechanism of the S-shaped nozzle under double culvert conditions is analyzed in depth to obtain the real and refined internal/external flow characteristics of the S-shaped nozzle under the influence of the complex intake structure.

2.3.1 Computational Models and Numerical Simulation Methods

Figure 2.21 give the overall geometric model of the S-shaped nozzle (M0) of a turbofan engine under real operating conditions, which consists of two parts: the exhaust mixing section (M1) and the S-shaped nozzle (M2). The inlet area of the inner/outer culvert channel of the exhaust mixing section model and the outlet area of the double S-shaped nozzle model are obtained by the engine overall program based on the overall performance of a certain type of mixed-exhaust turbofan engine, and the model of the tailcone and the flap mixer is obtained by the structure of this engine.

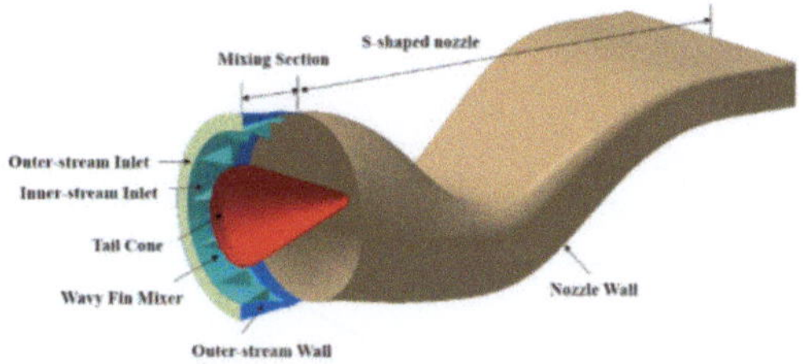

(a) Overall geometric model M0

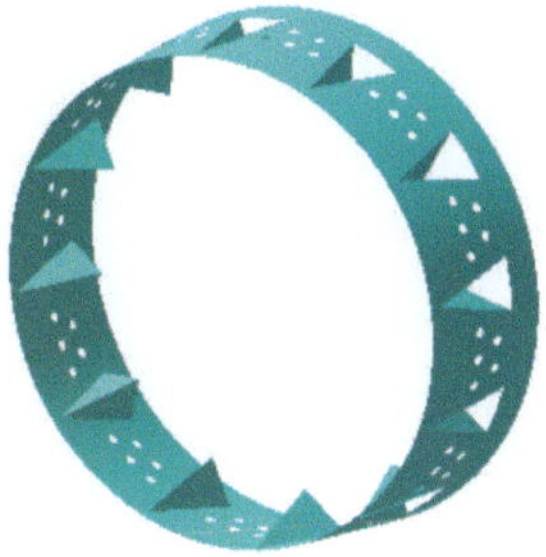

(b) Model M1 of the flap mixer

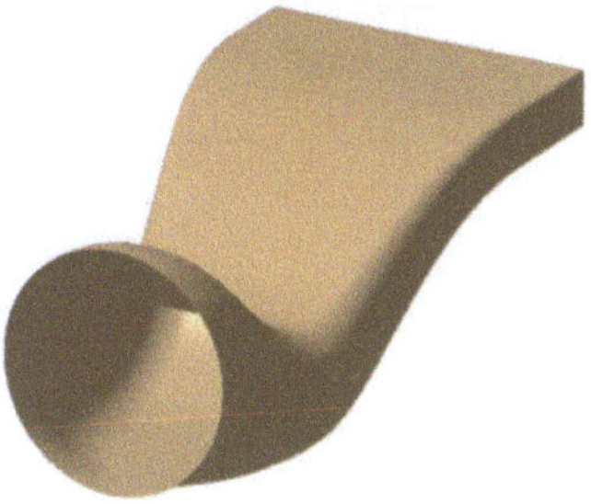

(c) Model M2 of the double S-shaped nozzle

Fig. 2.21 Geometric model of a double S-shaped nozzle based on a turbofan engine type. (**a**) Overall geometric model M0. (**b**) Model M1 of the flap mixer. (**c**) Model M2 of the double S-shaped nozzle

The main geometrical parameters involved in the overall geometrical model of the S-shaped nozzle are shown in Fig. 2.22, the coordinate origin is set at the inlet center position of the exhaust mixing section model, and the values of each parameter are shown in Table 2.1. The exhaust mixing section is based on a certain type of twin-axis mixed-exhaust turbofan engine, including the tail cone, the outer wall and the flap mixer. The flap mixer is different from the conventional mixer configuration, which is an expansive structure along the axial direction, and it divides the exhaust mixing section channel into the inner channel and outer channel, and the outer channel is a convergent channel, and the inner channel is an expansive channel. Its ring surface along the circumferential distribution of 12 "funnel-type" main mixing zone, and each main mixing zone on both sides of the trapezoidal shape of the equal spacing were arranged 5 mixing holes, the purpose is to further strengthen the connotation of high-pressure cold flow and connotation of low-pressure hot flow mixing effect. The tail cone is located in the center of the inner connotation channel, and its structure size is large, running through the exhaust mixing section and the S-shaped nozzle two sections of the model area, and the end is extended to the inside of the S-shaped nozzle, which reduces the design space of the bending surface of the S-shaped nozzle.

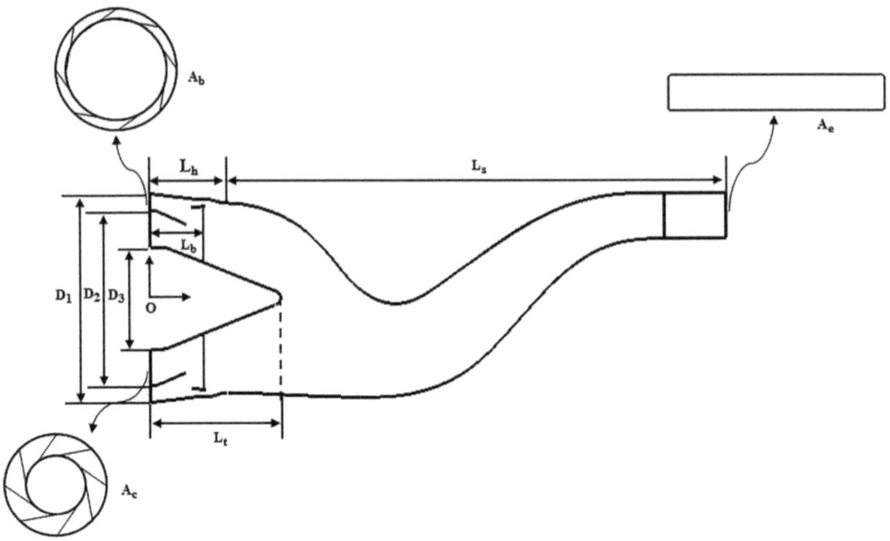

Fig. 2.22 S curved nozzle overall model main geometric parameters

Table 2.1 Values of the main geometrical parameters of the overall model

Geometric parameter	D_2/D_1	D_3/D_1	L_h/D_1	L_s/D_1	L_t/D_1	L_b/D_1
Value	0.838	0.486	0.370	2.362	0.622	0.265

Table 2.2 Valuation of key geometric parameters for the double S-shaped nozzle configuration

Geometric parameter	L/D	W_1/D	A_1/A_{in}	L_2/L_1	$\Delta Y_1/L_1$	W_e/H_e	L_3/D
Galue	2.6	1.015	0.6	1.5	−0.296	6	0.124

The S-shaped nozzle adopts the double S-shaped convergent nozzle configuration, and the nozzle outlet surface is rectangular-like. The S-shaped nozzle surface design adopts the multi-parameter coupled variable cross-section method introduced in Sect. 2.2, and based on the law of the common tangent, the flap mixer at the front end of the double S-shaped nozzle is completely blocked, and the geometrical characteristics of the S-shaped nozzle are as shown in Fig. 2.21. The inlet and outlet areas of the S-shaped nozzle are fixed, and the area contraction ratio is 0.436. The key geometrical parameters are taken in the design state as shown in Table 2.2. The values of key geometrical parameters are taken in the design condition as shown in Table 2.2.

ANSYS CFX commercial software is used to numerically simulate the flow field characteristics of an S-shaped exhaust system for turbofan engine with the flow type of compressible, viscous, three-dimensional turbulent flow. The full three-dimensional computational domain of the S-shaped exhaust system is discretized into a finite number of cells by the finite volume method to solve the Reynolds-averaged Navier-Stokes (N-S) equations. Among them, the convective and turbulent terms are solved in higher-order format, and the SST *k-w* model is selected as the turbulence model with reference to the results of the study on turbulence model selection for S-shaped nozzle for turbojet engine. The computational model of the S-shaped exhaust system for turbofan engine is meshed by ICEM commercial software. The full three-dimensional computational domain of the S-shaped exhaust system computational model has a size of 30D*10D*10D, which consists of the exhaust mixing domain, the S-shaped nozzle domain, and the far-field domain at the exit of the nozzle, as shown in Fig. 2.23, where D is the diameter of the nozzle inlet. The length of the far-field region along the jet direction is 30 times the diameter of the nozzle inlet, and the length perpendicular to the flow direction is 10 times the diameter of the nozzle inlet. The overall computational model of the S-shaped exhaust system is more complex, which can not be fully realized by structural mesh division, and the use of non-structural mesh will lead to the capture of the flow field characteristics of the inside and outside of the nozzle is not fine enough and will bring about a larger amount of computation, therefore, the overall computational domain using a mixture of mesh spliced together Therefore, the overall computational domain is stitched together by a hybrid mesh, and the mesh model of each region of the overall computational model of the S-shaped exhaust system is shown in Fig. 2.24. Due to the existence of complex geometrical structures such as the flap mixer and tail cone, the area near the flap mixer in the exhaust mixing section is meshed with a tetrahedral unstructured mesh as shown in Fig. 2.24a, while the other areas of the exhaust mixing section, the domain of the double S-curve nozzle and the surrounding far-field domains are meshed with a structured hexahedral mesh as shown in Fig. 2.24b, c, respectively. When splicing the computational domain, the number of nonstructural and structural meshes on both sides of the interface should be kept approximately equal in order to ensure the accuracy of data transfer on the mesh interface. In addition, the near-wall mesh of the S-shaped nozzle is encrypted

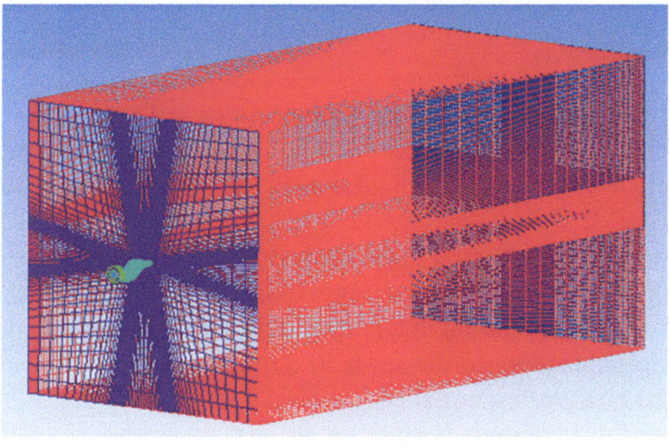

Fig. 2.23 Overall mesh model of S-shaped nozzle for turbofan engine

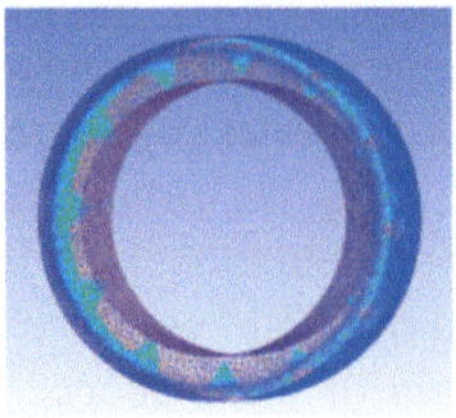

(a) Unstructured mesh model of the region near the flap mixer

(b) Structured mesh model of the exhaust mixing domain

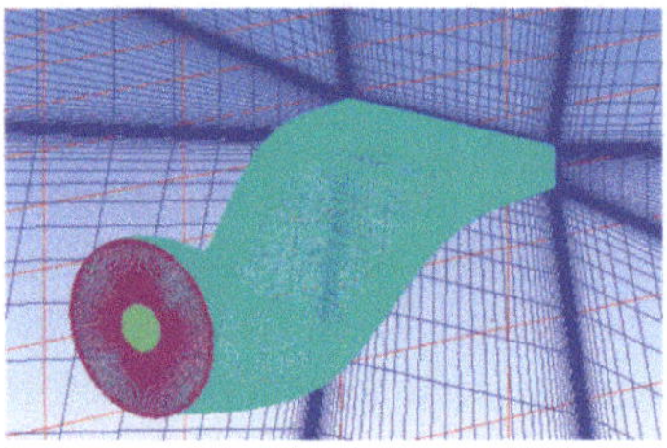

(c) Structural mesh model of the S-shaped nozzle and surrounding far-field region

Fig. 2.24 Mesh model of each region of S-shaped nozzle for turbofan engine. (**a**) Unstructured mesh model of the region near the flap mixer. (**b**) Structured mesh model of the exhaust mixing domain. (**c**) Structural mesh model of the S-shaped nozzle and surrounding far-field region

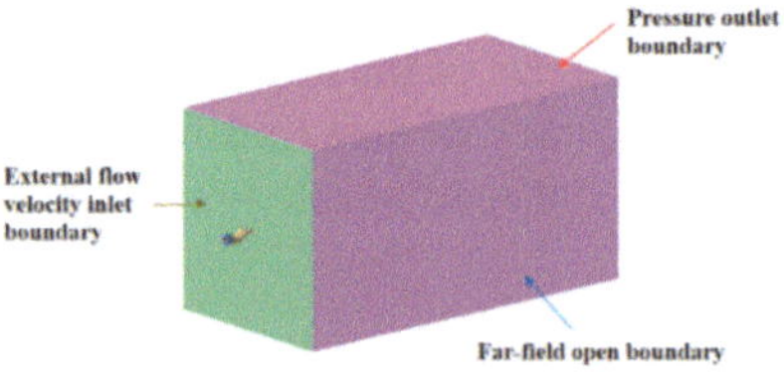

(a) Outflow areas

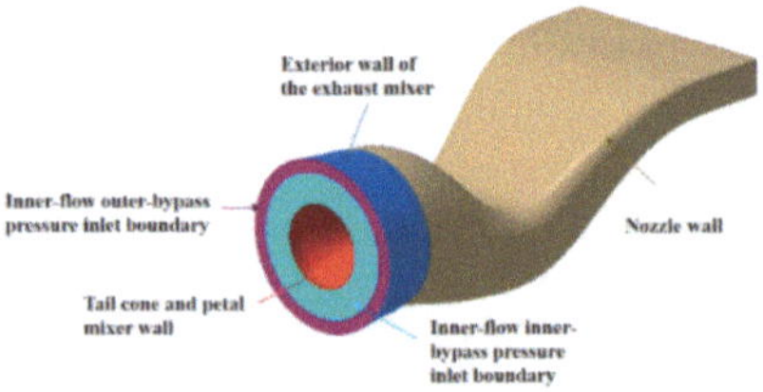

(b) Inflow areas

Fig. 2.25 Boundary condition settings for the calculation area. (**a**) Outflow areas. (**b**) Inflow areas

during meshing to ensure that the near-wall mesh of the nozzle corresponds to $y+<1$.

The computational domain of the S-shaped exhaust system is divided into the inner and outer domains, where the inner domain includes the exhaust mixing domain, the area near the wave flap mixer, and the double S-shaped nozzle domain, and the outer domain refers to the far-field domain, which involves four types of boundaries: inlet boundaries, outlet boundaries, far-field boundaries, and solid-wall boundaries, and the boundary conditions are set up as shown in Fig. 2.25. The engine design point is the ground condition, and the inlet boundary of the inner/

Table 2.3 S-shaped nozzle inner/outer culvert inlet boundary condition setting

Placement	Connotation imports	Connotation imports
Total pressure/atm	2.71	2.78
Total temperature/K	987.15	412.70

outer culvert runner is set as the pressure inlet boundary condition, i.e., given the total pressure, total temperature, and direction of the airflow inlet. The inlet conditions of the inner and outer culvert are obtained through the overall engine program, and the inlet boundary conditions of the inner/outer culvert for the ground condition are shown in Table 2.3, with the inlet incoming flow in the axial direction.

The inlet of the far-field domain around the S-shaped nozzle adopts the inlet boundary condition, with the inlet inlet velocity of 0. Based on the reasonableness of the numerical calculation, the inlet velocity is set to 0.05 Ma, and the inlet pressure is 1 atm; the outlet of the far-field domain adopts the pressure outlet boundary condition, with the outlet hydrostatic pressure of $P_0 = 1$ atm, and the boundary of the far-field is set as the open boundary condition; the outer wall surfaces of the double S-shaped nozzle and the exhaust mixer, the wave flap mixer and the wall surface of the tail cone are adopted no-slip adiabatic wall boundary conditions.

2.3.2 *Experimental Study of Flow Characteristics of S Curved Nozzle for Turbofan Engine*

In order to obtain the real internal flow field characteristics and aerodynamic performance of S-shaped nozzle for turbofan engine, and to provide experimental basis for the rationality and accuracy of numerical simulation method. In this section, the cold model blowing experiments of the S-shaped nozzle under the condition of double culverts are carried out based on the "Integrated Experimental Platform of Thrust Vector and Infrared Radiation for Dual-Flow Path Exhaust System", which is shown in Fig. 2.26. The experiments mainly measured the flow rate, thrust and wall static pressure distribution of the S-shaped nozzle under different drop pressure ratios.

2.3.2.1 Experimental Set-up

The experimental platform is mainly composed of the following parts: a six-component force balance system, a main flow system, a secondary flow system, and a flow field measurement system. The six-component balance is one of the core components of the experimental platform, with a static calibration accuracy of 0.5% F.S. It is used to measure the forces and moments in the three directions of the S-shaped nozzle, as shown in Fig. 2.27, in which its thrust along the axial direction is measured in the range of 0–6 kN. The main/secondary flow system is used for the

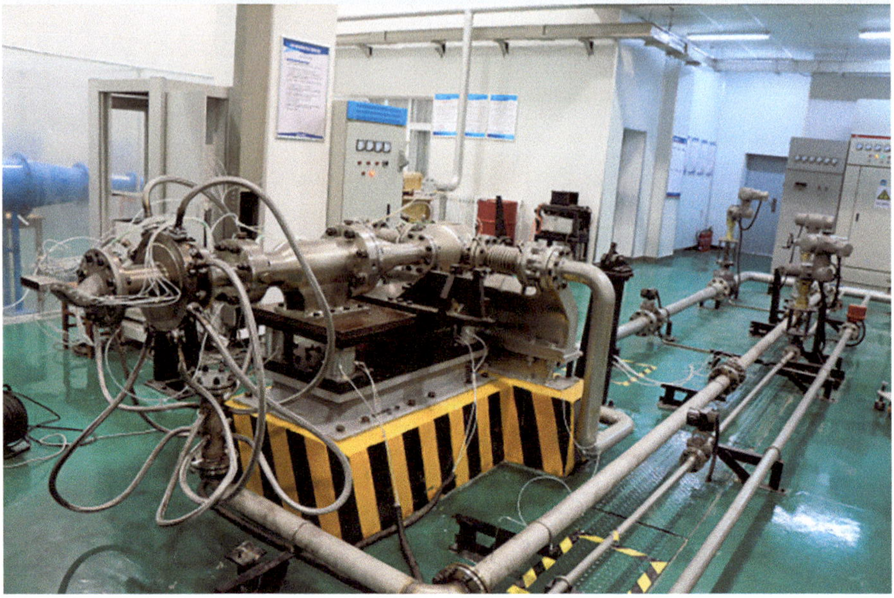

Fig. 2.26 Integrated Thrust Vectoring and Infrared Radiation Test Bed for Dual Flow Exhaust Systems

Fig. 2.27 Physical drawing of a six-component balance

adjustment of the nozzle's primary/secondary flows and pressures, respectively, in which the main flow ranges from 0 to 4.0 kg/s, and the secondary flow ranges from 0 to 2.0 kg/s. The flow field measurement system includes an electronic pressure scanning valve, a large diameter mirror color texture shader, as shown in Fig. 2.28.

(a) PSI Pressure Scanning Valve (b) Color Striker

Fig. 2.28 Experimental Measurements Instruments. (**a**) PSI Pressure Scanning Valve. (**b**) Color Striker

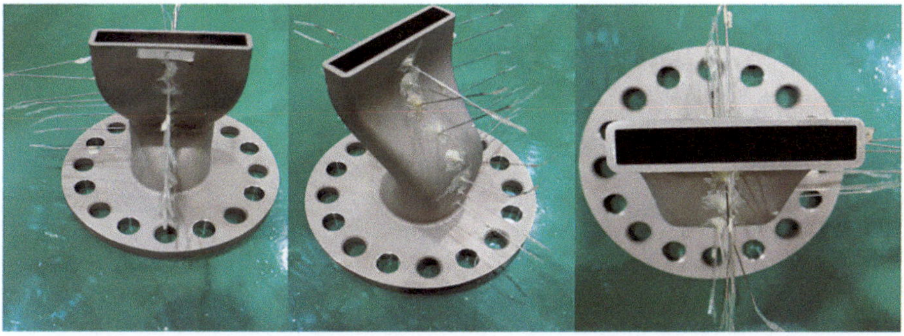

Fig. 2.29 Distribution of test pieces and wall hydrostatic pressure measurement points

The pressure scanning valve has a total of 96 channels with a measurement accuracy of 0.5%, which is used to measure the along-track wall pressure of the S-shaped nozzle under different operating conditions. The color ripple shader mainly consists of the main reflector system, the light source slit concentrator system and other structures, which can capture the complex flow characteristics in the flow field of high velocity airflow with Ma < 2.5, and is used to measure the structure of the surge system in the inner/outer flow field of the outlet of the S-shaped nozzle.

2.3.2.2 Experimental Models

The experimental model of the double S-shaped nozzle is shown in Fig. 2.29, which is obtained by downscaling the geometrical model of the S-shaped nozzle in the design state at Fig. 2.29. The model is set to be 10:1. Considering the similarity criterion and the experimental conditions, the model scale is set to 10:1. 3 hydro-static holes are opened along the longitudinal direction on the top, bottom and side

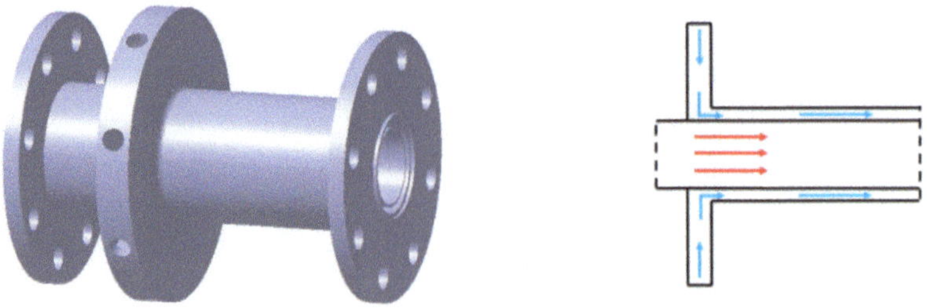

(a) Schematic of the three-dimensional structure of the parallel mixing chamber

(b) Schematic of the T-shaped longitudinal section of the parallel mixing chamber

Fig. 2.30 Schematic of Parallel Mixing Chamber Components. (**a**) Schematic of the three-dimensional structure of the parallel mixing chamber. (**b**) Schematic of the T-shaped longitudinal section of the parallel mixing chamber

wall of the nozzle, with 8 holes in each, totaling 24 holes. In order to more realistically simulate the inward/outward incoming flow conditions of the S-shaped nozzle under double culvert conditions, a flat mixer component was designed to connect the primary/secondary flow inducer with the S-shaped nozzle experimental parts, which is modeled as shown in the schematic diagrams of Fig. 2.30. The two flow channels of the parallel mixing chamber are used to simulate the inner/outer culvert channel of the exhaust system, the inner culvert channel is a circular pipe, and the main flow enters into the pipe in the axial direction; the outer culvert channel is designed as a radial air inlet device, which is designed to eliminate the error of the secondary incoming flow on the measurement of the axial thrust of the nozzle, i.e., the longitudinal cross-section of the outer culvert channel is an inverted T-shape (Fig. 2.30), and the secondary flow enters into the channel along the radial direction by eight circular tubes with same size and starts mixing with the main flow at the outlet position of the mixer. At the outlet position of the mixer the mixing with the main flow starts. A physical drawing of the model installation of the parallel mixing chamber and its connections to the secondary flow diversion piping is shown in Fig. 2.31. The installation of the S-shaped nozzle model on the test rig under double culvert conditions is shown in Fig. 2.32.

2.3.2.3 Experimental Measurement Program

Experiments were conducted to measure the wake jet flow field pattern images, along-travel pressure distributions on the upper, lower and side walls, total pressure parameters at the nozzle outlet, flow rate parameters and thrust parameters of the S-shaped nozzle under different Nozzle Pressure Ratio (*NPR*) and Second Nozzle Pressure Ratio (*SPR*) for the dual-container condition, respectively. Parameters, nozzle outlet total pressure parameters, flow parameters and thrust parameters. The experimental measurement program is shown in the Table 2.4.

(a) Physical drawing of the parallel mixing chamber installation

(b) Connection device of the parallel mixing chamber to the secondary flow induced gas piping

Fig. 2.31 Physical drawing of the parallel mixing chamber and its connections. (**a**) Physical drawing of the parallel mixing chamber installation. (**b**) Connection device of the parallel mixing chamber to the secondary flow induced gas piping

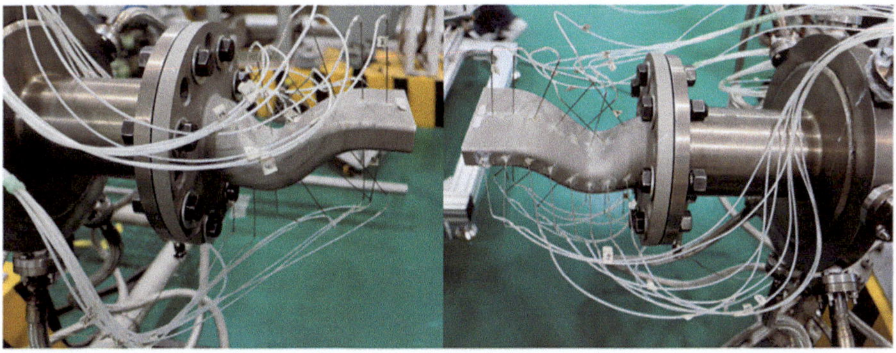

Fig. 2.32 Physical drawing of the installation of the S-shaped nozzle experimental piece on the experimental bench

2.3.2.4 Experimental Measurement Results

Figures 2.33, 2.34, and 2.35 give the wave system distribution of the S-shaped nozzle outlet under the double culvert condition with the internal drop pressure ratio of 2, 3 and 4 respectively. It can be seen that when the internal drop pressure ratio is 2, the S-shaped nozzle is in a slight under- expansion state, and the compression-expansion wave system generated downstream of the nozzle exit is weaker, and the ripple shadow map shows more fuzzy, and the two free boundaries of the nozzle exit are approximately parallel and along the axial direction. With the increase of connotation drop pressure ratio, the wave system distribution characteristics of the nozzle outlet do not change significantly. When the internal pressure drop ratio is 3, the under-expansion state of the S-shaped nozzle is intensified, and two expansion waves are generated at the nozzle exit, and the tail jet is deflected to the outside at a

Table 2.4 Experimental measurement program for S-shaped nozzle under double culvert conditions

Connotation *NPR*	External *SPR*	Total pressure, flow, thrust	Static pressure at the wall along the nozzle and wave system distribution at the outlet
2.0	1.8	√	√
	2.0	√	√
	2.2	√	√
	2.4	√	√
3.0	2.7	√	√
	3.0	√	√
	3.3	√	√
	3.6	√	√
4.0	3.6	√	√
	4.0	√	√
	4.4	√	√
	4.8	√	√

(a) *SPR*=1.8

(b) *SPR*=2.0

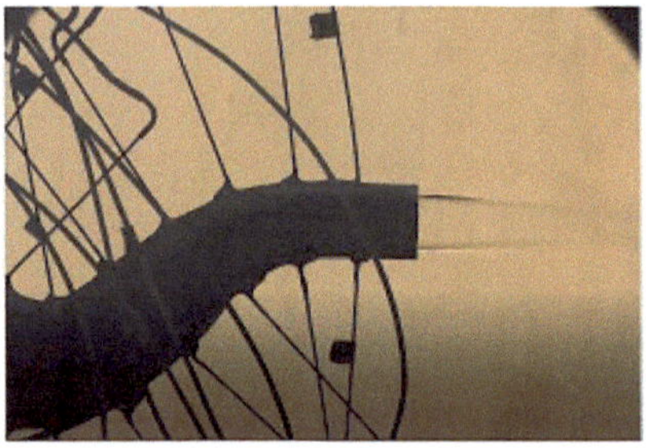

(c) *SPR*=2.2

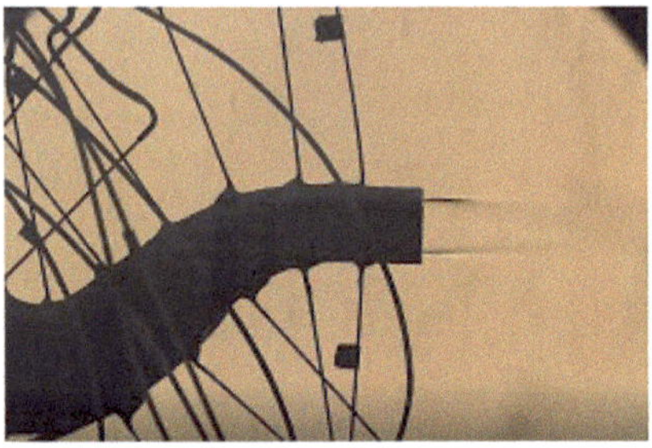

(d) *SPR*=2.4

Fig. 2.33 Wave system distribution at the exit of the S-shaped nozzle for the NPR = 2 condition. (**a**) *SPR* = 1.8. (**b**) *SPR* = 2.0. (**c**) *SPR* = 2.2. (**d**) *SPR* = 2.4

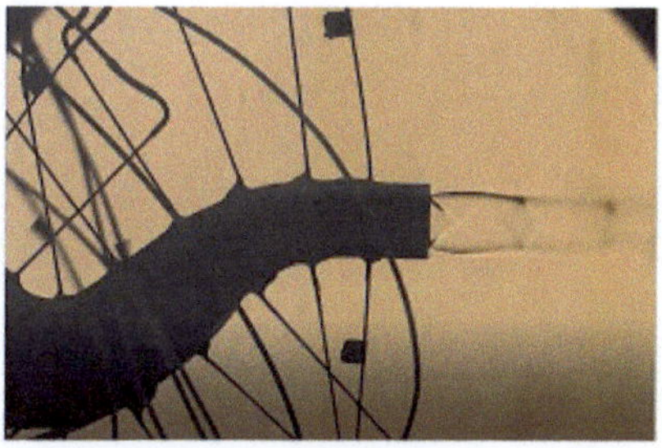

(a) *SPR*=2.7

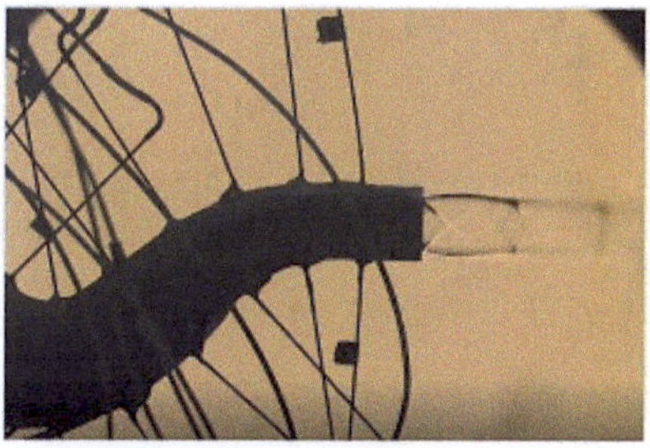

(b) *SPR*=3.0

(c) *SPR*=3.3

(d) *SPR*=3.6

Fig. 2.34 Wave system distribution at the exit of the S-shaped nozzle for the NPR = 3 condition. (**a**) *SPR* = 2.7. (**b**) *SPR* = 3.0. (**c**) *SPR* = 3.3. (**d**) *SPR* = 3.6

certain angle after passing through the expansion wave, that is to say, it produces a larger expansion angle of the airflow, and it is still an expansion wave after the intersection of the two expansion waves, and the expansion wave meets the free boundary and reflects two compression waves, and the airflow deflected inward at a certain angle after passing through the compression wave, and it is still a compression wave after the intersection of the two compression waves. The airflow is deflected inward by a certain angle after the compression wave, the two compression waves are still compression wave after the intersection, and the compression wave reflects two expansion waves at the airflow boundary, which is a typical compression-expansion wave system in alternating order. Due to the continuous expansion-compression-expansion-compression process, the two free boundaries at the nozzle exit show the shape of expansion-contraction-expansion-contraction. Compared with the implicit drop pressure ratio of 2, the distribution of the wave system shown by the ripple shadow is clearer under this drop pressure ratio condition. With the gradual increase of the implicit drop pressure ratio, the compression-expansion wave system has no obvious change. When the internal pressure drop ratio is 4, the structure of the basic wave system is basically the same as that of the pressure drop ratio of 3, but the expansion angle and compression angle are larger, the expansion- compression process of the airflow is more intense, and the distribution of the wave system shown in

(a) *SPR*=3.6

(b) *SPR*=4.0

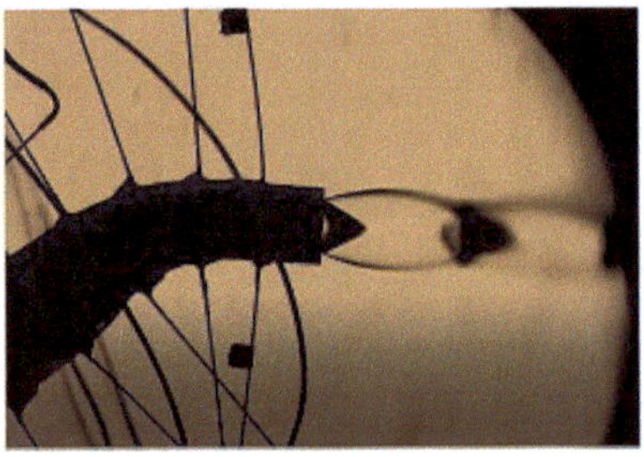

(c) *SPR*=4.4

(d) *SPR*=4.8

Fig. 2.35 Wave system distribution at the exit of the S-shaped nozzle for the NPR = 4 condition. (**a**) $SPR = 3.6$. (**b**) $SPR = 4.0$. (**c**) $SPR = 4.4$. (**d**) $SPR = 4.8$

the ripples is more significant. With the increase of the inner pressure drop ratio, the range of individual expansion-compression wave systems becomes larger, and the number of wave systems that can be accommodated in the texture film map decreases accordingly. With the gradual increase of the external drop pressure ratio, the compression-expansion wave systems remain unchanged.

Figures 2.36, 2.37, and 2.38 compare the distribution of dimensionless static pressure (dimensionless in terms of ambient pressure) on the upper and lower wall surfaces as well as on the side wall surfaces of the S-shaped nozzle model under different internal/external drop pressure ratios. It can be seen that the distribution trend of static pressure along the wall of the S-shaped nozzle under different drop pressure ratios is basically the same, and with the gradual increase of the drop pressure ratio, the value of static pressure at the wall point continues to increase. The static pressure on the upper wall of the nozzle decreases rapidly in the first bend and reaches a local minimum value near the first bend. As the air passes through the turn, the static pressure on the upper wall increases rapidly in the first part of the second turn and then decreases until the nozzle exit. The static pressure on the lower wall of the nozzle varies less in the region from the nozzle inlet to the front of the second bend passage, then begins to decrease rapidly and reaches a local minimum at the second bend turn, and continues to decrease in the isostatic section, but the

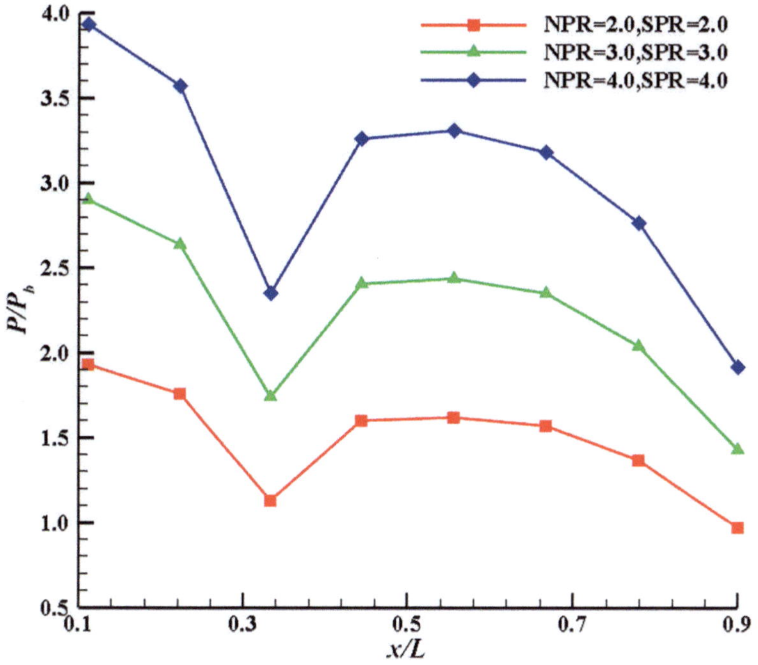

Fig. 2.36 Static pressure distribution on the upper wall of the nozzle at different drop pressure ratios

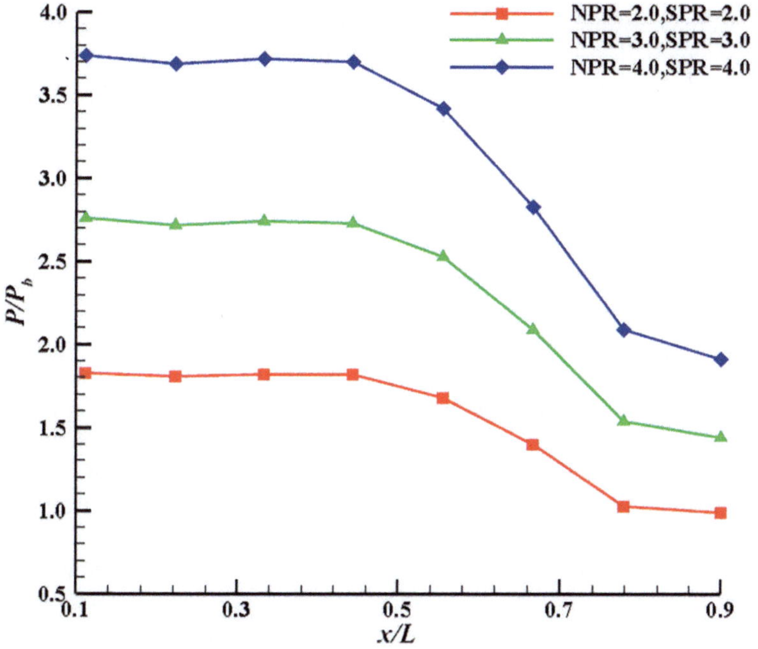

Fig. 2.37 Static pressure distribution on the lower wall of the nozzle at different drop pressure ratios

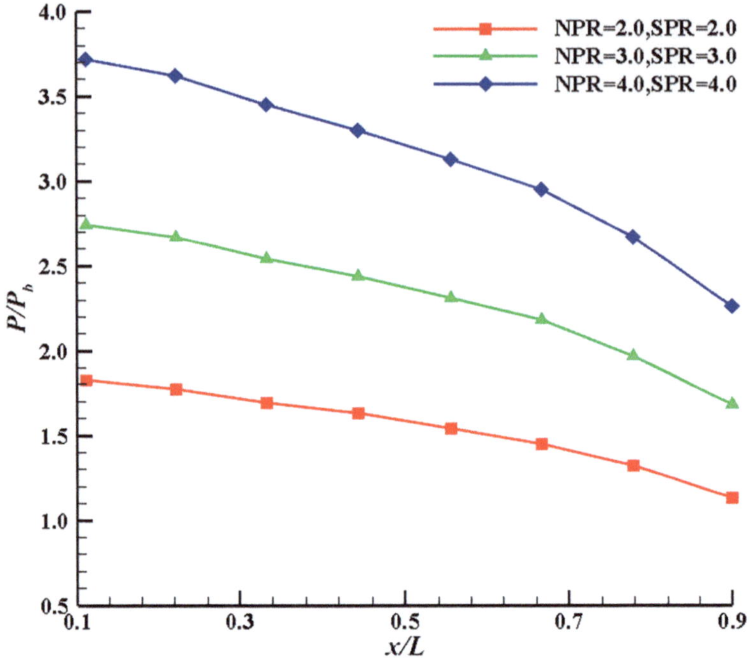

Fig. 2.38 Static pressure distribution on the side wall surface of S-shaped nozzle with different drop pressure ratios

rate of decrease slows down significantly. The static pressure along the sidewall surface gradually decreases, and the decrease gradually increases.

Figures 2.39, 2.40, and 2.41 compare the influence of different external drop pressure ratios on the dimensionless static pressure distribution on the upper and lower wall surfaces as well as on the sidewall surfaces of the S-shaped nozzle model under the conditions of three sets of internal drop pressure ratios. It can be seen that the trend of static pressure distribution along the wall of the S-shaped nozzle with the same internal pressure drop ratio and different external pressure drop ratios is basically the same, and the distribution law is consistent with that described earlier. With the gradual increase of the culvert pressure drop ratio, in addition to the nozzle and other straight sections of the region, along the course of the wall surface of the static pressure values continue to increase. When the internal pressure drop ratio $NPR = 2$ and 3, the external pressure drop ratio increases, and the magnitude of the increase in wall static pressure gradually decreases; when $NPR = 4$, the external pressure drop ratio increases, and the magnitude of the increase in wall static pressure firstly decreases and then increases. In the region of nozzle isostraight section, for the three states of connotation drop pressure ratio, with the increase of connotation drop pressure ratio, the side wall static pressure increases gradually, but the upper and lower wall static pressure appears to increase first and then decrease the trend. For NPR = 2, the trend of decreasing static pressure on the upper wall surface

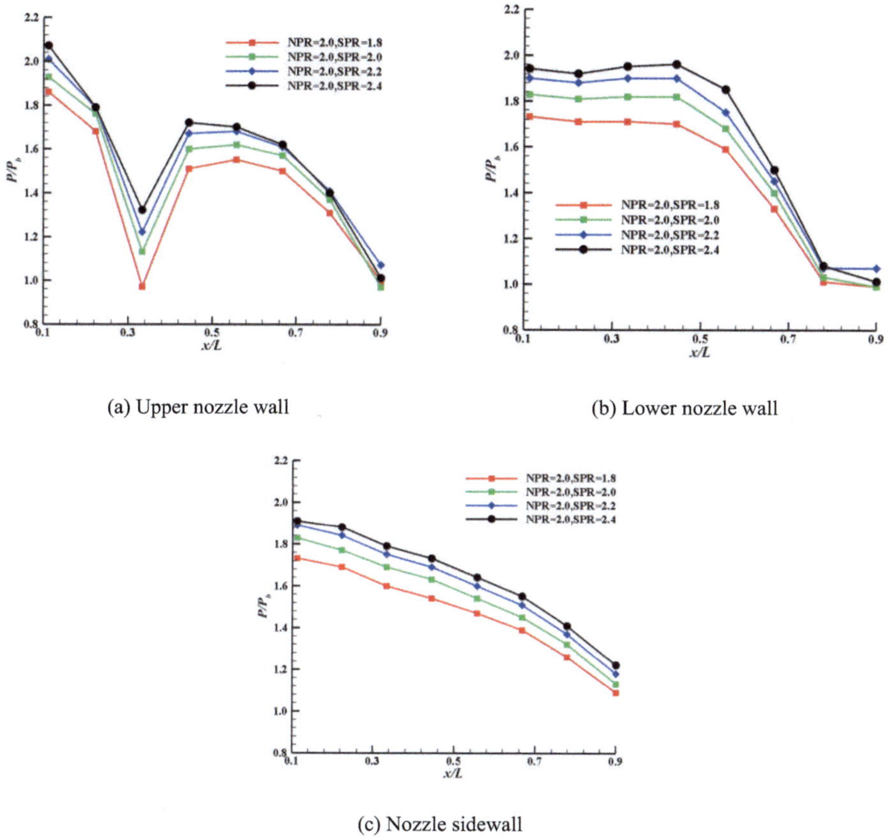

Fig. 2.39 Static pressure distribution on the side wall surface of S-shaped nozzle with different external drop pressure ratios (NPR = 2). (**a**) Upper nozzle wall. (**b**) Lower nozzle wall. (**c**) Nozzle sidewall

with the gradual increase of the outer culvert drop pressure ratio occurs when the outer culvert drop pressure ratio *SPR* changes from 2 to 2.2; for NPR = 3, the trend occurs when the *SPR* changes from 3.3 to 3.6; and for NPR = 4, the trend occurs when the *SPR* changes from 4.0 to 4.4. For the three states of the inner culvert drop ratio, the trend of decreasing static pressure on the lower wall surface with the gradual increase of the outer culvert drop pressure ratio The decreasing trend with increasing culvert drop pressure ratio occurs at the SPR change to the corresponding maximum culvert drop pressure ratio. The possible reason for this is the asymmetric static pressure distribution on the upper and lower wall surfaces due to the structural bending of the S-shaped nozzle along the longitudinal section.

Figures 2.42, 2.43, and 2.44 give the trends of total pressure recovery coefficient, culvert ratio and flow rate of the nozzle for different internal/external culvert drop pressure ratio conditions. Regarding *the* NPR = 2 condition, SPR = 1.8, 2.0, 2.2, 2.4 are defined as case1, case2, case3 and case4, respectively; accordingly, regarding

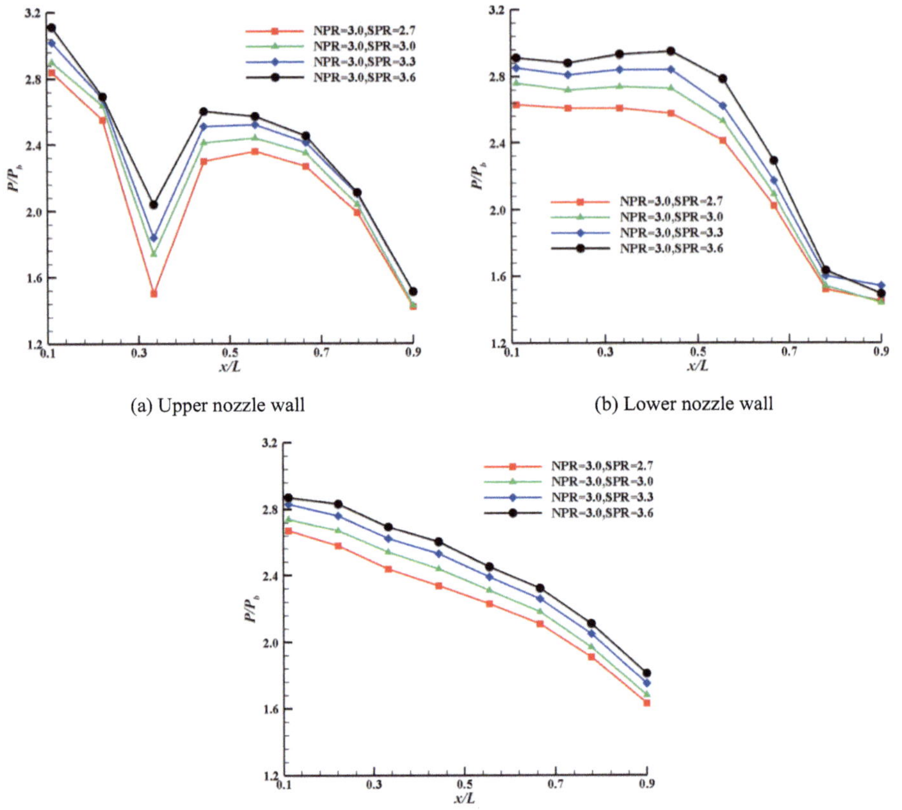

(a) Upper nozzle wall

(b) Lower nozzle wall

(c) Nozzle sidewall

Fig. 2.40 Static pressure distribution on the side wall surface of S-shaped nozzle with different external drop pressure ratios (NPR = 3). (**a**) Upper nozzle wall. (**b**) Lower nozzle wall. (**c**) Nozzle sidewall

the NPR = 3 condition, SPR = 2.7, 3.0, 3.3, 3.6 are defined as case1, case2, case3, and case4, respectively, and regarding the NPR = 4 conditions, SPR = 3.6, 4.0, 4.4, 4.8 are defined as case1, case2, case3 and case4, respectively. it can be seen that the total pressure recovery coefficient of the S-shaped nozzle decreases with the increase of the inner and outer culvert drop pressure ratios, and the culvert ratio of the S-shaped nozzle rises rapidly with the gradual increase of the outer culvert drop pressure ratio for the same inner culvert drop pressure ratio, while the inner culvert drop pressure ratio has a significant effect on the culvert ratio. The effect of the drop pressure ratio on the culvert ratio is relatively small. Regarding the trend of flow rate, when the inner culvert pressure drop ratio is certain, with the gradual increase of the outer culvert pressure drop ratio, the nozzle flow rate increases gradually, but the increase is small; when the outer culvert pressure drop ratio is certain, with the gradual increase of the inner culvert pressure drop ratio, the nozzle flow rate increases significantly. This reflects that the influence of the connotation drop

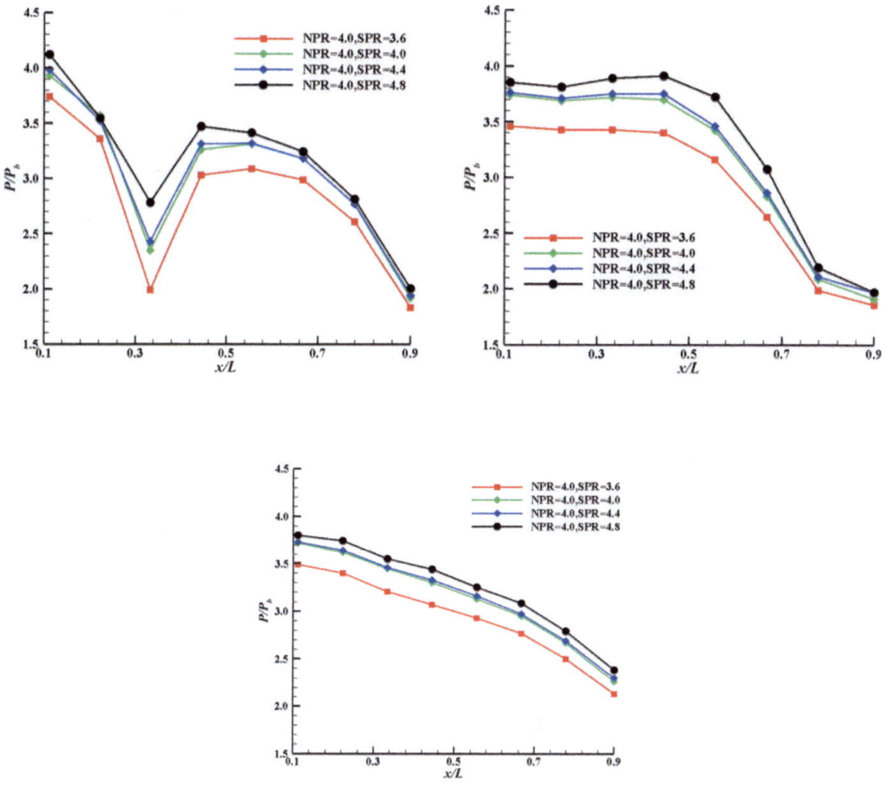

(c) Nozzle sidewall

Fig. 2.41 Static pressure distribution on the side wall surface of the S-shaped nozzle for different external drop pressure ratios (NPR = 4). (**a**) Upper nozzle wall. (**b**) Lower nozzle wall. (**c**) Nozzle sidewall

pressure ratio on the total flow of the S-shaped nozzle is much larger than the connotation drop pressure ratio.

2.3.3 Validation Study of Numerical Simulation Method for S-Shaped Nozzle under Double Culvert Condition

Figure 2.45 give the comparison between the numerical simulation results and experimental measurement data of the static pressure distribution on the upper and lower wall surfaces of the S-shaped nozzle under the conditions of different inner/outer culvert drop pressure ratios. As can be seen from the figure, the trend of static pressure distribution on the wall surface of the nozzle obtained by numerical calculation is basically consistent with the experimental results, and the change rule of

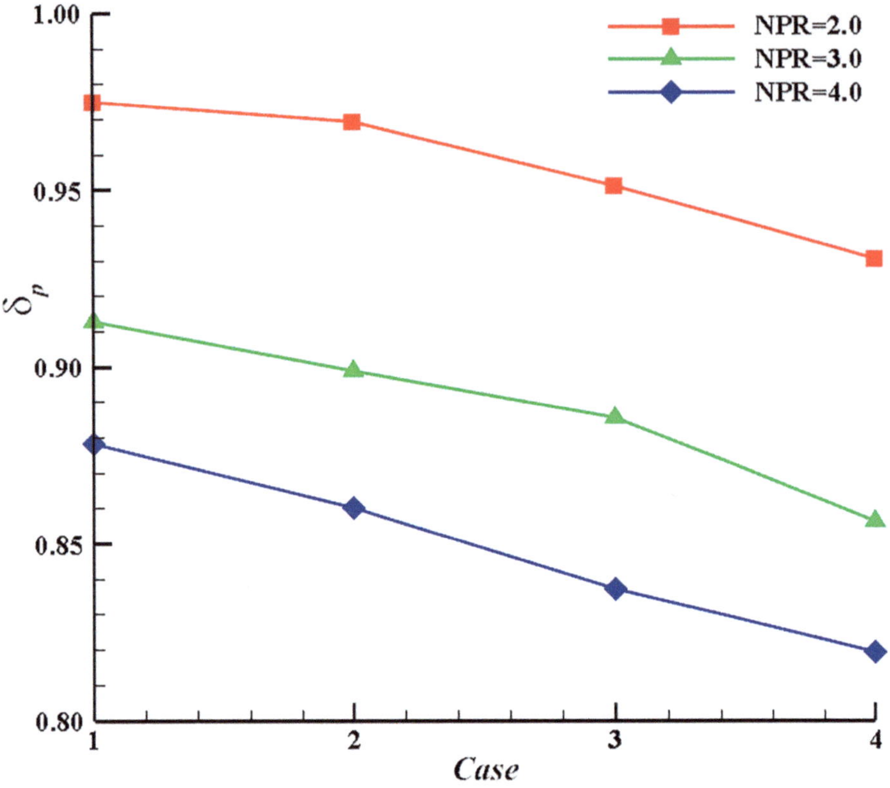

Fig. 2.42 Total pressure recovery coefficient of S-shaped nozzle with different drop pressure ratios

static pressure is as follows: the static pressure on the upper wall surface decreases rapidly in the first section of the S-shaped channel of the nozzle and achieves a local minimum near the first bend; the static pressure on the upper wall surface starts to rise sharply in the second section of the S-shaped channel before the static pressure starts to decrease rapidly after that, and then starts to decrease rapidly until the second bend; the rate of static pressure decrease slows down in the isostraight section of the nozzle. In the isostraight section of the nozzle, the rate of static pressure decrease slows down. The static pressure on the lower wall of the nozzle is essentially constant in the first S-shaped passage, and the static pressure decreases rapidly in the second S-shaped passage and reaches a localized minimum in the second turn. In addition, for different conditions of inner/outer culvert drop pressure ratios, the wall pressure values at each point obtained from experimental measurements mostly fall on the numerically predicted curves. The main differences between the numerically predicted results and the experimentally measured results are shown in the magnitude of the pressure values at the first bend measurement point on the upper wall surface as well as at the S-shaped nozzle exit measurement point.

This section also gives the formula for calculating the relative error between the predicted and experimental values of the aerodynamic parameters P_n and P_e at each

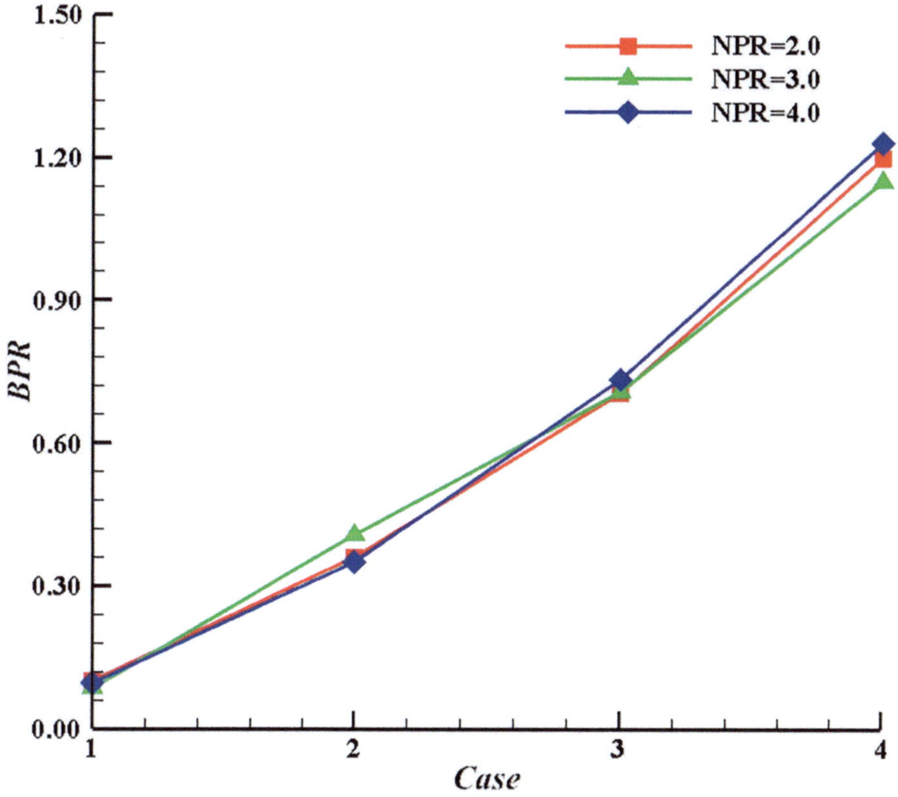

Fig. 2.43 S-shaped nozzle culvert ratio for different drop pressure ratios

measurement point: $|(P - P_{ex})/P_{ex}| * 100\%$, Fig. 2.46 show the relative error between the numerically predicted wall static pressure values and the experimentally measured values at each measurement point. As can be seen from the figure, under the conditions of NPR = 2, SPR = 2 and NPR = 3, SPR = 3, the relative error between experimental and numerical simulation at most of the measurement points on the upper and lower walls is less than 2%, and the relative error is larger only at the upper wall at the first turn and the lower wall at the second turn, but basically it is less than 4%; under the conditions of NPR = 4, SPR = 4, the relative error between experimental and numerical simulation is less than 2% at most of the measurement points on the upper and lower walls; under the conditions of NPR = 4 and SPR = 4, the relative error between experimental and numerical simulation is less than 2% at most of the measurement points on the lower wall. Under the condition of drop pressure ratio NPR = 4, SPR = 4, most of the relative errors between the experimental and numerical simulation results of the measurement points are almost less than 3%. The reason for the large errors of individual measurement points may be the deviation of the positioning of the experimental measurement points or the instability of the airflow drop pressure ratio in the experiments; in addition, the large error points all appear in the region with large static pressure gradient, which indirectly

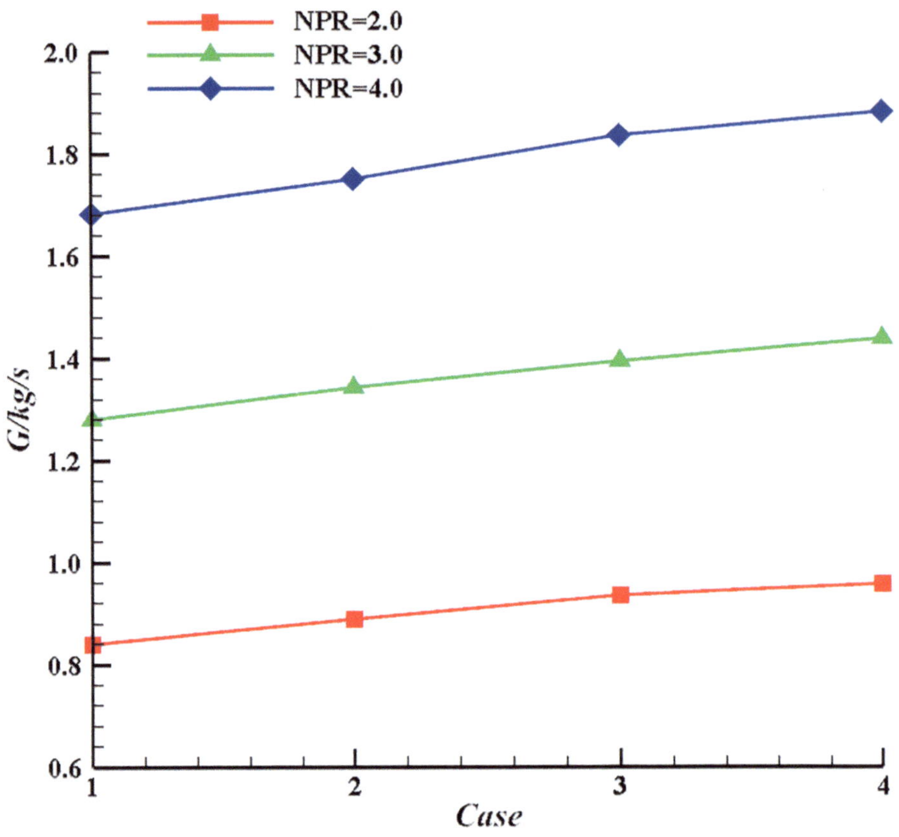

Fig. 2.44 S-shaped nozzle flow rate at different drop pressure ratios

increases the experimental measurement errors. In summary, under the conditions of different inner/outer culvert drop pressure ratios, the error between numerical simulation and experimental measurement of the static pressure on the wall of the S-shaped nozzle is generally small, and the numerical simulation method adopted in this section is able to accurately simulate the internal flow characteristics of the S-shaped nozzle under the conditions of double culvert.

2.3.4 Flow Mechanism Analysis of S Curved Nozzle for Turbofan Engine

The flow field characteristics of the S-shaped nozzle for turbofan engines are closely related to the geometrical configurations of the nozzle and the exhaust mixer. The standard geometrical model of the exhaust mixer structure and the S-shaped nozzle is shown in Fig. 2.21, and the key geometrical parameters and aerodynamic boundaries required for the flow field calculations have been introduced in Sect. 2.3.1 of

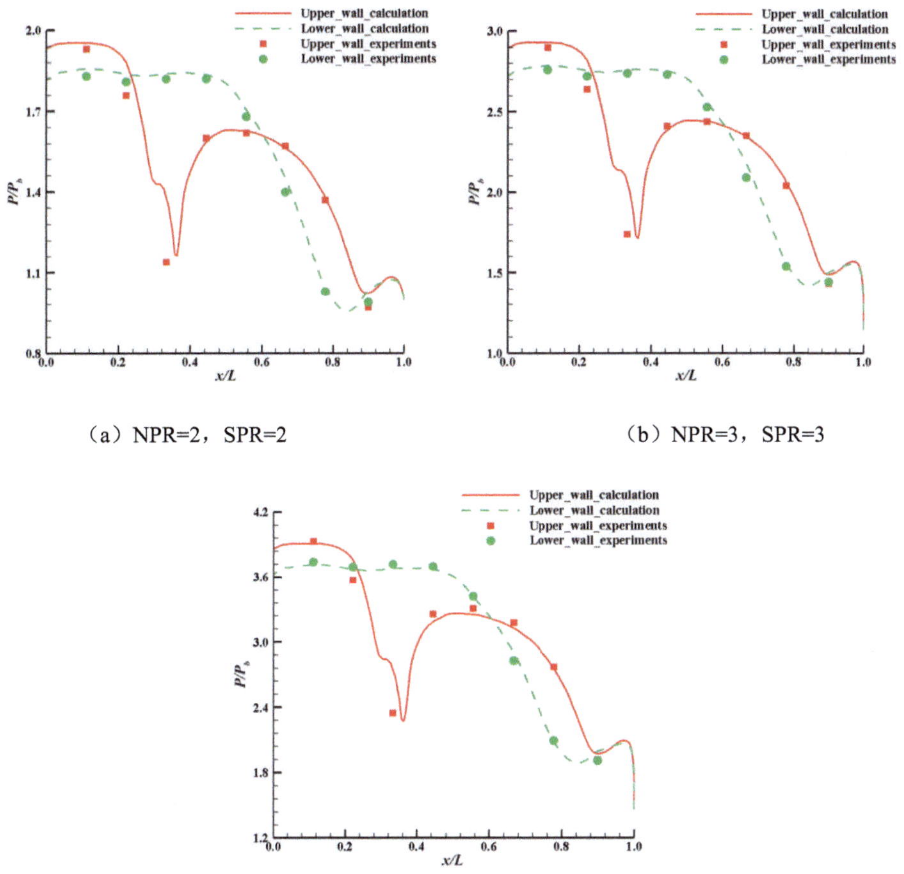

(a) NPR=2, SPR=2

(b) NPR=3, SPR=3

(c) NPR=4, SPR=4

Fig. 2.45 Comparison of numerical simulation and experimental measurements of static pressure distribution on the wall of a S-shaped nozzle. (**a**) NPR = 2, SPR = 2. (**b**) NPR = 3, SPR = 3. (**c**) NPR = 4, SPR = 4

this paper. On this basis, in order to facilitate the analysis of the specific flow field details inside the S-shaped nozzle, the dimensionless positional distribution of the transverse flow section of the nozzle along the axial direction is given here, as shown in Fig. 2.47, where: section A is the inlet cross section of the exhaust mixer; section B is the inlet cross section of the S-shaped nozzle; section C is located at the first turn; sections D and E are located in the middle of the first and the second turns; section F is located at the second turn that is, the inlet of the isostraight section of the nozzle and the outlet of the S-shaped converging section; cross-section G is the nozzle outlet cross-section. According to the longitudinal cross-section characteristics reflected in the main view, the centers of curvature of the nozzle internal circulation sections C and D are located on the upper side of the nozzle profile, and the

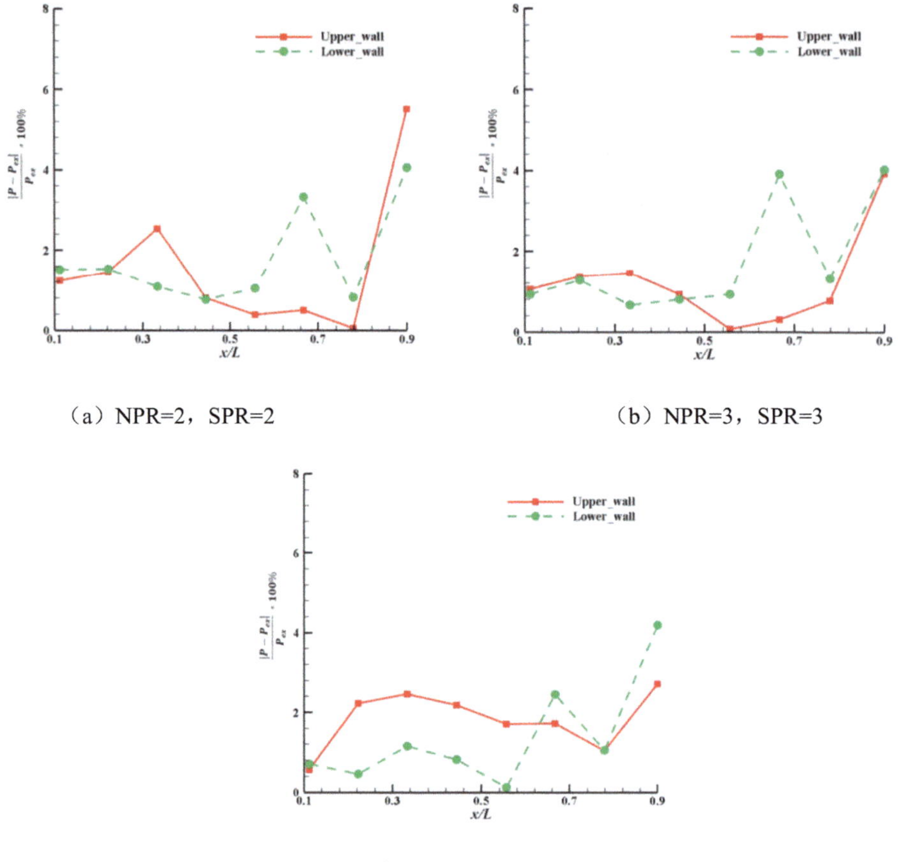

(a) NPR=2, SPR=2

(b) NPR=3, SPR=3

(c) NPR=4, SPR=4

Fig. 2.46 Relative error between numerical simulation and experimental results of static pressure distribution on the wall of the S-shaped nozzle. (**a**) NPR = 2, SPR = 2. (**b**) NPR = 3, SPR = 3. (**c**) NPR = 4, SPR = 4

center of curvature of section E is located on the lower side of the nozzle profile; according to the lateral cross-section characteristics reflected in the side view, the centers of curvature of the nozzle internal circulation sections C and D are located on the outer side of the nozzle profile, and the center of curvature of section E is located on the inner side of the nozzle profile.

In addition, the perimeter wall static pressure distribution characteristics of the along-travel section of the S-shaped nozzle are obtained by establishing a dimensionless coordinate system of the perimeter wall of the lateral circulation section, as shown in Fig. 2.48. The lateral circulation section is symmetric about the z = 0 plane, and the perimeter wall static pressure distribution on the right half of the lateral section is analyzed here. The lateral wall at Y = 0 is defined as the origin of the coordinate system, and the dimensionless arc length *s* is the ratio of the distance of the points on the perimeter wall from the origin to 1/2 the perimeter of the right perimeter wall. Thus, the midpoint of the upper wall of the perimeter wall of the

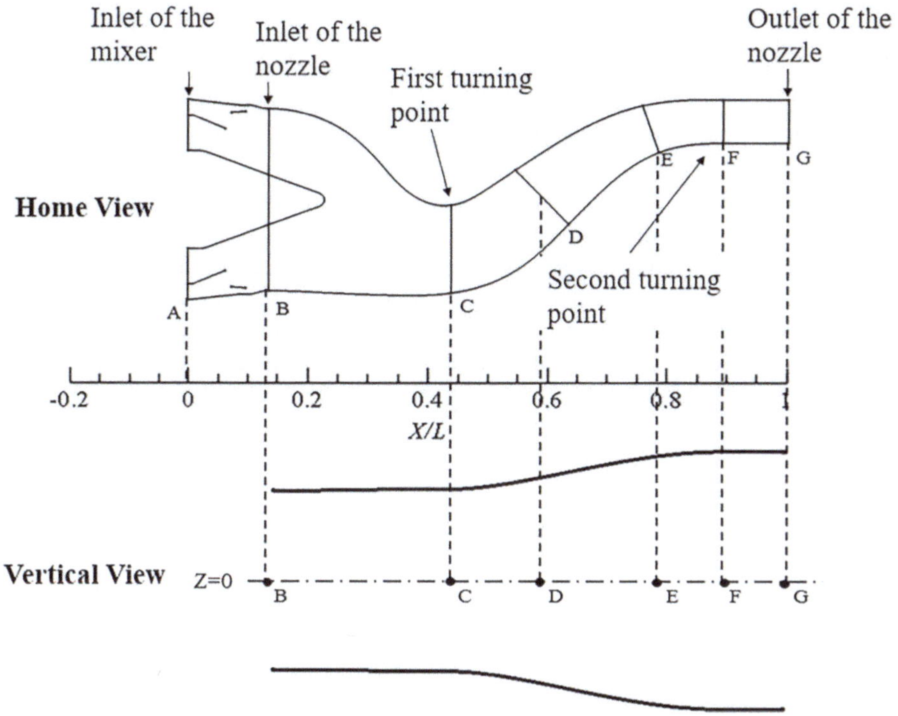

Fig. 2.47 Dimensionless positional distribution along the axial direction of the flow cross-section of the S-shaped nozzle

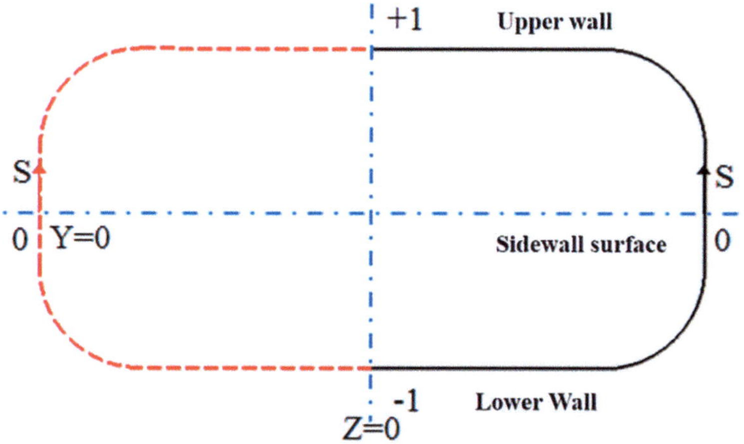

Fig. 2.48 Dimensionless coordinate system of the perimeter wall of the S-shaped nozzle

entire circulation cross-section corresponds to s = 1, and the midpoint of the lower wall corresponds to s = −1.

This section analyzes the complex flow characteristics and its generation mechanism inside the S-shaped nozzle under double culvert conditions for turbofan engines from the following aspects. The analysis includes: Ma, streamline and static pressure distribution on the symmetry plane wall; Ma, wall limit streamline distribution of the wave mixer structure, wall limit streamline distribution of the tail cone structure; x-direction vortex distribution, static temperature distribution and nozzle wall temperature distribution along the cross section; peripheral wall static pressure distribution and wall limit streamline distribution along the cross section; and total pressure distribution on the symmetry plane and along the cross section. Flow field characteristics.

Figures 2.49 and 2.50 show the distribution of Ma, streamlines and static pressure on the symmetric surface of the symmetric wall of the S-shaped nozzle,

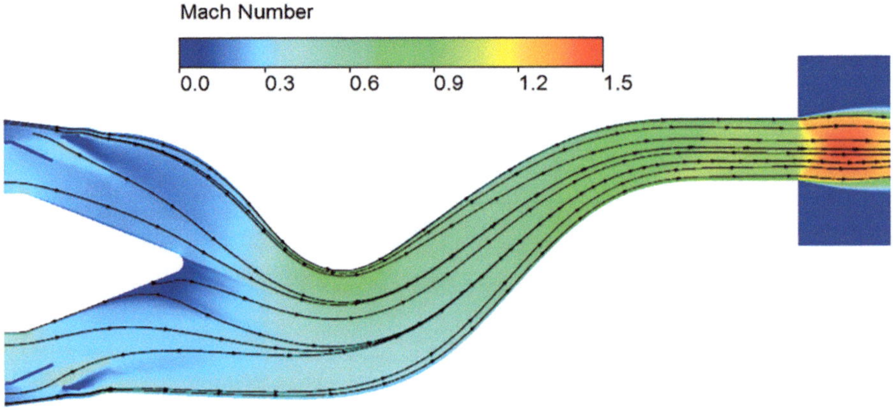

Fig. 2.49 Ma and flow distribution on the symmetry plane of the S-shaped nozzle

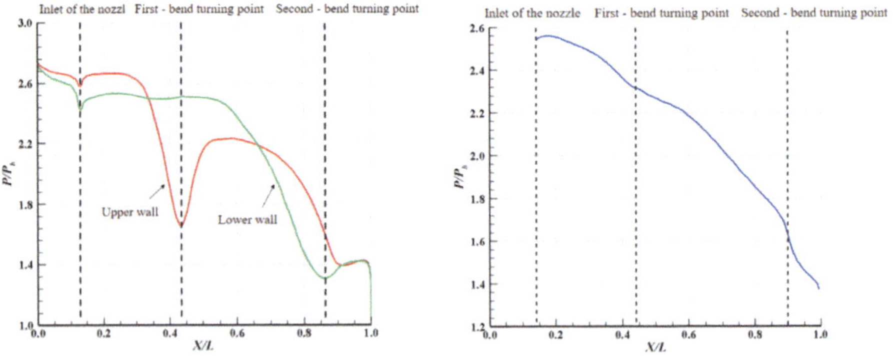

(a) Symmetric wall static pressure distribution (b) Side wall static pressure distribution

Fig. 2.50 S-shaped nozzle wall static pressure distribution. (**a**) Symmetric wall static pressure distribution. (**b**) Side wall static pressure distribution

respectively. From the Ma and streamline distribution, it can be seen that the S-shaped curved channel leads to obvious curvature and streamline convergence of the airflow inside the pipe; there are obvious non-uniform characteristics of the airflow velocity, and there are local high-speed zones in the region near the upper wall at the first turn and the region near the lower wall at the second turn; under the joint effect of the S-shaped curved pipe and the complex structures of the tail cone and the flap mixer, there are significant differences in the inner side of the tail cone and the inner side of the rear end of the flap mixer. Under the combined effect of the S-shaped curved pipeline and the complex structure of the tail cone and the flap mixer, significant flow separation occurs at the inner rear end of the flap mixer, forming a low-speed zone, and the low-velocity zone on the symmetry plane about the symmetry of the tail cone and the structure of the flap mixer shows a typical asymmetric distribution characteristics. From the pressure distribution on the symmetry surface, it can be seen that in the convergence section of the exhaust mixer, the static pressure on the upper and lower walls are gradually reduced. For the upper wall static pressure distribution characteristics: in the first section of the S-shaped channel, the wall static pressure is first stable, and then began to decline rapidly, to the first turn to reach a local minimum; in the second section of the S-shaped channel before the rapid rise in the static pressure, and then continued to decline until the second turn around the corner; in the isotropic section, the static pressure basically remains unchanged. For the lower wall static pressure distribution characteristics: in the first section of the S-shaped channel, the wall static pressure is basically unchanged; into the second bend channel, the static pressure continues to fall rapidly, and reaches a local minimum near the second turn; in the isostraight section, the static pressure rises slightly and then stays unchanged and is basically the same as the upper wall static pressure distribution. The static pressure on the side wall of the nozzle gradually decreases along the flow direction, and a steep pressure drop occurs near the first turn, while near the second turn, the pressure drop tends to slow down.

The curvature characteristics of the airflow flow is related to the corresponding wall pressure gradient in the normal direction, while the symmetric wall static pressure distribution is closely related to the distribution of air velocity, the difference between acceleration and deceleration of the airflow is related to the curvature of the geometric curvature of the surface and the direction of this section in accordance with the characteristics of the geometric configuration–differences in the acceleration and deceleration of airflow–static pressure distribution on the symmetric wall–pressure gradient in the normal direction of airflow–curvature of airflow streamlines–flow field distribution characteristics–flow field distribution characteristics of the idea of detailed analysis of the above turbofan engine S-shaped nozzle inside the generation of a variety of flow phenomena.

In the exhaust mixing section, the mixer structure presents symmetric features along the longitudinal direction, but the asymmetric geometric features of the downstream S-shaped curved pipe lead to an asymmetric pressure distribution, i.e., the upper wall pressure is greater than that on the lower wall, and this longitudinal asymmetric pressure distribution is transmitted forward in the subsonic flow field, thus changing the pressure distribution in the exhaust mixer, and thus the pressure value of the upper wall of the exhaust mixer is higher than that of the lower wall, and the wall static pressure of the exhaust mixer decreases gradually along the flow direction. In the S-shaped nozzle region, on the other hand, the center of curvature of the first bend channel is located on the outside of the upper wall surface of the nozzle, which leads to a rapid contraction of the airflow flow tube near the upper wall surface and a rapid decrease in the static pressure, which in turn creates a normal pressure gradient in an upward direction, and hence, the airflow streamline bends towards the outside of the upper wall surface of the nozzle. At the first bend turn, the curvature of the upper wall surface reaches a local maximum, and the local acceleration of the airflow is the strongest, forming a local high-speed zone. In the first section of the second bend channel, the curvature value gradually decreases along the flow direction, the airflow flow tube near the upper wall surface is expanding, the airflow begins to gradually decelerate, and thus the static pressure begins to rise, the normal pressure gradient gradually decreases, and the degree of curvature of the airflow continues to weaken. In the back section of the second bend channel, the center of curvature moves from the outside of the upper wall surface of the nozzle to the outside of the lower wall surface of the nozzle, and the airflow tubes near the lower wall surface contract rapidly, the static pressure decreases rapidly, and thus the direction of the normal pressure gradient is downward, and the airflow streamline bends to the outside of the lower wall surface of the nozzle. At the second turn, the curvature of the lower wall surface reaches a local maximum, and the local acceleration of the airflow is the strongest, forming a local high-speed zone. In the equal straight section, the upper and lower wall air velocity began to converge, the upper and lower wall static pressure also gradually converge to equal. Regarding the distribution of static pressure on the side wall of the S-shaped nozzle, the static pressure on the side wall gradually decreases due to the continuous acceleration of the airflow in the flow direction. However, downstream of the first turn, the transverse profile of the nozzle is curved and expanded, so the local acceleration of the sidewalls increases, and a steep local pressure drop occurs near the first turn. Upstream of the second turn, the lateral surface of the nozzle is curved and contracted, so the local airflow acceleration decreases, and the pressure drop is slowed down in the region upstream of the second turn.

Figure 2.51 show the near-wall Ma, streamline distributions on the inner/outer culvert side of the flap mixer structure within the exhaust mixer. It can be seen that there is an obvious low velocity region downstream of the "funnel-shaped" mixing zone in both the inner and outer impeller sides, which is due to the significant flow separation in this region. According to the previous description, the back pressure distribution downstream of the exhaust mixer is asymmetric, and thus the flow separation region in the outer impeller side is significantly smaller than that in the inner impeller side. Since the outer culvert channel is a converging channel, with the rapid compression of the outer culvert high-pressure gas flow, the outer culvert near-wall gas flow quickly re-adheres to the wall and flows through the wave flap mixer, and thus this is a closed separation region and the separation region is smaller. When the connotation of airflow along the wave valve mixer "funnel-shaped" surface into the

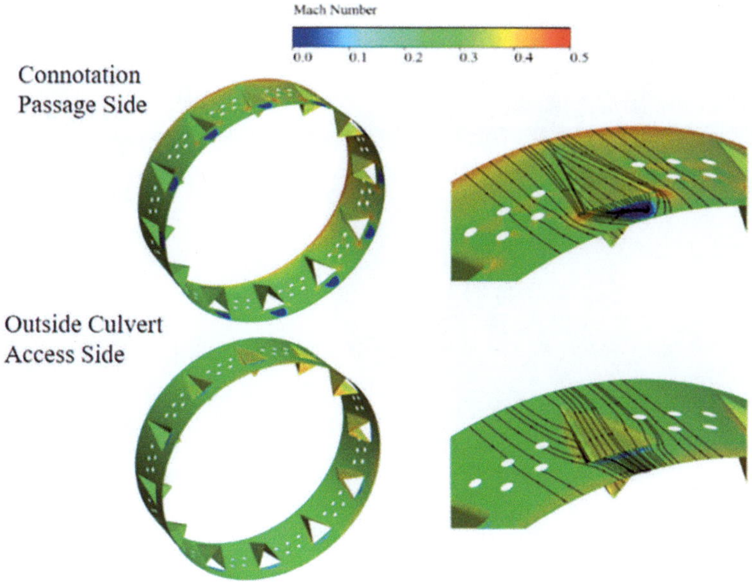

Fig. 2.51 Wavelength Mixer Structure Inside/Outside Culvert Side Near Wall Ma and Flow Distribution

connotation, impede and change the initial flow direction of the connotation of airflow, inside and outside the connotation of the mixed airflow along the "funnel-shaped" surface to the right of the bottom of the high-speed impact, the situation is similar to a high-speed jet, and thus in the valve mixer connotation side of the "funnel-shaped" side of the "funnel-shaped" side of the "funnel-shaped" side of the high speed impact. Mixer connotation side of the "funnel-shaped" doping area downstream of the emergence of a low-speed airflow area, but with the "high-speed jet" along the "funnel-shaped" surface of the development of the gradual weakening of the internal and external culvert of the flow of gas Gradually compressed, the low-velocity separation zone along the direction of the flow is gradually reduced, in a certain distance downstream of the wave mixer, the internal and external airflow to achieve a further second mixing. Therefore, this is also a closed separation zone, similar in shape to a "triangular cone", with a larger separation area compared to the outer culvert side separation zone, and the "triangular cone" separation zone has a relatively large deflection angle to the right along the axial direction.

As shown in the previous Fig. 2.49, the flow separation created by the tail cone structure results in two low Ma regions downstream, which come from the lower and upper wall surfaces of the tail cone. The large size of the tailcone and the short size of the exhaust mixer section make the flow characteristics near the tailcone wall surface more influenced by the S-shaped curved channel. Figure 2.52 give the wall limit flow distribution of the tail cone structure. It can be seen that flow separation occurs in two regions of the upper and lower wall surfaces of the tail cone. The flow separation at the upper wall occurs at the starting position of the upper hemisphere

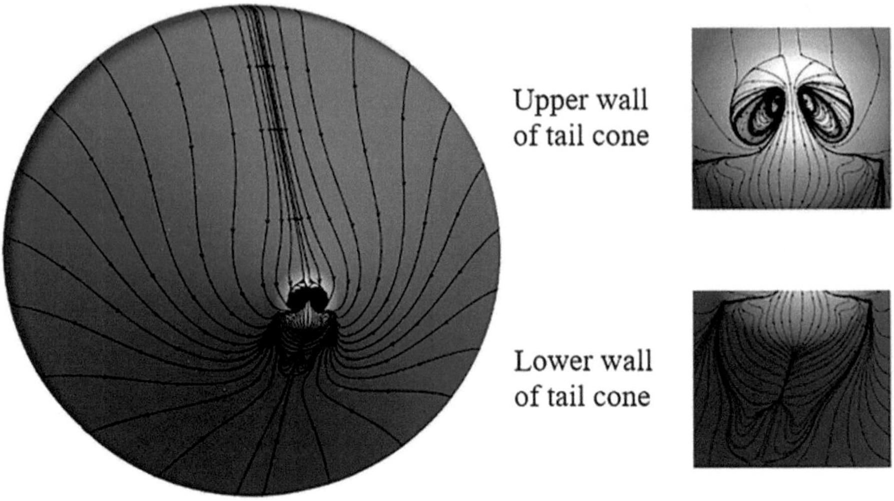

Fig. 2.52 Tail Cone Structure Wall Limit Flow Distribution

of the tail cone; the flow separation at the lower wall occurs after the middle position of the lower wall of the tail cone and before the starting position of the hemisphere of the lower wall. The characteristics of flow separation at these two positions are different, and the reasons for flow separation are also different. The flow separation at the starting position of the hemisphere on the upper wall surface is characterized by two laterally symmetric separation vortices, while the lower wall surface is a wider range of flow separation region, which is characterized by a separation flow line from the saddle point to the node, with the surrounding limit flow lines converging to the separation line. Since the tail-cone profile is a convergent structure, and thus the interior is an expansive channel, the interior flow exhibits decelerating pressurization, which generates an axial counterpressure gradient in the vicinity of the tail-cone wall. In addition, the effect of the downstream S-shaped curved channel causes a downward pressure gradient along the longitudinal direction in the vicinity of the tail-cone wall surface, as previously described. The combined effect of these two directional pressure gradients is completely opposite for the upper and lower wall surfaces of the tail cone. In the region of the upper wall of the tail cone, the angle between the longitudinal pressure gradient and the direction of air flow is acute, and thus the combined pressure gradient along the direction of flow for the smooth pressure gradient; in the region of the lower wall of the tail cone, the longitudinal pressure gradient and the direction of air flow is obtuse, and the combined pressure gradient along the direction of flow for the inverse pressure gradient, and, accordingly, the formation of the inverse pressure gradient along the direction of flow on the lower wall of the tail cone. Therefore, the flow separation near the hemisphere of the upper wall surface of the tail cone is caused by the larger curvature of the hemisphere of the upper wall surface of the tail cone, and the flow separation of the upper wall surface occurs at the starting position of the tail cone hemisphere,

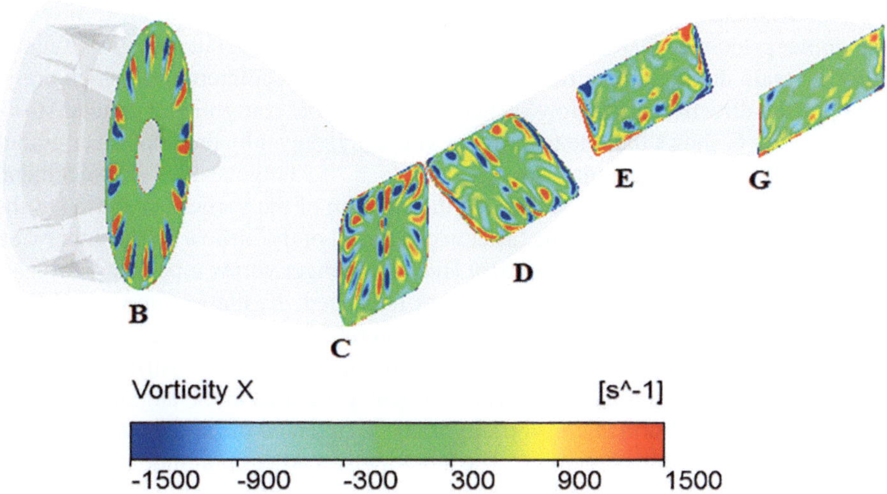

Fig. 2.53 Distribution of vorticity in x-direction along the course section of the S-shaped nozzle

and the separation area is smaller; whereas, the flow separation near the lower wall surface of the tail cone is caused by the counterpressure gradient along the direction of the streamline direction, which demonstrates that the flow separation of the lower wall surface occurs in the anterior side of the tail cone hemisphere, the middle of the position is behind the tail cone hemisphere, and the separation area is larger.

Figure 2.53 give the distribution of x-direction vortices on the along-travel cross section of the S-shaped nozzle. According to the literature, a flow vortex is a pair of vortices that are opposite to each other and develop in the flow direction. For the three-dimensional flow field of an S-shaped nozzle for a turbojet engine, the streamwise vortices with larger vortex magnitude mainly come from the wall boundary layer and the near-wall corner region. For the S-shaped nozzle 3D flow field studied in this section, the presence of the above vortices can be seen. In addition to this, in the core flow region, the introduction of the flap mixer and the tail cone structure lead to different streamwise vortex distributions, respectively.

As previously described, the complex flap mixer structure used in this section is an annular homogeneous set of 12 geometrically identical flap configurations, each of which consists of a "funnel-shaped" main mixing structure and a conventional annular structure. When the inner and outer culvert flow passes through the conventional annulus in the wave configuration, the fluid on the inner side of the annulus has a radially outward velocity component, while the fluid on the outer side of the annulus has a radially inward velocity component, resulting in a pair of reversed flow vortices downstream of the conventional annulus. Cross-section B shows that there are 12 pairs of flow vortices generated by the conventional annulus for the whole wave configuration, and they are distributed along the annulus with relatively weak flow vortices. The "funnel-shaped" structure can be approximated as an annulus with the opposite curvature direction of the conventional annulus. Based on the above principle, there are 12 pairs of flow vortices generated by the "funnel-shaped" main doping structure in the whole wavefront configuration, which are in the opposite direction of the flow vortices formed by the conventional annular structure. Section C shows that the two sets of vortices induced by the flap mixer gradually change from an annular arrangement to a "rectangular-like" arrangement. Under the normal pressure gradient, the distribution of the vortices is shifted to the upper side. In addition, due to the faster acceleration of the airflow on the upper wall of the bend, the larger velocity gradient leads to stronger vortex tube contraction, so the lateral vortices on the cross-section are enhanced. As the streamwise vortices continue downstream, due to the continued lateral expansion of the cross section, the streamwise vortices spread further outward and begin to be distributed against the wall and merge with the streamwise vortices generated in the angular region of the wall. In cross-section D, the vorticity in the region near the upper wall surface is weakened due to vortex stretching. The streamwise vortices continue to develop in cross-section E. The continued lateral expansion of the cross-section leads to a significant reduction of the vorticity in the core flow region of the cross-section and the distribution of vorticity becomes disorganized. The stronger streamwise vortices induced by the flap mixer are distributed close to the sidewalls. Since cross-section E is located near the second bend, the airflow tubes contract rapidly, the airflow velocity increases rapidly, and the vortices in the lower wall region are significantly enhanced. The vortices in the core flow region of cross-section G are also further weakened due to airflow mixing, and the vortex distribution appears to be more chaotic.

The tail cone is another important structure that causes the high vorticity distribution characteristics inside the S-shaped nozzle. The two low-Ma flows formed by the flow separation induced by the upper and lower wall surfaces of the tail cone form two pairs of counter-rotating vortices in cross-section B. Under the effect of the normal pressure gradient at the first bend of the nozzle, these two pairs of vortices are located away from the center of the cross-section, and are situated in the middle of the cross-section B in the upper part of the cross-section. The relatively upward position of the flow vortex is formed by the flow separation induced by the excessive curvature of the hemisphere on the upper wall, the vortex value is relatively small, and the vortex distribution area is also small; the relatively downward position of the flow vortex is formed by the flow separation induced by the counter-pressure gradient on the lower wall, the vortex value is relatively large, and the vortex distribution area is also relatively large. As the flow vortex continues to develop downstream, the flow vortex continues to force the inner and outer culvert flows to mix, the vortex volume of the flow vortex in the core region induced by the tail cone gradually decreases, and at cross-section D, the flow vortex generated by the tail cone and the flow vortex generated by the wave flap mixer are merged and the vortex volume is weakened. As the flow cross-section continues to expand laterally, the pair of streamwise vortices generated by the tail cone also spreads laterally.

Figures 2.54 and 2.55 show the static temperature distribution on the along-travel cross-section and the static temperature distribution on the wall surface of the

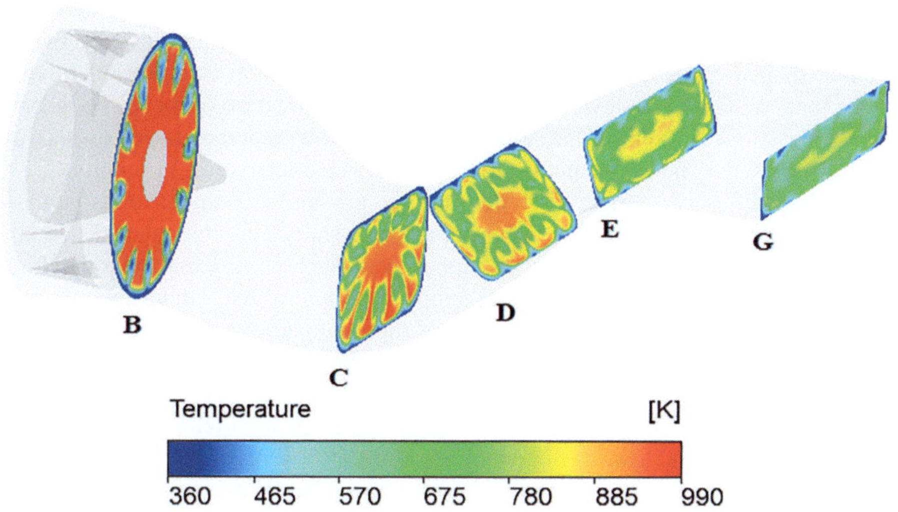

Fig. 2.54 Static temperature distribution along the cross-section of the S-shaped nozzle

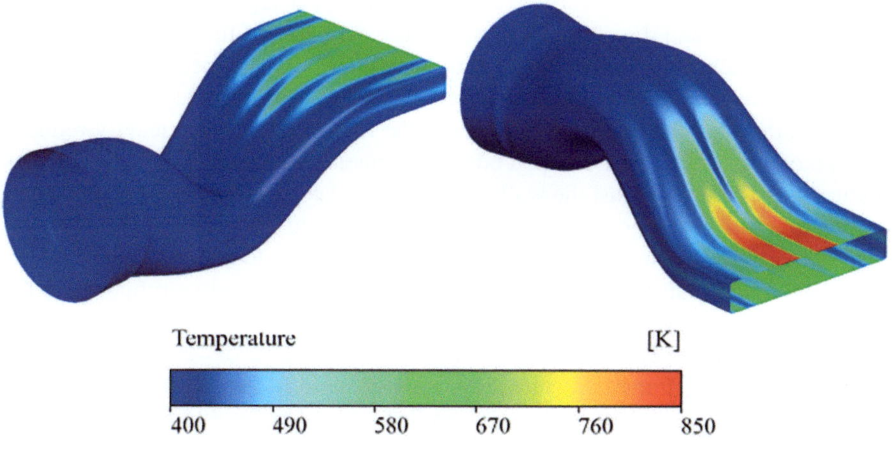

Fig. 2.55 Static temperature distribution on the wall of S-shaped nozzle

S-shaped nozzle, respectively. It can be seen that the static temperature distribution on the along-travel cross-section and on the wall of the nozzle mainly depends on the distribution of vortices (flow vortices) in the x-direction on the flow cross-section. Among them, the flow vortices induced by the flap mixer and the tail cone structure are the main drivers affecting the static temperature distribution, while the vortices in the boundary layer and the angular vortices induced by the S-shaped circular-turned-square configuration have little effect and are not mentioned in the following analysis.

In cross-section B, the flow vortices induced by the wave mixer can drive the inner heat flow to mix with the outer cold flow, but the flow vortices induced by the "funnel-shaped" structure of the wave mixer are the main motivation for the mixing of the inner and outer airflow. In the "funnel-shaped" structure of the vortex generated by the ring-shaped flow, the static temperature distribution on the cross-section B is similar to the "petal" shape of the distribution characteristics. In section C, the "petal-shaped" static temperature distribution gradually changes to a "spider-shaped" distribution. Due to the influence of the S-shaped curved channel, the airflow mixing in the region near the upper wall surface is enhanced, the high temperature region is reduced, and the temperature is lowered. Under the effect of normal pressure gradient, the distribution of high-temperature region on the cross-section moves to the upper side as a whole. With the lateral expansion of the cross-section, the high-temperature region gradually spreads to the outside. On cross-section D, the difference in the distribution of high-temperature regions between the upper and lower sides decreases as the airflow mixing in the region near the lower wall surface increases. With the continuous lateral expansion of the cross-section, the high-temperature core region stretches along the lateral direction. The flow vortex generated by the upper wall of the tail cone extends downward along the longitudinal direction and merges with the vortex generated by the flap mixer, which strengthens the doping effect of the airflow in the upper region of the core area of the cross-section, and thus the high-temperature core area of the cross-section D is changed from a "rectangle-like" to a "pea-shape". The high-temperature core region on cross-section D is changed from "rectangle-like" to "pea-shaped". In cross-section E, the high-temperature region near the lower wall is significantly reduced due to the influence of the S-shaped curved channel. With the flow direction vortex gradually distributed against the wall, the vortex sucked by the connotation of high-temperature gas flow impact by the connotation of cold flow wrapped in the low-temperature wall, the nozzle wall began to appear on the local "hot spot" feature, due to the lower wall near the region of the vortex is relatively stronger than the upper wall, so the lower wall of the thermal impact area is larger, the impact intensity is also greater. The high-temperature core on the cross-section is further stretched along the transverse direction, thus changing from a "pea shape" to a "whale tail shape". In cross-section G, the flow vortex in the middle region of the cross-section further spreads out, and the high-temperature gas flow continues to flare out to impact the nozzle wall. The high-temperature core in the center of the cross-section continues to be stretched laterally, and the "whale-tail-shaped" high-temperature core on the cross-section is significantly reduced in both scope and intensity.

In addition, the wall hot spots of the S-shaped nozzle are mainly concentrated at and downstream of the second bend of the nozzle. Since the vortex distribution inside the S-shaped nozzle is always symmetrical with respect to the longitudinal symmetry, the wall hot spot distribution of the nozzle is also symmetrical with respect to the longitudinal symmetry. The S-shaped nozzle has a typical asymmetric structure along the longitudinal curvature, and the vortex coverage and intensity of the lower wall is significantly higher than that of the upper wall in the second bend and its downstream area. Therefore, the intensity and range of the "hot spot" area on the lower wall of the S-shaped nozzle is significantly higher than that on the upper wall, and the high-intensity hot spot area corresponds to the corresponding high vortex area on the near-wall surface of the nozzle, which has a non-negligible impact on the structure of the S-shaped nozzle.

Figures 2.56 and 2.57 give the perimeter wall static pressure distribution and the limiting flow line and shear stress distribution on half of the wall of the S-shaped nozzle along the course section, respectively. It can be seen that the near-wall flow of the S-shaped nozzle has special phenomena such as longitudinal and transverse curvature, convergence, expansion, etc., which is closely related to the longitudinal pressure gradient and transverse pressure gradient of the nozzle perpendicular to the flow direction. And the distribution characteristics of the normal and transverse pressure gradient along the flow direction are determined by the curvature of the nozzle normal and transverse curvature size, that is, it depends on the geometrical

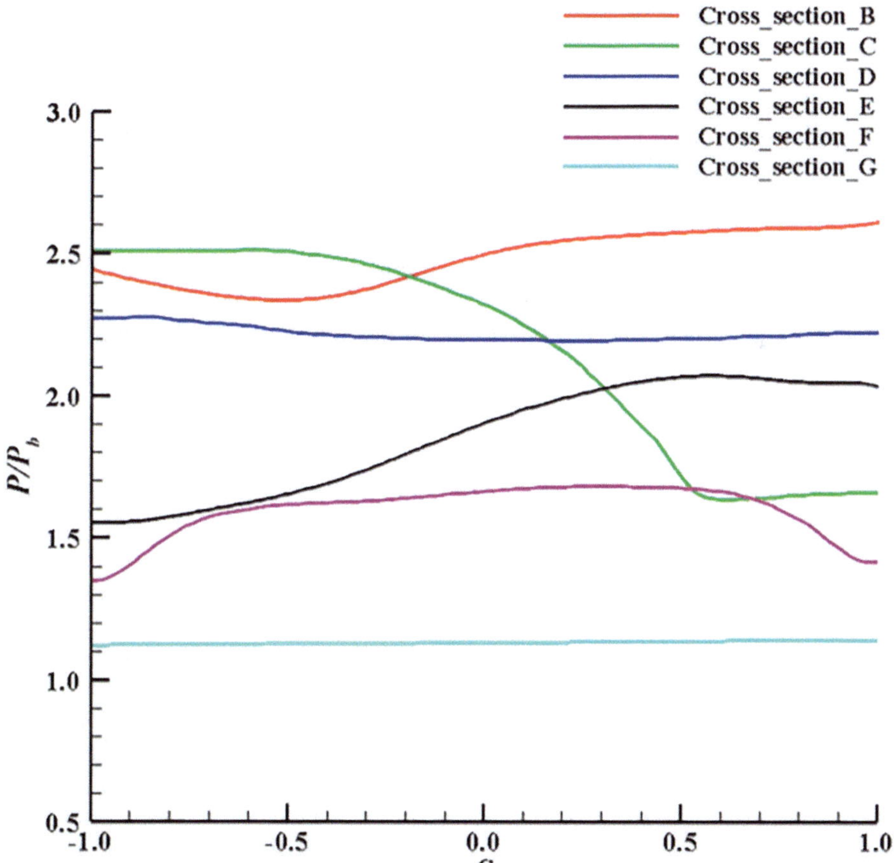

Fig. 2.56 S-shaped nozzle perimeter wall static pressure distribution along the course section

Fig. 2.57 S-shaped nozzle half wall extreme flow distribution

curvature configuration of the nozzle along the flow direction. In this section, according to the S-shaped nozzle curvature configuration–flow along the curvature distribution–pressure gradient trend–wall limit flow distribution of ideas focusing on analyzing the S-shaped nozzle special geometric configuration on its wall limit flow distribution. This section focuses on analyzing the influence mechanism of S-shaped nozzle's special geometric configuration on its wall ultimate streamline distribution characteristics.

In section B, as the nozzle bends downward along the longitudinal section, the center of longitudinal curvature is located in the lower side of the nozzle profile, so there is a longitudinal downward normal pressure gradient, so the nozzle upper wall static pressure is greater than the lower wall static pressure, and the distribution of peripheral wall static pressure throughout the section B is characterized by the upper wall static pressure > side wall static pressure > lower wall static pressure, and the wall limit flow line of the downstream region of the section bends downward. Section C is located in the first bend of the nozzle turn, the longitudinal center of curvature of the nozzle is located in the upper side of the profile, thus generating a longitudinal upward pressure gradient, the nozzle wall pressure decreases rapidly, the peripheral wall hydrostatic pressure distribution of section C is characterized by the lower wall hydrostatic pressure > side hydrostatic pressure > upper wall hydrostatic pressure, the wall limit flow line in the region downstream of the cross-section C curved upward. From cross-section C to cross-section D, the nozzle cross-section expands, and the tendency of the nozzle wall flow line to flow along the lateral to the outer bending expansion is more obvious. The larger bending curvature also leads to a sharp acceleration of the flow near the wall on section C and a significant increase in the wall shear stress. At section D, the longitudinal curvature of the upper side of the nozzle profile decreases and the static pressure on the upper wall surface increases. From section D to section E, the nozzle cross-section continues to expand laterally, and the curvature near section D is almost reduced to zero, so the static pressure on the upper, lower and side wall surfaces of the entire periphery of section D is almost the same, and the wall flow line continues to expand along the lateral expansion of the lateral curvature of the flow tendency is still more obvious. Section E is located in the second bend of the nozzle turn, the longitudinal curvature center of the nozzle moves to the lower side of the nozzle profile, and the pressure on the lower wall of the nozzle decreases rapidly. Under the effect of a large longitudinal back pressure gradient, the ultimate flow line of the wall near section E is bent in the axial direction. The larger curvature also leads to a sharp acceleration of the airflow near the wall under section E and a significant increase in the wall shear stress. From cross-section E to cross-section F, the nozzle cross-section expands, the transverse pressure gradient is directed toward the inner side of the nozzle, and the decrease in static pressure at the nozzle-side wall slows down. Therefore, the peripheral wall static pressure distribution of the entire section E is characterized by the upper wall static pressure > side wall static pressure > lower wall static pressure. The extreme flow lines on the upper and lower wall surfaces of the nozzle begin to converge in the inward curvature of the flow lines under the action of the lateral reverse pressure gradient. Section F is located at the end of the S-shaped converging section of the S-shaped nozzle, the normal pressure gradient along the longitudinal downward direction is significantly reduced, and the difference between the upper and lower wall static pressures is significantly reduced. Under the effect of the continuous longitudinal reverse pressure gradient, the wall limit flow line at cross-section E is bent to continue bending downward along the axial direction. The transverse curvature near cross-section F continues to increase compared to cross-section E. The extreme flow lines on the upper and lower walls of the nozzle further converge to the inward direction under the continuous transverse counterpressure gradient. The effect of lateral curvature at this cross-section is very prominent, making the lateral wall pressure greater than the upper and lower wall pressure. Cross-section F to the nozzle outlet cross-section G is straight section, the center of curvature of the nozzle is zero, the nozzle wall static pressure gradually converge, the wall limit flow direction is basically unchanged, the upper and lower wall limit flow line continues to converge to the inner curvature. As the airflow reaches supersonic speed in the isostraight section, the wall shear stress in the isostraight section increases significantly.

Figure 2.58 gives the distribution of total pressure recovery coefficients on the symmetry plane and the along-travel cross section of the S-shaped nozzle. It can be seen that in addition to the large total pressure loss due to viscous and vortex losses

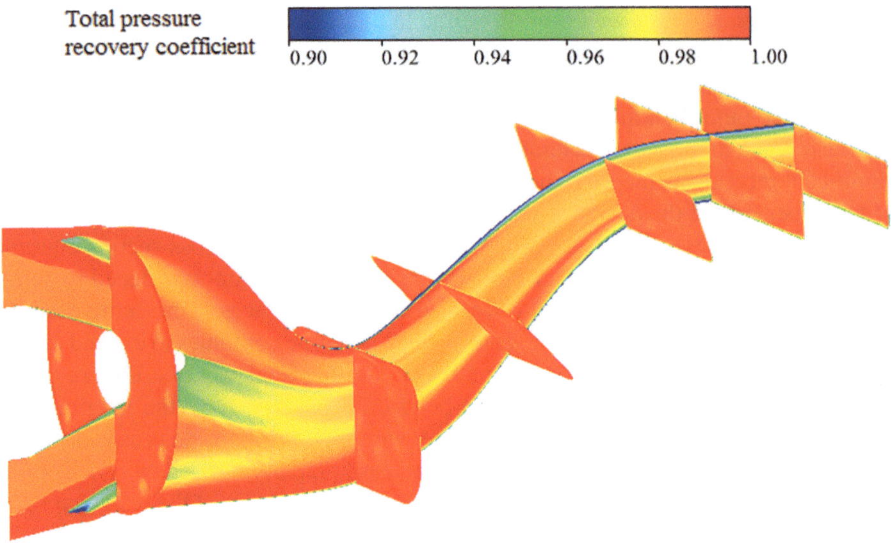

Fig. 2.58 Distribution of total pressure recovery coefficients on the symmetry plane and along-range cross section of the S-shaped nozzle

in the near-wall boundary layer, the total pressure loss generated by the flap mixer and the tail cone structure should not be neglected. The conventional annular structure in the flap mixer on the one hand produces flow vortex, which brings vortex loss; on the other hand, the inner and outer culvert gas flow through the annular structure of the formation of the shear layer between the doping, thus increasing the viscous loss and doping loss in the nozzle. In addition to the above losses, the most important thing is that the mixing of internal and external airflow induced by this structure leads to serious flow separation downstream, which brings large vortex loss. The tail cone, with its large curvature hemisphere and large convergence angle at the end of the structure, induces two serious flow separation regions, which also generate large vortex losses. The vortex loss generated by the flap mixer and the tail cone runs through the entire S-shaped channel, which interacts with the round-to-square bending configuration of the S-shaped nozzle, intensifying the mixing effect of the inner/outer airflow and increasing the mixing loss of the cold/hot flow in the entire nozzle channel. In addition, according to the research results of the S-shaped nozzle for turbojet engine, the S-shaped nozzle produces significant acceleration loss, secondary flow loss, collision loss and other local losses due to the curved channel structure of the round-to-square transition surface, which also exists in the S-shaped nozzle under the double-connotation condition for turbofan engine. Therefore, the aerodynamic performance losses of the S-shaped nozzle under double-connotation conditions mainly include vortex losses, doping losses, viscous losses and localized losses, and the sources of the losses are mainly the wall boundary layer, the inner/outer-connotation two-streams, the flap mixer and the tail cone structure.

From the above analysis, it can be seen that the existence of internal/external double-stranded flow in the exhaust mixer, the flap mixer, and the tail cone, and their interaction with the round-to-square bending configuration of the S-shaped nozzle are the most essential features of the development and evolution of the flow field of the S-shaped nozzle for the turbofan engine. The introduction of the flap mixer and tail cone induces some new special flow phenomena, such as annular flow vortices, separation vortices downstream of the tail cone, and internal/external shear layer, etc. These vortex distributions develop and evolve under the constraints of the S-shaped circularly curved channel, which significantly affect the symmetric plane Ma number distribution, vortex distribution along the cross-section, and static temperature distribution, and are in contrast to typical flow features inherent to S-curved nozzle configuration, such as local acceleration, transverse acceleration, and transversal temperature distribution. They are intertwined with the typical flow characteristics inherent in the S-shaped nozzle configuration, such as local acceleration, transverse secondary flow, and flow vortices at the corners, which makes the flow field inside the S-shaped nozzle more special and complex, and gives rise to such features as asymmetric distribution of vortices due to vortex stretching, vortex-volume absorbing heat flow impinging on the wall surface to form a "hot spot", and core flow with inner/outer connotations. S-shaped nozzle round to square bending geometric configuration affects the flow along the lateral, normal curvature distribution, which in turn affects the symmetry surface wall pressure distribution and the perimeter of the cross-section along the course of the nozzle. The S-shaped nozzle geometry affects the lateral and normal curvature distribution along the flow, which in turn affects the symmetry surface of the nozzle symmetry surface wall pressure distribution and the wall pressure distribution along the periphery of the cross-section, which determines the trend of the transverse and normal pressure gradient, and thus affects the distribution characteristics of the wall limit flow. These flow field characteristics generate the main flow losses such as vortex loss, mixing loss, viscous loss and localized loss, which ultimately have a very important impact on the aerodynamic performance of the S-shaped nozzle.

2.4 Effect of Geometrical Parameters on the Flow Characteristics of S-Shaped Nozzle for Turbofan Engines

There are many geometric design parameters of S-shaped nozzle for turbofan engine, and different parameters have different effects on its aerodynamic performance. According to the key geometric parameters of the S-shaped nozzle and the research results on the influence law of the geometric parameters of the S-shaped nozzle for turbojet engine shown in Fig. 2.3, the key geometric parameters that have a significant influence on the aerodynamic performance of the S-shaped nozzle are selected in this section, which include: the axial lengths of the first and the second bends of the nozzle, L_1 and L_2, the nozzle exit width, W_e, and the nozzle exit height,

H_e, and the longitudinal deviation, ΔY, of the second bend$_2$. Taking the nozzle inlet diameter D as the reference parameter, the above five geometrical parameters are converted into four dimensionless parameters through dimensionless processing, including the exit width/height ratio W_e/H_e, the length/diameter ratio L/D, the shading ratio $\Delta Y_2/L_2$, and the axial length ratio of the two bends L_2/L_1, with the axial lengths of the nozzles $L = L_1 + L_2 + L_3$, and the shading ratios $\Delta Y_2/L_2 = 0.513$ means that the shading rate is 100%, i.e., the nozzle achieves complete shading of the inlet high-temperature parts. Aiming at the four geometrical parameters mentioned above, the study on the influence of key geometrical parameters on the flow characteristics of the S-shaped nozzle for turbofan engine is carried out to analyze the development and evolution of the flow field characteristics of the S-shaped nozzle under the action of key parameters, and to obtain the trend of the aerodynamic performance of the S-shaped nozzle under the different parameter ranges. Table 2.5 gives the values of each key dimensionless geometric parameter in the design state and its variation range during the parametric study. Except for the study of the obscuration rate parameter, the study of the remaining parameters must be constrained by the criterion of fully obscuring the high-temperature components, i.e., the longitudinal deflection of the second bend will be varied.

2.4.1 *Effect of Outlet Aspect Ratio on the Flow Characteristics of S-Shaped Nozzle*

In this section, the flow field characteristics of four different S-shaped nozzles with different exit aspect ratios W_e/H_e for turbofan engines are numerically investigated under ground conditions. The geometrical configurations of the different S-shaped nozzles are shown in three views in Fig. 2.59. The values of the exit aspect ratios and the corresponding dimensionless longitudinal offsets of the second bend are shown in Table 2.6. The dimensionless longitudinal offsets of the second bend reflect the bending curvature of the S-shaped nozzle at the longitudinal turn, and the other geometrical parameters are the values in the design condition. It can be seen that, under the constraint of the criterion of complete shielding of high-temperature components, the increase of the nozzle outlet aspect ratio will reduce the difficulty of shielding the S-shaped nozzle from the high-temperature components such as the wave mixer and the tail cone, which leads to the gradual decrease of the second-bend dimensionless longitudinal deviation, and the transverse expansion range of the nozzle profile increases, which results in the gradual decrease of the bending

Table 2.5 Values of key geometrical parameters of double S-shaped nozzle in the design state and their range of variation

Geometric parameter	L/D	$\Delta Y_2/L_2$	L_2/L_1	W_e/H_e
Design state	2.6	0.51	1.5	6
Scope of change	2.2 ~ 2.8	0.13 ~ 0.51	0.67 ~ 2	3 ~ 6

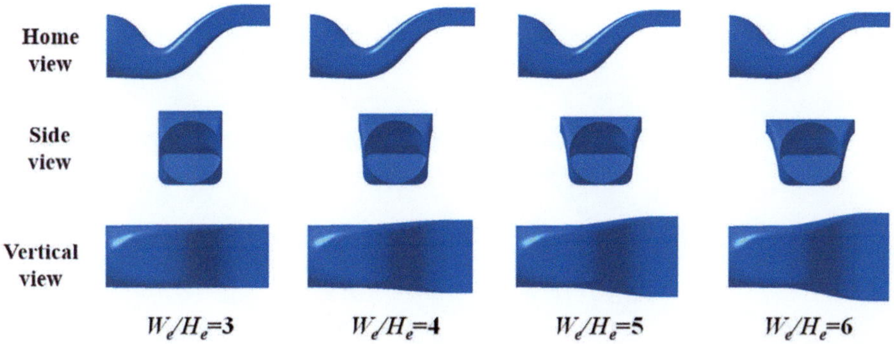

Fig. 2.59 Three views of geometric configuration of S-shaped nozzle with different outlet aspect ratios

Table 2.6 Exit aspect ratio of the S-shaped nozzle geometry model and the corresponding dimensionless longitudinal offset taken for the second bend

W_e/H_e	3	4	5	6
$\Delta Y_2/L_2$	0.56	0.54	0.52	0.51

curvature of the profile along the longitudinal direction of the S-shaped nozzle, and the gradual increase of the bending curvature of the profile along the transverse direction of the S-shaped nozzle. The curvature of the S-shaped nozzle curvature decreases in the longitudinal direction and increases in the transverse direction. Therefore, with the increase of the outlet aspect ratio, the entire S-shaped nozzle geometry becomes more gentle along the longitudinal bending flow path, and the expansion angle along the transverse direction gradually increases.

Figures 2.60 and 2.61 show the symmetric surface Mach number distribution and the wall static pressure distribution of the S-shaped nozzle for different outlet aspect ratios, respectively. With the increase of outlet aspect ratio, the upper wall static pressure near the first bend of the S-shaped nozzle gradually decreases, and the airflow velocity near this region gradually increases; the lower wall static pressure near the second bend gradually increases, and the airflow velocity near this region gradually decreases. The reason is that the increase in the exit aspect ratio leads to a reduction in the longitudinal deviation of the second bend under the constraints of the complete shading criterion, and the longitudinal curvature of the nozzle near the first and second bends is reduced, and accordingly, the degree of contraction of the airflow tubes in the two regions is reduced, the degree of acceleration of the airflow is weakened, and the hydrostatic pressure is increased. In addition, with the increase of the exit aspect ratio, the sidewall static pressure near the first bend gradually decreases, and the sidewall static pressure near the second bend gradually increases. This is because the increase in the exit aspect ratio leads to an increase in the lateral expansion angle of the nozzle, which in turn makes the first and second bend near the lateral curvature increases, but because the first bend near the center of the

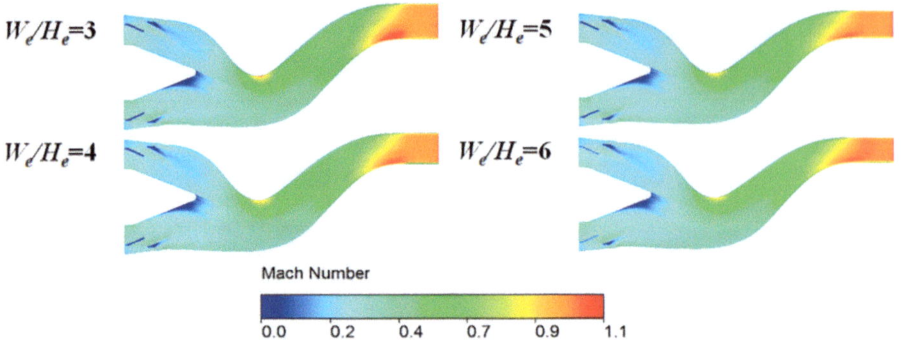

Fig. 2.60 Distribution of Ma number in the symmetry plane of the S-shaped nozzle for different outlet aspect ratios

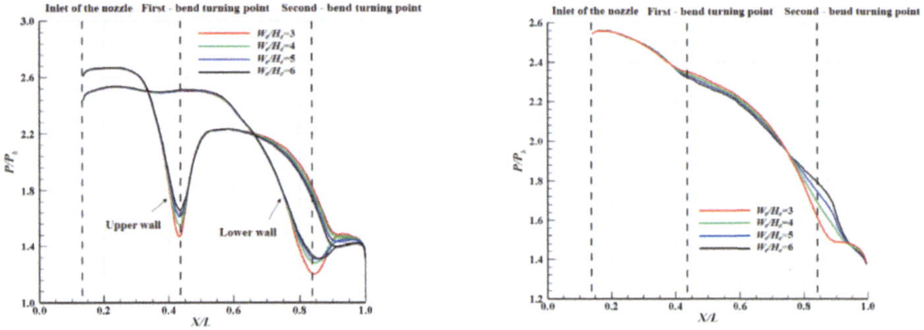

(a) Symmetric wall static pressure distribution

(b) Side wall static pressure distribution

Fig. 2.61 Static pressure distribution on the wall of S-shaped nozzle for different outlet aspect ratios. (**a**) Symmetric wall static pressure distribution. (**b**) Side wall static pressure distribution

lateral curvature is located in the outside of the wall surface of the nozzle, the lateral curvature increase makes the first bend near the side wall surface of the airflow tube contraction degree is reduced, the degree of airflow acceleration weakened, the static pressure rises; and the second bend near the center of the lateral curvature is located in the wall surface of the nozzle On the other hand, the center of the transverse curvature near the second bend is located on the inner side of the nozzle wall, and the increase of the transverse curvature increases the contraction degree of the airflow tubes on the side wall near the second bend, the acceleration degree of the airflow is enhanced, and the static pressure is reduced.

According to the nozzle symmetry surface Ma distribution and wall static pressure distribution, it can be seen that the differences in the flow field characteristics of the S-shaped nozzle under different outlet aspect ratios are mainly located in the region between the first bend turn and the nozzle outlet position, and the differences in the flow field characteristics in the exhaust mixer and the first section of the

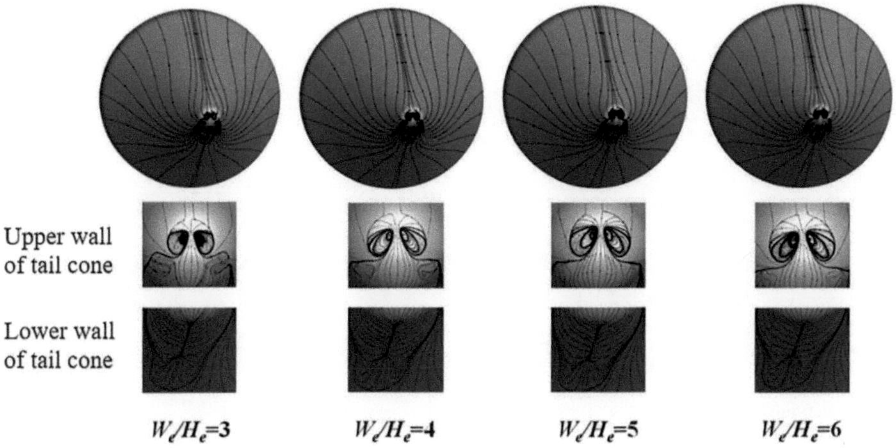

Fig. 2.62 Ultimate flow distribution of the tail cone surface for different outlet aspect ratios

S-shaped channel are smaller. With the increase of the outlet aspect ratio, the flow separation generated by the upper and lower side structures of the flap mixer has basically the same range of influence on the flow field downstream of the nozzle. Figure 2.62 give the limiting streamline distributions on the surface of the wake cone structure in the exhaust mixer for different aspect ratio conditions. The results show that the variation of aspect ratio has less effect on the flow separation characteristics near the tail cone wall. The larger flow vortices generated in the upper wall hemisphere and the flow separation zone in the lower wall both change less with the increase of aspect ratio. On the other hand, the range of the flow vortices in the flow separation region induced by the lower wall surface decreases, and the pair of flow vortices disappears completely when the aspect ratio is increased to 5, and the region where the flow vortices are located evolves into a separated flow line from saddle point to node. The reason is that the increase in aspect ratio causes the pressure gradient upward in the longitudinal direction at the first turn to decrease, and this flow field disturbance is transmitted forward, causing the pressure gradient downward in the longitudinal direction downstream of the tail cone to decrease slightly, and it manifests itself as a slightly decreasing inverse pressure gradient in the region of the lower wall surface of the tail cone, which reduces the degree of airflow separation near the lower wall surface.

Figure 2.63 give the x-direction vortex distribution on the along-travel cross section of the S-shaped nozzle for different outlet aspect ratios. Combined with the above analysis, the downstream flow field perturbation of the nozzle caused by the difference in aspect ratios has less influence on the flow characteristics near the upstream exhaust mixer, and thus the vortex distribution on the inlet section B of the S-shaped nozzle with different outlet aspect ratios is basically the same. However, when the flow passes through the first bend turn, the effect of downstream flow field perturbation starts to appear. In cross-section C, the increase of aspect ratio makes

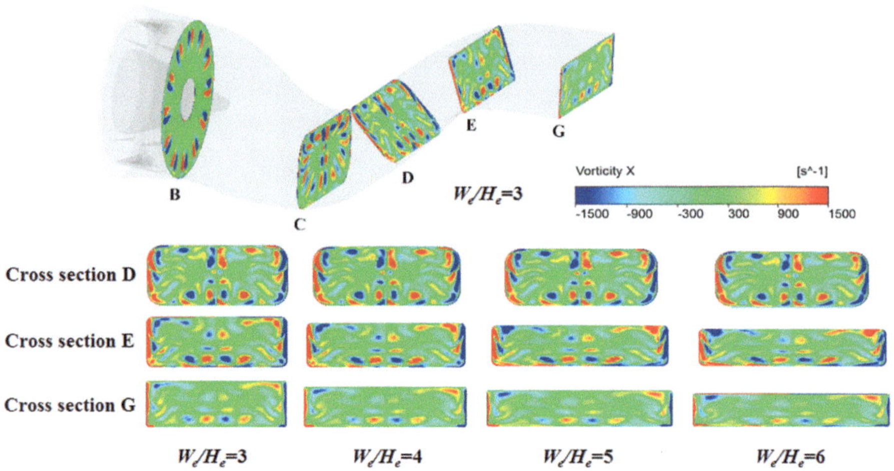

Fig. 2.63 Distribution of x-direction vorticity on the along-travel cross section of the S-shaped nozzle for different outlet aspect ratios

the longitudinal curvature decrease, the pressure gradient weaken, and the tendency of the vortex distribution to move upward slows down, which ultimately leads to the weakening of the asymmetric vortex distribution characteristics along the longitudinal direction in this cross-section; while the transverse expansion angle of the nozzle increases gradually with the increase of the aspect ratio, which results in the increase of the transverse pressure gradient, and the enhancement of the transverse stretch of the vortex near the sidewall surface. As the flow moves to section D, the longitudinal pressure gradients of the nozzle at different aspect ratios are basically equal, so the distribution characteristics of the vortices along the longitudinal direction in this section are basically the same. When the flow develops to section E, with the increase of aspect ratio, the transverse curvature near the second bend gradually increases, and the induced counterpressure gradient inhibits the transverse flow of the airflow, however, the increase of aspect ratio leads to the continuous expansion of the transverse flow cross-section, and the two together lead to the vortex near the side wall surface has a tendency to move toward the center region, but the distance of the flow to the vortex from the center position gradually increases.

Figures 2.64 and 2.65 give the static temperature distributions on the along-travel cross section and on the wall surface of the S-shaped nozzle for different outlet aspect ratios, respectively. According to the above analysis, the static temperature distribution along the cross-section and on the wall of the nozzle mainly depends on the distribution of vortices (flow vortices) in the x-direction on the circulation cross-section, and the temperature distribution at the inlet cross-section B of the nozzle with different aspect ratios is basically the same. At cross-section C, the high temperature region near the upper wall surface is gradually weakened by the vortex enhancement as the outlet aspect ratio increases. The longitudinal asymmetry of the temperature also diminishes gradually due to the reduced upward movement of the vortex. The increase in aspect ratio exacerbates the degree of transverse diffusion of

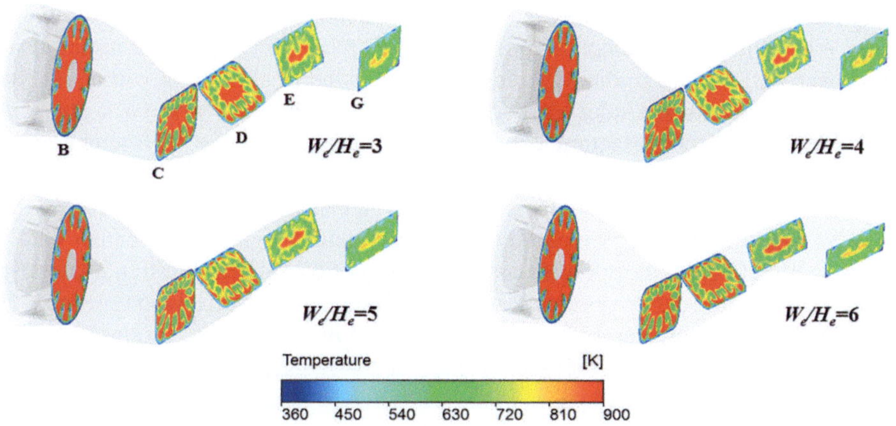

Fig. 2.64 Static temperature distribution on the along-travel cross-section of the S-shaped nozzle for different outlet aspect ratios

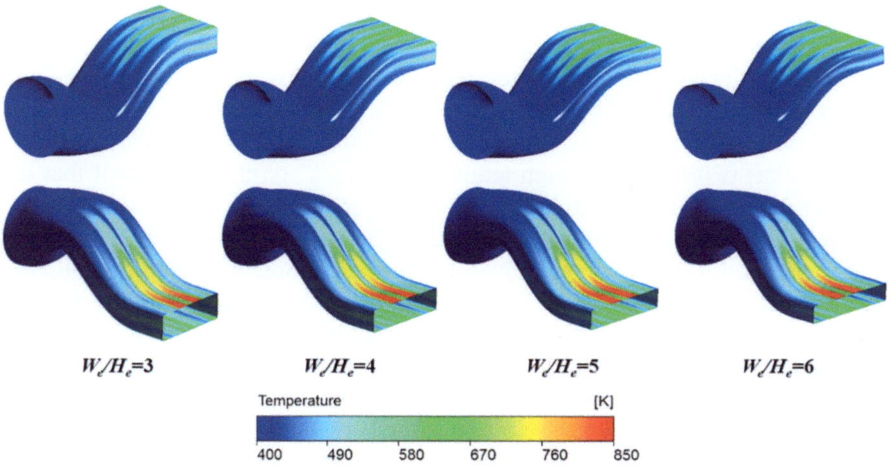

Fig. 2.65 Static temperature distribution on the wall of the S-shaped nozzle for different outlet aspect ratios

the vorticity in this cross-section, resulting in a more pronounced transverse distribution characteristic of the high temperature region. It can be seen that the high-temperature region of the nozzle with aspect ratio 6 is characterized by flatter and the longitudinal position of the high-temperature core is relatively lower compared to that of aspect ratio 3. In cross-section D, the temperature distribution of the S-shaped nozzle with different aspect ratios is consistent because the pressure gradients at the upper and lower walls of the nozzle are basically equal. However, the increase of aspect ratio further aggravates the lateral diffusion of the flow into the vortex, and thus the high temperature region further spreads along the lateral direction. When the flow develops to cross-section E, the high temperature region near the lower wall surface gradually strengthens due to the weakening of vortices as the aspect ratio increases. Under the combined effect of the transverse back pressure gradient and the cross-section expansion, the distance of the high-temperature region near the side wall surface of the second bend from the side wall gradually increases, but its distance from the center position also continues to increase. In the exit cross-section G, the "whale-tail" high-temperature region at the center of the cross-section G of the nozzle with an aspect ratio of 6 continues to stretch laterally and its temperature decreases compared to that of the nozzle with an aspect ratio of 3.

Regarding the static temperature distribution on the wall surface of the S-shaped nozzle, the "hot spots" on the wall surface of the S-shaped nozzle with different outlet aspect ratios mainly appear in the wall area between the downstream of the first bend and the outlet of the nozzle. With the increase of aspect ratio, due to the nozzle cross-section along the longitudinal direction gradually contracted, under the action of the longitudinal pressure gradient, by the heat flow impact of the S-shaped nozzle on the upper and lower wall surface of the transverse area increases, and the flow direction of the vortex sucked by the core area of the heat flow on the upper and lower wall surface of the stronger degree of impact, it can be seen that, compared with the aspect ratio of 3 of the nozzle, the aspect ratio of 6 of the nozzle under the wall surface of two localized "high temperature band" is wider, and the temperature band is wider, and the heat band is wider, and the heat band is wider. It can be seen that compared with the nozzle with an aspect ratio of 3, the nozzle with an aspect ratio of 6 has two localized "high temperature zones" on the lower wall that are wider, and four localized "sub-high temperature zones" on the upper wall that are relatively low in temperature value and narrow in width. However, the increase in aspect ratio leads to a reduction in the longitudinal pressure gradient of the nozzle, the longitudinal area of the upper and lower walls of the S-shaped nozzle impacted by the heat flow is reduced, compared to the nozzle with an aspect ratio of 6, the nozzle with an aspect ratio of 3 has a longer longitudinal range of the "hot spot" on the upper and lower walls of the nozzle. It can also be seen that, with the increase in aspect ratio, the nozzle flow cross-section of the lateral width increases, which leads to the lateral flow of the air flow under the counterpressure gradient increases, and the lateral flow distance is longer, so the flow direction of the vortex sucked by the core of the heat flow on the side wall surface of the impact of the ability to be weakened. Compared to the nozzle with an aspect ratio of 6, the nozzle with an aspect ratio of 3 has a larger range of "hot spots" on the sidewall surface and higher temperature values.

Figures 2.66 and 2.67 show the perimeter wall static pressure distribution and the limiting flow line and shear stress distributions on half of the wall for S-shaped nozzles with different outlet aspect ratios along the course section, respectively. There are significant differences in the near-wall flow characteristics of the S-shaped nozzle at different outlet aspect ratios. As the outlet aspect ratio increases, the pressure on the lower side of section B decreases, and the resulting longitudinal pressure gradient in the section increases. In contrast, the increase in aspect ratio leads

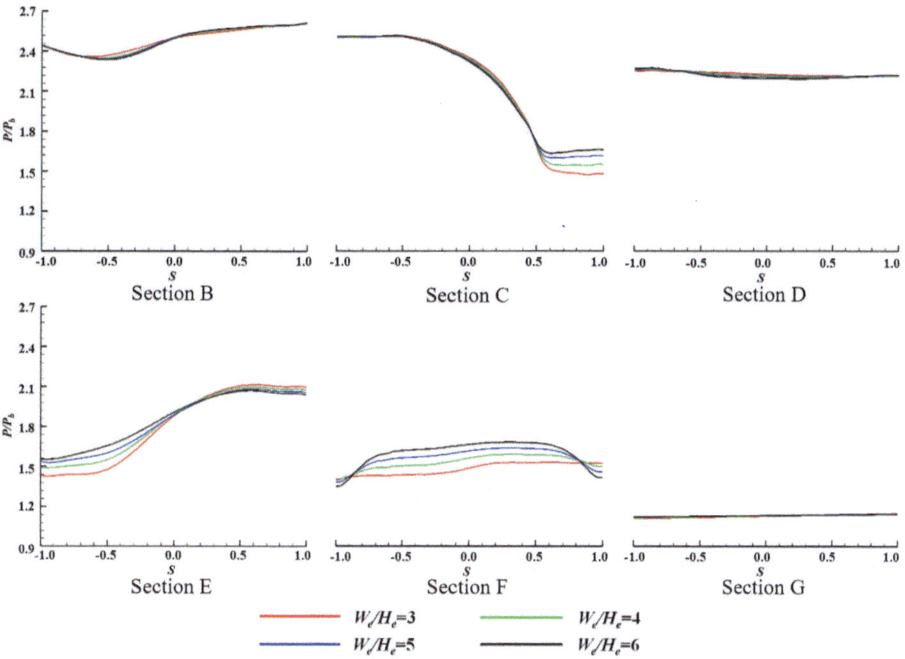

Fig. 2.66 Perimeter wall static pressure distribution along the course section of S-shaped nozzle for different outlet aspect ratios

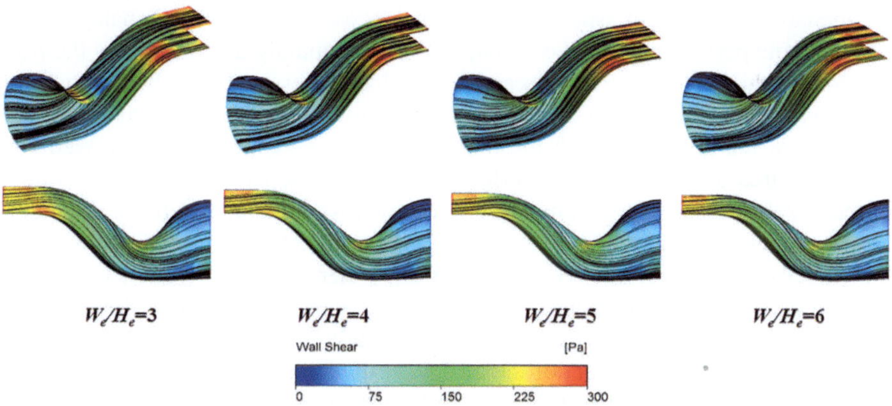

Fig. 2.67 Ultimate flow distribution of S-shaped nozzle wall for different outlet aspect ratios

to an increase in the upper wall pressure at cross-section C due to a decrease in the longitudinal curvature, and the wall limit flow line near this cross-section increases in curvature outward along the transverse direction and decreases in curvature upward along the longitudinal direction. In addition, the reduced longitudinal curvature in the vicinity of this cross-section results in weakened airflow acceleration, and the upper wall shear stress in this region is significantly higher for a nozzle with aspect ratio 3 than for a nozzle with aspect ratio 6. At cross-section D, the longitudinal curvature of the S-shaped nozzles with different aspect ratios is essentially the same, and thus the perimeter wall pressure distributions at this cross-section are approximately equal. At cross-section E at the second bend turn, the longitudinal curvature at this cross-section decreases as the outlet aspect ratio increases, the lower wall hydrostatic pressure is relatively high, and the longitudinal pressure gradient formed decreases, so that the nozzle with an aspect ratio of 6 has a weaker degree of flow curvature on the side wall surfaces, whereas the nozzle with an aspect ratio of 3 has a relatively high shear stress on the lower wall surface in this region. The airflow develops up to cross-section F where the lateral pressure gradient of the S-shaped nozzle reaches its maximum. As the exit aspect ratio increases, the lateral pressure gradient of the nozzle gradually increases, the static pressure on the side wall surface gradually increases, and compared with the nozzle with an aspect ratio of 3, the flow lines on the upper and lower wall surfaces of the nozzle with an aspect ratio of 6 rapidly converge to the inner side of the nozzle with a larger bending angle. At the outlet cross section G, the wall pressures of the nozzle tend to equalize. As the exit aspect ratio increases, the reduction of longitudinal curvature leads to a weaker acceleration of the airflow in the region of the isostraight section, and the shear stresses on the upper and lower wall surfaces of the nozzle are reduced.

Figure 2.68 gives a comparison of the aerodynamic performance of the S-shaped nozzle at different outlet aspect ratios. It can be seen that the total pressure recovery coefficient, flow coefficient and thrust coefficient all show a decreasing trend with the increase of the outlet aspect ratio, but the decrease is small. The change of the aerodynamic performance of the S-shaped nozzle is mainly determined by the friction loss, localized loss, vortex loss and doping loss generated by the boundary layer of the wall of the nozzle, the inner/outer culvert flow, the wavelength mixer and the structure of the tail cone. With the increase of the exit aspect ratio, the wall wet perimeter area of the S-shaped nozzle increases, and the lateral bending curvature of the nozzle increases, which makes the friction loss generated by the wall boundary layer increase; and the increase of the wet perimeter area of the nozzle makes the velocity shear layer between the inner and outer culvert flow increase, which leads to the increase of the doping loss of the inner/outer culvert flow. However, the increase of the outlet aspect ratio reduces the longitudinal curvature of the nozzle, reduces the acceleration effect of the airflow at the longitudinal turn of the nozzle, reduces the wall shear stress at the turn and the degree of bending of the airflow at the side wall, and reduces the localized loss and friction loss inside the nozzle. Under different outlet aspect ratios, the difference in the amount of flow vortex produced by the wave mixer and tail cone structure is small, and the vortex loss and internal and external doping loss are basically the same. The combination of the above factors ultimately leads to little change in the aerodynamic performance of the S-shaped nozzle.

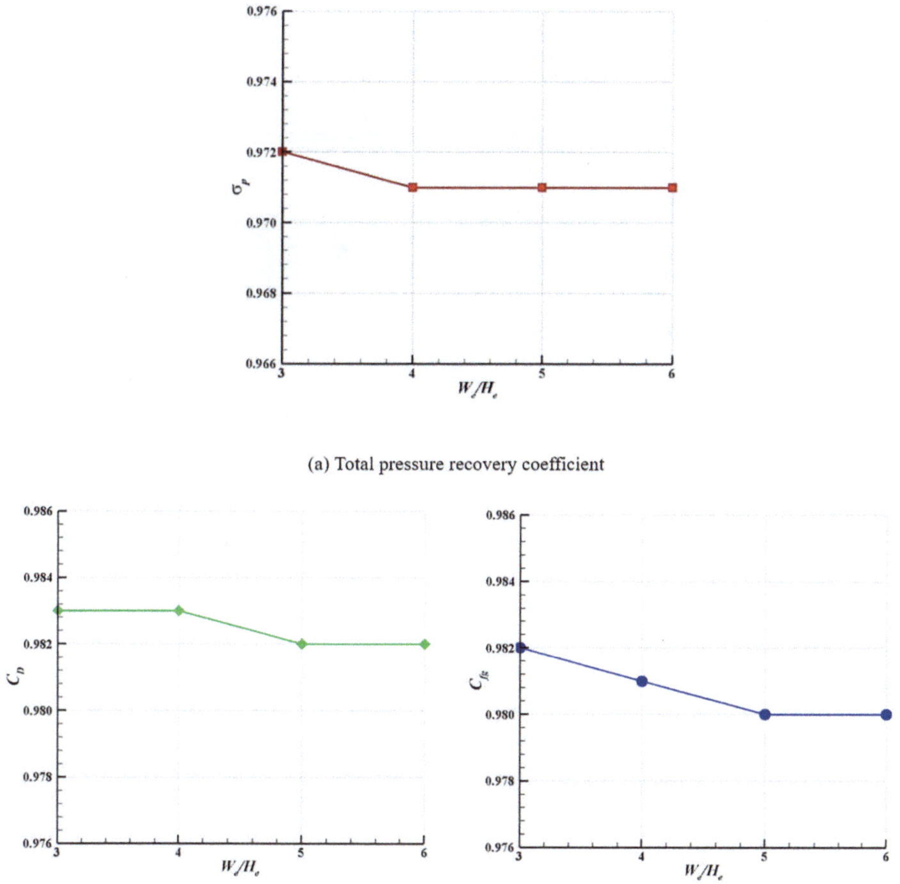

(a) Total pressure recovery coefficient

(b) Discharge coefficient

(c) Thrust coefficient

Fig. 2.68 Trend of aerodynamic performance of S-shaped nozzle with different outlet aspect ratios. (**a**) Total pressure recovery coefficient. (**b**) Discharge coefficient. (**c**) Thrust coefficient

2.4.2 Effect of Length-to-Diameter Ratio on the Flow Characteristics of S-Shaped Nozzles

In this section, the flow field characteristics of four different S-shaped nozzles with different length-to-diameter ratios are numerically investigated under ground conditions. The geometrical configurations of the different S-shaped nozzles are shown in three views at Fig. 2.69. The values of the length-to-diameter ratio and the corresponding longitudinal offsets of the second bend are shown in Table 2.7. The dimensionless longitudinal offsets reflect the magnitude of the bending curvature at the longitudinal turn of the S-shaped nozzle. The other geometric parameters are the

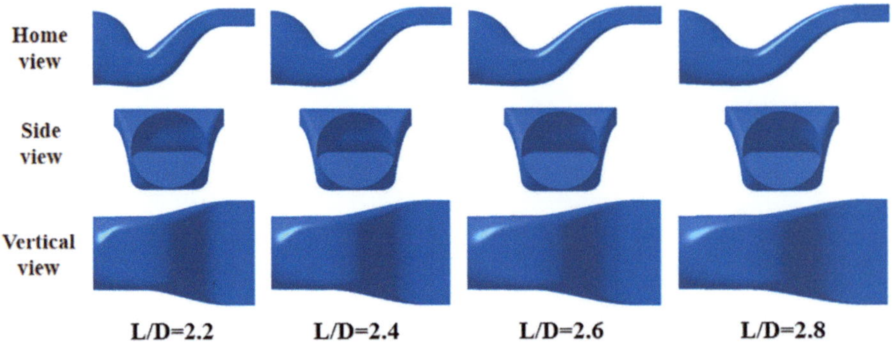

Fig. 2.69 Three views of geometric configuration of S-shaped nozzle with different length-to-diameter ratios

Table 2.7 The length-to-diameter ratio of the geometric model of the S-shaped nozzle and the corresponding longitudinal offset of the second bend

L/D	2.2	2.4	2.6	2.8
ΔY_2 (mm)	530	540	545	555
$\Delta Y_2/L_2$	0.61	0.56	0.51	0.48

values in the design state. It can be seen that an increase in the nozzle length-to-diameter ratio leads to a gradual increase in the longitudinal deflection of the second bend under the constraint of the criterion of complete shielding of the high temperature components. However, the longitudinal curvature of the profile in the longitudinal direction of the S-shaped nozzle gradually decreases due to the increase in the axial length of the second bend of the nozzle as a result of the increase in the length-to-diameter ratio, which leads to a gradual decrease in the dimensionless longitudinal deflection of the first and second bends that affect the longitudinal curvature of the S-shaped nozzle. The increase in the length-to-diameter ratio makes the nozzle along the transverse surface expansion of the degree of urgency is weakened, the transverse bending curvature gradually decreases. Therefore, with the increase of the length-to-diameter ratio, the geometric configuration of the S-shaped nozzle becomes more gentle along the longitudinal bending flow path, and the expansion angle along the transverse direction gradually decreases.

Figures 2.70 and 2.71 give the symmetric surface Ma number distribution and wall static pressure distribution of the S-shaped nozzle at different length-to-diameter ratios, respectively. It can be seen that, when the length-to-diameter ratio is 2.2, due to the low static pressure at the upper wall of the first turn, a large counterpressure gradient is generated downstream, which leads to serious flow separation; the gas flow is accelerated to supersonic speed at the entrance of the isostraight section after the second turn. With the increase of the length-to-diameter ratio, the static pressure on the upper wall of the first bend and the lower wall of the second bend of the S-shaped nozzle are gradually increased, and the air velocity near the region gradually decreases, and the flow separation zone downstream of the upper wall of the first bend and the supersonic zone of the airflow near the downstream of

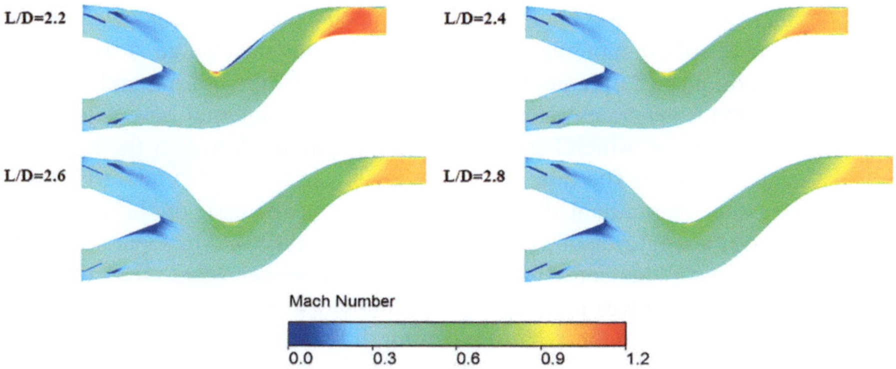

Fig. 2.70 Distribution of Ma number in the symmetry plane of the S-shaped nozzle for different aspect ratios

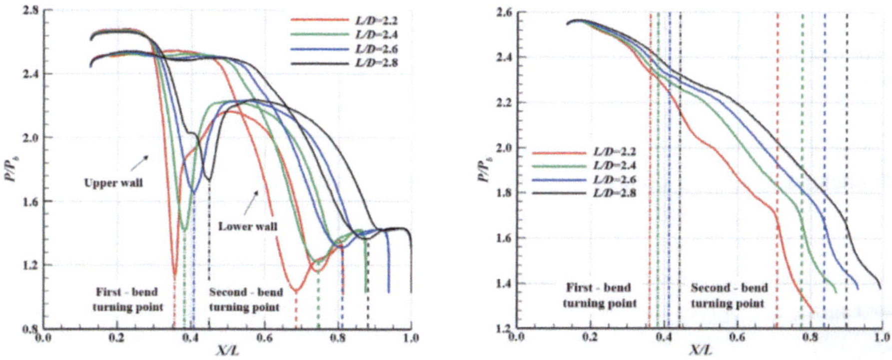

(a) Symmetric wall static pressure distribution

(b) Side wall static pressure distribution

Fig. 2.71 Static pressure distribution on the wall of S-shaped nozzle with different aspect ratios. (**a**) Symmetric wall static pressure distribution. (**b**) Side wall static pressure distribution

the lower wall of the second bend (the entrance of the isostraight section) disappears. The main reason is that the increase of the aspect ratio leads to the decrease of the dimensionless longitudinal deviation of the first and second bends under the constraints of the code of the completely shielded high-temperature parts, and the longitudinal bending curvature is reduced, and thus the acceleration of the airflow in these two regions is weakened and the static pressure is increased. As the length-to-diameter ratio increases, the static pressure on the sidewall surface near the first and second bends of the S-shaped nozzle remains basically unchanged. This is due to the increase in the length-diameter ratio leads to a decrease in the lateral curvature of the nozzle, so that the first two bends near the side wall of the airflow acceleration weakened, the static pressure rises. However, the increase in length leads to an increase in the flow loss in the nozzle channel, which in turn leads to a decrease in the static pressure on the sidewalls near the first and second bends. The two factors have opposite effects, and together they make the static pressure on the sidewalls near the first and second bends of the S-shaped nozzle basically unchanged.

According to the Ma distribution on the symmetry plane of the nozzle and the static pressure distribution on the wall, it can be seen that the change of the length-to-diameter ratio changes the geometric configuration of the nozzle, and the flow field perturbation it generates significantly spreads to the internal flow field of the entire S-shaped nozzle, which has a greater impact on the flow near the upstream exhaust mixer structure. The flow separation generated by the flap mixer is asymmetric along the longitudinal direction, and with the increase of the length-to-diameter ratio, the angle of downward deflection of the low Ma separation flow induced by the upper side configuration of the flap mixer decreases along the axial direction. Figure 2.72 gives the limiting streamline distributions on the surface of the wake cone structure in the exhaust mixer for different length-to-diameter ratios. The results show that the change in *L/D* leads to a change in the flow separation characteristics and generation location near the upper and lower wall surfaces of the tail cone. As the aspect ratio increases, the counter-rotating separation vortex generated on the upper wall surface gradually decreases, and the separation location gradually moves downstream of the tail cone; the counter-rotating separation vortex generated on the lower wall surface gradually disappears, and the flow separation on the lower wall surface as a whole is manifested as a separation streamline from saddle to node, and the degree of separation weakens, and the separation location gradually moves downstream of the tail cone. Therefore, the difference in length-to-diameter ratio has a greater influence on the flow characteristics near the exhaust mixer structure upstream of the nozzle. This is evidenced by the symmetry plane Ma distribution shown in Fig. 2.18, where the low velocity region of the flow downstream of the tail cone is significantly reduced at larger aspect ratios.

Figure 2.73 gives the x-direction vortex volume distribution on the along-travel cross section of the S-shaped nozzle for different aspect ratios. On cross section C, the two pairs of flow vortices generated by the upstream tail cone structure gradually weaken with the increase of length-to-diameter ratio, and the flow vortices induced by the upper wall of the tail cone disappear at length-to-diameter ratio of

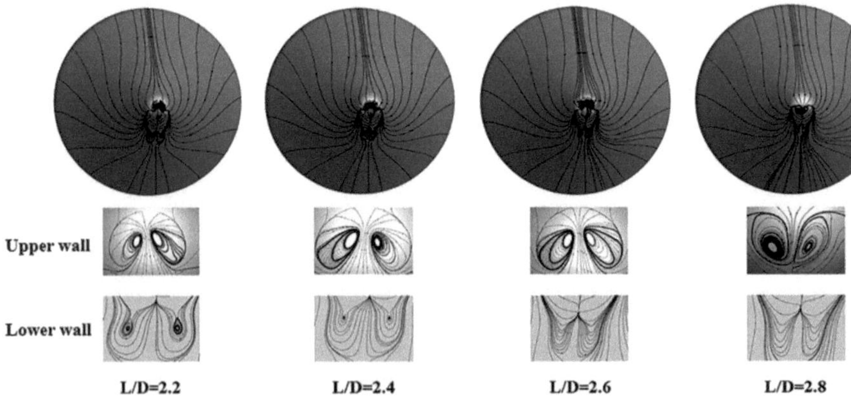

Fig. 2.72 Ultimate flow distribution on the tail cone surface for different aspect ratios

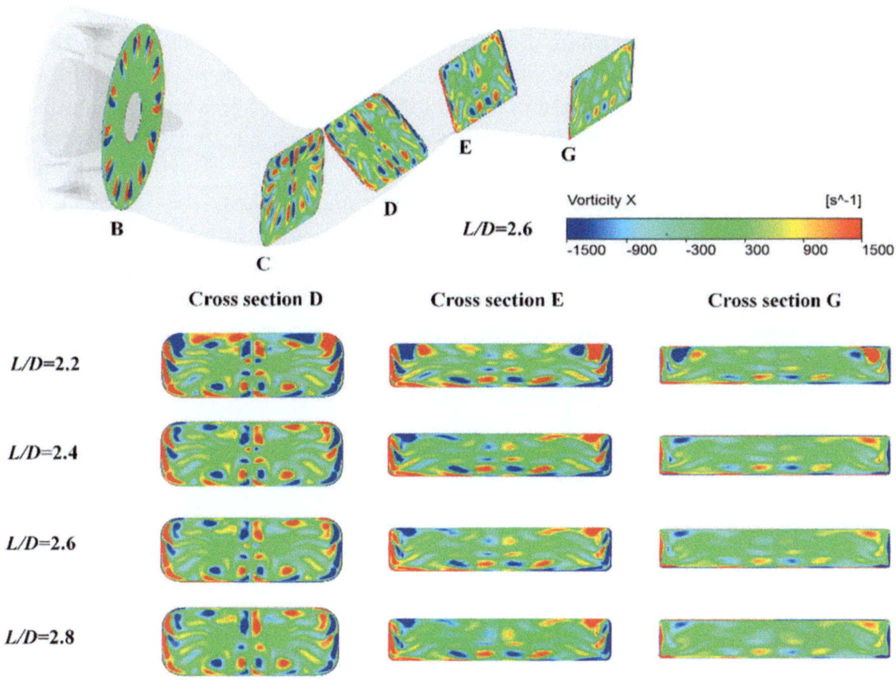

Fig. 2.73 Distribution of x-direction vorticity on the along-travel cross section of S-shaped nozzle with different aspect ratios

2.8. This is due to the weakening of flow separation generated at the tail of the tail-cone structure as a result of the reduction of the longitudinal pressure gradient. As the longitudinal curvature near the cross-section gradually decreases due to the increase of the aspect ratio, the vortex stretching effect in the region on the upper side of the cross-section gradually decreases, and thus the amount of flow vortices on the upper side of the cross-section gradually decreases. In cross-section D, the vortex in the upper region of cross-section D increases significantly when the length-to-diameter ratio is 2.2, which is due to the severe flow separation on the upper wall of the nozzle near this region due to the large curvature of the profile, and with the gradual increase of the length-to-diameter ratio, the flow separation disappears and the vortex starts to decrease. In sections E and G, the increase in the aspect ratio reduces the degree of curvature of the profile at the second bend, resulting in a gradual decrease in the amount of flow vortices near the nozzle side wall surface and the lower wall surface, and the gradient of backpressure to which the flow vortices are subjected by the transverse motion decreases, ultimately resulting in the enhancement of the degree of adherence of the flow vortices to the wall near the nozzle side wall surface. In addition, with the increase of the aspect ratio, the streamwise vortex induced by the tail cone extends upward, and gradually gets closer to the streamwise vortex generated on the upper side of the wave flap mixer, and the two pairs of streamwise vortices merge with each other when the aspect ratio is 2.8.

Figures 2.74 and 2.75 give the static temperature distribution on the along-travel cross section and the temperature distribution on the wall of the S-shaped nozzle for different aspect ratios, respectively. In cross-section C, the decrease in length-to-diameter ratio leads to an increase in the longitudinal upward pressure gradient, and thus the high temperature core in the middle region of the nozzle gradually moves upward. And the reduction of the length-to-diameter ratio shortens the mixing process of the inner and outer culvert gas streams and weakens the mixing effect between the hot and cold gas streams, which leads to a larger range of high-temperature core area. On the cross-section D, for the S-shaped nozzle with an length-to-diameter ratio of 2.2, the flow separation leads to a larger range of high vorticity zone in the upper side region of the cross-section, which strengthens the

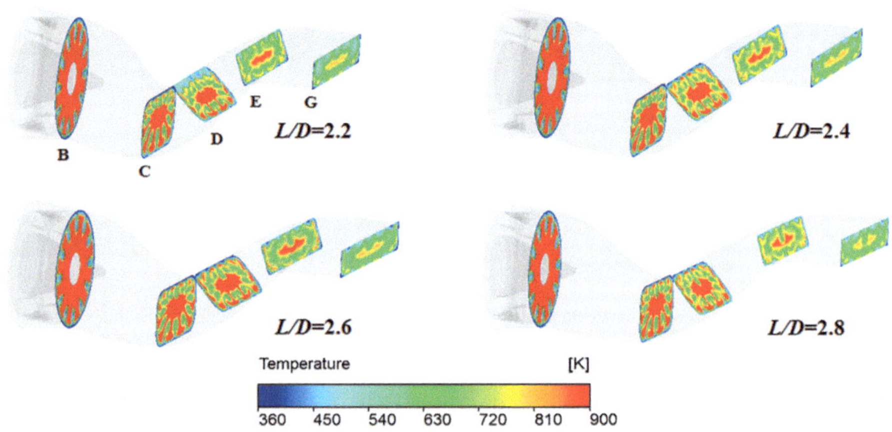

Fig. 2.74 Static temperature distribution on the along-travel cross-section of the S-shaped nozzle for different aspect ratios

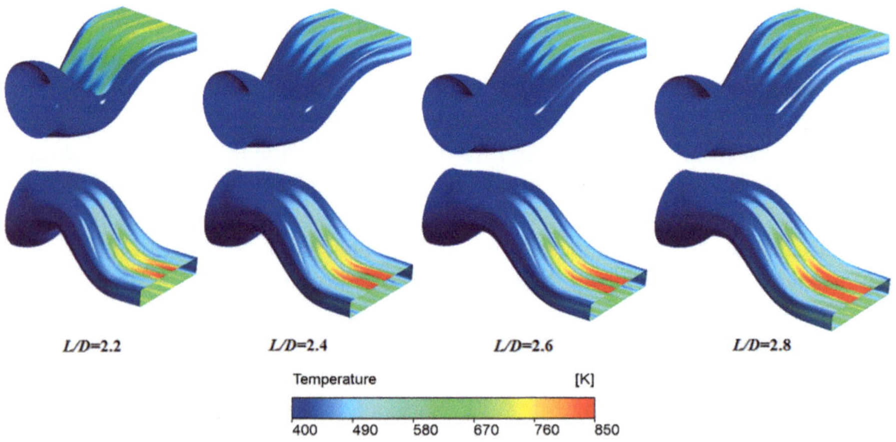

Fig. 2.75 Static temperature distribution on the wall of S-shaped nozzle with different aspect ratios

doping effect between the hot and cold streams, and thus the temperature is lower. In addition, with the increase of the aspect ratio, the vortex volume of the flow vortex generated on the upper and lower walls of the tail cone gradually decreases, and the position of the flow vortex gradually moves downward, which results in the high-temperature core area gradually changing from a "rectangle-like" to a "pea shape". In cross-section E and G, with the increase of the length-to-diameter ratio, the high temperature core area from the flatter "pea-shaped" gradually changed to the "whale tail" shape, when the length-to-diameter ratio of 2.8, the flow vortex generated by the wave mixer and the tail cone generated by the flow vortex merge, in the high temperature core area produces a strong cold and heat. The core area produces strong mixing of hot and cold flows, and the high-temperature core area in the shape of "whale tail" is divided into two parts from the center of the cross-section.

Regarding the static temperature distribution on the wall surface of the S-shaped nozzle, the "hot spots" on the wall surface of the S-shaped nozzle with different aspect ratios mainly appeared in the wall area of the second S-shaped channel. When the length-to-diameter ratio is 2.2, due to the serious flow separation in the upper wall region downstream of the first bend, it generates a high vortex region that increases the impact of the internal heat flow on the upper wall, while the flow separation disappears and the impact of the heat flow is reduced when the length-to-diameter ratio is increased to 2.4. In addition, compared with the nozzle with an length-to-diameter ratio of 2.4, the nozzle with an length-to-diameter ratio of 2.2 has a larger longitudinal curvature, which induces the formation of a larger normal pressure gradient, and thus the upper and lower wall surfaces of the nozzle are subjected to a stronger impact effect of the heat flow. Taken together, when the length-to-diameter ratio increases from 2.2 to 2.4, the "hot spot" range of the second bend of the S-shaped nozzle channel on the upper and lower wall area is greatly reduced, and the temperature value is significantly reduced. Length-to-diameter ratio increased from 2.4 to 2.8, due to the increase in the wet perimeter of the wall of the nozzle, the nozzle on the upper and lower walls of the "hot spot" distribution range is wider, located in the lower wall of the two localized "high-temperature zone" becomes longer, the upper wall of the new number of relatively low temperature values and the narrower width of the localized "hot spot". A number of localized "sub-high temperature zones" with relatively low temperature values and narrow widths are added on the upper wall. It can also be seen that, with the increase of the length-to-diameter ratio, the lateral curvature of the side wall of the nozzle gradually decreases, which leads to a weakening of the back pressure gradient of the lateral motion of the flow vortex, and the flow vortex absorbs the heat flow in the core area and increases the impact ability of the side wall. As a result, the temperature of the "hot spot" on the side wall of the S-shaped nozzle is higher.

Figure 2.76 gives the perimeter wall static pressure distributions along the course cross section of an S-shaped nozzle for different aspect ratios. At cross-section B, an increase in the aspect ratio leads to a decrease in the longitudinal curvature of the nozzle, which results in a decrease in the longitudinal pressure gradient downstream of the nozzle inlet, and the upper wall pressure > side wall pressure > lower wall

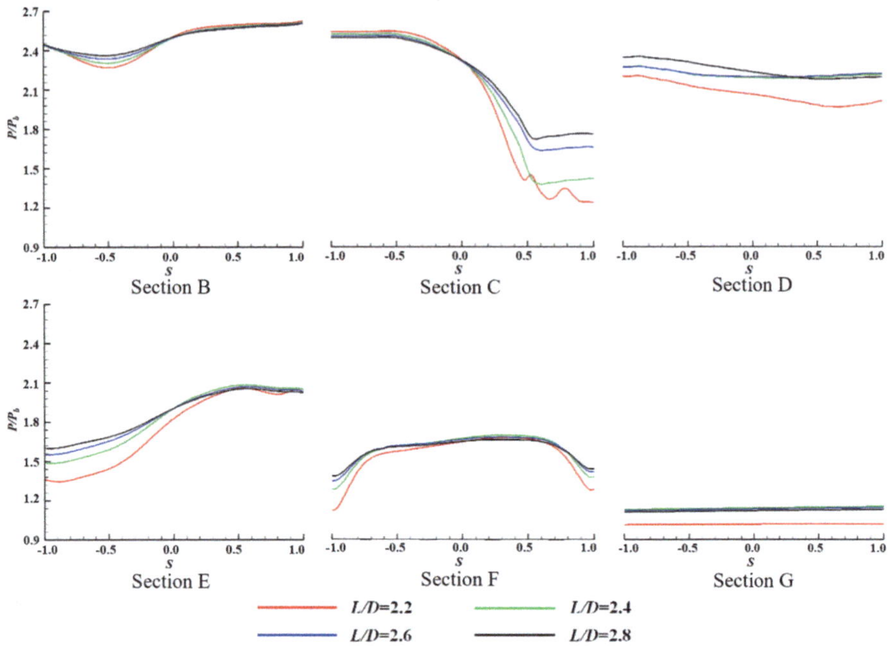

Fig. 2.76 Perimeter wall static pressure distribution along the course section of S-shaped nozzle with different length-to-diameter ratios

pressure. At section C, the longitudinal bending curvature decreases as the aspect ratio increases, and thus the upper wall static pressure at the perimeter of this section continues to decrease. For the nozzle with an length-to-diameter ratio of 2.2, the pressure fluctuations occurring at the upper wall of the perimeter are mainly caused by the flow separation zone downstream. The flow develops to section D. When the length-to-diameter ratio is 2.2, a localized low-pressure zone is generated in the upper side region of this section due to the flow separation. Accompanied by the increase of the length-to-diameter ratio to 2.4, the flow separation disappears, and this localized low-pressure zone disappears, and the peripheral upper-wall static pressure increases significantly. At section E near the second bend, the center of longitudinal curvature is shifted to the lower side of the bending channel, and the increase of the length-to-diameter ratio leads to a gradual decrease of the bending curvature and a gradual increase of the lower wall static pressure. At section F, the lateral wall static pressure is the largest, and the increase of the length-to-diameter ratio leads to a weakening of the lateral curvature of the nozzle and a gradual decrease of the differential pressure on the circumferential wall of this section.

Figure 2.77 gives the limiting flow lines and shear stress distributions on the wall half of the S-shaped nozzle for different aspect ratios. The increase in length-to-diameter ratio reduces the degree of profile bending of the S-shaped nozzle. Under the effect of transverse pressure gradient, the diffusion degree of the extreme

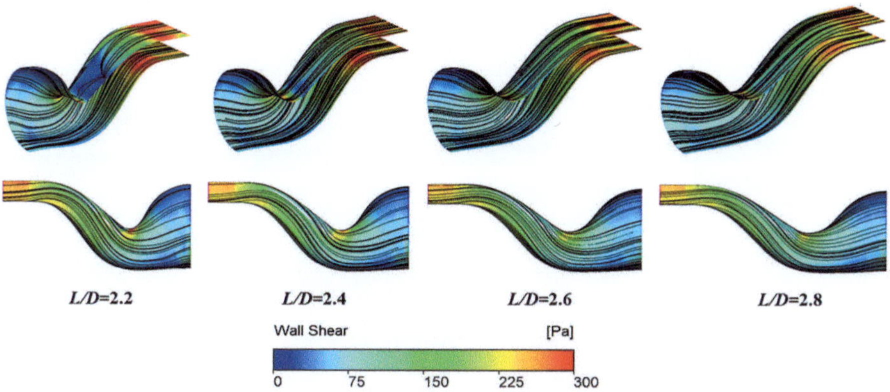

Fig. 2.77 Ultimate flow distribution on the wall of S-shaped nozzle with different aspect ratios

streamlines on the upper and lower wall surfaces of the nozzle downstream of the first bend and the convergence degree near the second bend are gradually reduced; under the effect of longitudinal pressure gradient, the curvature degree of the streamlines on the side wall surfaces of the nozzle is significantly weakened, and the axial deflection angle of streamlines in the outlet region of the nozzle is also gradually reduced. With the increase of the aspect ratio, the local acceleration of the airflow in the curved channel is slowed down, and the wall shear stresses on the upper wall at the first turn, the lower wall at the second turn and the nozzle exit region are gradually reduced. For the nozzle with an length-to-diameter ratio of 2.2, a large range of flow separation occurs in the upper wall area downstream of the first bend due to a large counterpressure gradient, and the flow separation as a whole is shown as a separating limit streamline from the saddle point to the node. The flow separation gradually disappears when the length-to-diameter ratio increases to 2.4.

Figure 2.78 compares the aerodynamic performance of the S-shaped nozzle at different aspect ratios. The total pressure recovery coefficient, flow coefficient and thrust coefficient of the nozzle firstly increase and then slightly decrease with the increase of length-to-diameter ratio. Increasing the length-to-diameter ratio effectively increases the wetted perimeter area of the S-shaped nozzle, which leads to an increase in the friction loss generated by the wall boundary layer, and at the same time increases the area of the velocity shear layer between the inner and outer culvert streams, which leads to an increase in the doping loss of the inner/outer culvert streams. In addition, the vortex loss generated by the flap mixer and the tail cone structure and the resulting airflow doping loss also gradually increase. However, increasing the aspect ratio reduces the degree of nozzle surface bending, which on the one hand reduces the vortex loss due to flow separation in the longitudinal direction, and on the other hand reduces the flow acceleration effect at the longitudinal turn of the nozzle, and reduces the wall shear stress and the degree of flow bending at the turn, so that the localized loss and friction loss inside the nozzle are reduced. To summarize, when the length-to-diameter ratio increases from 2.2 to 2.4, the vortex loss generated by flow separation is the main factor affecting the aerodynamic

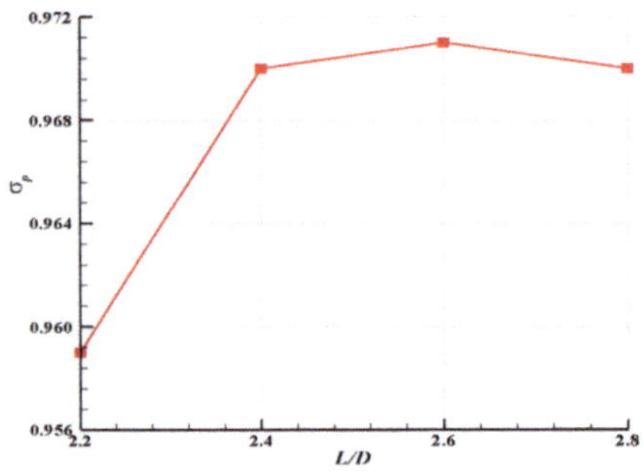

(a) Total pressure recovery coefficient

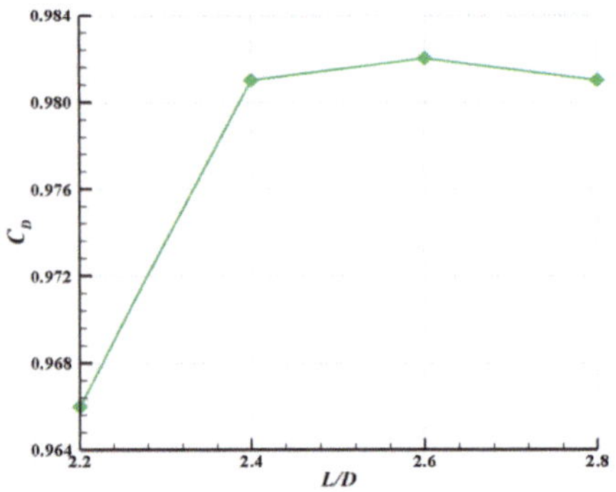

(b) Discharge coefficient

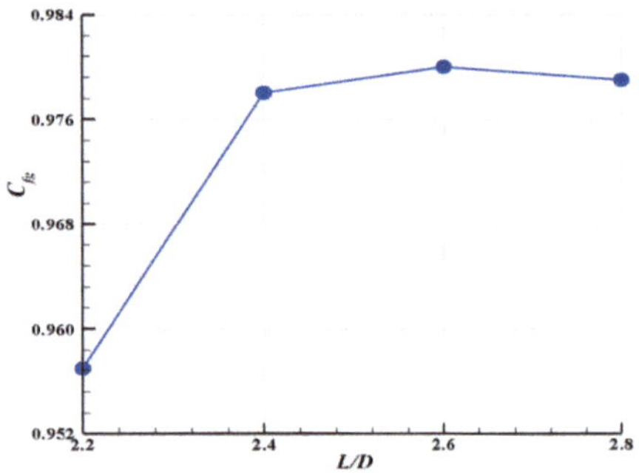

(c) Thrust coefficient

Fig. 2.78 Trend of aerodynamic performance of S-shaped nozzle with different aspect ratios. (**a**) Total pressure recovery coefficient. (**b**) Discharge coefficient. (**c**) Thrust coefficient

performance of the nozzle, and the disappearance of the flow separation significantly improves the aerodynamic performance of the S-shaped nozzle; while when the length-to-diameter ratio increases from 2.4 to 2.8, the various influencing factors interact with each other, and thus the aerodynamic performance of the nozzle remains basically unchanged.

2.4.3 Effect of Obscuration Rate on the Flow Characteristics of S-Shaped Nozzle

In this section, the flow field characteristics of four S-shaped nozzles with different shading rates ΔY are numerically investigated under ground conditions. The second bend longitudinal deviation ΔY_2 of the double S-shaped nozzle is defined as the shading rate ΔY, as shown in Fig. 2.79, i.e., the shading rate is $\Delta \mathrm{Y} = 1$ when the nozzle profile achieves complete shading of the high temperature components, and

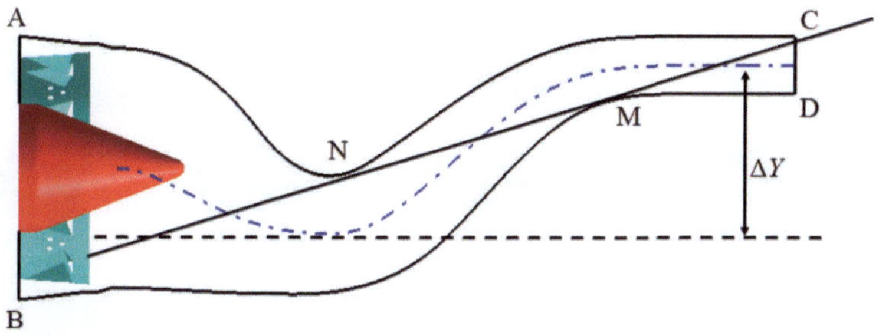

(a) Occlusion rate ΔY=1

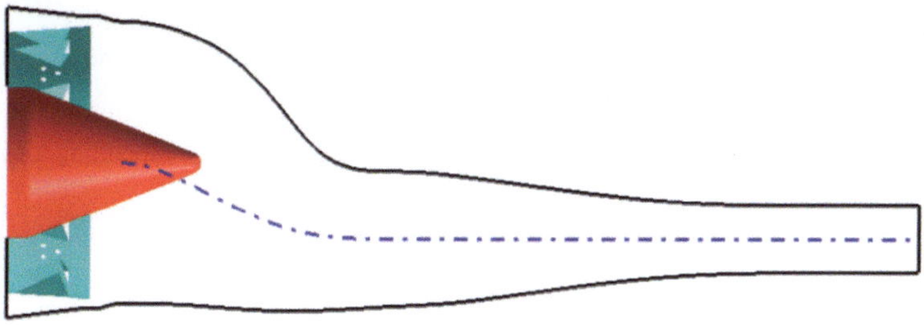

(b) Occlusion rate ΔY=0

Fig. 2.79 Schematic of double S-shaped nozzle obscuration rate for turbofan engine. (**a**) Occlusion rate $\Delta Y = 1$. (**b**) Occlusion rate $\Delta Y = 0$

$\Delta Y = 0$ when the longitudinal deviation of the second bend is zero. The other shading rates are the ratios of the longitudinal deviation of the second bend of the nozzle to the longitudinal deviation of the second bend with the shading rate $\Delta Y = 1$. The shading rate $\Delta Y = 0.5$ represents the ratio of the second-bend deviation of the S-shaped nozzle to that of the second-bend deviation with shading rate $\Delta Y = 1$, which is the same as that of the second-bend nozzle with the second bend. The masking rate $\Delta Y = 0.5$ means that the deflection of the S-shaped nozzle is 0.5 times the deflection of the S-shaped nozzle when the masking rate is $\Delta Y = 1$. The geometrical configuration of the S-shaped nozzle with different shading rates is shown in three views at Fig. 2.80. The values of shading rate and the corresponding dimensionless longitudinal deflection of the second bend are shown in the Table 2.8. The dimensionless longitudinal deflection of the second bend reflects the curvature of the S-shaped nozzle at the longitudinal turn, and the length-to-diameter *(L/D)* ratio is 2.2, while the other geometrical parameters are all the same as those in the design state. It can be seen that reducing the shading rate of the S-shaped nozzle leads to a gradual decrease in the dimensionless longitudinal offset of the second bend, which gradually reduces the longitudinal curvature of the nozzle; and the reduction of the shading rate has no effect on the degree of bending urgency of the profile along the

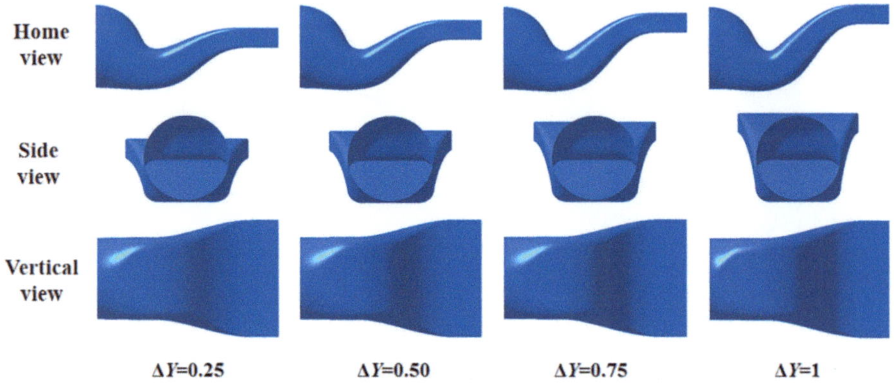

Fig. 2.80 Three views of geometric configuration of S-shaped nozzle with different shading rates

Table 2.8 Occlusion rate of S-shaped nozzle geometry model and the corresponding longitudinal offset taken for the second bend

ΔY	1	0.75	0.50	0.25
$\Delta Y_2/L_2$	0.513	0.385	0.257	0.128

transverse direction of the S-shaped nozzle. Therefore, the S-shaped nozzle profile becomes smoother in the longitudinal direction as the shading ratio decreases, and remains basically unchanged in the transverse direction.

Figures 2.81 and 2.82 show the symmetric surface Ma distribution and the wall static pressure distribution of the S-shaped nozzle with different shading ratios, respectively. With the increase of the shading ratio, the longitudinal curvature of both bends of the S-shaped nozzle gradually increases, which leads to more intense acceleration of the airflow near the upper wall at the first bend and the lower wall at the second bend, and the local static pressure gradually decreases. When the masking ratio is 0.75, the airflow near the upper wall of the first turn accelerates to supersonic speed, and slight flow separation begins to occur downstream of the region due to the large reverse pressure gradient. And as the shading rate is further increased to 1, the counterpressure gradient continues to increase, the flow separation becomes more severe, and the static pressure at the upper wall surface in the front section of the second turn channel is significantly lower than the other three shading rate cases. In addition, when the shading ratio is 0.25, the flow is accelerated to a supersonic state at the exit of the second bend and a slanting sonic surface is generated, and the increase of the shading ratio leads to a more severe supersonic state of the flow in this region, and the static pressure at the upper wall of the exit of the second bend decreases significantly when the shading ratio is 1, and the flow is in a stronger over-expansion state, and thus a stronger compression wave is generated. It can also be seen that the distribution of static pressure on the side wall of the three groups of S-shaped nozzles with the shading ratio of 0.25, 0.5 and 0.75 is basically the same. However, the sidewall static pressure distribution of the nozzle with the shading

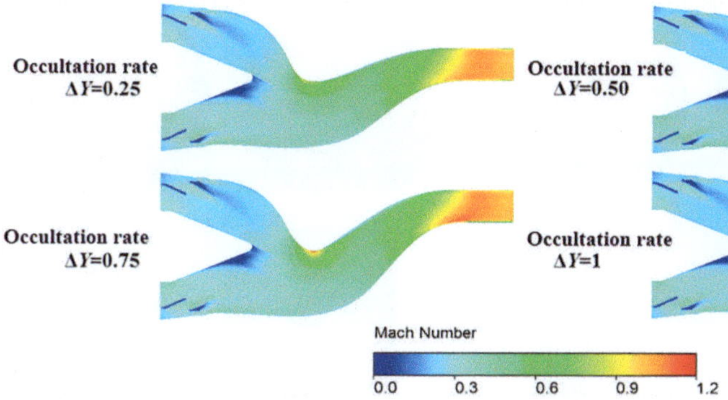

Fig. 2.81 Distribution of Ma number in symmetry plane of S-shaped nozzle for different shading rates

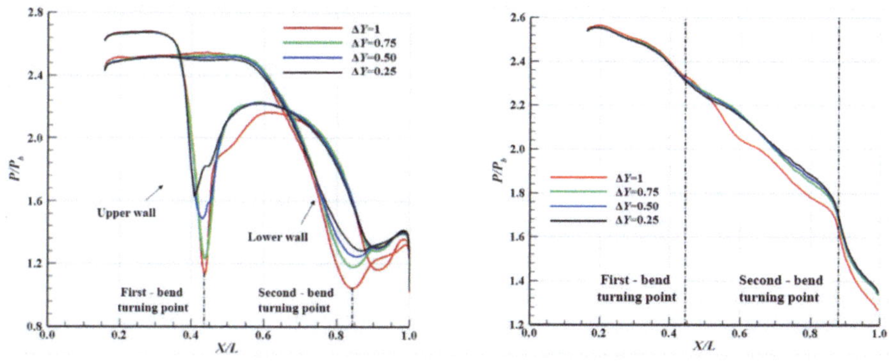

(a) Symmetric wall static pressure distribution (b) Side wall static pressure distribution

Fig. 2.82 Static pressure distribution on the wall of S-shaped nozzle with different shading rates. (**a**) Symmetric wall static pressure distribution. (**b**) Side wall static pressure distribution

ratio of 1 is quite different from that of the other three shading ratio conditions. The main reason is that the nozzle with the shading ratio of 1 has a serious flow separation downstream of the first bend, which significantly reduces the sidewall static pressure of the nozzle.

According to the above analysis, there are large differences in the geometrical configurations of the S-shaped nozzle based on different shading ratios, and the flow field perturbation it generates has some influence on the flow characteristics near the upstream exhaust mixer, leading to asymmetrical flow separation along the longitudinal direction generated by the wave flap mixer, as shown in Fig. 2.83. The limiting streamline distributions on the surface of the wake cone structure in the exhaust mixer are given for different shading rate conditions. The results show that the separation vortex features on the upper wall surface of the tail cone change less and their positions are relatively stable as the shading ratio increases. At a shading ratio of 0.25, multiple pairs of smaller separation vortices appear in the flow separation

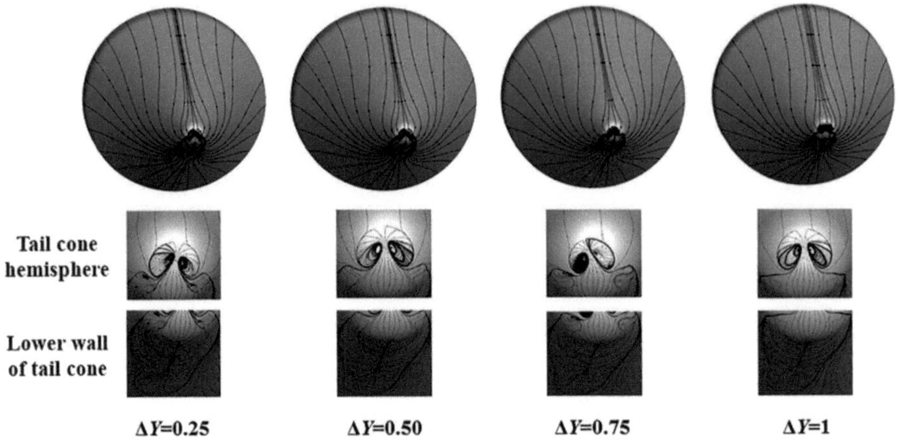

Fig. 2.83 Ultimate flow distribution on the tail cone surface for different shading ratios

region on the lower wall of the tail cone near the hemisphere. As the masking ratio increases, the separation vortices in this region gradually decrease and move toward the back side of the hemisphere on the lower wall. When the masking rate is 1, the separation vortices disappear completely, and the flow separation region on the lower wall of the tail cone exhibits a separation flow line from the saddle point to the node. The reason for this is that the increase of the shading rate causes the longitudinal pressure gradient in the curved channel downstream of the tail cone to decrease slightly, and it manifests itself as a gradually decreasing counterpressure gradient in the region of the lower wall surface of the tail cone, which makes it more difficult for flow separation of the airflow to occur. Therefore, the increase of the shading rate slows down the flow separation at the tail cone surface to some extent, but this slowing down effect has less effect on the flow characteristics downstream of the tail cone. From the symmetry plane Ma distributions, it can be seen that for the S-shaped nozzle configurations with different shading ratios, the differences in the low Ma regions downstream of the tail cone due to flow separation are small.

Figure 2.84 gives the x-direction vortex distribution on the along-travel cross section of the S-shaped nozzle for different shading rates. It can be seen that the vortex distribution on cross-section B is basically the same under different shading rate conditions. On cross-section C, the longitudinal curvature near the first bend gradually increases with the increase of the shading rate, and thus the flow vorticity on the upper side of the cross-section is gradually enhanced; at the same time, the longitudinal pressure gradient also increases, and the vortex distribution as a whole shifts along the longitudinal direction and upward side. On the cross-section D, when the shading ratio increases from 0.25 to 0.75, the extension length of the flow-oriented vortices located at the middle position of the upper side region increases along the longitudinal direction, which is due to the fact that the increase of the shading ratio leads to the decrease of the longitudinal upward pressure gradient at the cross-section, which reduces the obstruction to the longitudinal extension of the flow-oriented vortices. And when the shading ratio is further increased to 1, flow

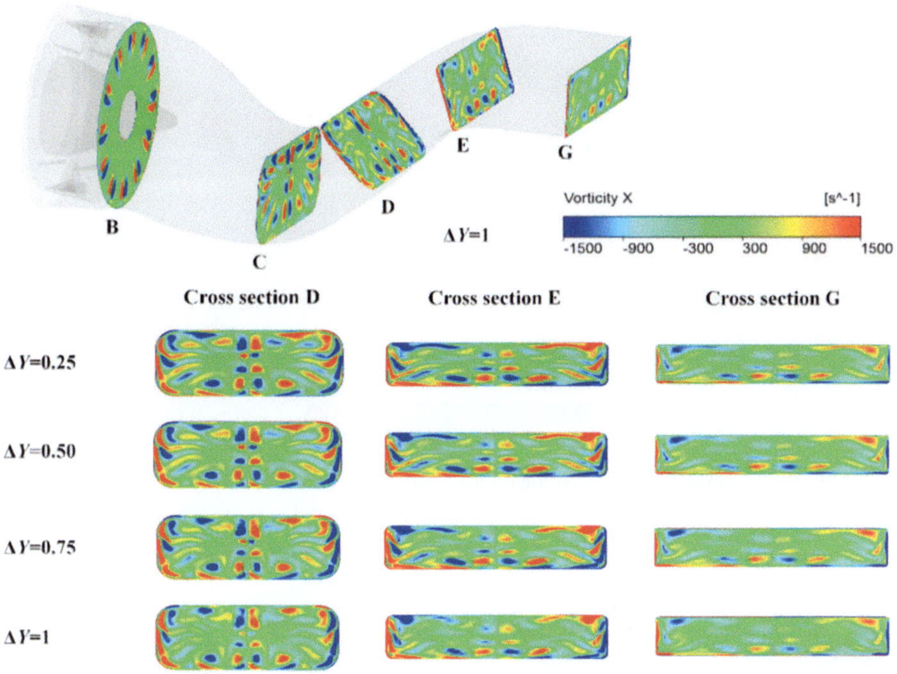

Fig. 2.84 Distribution of x-direction vorticity on the along-travel cross-section of the S-shaped nozzle for different shading rates

separation occurs in the region near the upper wall surface, and the larger vortex volume it generates squeezes the extension space of the flow-oriented vortex, effectively shortening the longitudinal length of this flow-oriented vortex. On the cross sections E and G, the increase of the shading ratio causes the longitudinal curvature of the second bend to increase significantly, which leads to a gradual increase of the vortex volume in the region of the lower wall surface. And two pairs of counter-rotating streamwise vortices are formed near the upper side corner region of the cross-section. When the shading ratio increases from 0.25 to 0.75, the two pairs of vortices merge in the two side corners, and their shapes are similar to the "seagull shape"; and when the shading ratio increases to 1, the two pairs of vortices are separated by the influence of the upstream flow separation, and their shapes are the "elongated strip". The two pairs of flow vortices are separated by the upstream flow separation and have the shape of "long strip".

Figures 2.85 and 2.86 give the static temperature distributions on the along-travel cross-section and on the wall surface of the S-shaped nozzle for different shading ratios, respectively. On section C at the first bend turn, since the center of curvature is located at the upper side of the nozzle, the increase of the shading ratio leads to a significant increase of the vortex near the upper wall surface, which enhances the hot and cold flow mixing in this region, and thus the high-temperature region at the upper side of the section is narrowed, with a lower temperature, and exhibits an

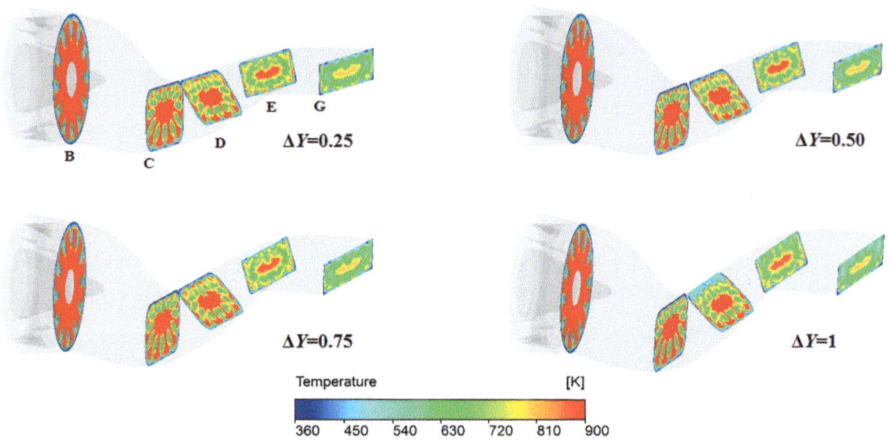

Fig. 2.85 Static temperature distribution on the along-travel cross-section of the S-shaped nozzle for different shading rates

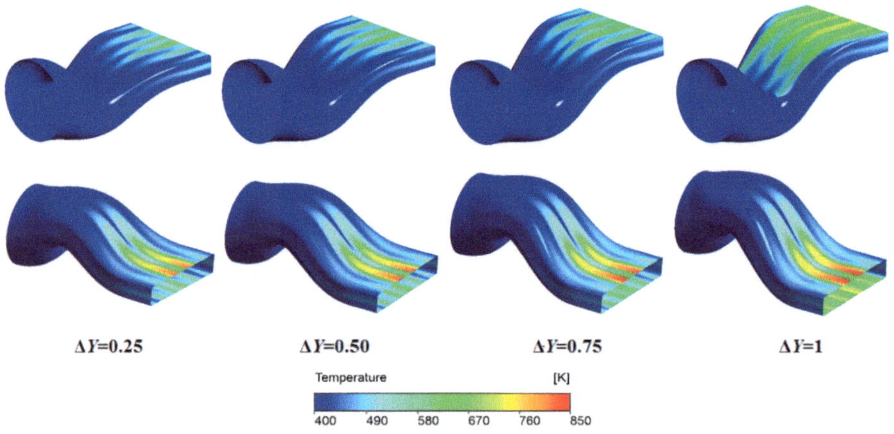

Fig. 2.86 Static temperature distribution on the wall of the S-shaped nozzle for different shading rates

asymmetric feature along the longitudinal direction. On section D, when the shading ratio increases from 0.25 to 0.75, the flow vortex located on the upper side of the section gradually penetrates into the high-temperature core region, which gradually changes the shape of the high-temperature core region from a "rectangle-like" to a "pea shape". When the shading ratio increases to 1, the larger vortex generated by the flow separation squeezes the spatial position of the original flow vortex, and a larger range of low-temperature zone appears in the upper region of the cross-section, and the extension length of the squeezed flow vortex along the longitudinal direction is shortened, which results in the high-temperature core zone changing from "pea-like" to "rectangle-like" in shape. The shape of the high-temperature core changes from "pea-shaped" to "rectangle-like". The continuous expansion of the nozzle cross-section along the flow direction causes the high-temperature core region to develop a "whale shape" in cross-section E. For a nozzle with a shading ratio of 1, the "whale shape" feature is significantly weakened due to the effect of flow separation. In addition, the vorticity in the lower region of the cross-section increases and the temperature decreases with the increase of the shading ratio. The temperature distribution characteristics on cross-section G are basically the same as those on cross-section E, but the range and temperature of the high-temperature region are further reduced.

Regarding the static temperature distribution on the wall surface of the S-shaped nozzle, the "hot spots" on the wall surface of the S-shaped nozzle with different shading ratios mainly appeared in the wall area of the second S-shaped channel. With the increase of the shading rate, the longitudinal curvature of the second turn increases, and the resulting pressure gradient gradually increases, which leads to the enhancement of the impact of the heat flow in the core area of the flow vortex suction on the upper and lower wall surfaces, and the impact range increases along the transverse direction. It can be seen that the second bend channel on the upper and lower walls of the "hot spot" range increases, the temperature increases. In particular, the two localized "high-temperature zones" on the lower wall gradually widen, the temperature value gradually increases, and the low-temperature region between them gradually narrows; whereas four localized "sub-high-temperature zones" with relatively low temperature values and narrow widths gradually appear on the upper wall. In addition, for the nozzle with a shading ratio of 1, due to the serious flow separation in the upper wall region downstream of the first bend, the high vortex generated by it increases the impact of the inner heat flow on the wall, thus increasing the distribution range of the upper wall "hot spots" along the flow direction, and compared with the other three types of nozzles with shading ratios, the upper wall surface of the nozzle with a shading ratio of 1 has the following characteristics The range of "hot spots" on the upper wall of the nozzle with a shading ratio of 1 extends to the first turn.

Figure 2.87 gives the perimeter wall static pressure distributions of the S-shaped nozzle along the course cross section for different shading rates. At cross-section B, the disturbance of the flow field inside the nozzle caused by the increase of the shading ratio is transmitted forward, resulting in a gradual decrease of the static pressure at the lower side of this cross-section. At cross-section C, the center of curvature is located on the upper side of the curved channel, and with the increase of the shading ratio, the longitudinal curvature of the nozzle increases, the airflow velocity near the upper wall surface rises, and the static pressure on the upper wall surface of the perimeter gradually decreases. For the nozzle with a shading ratio of 1, the peripheral upper wall pressure of the cross-section shows fluctuations due to severe flow separation in the downstream region of the cross-section due to a large back pressure gradient. The flow develops to section D. When the shading ratio increases from 0.25 to 0.75, the pressure distribution on the perimeter wall is basically the same. When the masking ratio continues to increase to 1, the flow separation leads to the creation of an airflow low-energy zone in the upper side region of the cross-section, and the perimeter upper wall pressure decreases significantly. At cross section E near the second bend, the center of longitudinal curvature shifts to the lower side of the bend channel. As the shading rate increases, the longitudinal curvature

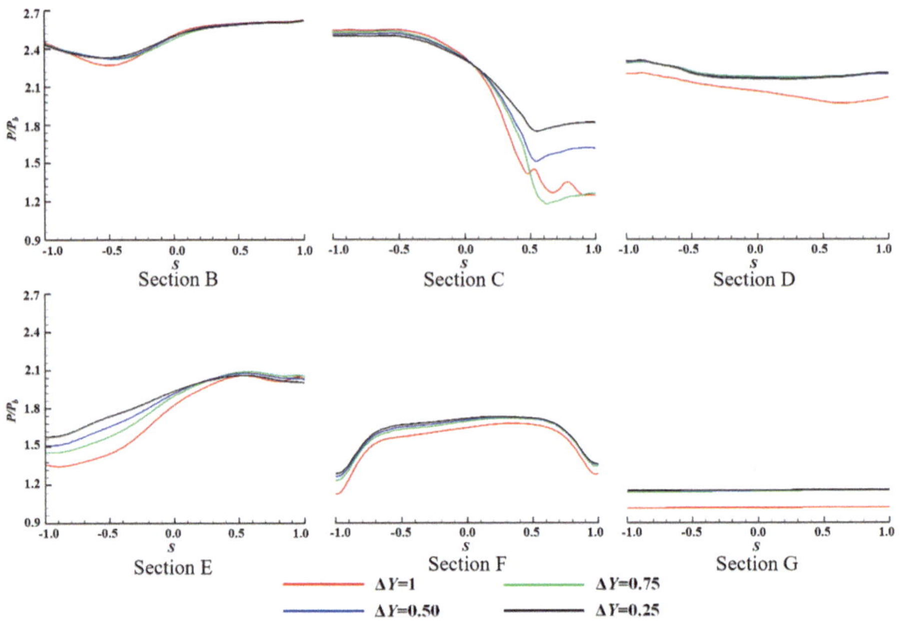

Fig. 2.87 Perimeter wall static pressure distributions of the S-shaped nozzle along the course section for different shading rates

gradually increases, the lower wall static pressure gradually increases, and the pressure gradient formed gradually decreases. At cross-section F, the perimeter wall pressure distribution of cross-section F is basically the same for the three types of nozzles with 0.25, 0.5 and 0.75 shading ratio, respectively, due to the small influence of shading ratio on the transverse profile characteristics of the S-shaped nozzle. When the shading ratio is increased to 1, the flow field perturbation caused by flow separation decreases the perimeter wall static pressure in this cross-section.

Figure 2.88 gives the limiting flow lines and shear stress distributions on the wall half of the S-shaped nozzle for different shading rates. The increase of the shading rate significantly increases the longitudinal bending of the S-shaped nozzle at both bends, while the effect on the transverse profile characteristics is small. For the three groups of nozzles with shading ratios of 0.25, 0.5, and 0.75, the extent of diffusion of the extreme flow lines on the upper and lower walls downstream of the first bend and the extent of convergence near the second bend are basically the same. For the S-shaped nozzle with a shading ratio of 1, due to the large extent of flow separation on the upper wall surface downstream of the first bend, the region exhibits separated limit flow lines from the saddle point to the node, with the surrounding limit flow lines converging toward the separation line. It can also be seen that the degree of longitudinal bending of the nozzle profile intensifies with the increase of the shading ratio, and the degree of airflow acceleration at the two turns rises, which leads to a gradual increase in the shear stresses on the upper wall surface at the turn of the first turn, on the lower wall surface at the turn of the second turn and on the exit wall surface of the nozzle, and a significant increase in the bending degree of the flow

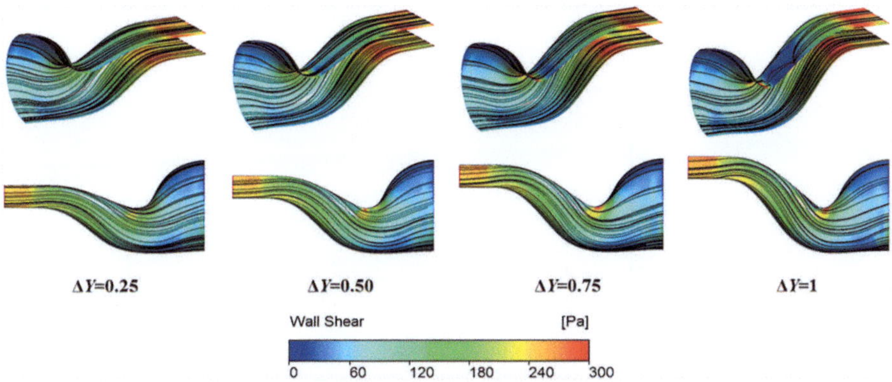

Fig. 2.88 Ultimate flow distribution on the wall of the S-shaped nozzle for different shading rates

line on the side wall surface of the nozzle, with a gradual increase in the angle of the flow line deflection and a rise in the degree of the convergence of the flow lines in the exit region.

Figure 2.89 compares the aerodynamic performance of the S-shaped nozzle at different shading rates. The total pressure recovery coefficient, flow coefficient and thrust coefficient of the nozzle gradually decrease with the increase of the shading ratio. When the shading ratio increases from 0.25 to 0.75, the longitudinal curvature of the two turns increases, the corresponding wall shear stress increases, and the degree of flow bending increases, thus the friction loss and localized loss generated by the nozzle wall increases, while the difference in the amount of flow vortices generated by the wave mixer and the tail-cone structure is relatively small, and the resultant vortex loss and internal and external culvert doping loss do not change much. As a result, the S-shaped nozzle aerodynamic performance is slightly reduced. When the masking rate is further increased from 0.75 to 1, significant flow separation occurs on the upper wall downstream of the first bend of the nozzle, which generates a larger vortex volume and strengthens the doping effect of the inner/outer culvert flow, and the resulting vortex loss and doping loss become the main factors affecting the aerodynamic performance of the nozzle, which results in a significant decrease in the aerodynamic performance of the S-shaped nozzle. Taken together, the aerodynamic performance of the S-shaped nozzle decreases gradually with the increase of the shading rate.

2.4.4 Effect of Axial Length Ratio of Two Bends on the Flow Characteristics of S-Shaped Nozzle

In this section, the flow field characteristics of four different S-shaped nozzles with different two-bend axial length ratios $L_1/L_2$are numerically investigated under ground conditions. The geometrical configurations of the different S-shaped nozzles are shown in three views in Fig. 2.90, and the values of the two-bend axial

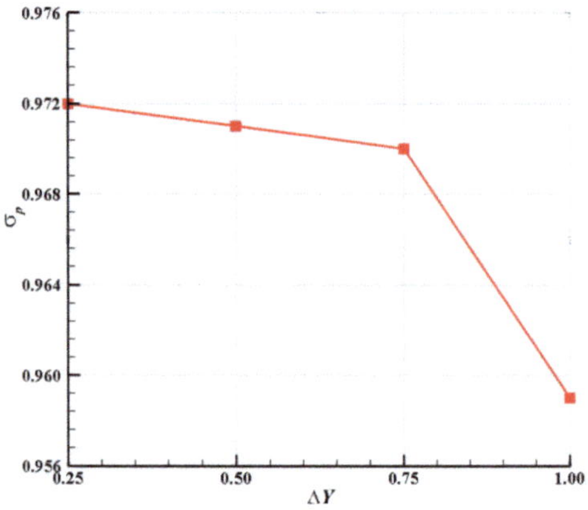

(a) Total pressure recovery coefficient

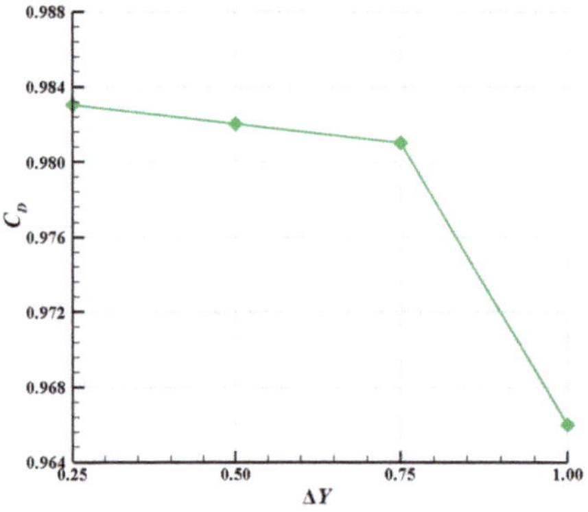

(b) Discharge coefficient

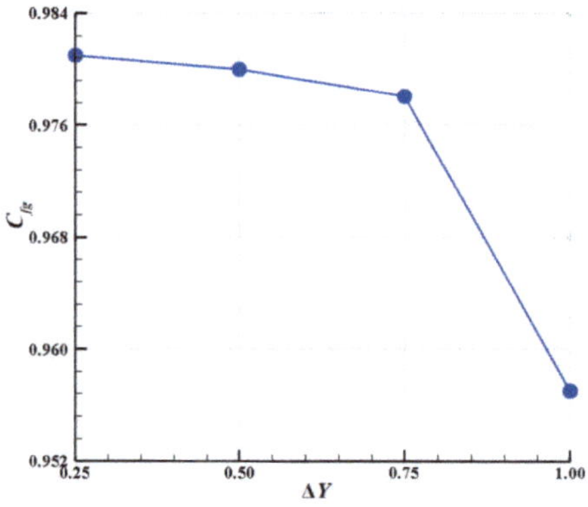

(c) Thrust coefficient

Fig. 2.89 Trend of aerodynamic performance of S-shaped nozzle with different shading rates. (**a**) Total pressure recovery coefficient. (**b**) Discharge coefficient. (**c**) Thrust coefficient

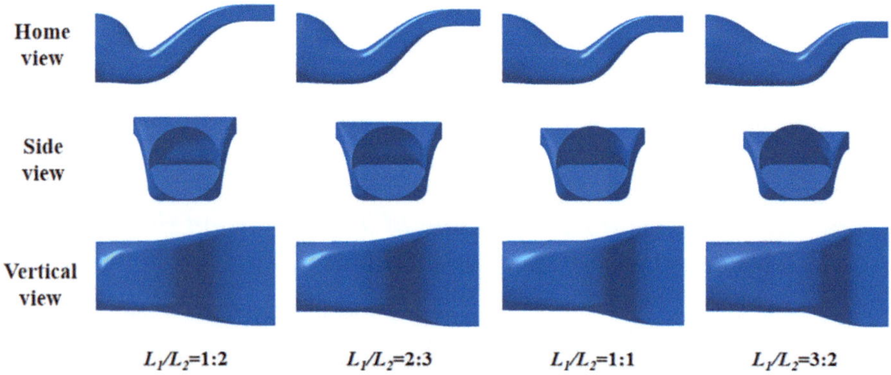

Fig. 2.90 Three views of geometric configuration of S-shaped nozzle with different two-bend axial length ratios

Table 2.9 The axial length ratio of the two bends and the corresponding longitudinal offsets of the first and second bends of the geometric model of the S-shaped nozzle

Parameters	Value			
L_1/L_2	1:2	2:3	1:1	3:2
$\Delta Y_1/L_1$	0.356	0.296	0.237	0.198
$\Delta Y_2/L_2$	0.512	0.513	0.548	0.607

length ratios and the corresponding longitudinal offsets of the first and second bends are shown in Table 2.9, while the other geometrical parameters are the values in the design condition. It can be seen that, under the constraint of the criterion of completely covering high-temperature parts, the increase of the axial length ratio of the two bends of the nozzle will improve the difficulty of covering the high-temperature parts such as the wave mixer and the tail cone of the S-shaped nozzle, which results in the gradual decrease of the dimensionless longitudinal deflection of the first bend and the gradual increase of the dimensionless longitudinal deflection of the second bend. The longitudinal curvature at the turn of the first bend is jointly determined by the dimensionless longitudinal offset of the two bends, and it tends to decrease and then increase with the increase of the axial length ratio of the two bends, whereas the longitudinal curvature at the turn of the second bend is only determined by the dimensionless longitudinal offset of the second bend, and it gradually increases with the increase of the axial length ratio of the two bends. In addition, the increase of the ratio of the axial lengths of the two bends exacerbates the degree of transverse curvature of the second bend channel of the nozzle, resulting in a gradual increase of the transverse curvature near the second bend. Therefore, with the increase of the ratio of the axial lengths of the two bends, the geometrical configuration of the entire S-shaped nozzle in the longitudinal direction is first gentle and then steep in the first bend, and then continues to become steeper in the second bend, while the profile characteristics of the nozzle in the transverse direction are also progressively steeper in the second bend channel.

Figures 2.91 and 2.92 give the symmetry plane Ma distributions and wall static pressure distributions of the S-shaped nozzle for different two-bend axial length ratios, respectively. It can be seen that with the increase of the ratio of the axial length of the two bends, the longitudinal curvature of the first bend decreases and then increases, and the center of the longitudinal curvature at the bend gradually moves backward, which results in the contraction of the airflow tube near the upper wall surface decreases and then increases, the acceleration effect of the airflow weakens and then strengthens, and the value of the static pressure in the local low-pressure zone increases and then decreases, and the location of the low-pressure zone is constantly moving backward. When the ratio of the axial length of the two bends is 3:2, a slight flow separation occurs downstream of the region due to the large reverse pressure gradient. The longitudinal curvature of the second bend increases with the increase of the ratio of the axial lengths of the two bends, resulting in more intense acceleration of the airflow on the lower wall of the second bend,

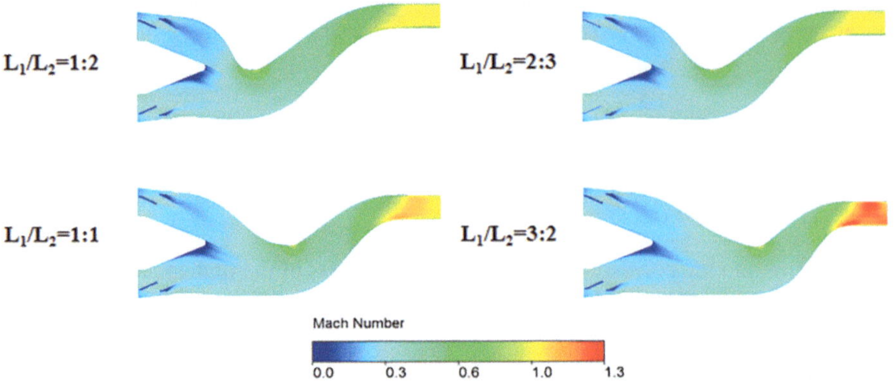

Fig. 2.91 Distribution of Ma number in the symmetry plane of the S-shaped nozzle for different two-bend axial length ratios

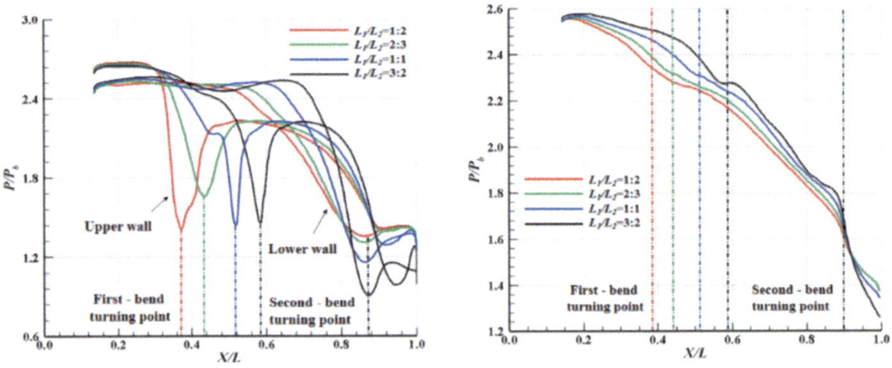

(a) Symmetric wall static pressure distribution (b) Side wall static pressure distribution

Fig. 2.92 Static pressure distribution on the wall of S-shaped nozzle with different axial length ratios of the two bends. (**a**) Symmetric wall static pressure distribution. (**b**) Side wall static pressure distribution

and the value of the local static pressure gradually decreases. When the axial length ratio of the two bends increases to 3:2, the nozzle in the second bend of the channel exit due to the intense acceleration of the airflow and a serious over-expansion state, thus forming a slanting surge in the isotropic section. Under the influence of the oblique shock wave, the static pressure distribution on the upper wall of the isostatic section first decreases and then increases. The increase of the axial length ratio of the two bends also significantly changes the profile characteristics of the nozzle in the transverse direction, and the flow field perturbation brought by it causes the static pressure on the side wall of the nozzle to rise continuously.

The two-bend axial length ratio changes the geometrical configuration of the S-shaped nozzle, which significantly affects the flow characteristics downstream of the tail cone and the flap mixer. With the increase of the two-bend axial length ratio, the flow separation generated by the flap mixer and tail cone increases the range of influence on the flow field downstream of the nozzle. Under the effect of nozzle geometry, the longitudinal pressure gradient downstream of the exhaust mixer decreases with the increase of the ratio of the axial lengths of the two bends, the static pressure on the upper side of the nozzle decreases, and the static pressure on the lower side increases, and thus, the angle of the low-Ma separation flow induced by the upper side of the wave flap mixer is gradually reduced by the axial downward deflection.

Figure 2.93 gives the limiting streamline distributions on the surface of the tail cone structure for different two-bend axial length ratios. When the two-bend axial length ratio is gradually increased from 1:2 to 1:1, the separation vortex characteristics near the upper wall surface of the tail cone change less. When the axial length ratio of the two bends continues to increase to 3:2, the degree of flow separation near this region increases, the separation vortex range is larger, and the location of the separation vortex moves to the lower side of the hemisphere. And the limiting flow line change trend on the lower wall of the tail cone is different from that on the upper wall. When the ratio of the axial lengths of the two bends is 1:2, a larger range of flow separation occurs on the lower wall of the tail cone, which is manifested in the separation flow line from the saddle point to the node, and a pair of smaller separation vortices is formed on the anterior side of the separation zone. As the axial length ratio of the two bends increases to 2:3, the separation vortex disappears. When the length ratio of the two bends is increased to 3:2, the flow separation zone on the front side of the hemisphere on the lower wall surface of the tail cone disappears completely, and the flow separation zone is confined to the inner side of the hemisphere only. The reason is that, with the increase of the axial length ratio of the

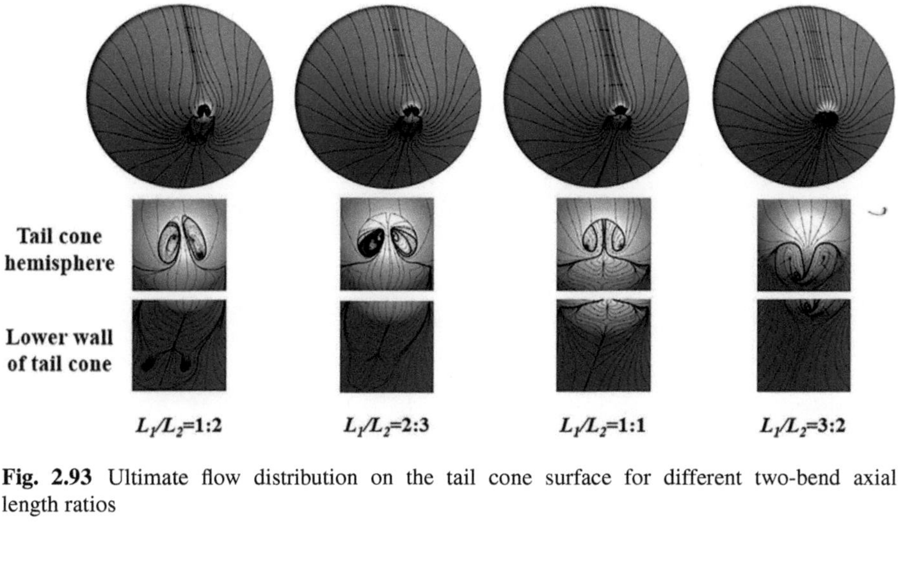

Fig. 2.93 Ultimate flow distribution on the tail cone surface for different two-bend axial length ratios

two bends, the pressure gradient along the longitudinal downstream of the tailcone downstream of the curved channel decreases significantly, and it manifests itself as a gradually decreasing down-pressure gradient in the upper wall region, and thus the separation vortex generated by the hemisphere on the upper wall surface moves downward to the position of a larger curvature of the hemisphere; however, it manifests itself as a gradually decreasing counter-pressure gradient in the lower wall region, and the flow characteristics in the vicinity of the tailcone's lower wall surface disappear from the flow field characteristics near the lower wall of the tail cone change gradually from a state in which a larger range of separating vortices and separating streamlines coexist to a state in which a smaller range of separating streamlines exists. Therefore, the increase of the axial length ratio of the two bends reduces the flow separation generated by the tail cone structure to a certain extent.

Figure 2.94 gives the x-direction vortex distributions on the along-travel cross section of the S-shaped nozzle for different two-bend axial length ratios. On the cross-section C at the first turn, when the two-bend axial length ratio is 1:2, there are two pairs of counter-rotating flow vortices at the center of the cross-section, and the vorticity of the flow vortices at the lower side of the tailcone is significantly higher than that at the upper side due to the larger flow separation zone generated at the lower wall of the tailcone compared to that at the upper wall. As the axial length ratio increases to 3:2, due to the separation vortex formed on the upper wall of the

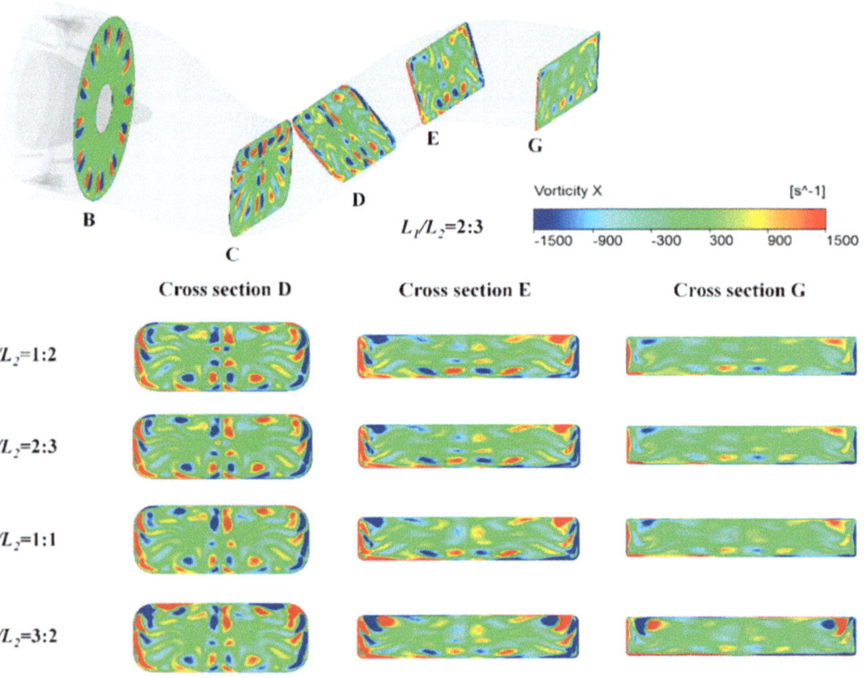

Fig. 2.94 Distribution of x-direction vorticity in the along-travel cross-section of an S-shaped nozzle for different two-bend axial length ratios

tail cone shifts downward, the flow separation feature generated on the lower wall of the tail cone is weakened, and only one pair of flow vortices exists in the center region of the cross-section, and the location of the flow vortices is located on the lower side of the center of the cross-section. In cross-section D, for the three groups of nozzles with two bends and axial length ratios of 1:2, 2:3 and 1:1, the vortex distribution characteristics induced by the wave mixer are basically the same. When the axial length ratio of the two bends is increased to 3:2, flow separation occurs in the upper wall region downstream of the first bend, and thus a pair of counter-rotating flow vortices are formed on the upper side of this cross-section. In cross-section E, all four groups of nozzles formed a pair of streamwise vortices near the corner region of the upper wall surface, and for the two groups of nozzles with the axial length ratios of the two bends of 1:1 and 3:2, the streamwise vortices generated by the flow separation of the wall surfaces moved laterally to the vicinity of the vortices of the corner region and the vortices in the middle region of the cross-section increased compared with those of the other two groups of nozzles. In addition, as the axial length ratio of the two bends increases, the transverse curvature at the turn of the second bend increases, which effectively hinders the transverse movement of the flow-oriented vortices, resulting in the enhancement of the degree of transverse stretching of the flow-oriented vortices in the vicinity of the side wall surfaces of the nozzle, and the weakening of the degree of adherence to the wall.

Figures 2.95 and 2.96 give the static temperature distributions on the along-travel cross section and on the wall of the S-shaped nozzle at different two-bend axial length ratios, respectively. In the cross-section C at the first turn, when the axial length ratio of the two bends increases to 1:1, the flow vortex induced by the tail cone gradually moves downward, which sucks the cold flow into the high-temperature core area for mixing, resulting in a "notch" in the middle of the

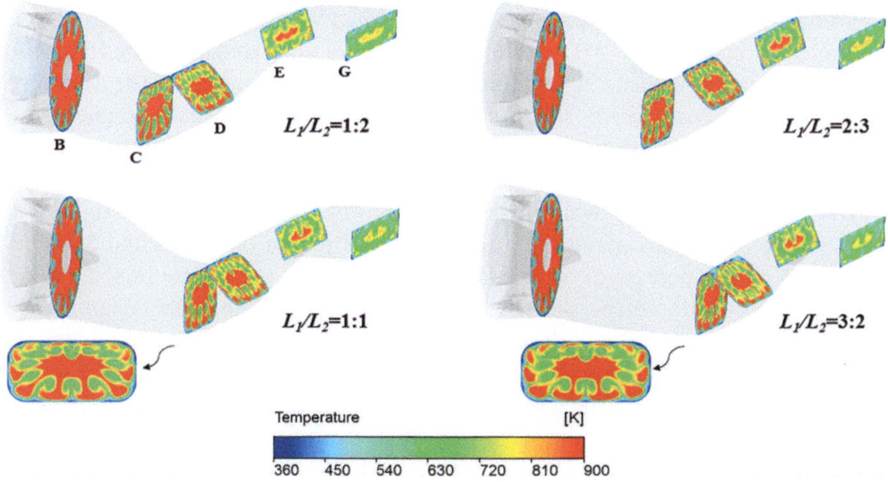

Fig. 2.95 Static temperature distribution on the along-travel cross-section of the S-shaped nozzle for different two-bend axial length ratios

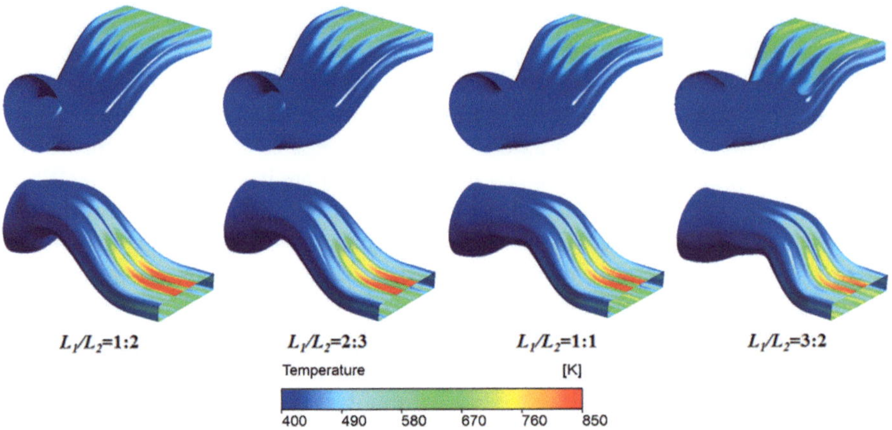

Fig. 2.96 Static temperature distribution on the wall of the S-shaped nozzle for different two-bend axial length ratios

high-temperature core area. With the further increase of the axial length ratio of the two bends, the "notch" in the high-temperature core region becomes larger. In cross-section D, for the three sets of nozzles with the two-bend axial length ratios of 1:2, 2:3, and 1:1, the local high-temperature zones around the cross-section are basically the same due to the similar vortex distributions induced by the flap mixer. When the ratio of the axial lengths of the two bends is 3:2, the vortices generated on the upper side of the cross-section exacerbate the mixing of the hot and cold flows in this region, resulting in a significant reduction in the area of the local high-temperature zone. On section E, the lateral motion of the airflow causes the high-temperature core area to develop into a "whale tail" shape. As the axial length ratio of the two bends increases, the vortex in the middle region of the cross-section increases, and the "whale-shaped" feature becomes more obvious. When the axial length ratio of the nozzle is increased to 1:1, the high-temperature core area is divided into two parts. The increase in the axial length ratio of the two bends increases the transverse curvature of the S-shaped nozzle, forming a stronger back pressure gradient, and thus the high-temperature core area in the shape of a "whale tail" shrinks along the transverse direction.

The "hot spots" on the wall of the S-shaped nozzle with different axial length ratios of the two bends mainly appear in the wall region of the second S-shaped channel. With the increase of two-bend axial length ratio, the longitudinal curvature of the second bend increases, which leads to the increase of the impact degree of the heat flow in the core region of the flow vortex absorption on the upper wall, and the impact range increases along the transverse direction. It can be seen that the second turn of the channel wall surface gradually appeared four relatively low temperature values and narrow width of the local "sub-high temperature zone". For the lower wall of the nozzle, the larger bending curvature leads to an increase in the upward deflection angle of the airflow near the second bend, which weakens the impact effect of the heat flow on the lower wall of the nozzle, and the two localized "high-temperature zones" located on the lower wall gradually become narrower, with gradually decreasing temperatures, and the low-temperature region between them becomes wider and wider. In addition, for the two-bend axial length ratio of 3:2 nozzle, due to the first bend downstream of the upper wall area of the flow separation, which formed a high vortex region increases the impact of the inner heat flow on the wall, so that the second bend channel wall "hot spot" range extends to the first turn. It can also be seen that the increase in the ratio of the axial length of the two bends leads to a gradual increase in the lateral curvature of the second bend of the nozzle, the flow to the vortex transverse movement of the counterpressure gradient to enhance the impact of the heat flow on the side of the wall of the nozzle continues to weaken. Thus, with the increase of the ratio of the axial length of the two bends, the "hot spot" produced by the side wall of the S-shaped nozzle range and temperature values are gradually reduced.

Figure 2.97 gives the static pressure distributions on the perimeter wall surface of the S-shaped nozzle along the cross section for different two-bend axial length ratios. At cross-section B, the flow field perturbation inside the nozzle caused by the increase of the two-bend axial length ratio is transmitted forward, resulting in a significant increase of the static pressure at the lower perimeter wall of this cross-section. In section C, the center of longitudinal curvature is located in the upper side of the curved channel, with the increase of the ratio of the two curved axial lengths, the longitudinal curvature of the nozzle decreases and then increases, resulting in the

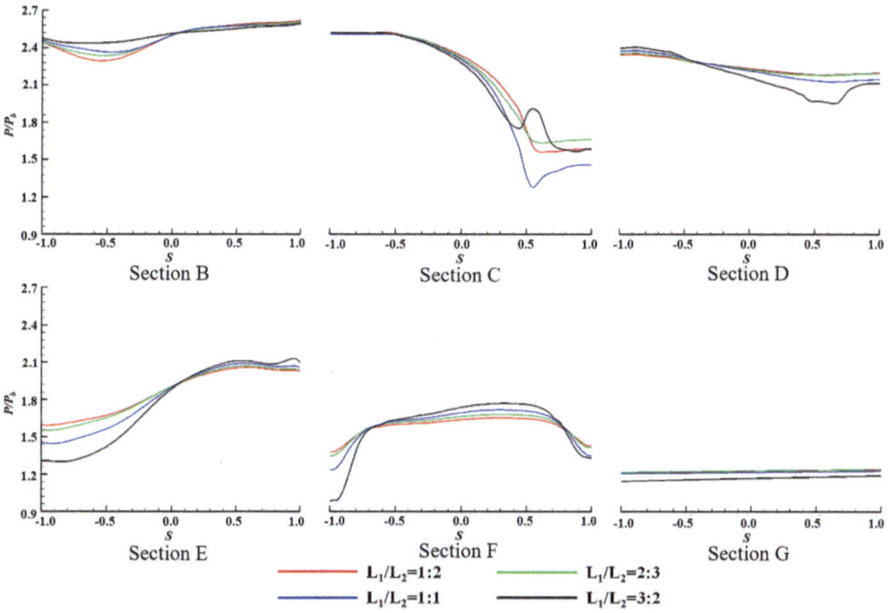

Fig. 2.97 Perimeter wall static pressure distributions along the course section of S-shaped nozzle for different two-bend axial length ratios

static pressure of the upper peripheral wall increases and then decreases. For the nozzle with two bends axial length ratio of 3:2, the peripheral upper wall pressure shows fluctuation due to the flow separation in the upper wall region downstream of this cross-section. The flow develops to cross-section D, and the distribution of peripheral wall static pressure is basically the same for the three groups of nozzles with two-bend axial length ratios of 1:2, 2:3, and 1:1, respectively. When the two-bend axial length ratio is 3:2, the flow separation creates a flow low-energy zone in the upper region of the cross-section, and the peripheral upper-wall pressure decreases significantly. At section E near the second bend, the center of longitudinal curvature is shifted to the lower side of the curved channel, and with the increase of the axial length ratio of the two bends, the longitudinal curvature gradually increases, and the static pressure at the lower wall surface continues to decrease, and the pressure gradient gradually increases. At section F, the lateral wall surface static pressure is the largest due to the larger transverse curvature and pointing to the inner side of the nozzle. The increase of the axial length ratio of the two bends makes the transverse curvature at the turn of the second bend gradually increase, the static pressure at the side wall surface gradually increase, and the pressure gradient at the circumferential wall surface continues to increase. On section G, for the nozzle with a 3:2 ratio of the axial lengths of the two bends, the value of the perimeter wall static pressure in this section is much lower than that of the other three groups of nozzles due to the occurrence of a strong surge upstream and the generation of a large surge loss.

Figure 2.98 gives the limiting flow lines and shear stress distributions on half of the wall of the S-shaped nozzle for different two-bend axial length ratios. With the increase of two-bend axial length ratio, the longitudinal curvature at the first bend decreases and then increases, while the longitudinal curvature at the second bend increases continuously. This results in the upper wall shear stress at the first bend first decreasing and then increasing, and the degree of streamline curvature at the sidewall surface first decreasing and then increasing; whereas the shear stress at the lower wall and nozzle exit region at the second bend gradually increases, the degree

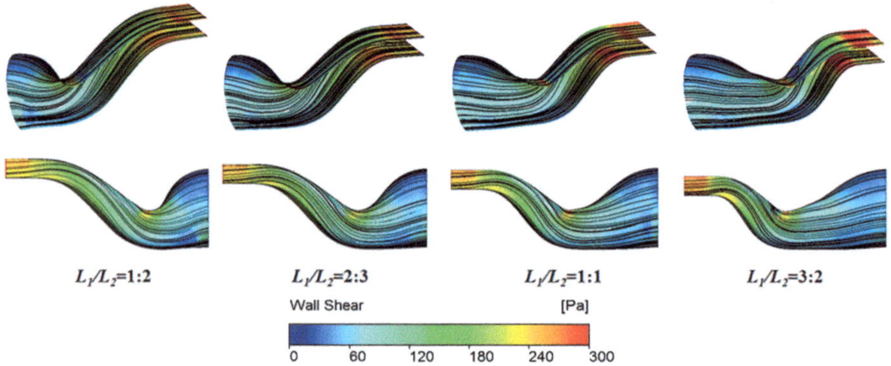

Fig. 2.98 Ultimate flow distribution of S-shaped nozzle wall for different two-bend axial length ratios

of streamline curvature at the sidewall surface of the region is significantly enhanced, and the deflection angle of the streamline at the nozzle exit region along the axial downward direction gradually increases, and the degree of streamline convergence rises. In addition, the increase in the ratio of the axial lengths of the two bends causes the lateral curvature of the second bend channel of the nozzle to continuously increase. This leads to a gradual increase in the degree of convergence of the extreme streamlines of the upper and lower wall surfaces of the nozzle in the vicinity of the second bend. It can also be seen that for the nozzle with a 3:2 axial length ratio of the two bends, flow separation occurs in the region of the upper wall surface downstream of the first bend, which as a whole manifests itself as a separating flow line from the saddle point to the node, with the limiting flow lines around it converging towards the separating line.

Figure 2.99 compares the aerodynamic performance of the S-shaped nozzle for different two-bend axial length ratios. The total pressure recovery coefficient, flow

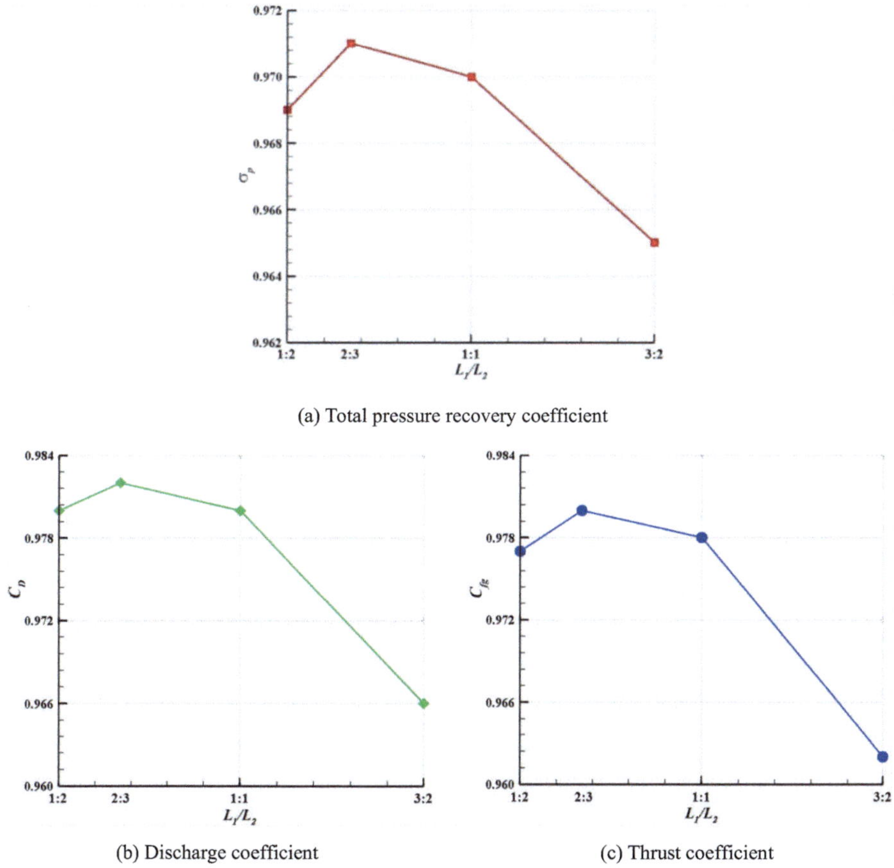

(a) Total pressure recovery coefficient

(b) Discharge coefficient

(c) Thrust coefficient

Fig. 2.99 Trend of aerodynamic performance of S-shaped nozzle with different two-bend axial length ratios. (**a**) Total pressure recovery coefficient. (**b**) Discharge coefficient. (**c**) Thrust coefficient

coefficient and thrust coefficient of the nozzle firstly increase and then significantly decrease with the increase of the two-bend axial length ratio. With the increase of two-bend axial length ratio, the amount of flow vortex generated by the tail cone and the flap mixer is reduced, and the resulting vortex loss and internal and external mixing loss are reduced. When the axial length ratio of the two bends increases from 1:2 to 2:3, the longitudinal curvature of the first bend decreases, the corresponding upper wall shear stress decreases, and the bending degree of the side wall flow line decreases, resulting in a reduction of friction loss and local loss of the wall; while the longitudinal curvature at the turn of the second bend increases, the corresponding lower wall shear stress increases, and the degree of bending of the side wall flow line increases, resulting in the generation of the friction loss and local loss of the wall. Losses and localized losses increase. The combined effect of the above factors makes the aerodynamic performance of the S-shaped nozzle increase. When the axial length ratio of the two bends increases from 2:3 to 1:1, the longitudinal curvature of the two bends increases, the corresponding wall shear stress increases, and the bending degree of the flow line intensifies, which leads to an increase in friction loss and local loss on the wall. This ultimately leads to a downward trend in the aerodynamic performance of the nozzle. The above results show that when the axial length ratio of the two bends is increased from 1:2 to 1:1, the main drivers affecting the aerodynamic performance of the nozzle are the friction loss and localized loss on the wall surface of the nozzle. When the axial length ratio of the two bends is increased to 3:2, flow separation occurs at the upper wall downstream of the first bend, forming a high vorticity region, which strengthens the doping effect of the hot and cold flows; while the intense airflow acceleration near the lower wall at the turn of the second bend makes the airflow over-expanded severely, resulting in a slanting surge in the isotropic section of the nozzle. The vortex loss, doping loss, and surge loss caused by the flow separation at the nozzle wall become the main factors affecting the aerodynamic performance of the nozzle, which ultimately leads to a significant decrease in the aerodynamic performance of the S-shaped nozzle. Therefore, with the increase of the axial length ratio of the two bends, the aerodynamic performance of the S-shaped nozzle first increases and then decreases significantly.

2.4.5 *Effect of Inlet Cyclone on the Flow Characteristics of an S-Shaped Nozzle*

In a conventional turbofan engine, the airflow passing through the low-pressure turbine will generate a certain degree of circumferential cyclonic flow in the inlet channel of the exhaust system, which can significantly change the flow field characteristics inside the S-shaped nozzle and is an important parameter affecting the aerodynamic performance of S-shaped nozzle. In this section, the flow field characteristics of the double-connotation S-shaped nozzle with different inlet swirl angles are numerically investigated under ground conditions by selecting the geometric model of the S-shaped nozzle in the design state. In the boundary condition setting, the inlet rotation angle is obtained by setting the circumferential velocity of the inlet flow. The inlet rotation angles are 0°, 5°, 10°, 15°, and 20°, and the direction of rotation is clockwise (along the X-direction), where 0° rotation is the no-rotation condition.

Figures 2.100 and 2.101 show the *Ma* number distribution on the symmetry plane of the S-shaped nozzle and the Ma number and streamline distribution on the surface of the exhaust mixer for different inlet rotation angles, respectively. It can be seen that the differences in the Ma number distribution of the symmetry plane of the S-shaped nozzle under different rotational angles are mainly shown in the flow field near the tail cone and its downstream. Under the no-spin condition, a large flow separation occurs near the tail cone surface, and the flow separation is manifested as a pair of counter-rotating separating vortices in the hemispherical region of the tail cone surface, which is manifested as a separating flow line from saddle point to node before the hemispherical position of the tail cone lower wall surface, and thus a large range of low Ma region is formed downstream of the tail cone. Under the 5° spin angle condition, the separation vortex in the hemispherical region of the tail cone and the separation line on the lower wall surface disappear, while the tail cone interacts with the spin flow and induces a vortex rotating in the clockwise direction on the lower side of the hemispherical body (along the X-direction), and thus the range of the low-Ma region downstream of the tail cone is significantly reduced. The low-Ma flow is the flow inside the vortex, which has a lower velocity and static pressure than the surrounding fluid. Under the action of the vortex, the velocity of the surrounding airflow rises significantly, forming a high Ma number region. This vortex develops downstream in the S-shaped curved flow path and eventually changes the velocity distribution within the isostraight section of the nozzle exit. With the increase of the swirl angle, the swirl range in the tail cone hemisphere region gradually increases, and the velocity of the airflow around the swirl further rises. When the swirl angle increases to 15°, two high Ma-number airflow bands

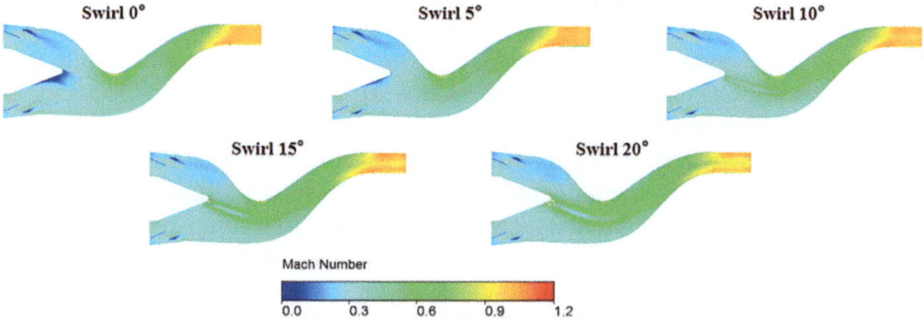

Fig. 2.100 Distribution of Ma number in the symmetry plane of the S-shaped nozzle for different spin angles

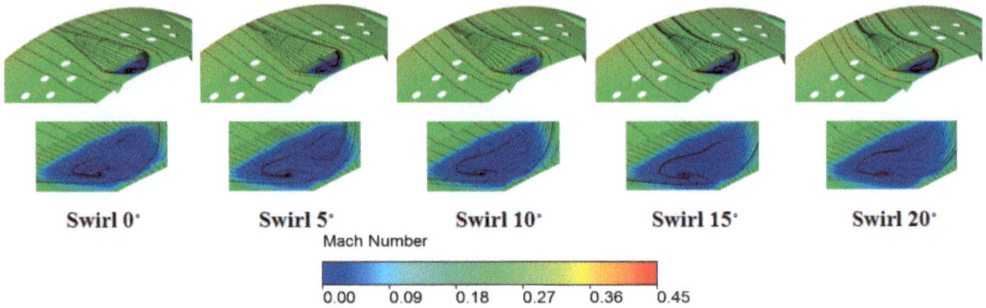

(a) Surface Ma and streamline distributions of the flap mixer at different spin angles

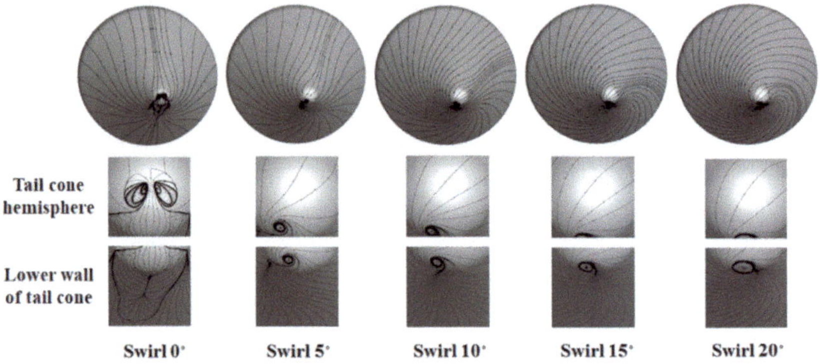

(b) Ultimate streamline distribution on the surface of the tail cone for different spin angles

Fig. 2.101 Surface Ma and streamline distribution of exhaust mixer at different spin angles. (**a**) Surface Ma and streamline distributions of the flap mixer at different spin angles. (**b**) Ultimate streamline distribution on the surface of the tail cone for different spin angles

appear downstream of the tail cone, and the influence of the swirl on the velocity distribution in the isotropic section of the nozzle exit is very significant.

The internal centerline Ma distribution of the S-shaped nozzle can reflect the evolutionary characteristics of the vortex flow direction generated by the interaction between the tail cone and the cyclonic flow more clearly, as shown in Fig. 2.102. Under the no-whirl condition, the airflow velocity downstream of the tail cone is low due to the existence of flow separation. The airflow is accelerated inside the S-shaped curved channel, and a compression wave is generated in the isostraight section due to violent acceleration, which leads to local fluctuations in the airflow velocity. In the 5° swirling condition, the vortex generated on the surface of the tail cone hemisphere leads to slight fluctuations in the downstream airflow velocity, and the airflow velocity inside the first curved channel increases compared to the non-swirling condition. As the vortex generated on the surface of the tail cone continues to develop to the nozzle exit, the airflow velocity in the isostraight section is lower compared to the no-spin condition. With the increase of the spin angle, the vortex induced by the surface of the tail cone hemisphere is gradually enhanced, which leads to more drastic fluctuations of the airflow velocity on the downstream

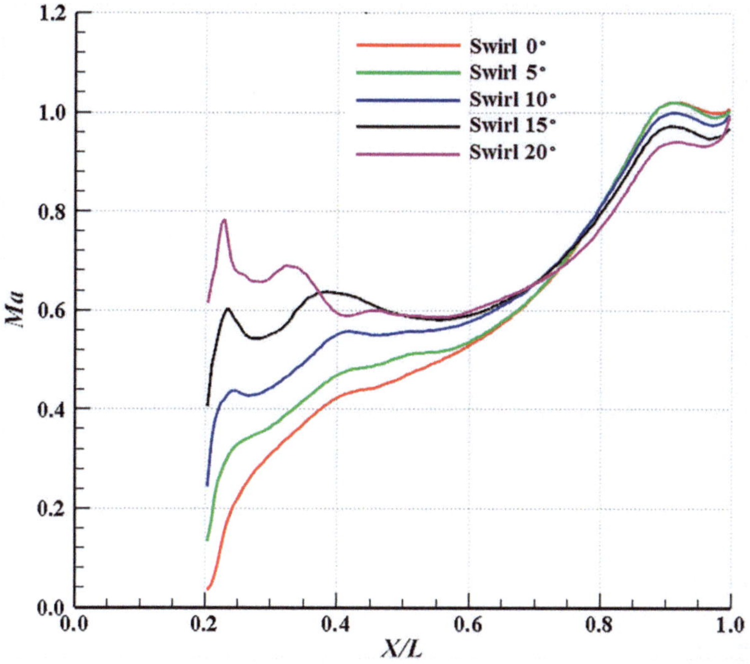

Fig. 2.102 Distribution of Ma number in the centerline of S-shaped nozzle for different spin angles

centerline, and has a greater impact on the flow characteristics in the isostream region, and thus the airflow velocity in the isostream section continues to decrease. When the vortex angle increases to 20°, due to the stronger vortex on the surface of the tail cone, the airflow velocity around the vortex is higher, the magnitude of the airflow velocity fluctuation downstream of the tail cone is larger, and the trend of the velocity change on the centerline of the nozzle is shown as decreasing firstly and then increasing.

There is also a certain degree of difference in the surface flow characteristics of the flap mixer at different spin angles, as shown in Fig. 2.101a, which is mainly in the low Ma region at the back of the "funnel-shaped" structure of the flap mixer. In the no-spin condition, there are two separation zones in this region, the left side of the separation zone is characterized by a long flow separation line, while the right side of the separation zone is characterized by a large separation vortex. At 5° cyclonic flow condition, the separation zone on the left side (along the X direction) changes to two shorter flow separation lines, which indicates that the flow separation under cyclonic flow is weakened. With further increase of the rotation angle, the flow separation lines on the left side move to the right side and gradually merge with the flow lines near the separation vortex. When the rotation angle is increased to 20°, the flow separation in the left region disappears completely, and all the airflow flows to the right region to form a separation vortex. In addition, with the increase of the spin angle, the bending degree of the flow lines on the surface of the flap mixer along the circumferential direction is increasing.

Figure 2.103 gives the static pressure distribution on the wall of the S-shaped nozzle at different inlet cyclone angles. It can be seen that the wall static pressure distribution law is basically the same for different cyclone angles, but the existence of the cyclone effectively changes the static pressure value of the nozzle wall. As the cyclone angle increases, the static pressure on the symmetric surface of the S-shaped nozzle wall gradually decreases. The reason is that, under the action of the longitudinal pressure gradient induced by the S-shaped curved channel, the vortex produced by the whirlpool sucks the connotation of low-pressure airflow outward to the upper and lower wall of the nozzle, which reduces the value of the static pressure on the wall. The larger the spin angle, the stronger the vortex, the greater the impact of low-pressure airflow on the nozzle wall. However, the static pressure value of the upper wall surface at the first bend turn increases with the gradual increase of the swirl angle. This is due to the fact that this region is located in the inner side of the longitudinal bending channel, where the effect of the internal airflow is very significant, and the increase of the swirl angle makes the airflow velocity in this region decrease and the static pressure increase. In addition, as the spin angle increases from 0° to 15°, the static pressure on the side wall surfaces of the S-shaped nozzle tends to be consistent on the whole. The fundamental reason is that the lateral width of the nozzle is larger, and the curvature of the lateral surface is smaller, in the smaller spin angle conditions, the vortex sucks the inner low-pressure airflow to the side wall surface of the influence is smaller. When the swirl angle increases to 20°, the vortex generated by the interaction between the swirl and the tail cone is stronger, so it sucks the low-pressure airflow with the connotation of the side wall of the nozzle, and the static pressure of the side wall surface is reduced.

Figure 2.104 gives the vortex distributions in the x-direction on the along-travel cross section of the S-shaped nozzle for different spin angles. There is a clear difference between the vortex distributions in the cyclonic and no-cyclonic conditions. In the no-spin condition, the flow separation induced by the tail cone forms two pairs of counter-rotating flow vortices in the center region of cross-section C. In the

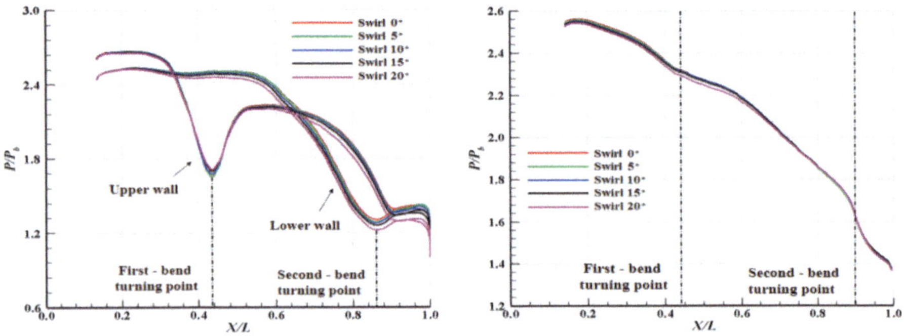

(a) Symmetric wall static pressure distribution (b) Side wall static pressure distribution

Fig. 2.103 Static pressure distribution on the wall of S-shaped nozzle at different spin angles. (**a**) Symmetric wall static pressure distribution. (**b**) Side wall static pressure distribution

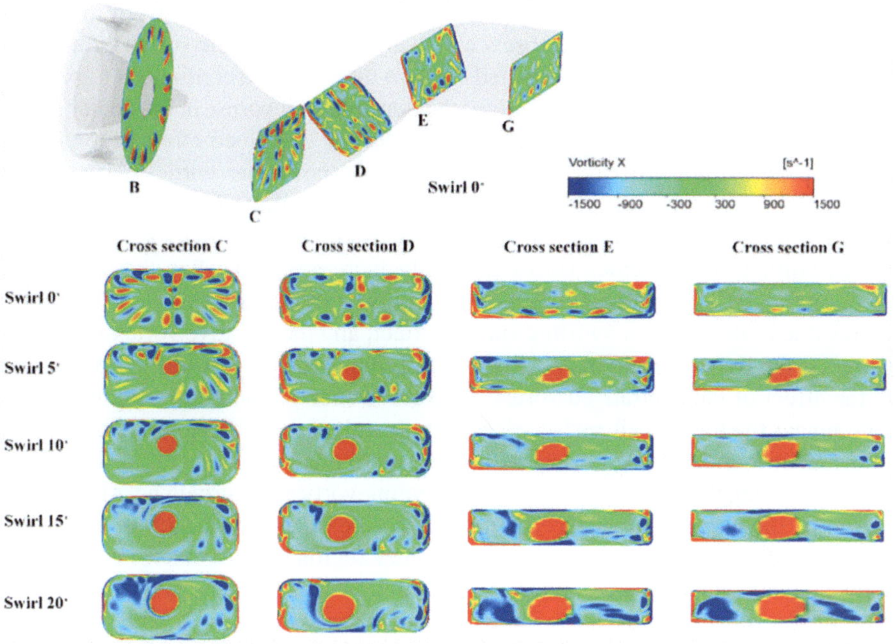

Fig. 2.104 Distribution of x-direction vorticity on the along-travel cross section of an S-shaped nozzle for different spin angles

no-spin condition, the flow separation induced by the tail cone forms two pairs of counter-rotating flow vortices. The flow separation induced by the flap mixer forms a ring-shaped distribution of flow vortices in cross-section B, which spread to the downstream cross-section C near the wall of the nozzle and show a "rectangle-like" distribution characteristics, due to the influence of the curved channel, the upper region of the airflow accelerated more strongly, and thus the amount of flow vortices on the upper side of the cross-section is larger than that on the lower side. Under the 5° rotating flow condition, the clockwise direction (along the X-axis) of the rotating flow leads to an increase in the negative vortex region and a decrease in the positive vortex region generated by the wave mixer, and the negative vortex on the cross-section B is characterized by clockwise circumferential bending. When the flow develops to cross-section C, the bending feature of negative vortex becomes more and more obvious. And the spinning flow and the tail cone work together to form a concentrated vortex with clockwise rotation and high positive vorticity in the center region of section C. As the rotation angle increases, the vortex volume of the concentrated vortex continues to increase, and the bending degree of the vortex induced by the wave mixer becomes more and more significant and begins to merge in a wide range. At a rotation angle of 20°, two pairs of counter-rotating vortices are formed around the concentrated vortex.

In cross-section D, the vortex in the upper region extends downward to form a "long" vortex distribution in the absence of cyclonic flow, and the vortex induced by the wave mixer begins to be distributed close to the wall by a continuous transverse motion. Under the 5° cyclone condition, the concentrated vortex in the center of the cross-section moves to the lower right side under the combined effect of the longitudinal pressure gradient and the cyclone. Due to the higher vorticity, it starts to entrain the flow vortices in the neighboring region, resulting in an increase in the vorticity in the surrounding region. With the increase of the spin angle, the positive vortex volume of the center vortex gradually increases, and when the spin angle increases to 20°, only a pair of counter-rotating vortices exists around the center vortex due to the stronger swirling suction effect, and the negative vortex volume is larger on the right side and the positive vortex volume on the left side is smaller due to the effect of the centralized vortex. In cross-section E, the vortex volume in the region near the lower wall surface rises significantly due to the violent acceleration of the airflow under the no-swirl condition. Due to the continuous expansion of the cross-sectional channel, a pair of counter-rotating vortices is formed near the corner region of the upper wall surface. At 5°, due to the lateral expansion of the cross-section, the concentrated vortices in the center region are stretched laterally and show a "rectangular" characteristic. As the rotation angle increases, the vortex volume of the concentrated vortex increases and more vortices with higher negative vorticity are sucked out of the cross-section, and the vortex near the corner area of the upper wall is stretched along the circumference of the concentrated vortex. When the vortex angle increases to 20°, all the vortices with higher negative vorticity surround the centralized vortex due to the stronger swirling effect, and the streamwise vortices existing near the corner zone of the upper wall surface disappear due to the strong swirling effect. The vortex distribution on section G is basically the same as that on section E.

Figures 2.105 and 2.106 give the static temperature distributions along the cross section and on the wall of the double culvert S-shaped nozzle for different cyclonic flow angles, respectively. There is a clear difference between the temperature

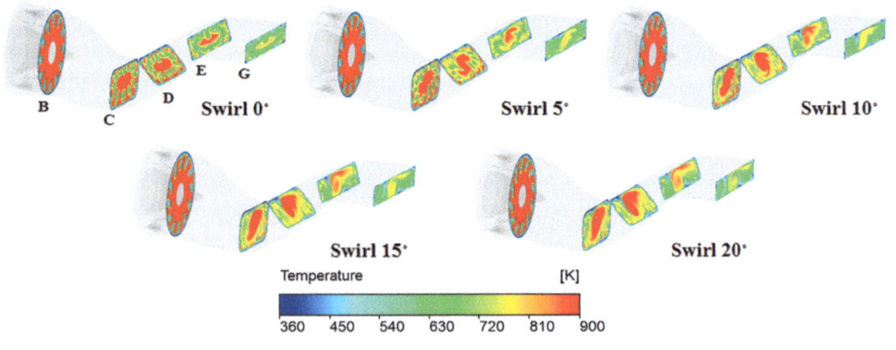

Fig. 2.105 Static temperature distribution on the along-travel cross-section of the S-shaped nozzle for different spin angles

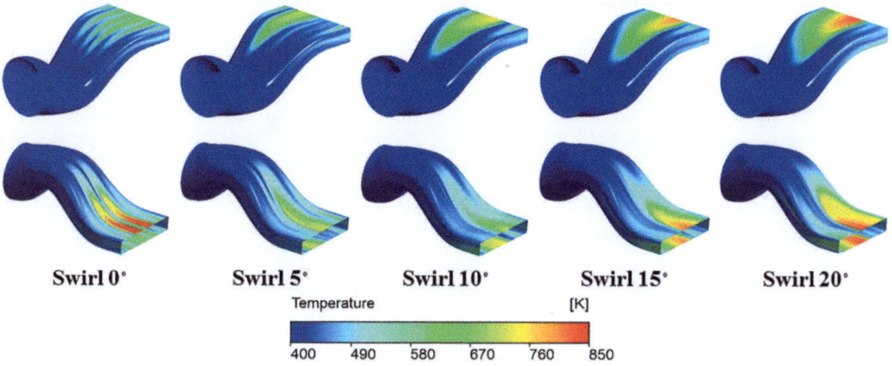

Fig. 2.106 Static temperature distribution on the wall of the S-shaped nozzle at different spin angles

distributions for the cyclonic and no-cyclonic conditions. In section B, under the no-rotation condition, the flow vortex induced by the wave mixer strengthens the mixing between the inner and outer culverts, resulting in the formation of a series of "square low-temperature grooves" along the circumference of the annular high-temperature zone, and the cross-sectional temperature distribution is characterized by the "petal shape". Under the 5° spin condition, the "petal-shaped" high-temperature zone bends clockwise along the circumference, and the degree of bending of the high-temperature zone increases with the increase of the spin angle. At section C, the temperature distribution is characterized by a "spider shape" without spinning, and the high-temperature zone at the upper side of the section is smaller than that at the lower side due to the higher vorticity at the upper side of the section. Under the 5° spin condition, the concentrated vortex in the center of the cross-section sucks the cold flow from the outer connotation into the inner connotation, which reduces the area of the high-temperature core area, and presents a tilted "peanut-shaped" distribution under the action of the spin. The local high-temperature area around the cross-section bends along the circumference under the action of the cyclone, and the local high-temperature area is obviously reduced due to the increase of the vortex volume in the flow direction by the cyclone. With the increase of the cyclone angle, the "peanut-shaped" high-temperature core area in the center of the cross-section is further reduced and gradually changed to "torch-shaped", and the local high-temperature area around the cross-section bends along the circumferential direction, and the area also continues to decrease. When the rotation angle is 20 °, the local high-temperature area around the cross-section basically disappears.

At cross-section D, the flow vortex on the upper side of the cross-section extends downward in the absence of rotation, resulting in the creation of a "low-temperature trough" in the upper region of the core. Due to the continuous mixing of the inner/outer air streams along the flow direction, the high-temperature core area is further reduced in size and is characterized by a "pea shape". At 5° rotation, the concentrated vortex in the center of the cross-section continues to suck the outer cold flow to mix the high-temperature core area, thus presenting an "S-shaped" distribution on the cross-section, and the local high-temperature area around the cross-section is also further reduced. With the increase of the rotation angle, the centralized vortex in the center of the cross-section increases, and the "S-shaped" high-temperature core area gradually changes to a "triangle" feature. When the rotation angle increases to 10°, the localized high-temperature zone around the cross-section disappears. At section E, under the no-rotation condition, due to the transverse expansion of the section, the "pea-shaped" high-temperature core region is stretched along the transverse direction, and gradually evolves into the "whale-tail" feature, with the localized high-temperature region around the core region distributed close to the wall. Under the 5° cyclonic flow condition, the concentrated vortices move to the lower right side of the cross-section, and the vortices on the lower side of the cross-section are larger, which together lead to the evolution of the "S-shape" high-temperature core area into the "hoe-shaped" feature, and the local high-temperature zone around the core area disappears. With the increase of rotational angle, the high temperature core area on section D is aggregated, and the area of high temperature area on the lower side of section E continues to shrink under the effect of hot and cold flow mixing but does not disappear completely, and stretches along the transverse direction. When the rotation angle is 20°, the distribution of high-temperature core area shows an inverted "L shape".

The "hot spots" on the wall of the nozzle at different cyclone angles mainly appear in the nozzle exit region. There is a significant difference in the temperature distribution on the wall of the S-shaped nozzle with and without cyclone. In the no-rotation condition, two "high-temperature zones" with relatively high temperature values and large widths appear in the middle of the lower wall in the nozzle exit region. In the middle of the upper wall of the nozzle exit area, there are four localized "sub-high-temperature zones" with relatively low temperature values and narrow widths. Compared with the upper wall, the lower wall of the nozzle has a larger range of "hot spot" areas with higher temperature values. In the 5° spin condition, there is a concentrated vortex with a large positive vortex in the center of the cross-section of the nozzle exit area, which continues to suck other vortices in the surrounding area, so that the vortex in the lower part of the cross-section is significantly reduced compared to the non-spin condition, and the impact of the heat flow on the lower wall surface is significantly weakened. As a result, a local "sub-high temperature zone" is formed in the middle of the lower wall of the nozzle exit area, and the range and temperature of the "hot spot" are significantly reduced; however, under the effect of the longitudinal pressure gradient formed by the curved channel, the concentrated vortex is located in the lower side of the cross-section, and thus its influence on the flow vortex on the upper side of the cross-section is significantly reduced. However, under the effect of the longitudinal pressure gradient formed by the curved channel, the concentrated vortex is located at the lower side of the cross-section, and thus it has a relatively weaker degree of attraction to the vortex in the upper side of the section. Under the effect of cyclonic flow, the amount of flow vortices on the upper side of the cross-section increases compared to the non-cyclonic condition, which leads to a significant increase in the degree of impact of the heat flow on the upper wall. As a result, a localized "sub-high temperature zone" is formed on the upper wall of the nozzle exit area, and the range and temperature of the "hot spot" increase.

With the increase of the spin angle, the range and temperature of the "hot spot" on the lower wall of the nozzle decreases and then increases, and when the spin angle is 10°, the range and temperature of the "hot spot" reaches the minimum value. The reason is that increasing the rotation angle leads to an increase in the concentrated vortex volume along the cross-section, which increases the degree of vortex absorption on the lower side of the cross-section, and the vortex volume in this region decreases, thus reducing the range and temperature of the "hot spot" on the lower wall. However, when the spin angle continues to increase, the concentration of vortex and the distance between the lower wall surface is gradually close to the connotation of high-temperature gas flow it sucked violently impact on the lower wall surface of the nozzle, so that the lower wall surface of the "hot spot" range and temperature instead of increasing. With the increase of the spin angle, the nozzle on the wall of the "hot spot" range and temperature gradually increased. When the rotation angle increases to 10°, due to the centralized vortex is located in the lower side of the cross-section, its vortex on the upper side of the cross-section of the vortex of the relative weak, and increase the rotation angle leads to the vortex on the upper side of the cross-section of the enhancement, so that the upper wall of the "hot spot" range and temperature increase. When the rotation angle is 20°, the vortex volume of the concentrated vortex is the highest, and it is almost close to the upper and lower walls, and the range and temperature of the "hot spot" on the nozzle wall are the highest. It can also be seen that in the absence of spinning conditions, the "hot spot" area on the upper and lower walls of the nozzle is mainly located in the middle position. However, under the effect of cyclone, the high temperature region in the upper part of the cross-section along the course is located in the left side, and the high temperature region in the lower part is located in the right side, so the "hot spot" region is mainly located in the left side of the upper wall and the right side of the lower wall of the nozzle.

Due to the large width-to-height ratio of the S-shaped nozzle and the pressure gradient toward the inside of the nozzle induced by the bending feature of the transverse profile, the range and temperature of the "hot spot" on the side wall of the nozzle exit area are relatively low. The existence of the rotating flow effectively increases the range of hot spots on the side wall of the S-shaped nozzle, and when the rotating flow angle is 20°, the side wall of the nozzle exit area is almost covered by the "hot spots" with low temperature values.

Figure 2.107 gives the perimeter wall static pressure distributions on the along-range cross-section of the S-shaped nozzle with double culvert channel for different spin angles. The region where the cyclonic flow has a greater effect on the perimeter wall static pressure of the nozzle is concentrated in the downstream channel of the first bend, while in the other regions of the nozzle, the effect of the cyclonic flow is mainly located in the middle of the upper and lower walls because the concentrated vortex is located in the center of the along-travel cross-section of the nozzle. On section B, the distribution of static pressure on the perimeter wall at different cyclonic flow angles is basically the same, and the presence of the cyclonic flow

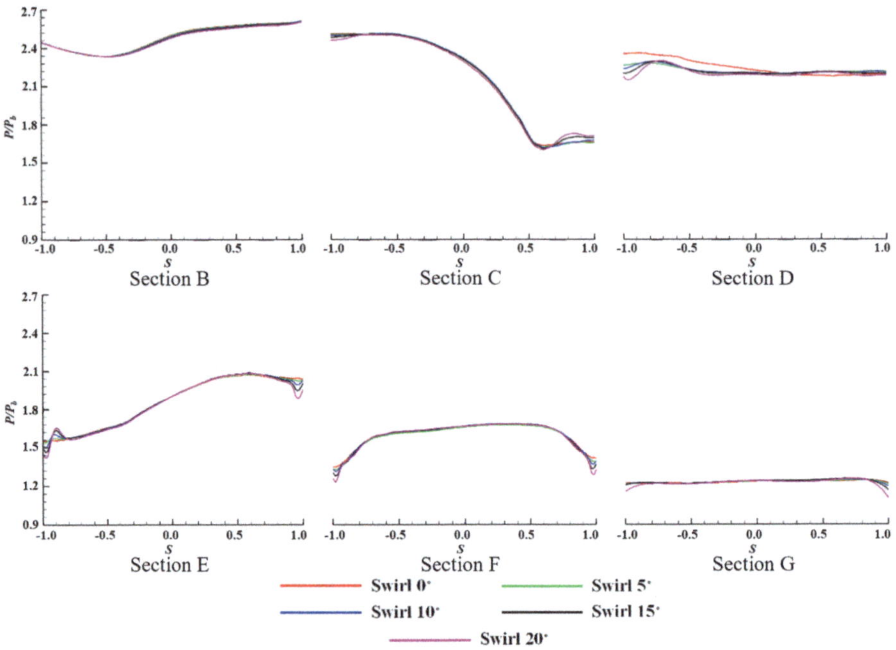

Fig. 2.107 Perimeter wall static pressure distributions of the S-shaped nozzle along the course section for different spin angles

slightly reduces the static pressure on the upper wall of the nozzle. In section C, the increase of the spin angle leads to a significant increase of the static pressure at the middle of the upper perimeter wall and a slight decrease of the static pressure at the middle of the lower wall. The reason for this is that the center of curvature of the curved flow path is located on the upper side of the pipe, inducing a longitudinal upward pressure gradient in this cross-section, with a larger vortex and lower static pressure on the upper wall surface, a smaller vortex and higher static pressure on the lower wall surface, and airflow pressures in the other regions of the cross-section falling in-between. The vortex exacerbates the airflow mixing near the upper and lower wall surfaces, resulting in a slight decrease in the static pressure at the lower perimeter wall surface and a significant increase in the static pressure at the upper perimeter wall surface.

In cross-section D downstream of the first bend, the trend of the perimeter wall static pressure distribution with the spin angle is basically the same as that of cross-section C. Due to the reduction of the longitudinal pressure gradient in the cross-section, the vortex volume moves down the wall, and the difference in the perimeter down-wall static pressure distribution is greater at different spin angles compared to cross-section C. The perimeter down-wall static pressure distribution at different spin angles varies more than that of cross-section C. The magnitude of static pressure change is the largest when the spin angle is 20°. In section E, due to the

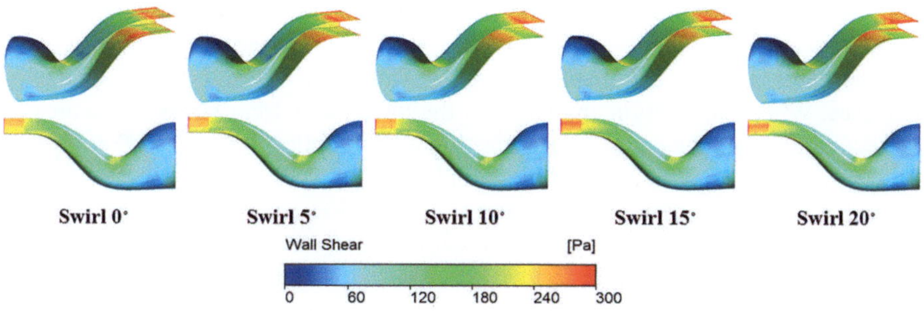

Fig. 2.108 Shear stress distribution on the wall of S-shaped nozzle at different spin angles

curvature center of the curved runner is located in the lower side of the pipe, the peripheral wall static pressure distribution of the cross-section with the spin angle of the trend of change with the cross-section C is opposite, but due to the reduced height of the cross-section, the center of the cross-section region of the centralized vortex is larger, so that the cross-section peripheral wall static pressure changes in the magnitude of the upper and lower wall is basically the same. Therefore, with the increase of the spin angle, the static pressure at the center of the upper wall of the periphery increases, and the static pressure at the center of the lower wall decreases. Sections F and G are located in the straight section of the S-shaped nozzle, because the static pressures on the upper and lower perimeter walls are basically equal and smaller than the other regions of the section, and the existence of concentrated vortices exacerbates the mixing of the inner/outer culvert flow, so the static pressures in the middle of the upper and lower perimeter walls decrease with the increase of the swirl angle.

Figure 2.108 gives the wall shear stress distributions of the S-shaped nozzle of the double culvert channel for different spin angles. Inside the first bend channel, due to the wrapping of the cold flow in the outer culvert, the cyclonic flow has less influence on the shear stress distribution on the nozzle wall. As the airflow is deflected several times in the S-shaped channel, the flow vortex sucks the inner airflow through the outer cold flow layer and impacts the nozzle wall, which leads to the decrease of the static pressure near the wall, the increase of the airflow velocity, and the consequent increase of the wall shear stress. Compared with the no-spin condition, the upper wall shear stress increases and the lower wall shear stress decreases at the second turn of the S-shaped nozzle and the isostraight section under the 5° spin condition. The reason is that, as shown in Figs. 2.102 and 2.105, the cyclone induces the formation of concentrated vortices on the surface of the tail cone, which weakens the impact of the internal airflow on the lower wall surface, resulting in the weakening of the strength of the "hot spot" on the lower wall surface, the increase of the static pressure on the wall surface, and the corresponding decrease of the wall surface shear stress. On the other hand, the degree of impact of the inner flow on the upper wall increases, and the shear stress on the upper wall increases. With the increase of the rotation angle, the intensity of the "hot spot" on the lower wall of the second turn and the equal straight section decreases and then increases, while the intensity of the "hot spot" on the upper wall continues to increase, so the corresponding region of the lower wall shear stress decreases and then increases, and the upper wall shear stress The upper wall shear stress gradually increases. When the rotation angle is 10°, the lower wall shear stress in the area of the second turn and the equal straight section is the smallest.Figure 2.109 compares the aerodynamic performance of the double culvert S-shaped nozzle at different spin angles. The total pressure recovery coefficient, flow coefficient and thrust coefficient of the nozzle decrease gradually with the increase of the swirl angle. The changes in the aerodynamic performance of the double culvert S-shaped nozzle are mainly determined by the friction loss, localized loss, vortex loss, and doping loss generated by the boundary layer of the nozzle wall, the inner/outer culvert flow, the flap mixer, and the tail cone structure. Compared with the no-spin condition, in the 5° spin condition, the flow separation on the tail cone surface is weakened, the vortex generated by the flow separation is reduced, the mixing effect of the internal/external airflow is reduced, and the resulting vortex loss and mixing loss are reduced. However, the presence of the rotating flow makes the surface of the tail cone

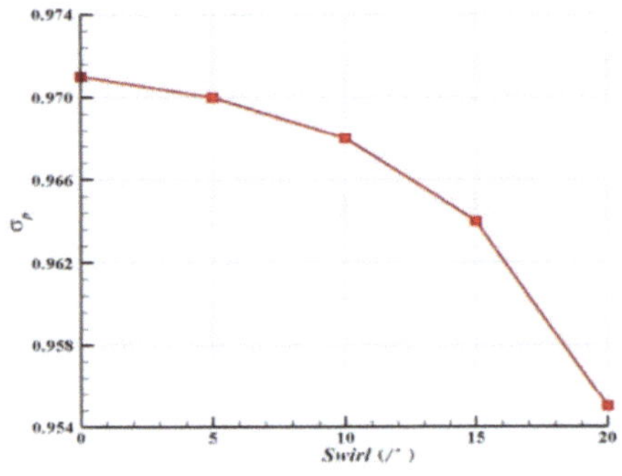

(a) Total pressure recovery coefficient

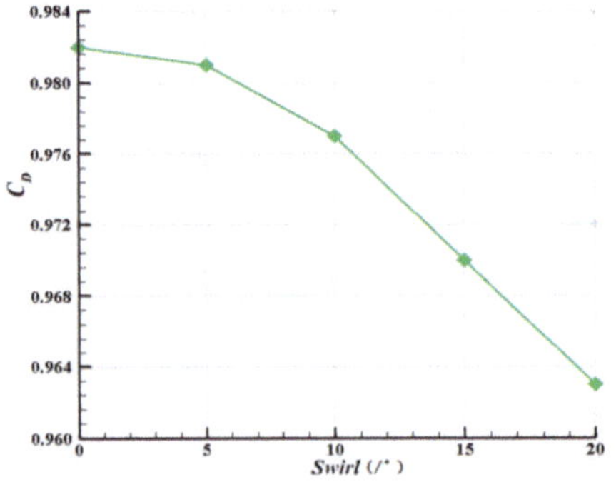

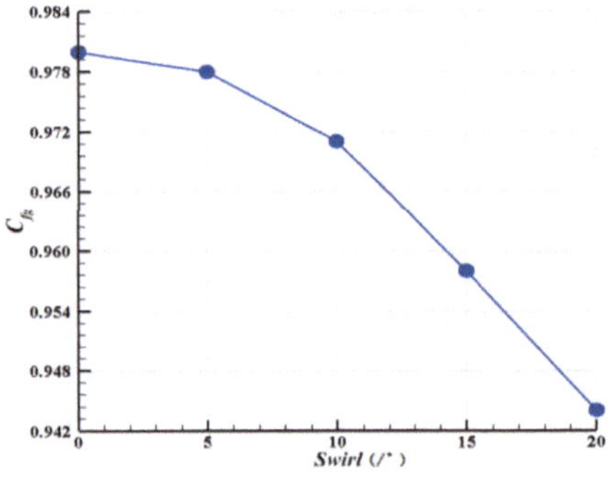

(b) Discharge coefficient

(c) Thrust coefficient

Fig. 2.109 Trend of aerodynamic performance of S-shaped nozzle at different spin angles. (**a**) Total pressure recovery coefficient. (**b**) Discharge coefficient. (**c**) Thrust coefficient

hemisphere generate concentrated vortices with larger vortices, and the vortices of the flow vortices induced by the wave mixer also increase, so the vortex loss and the doping loss of the nozzle internal flow field are elevated. As a result, the presence of the vortex reduces the aerodynamic performance of the S-shaped nozzle. With the increase of the swirl angle, the amount of concentrated vortex induced by the surface of the tail cone is stronger, and the amount of flow-oriented vortex generated by the flap mixer continues to increase, resulting in increasing vortex loss and doping loss, while the increase of the swirl angle further increases the wall shear stress of the nozzle, which makes the wall friction loss and the localized loss rise gradually. Therefore, the aerodynamic performance of the S-shaped nozzle decreases with the increase of the swirl angle, and the rate of decrease increases continuously.

Chapter 3
Study on the Influence of Infrared Radiation Characteristics of Serpentine Nozzles

3.1 Introduction

Infrared stealth capability is also one of the key focuses for future advanced stealth fighters. Relevant studies on aircraft infrared radiation characteristics (Mahulikar et al. 2007; Sang and Zhang 2013; Rao and Mahulikar 2002) show that the engine exhaust system is the main source of infrared radiation on the aircraft, producing more than 95% of the infrared signature in the tailward direction of the aircraft. Serpentine nozzles can effectively suppress the infrared radiation intensity of the exhaust system, thereby improving the aircraft's infrared stealth capability. Serpentine nozzle technology has been listed as one of the key technologies for advanced stealth fighters (Sang and Zhang 2013; Zhang et al. 2014). The design of serpentine nozzles involves many geometric parameters, and changes in geometric parameters will alter the profile structure of the serpentine nozzle, thus affecting the flow field characteristics and infrared radiation characteristics of the serpentine nozzle. Therefore, in the design of serpentine nozzles, both aerodynamic performance and infrared radiation characteristics need to be considered simultaneously in order to achieve the goal of significantly suppressing infrared radiation characteristics while maintaining high aerodynamic performance of the nozzle. This chapter uses numerical simulation methods to analyze the distribution characteristics of infrared radiation intensity of serpentine nozzles and the mechanisms generating it, studies the influence rules of geometric design parameters on the infrared radiation characteristics of serpentine nozzles, compares the differences in infrared radiation characteristics of serpentine nozzles under different inlet conditions; from the perspective of infrared stealth, carries out research on low-infrared radiation design criteria for serpentine nozzles; based on approximate modeling techniques and multi-objective optimization algorithms for serpentine nozzles, performs aerodynamic/infrared multi-objective optimization design with the goals of high aerodynamic performance and low infrared radiation intensity.

J. Shi et al., *Special Purpose Nozzles*,
https://doi.org/10.1007/978-981-95-2939-1_3

3.2 Calculation Method of Infrared Radiation Characteristics

All objects in nature continuously emit and absorb electromagnetic waves, and the intensity of radiation is related to the wavelength, as well as the properties and temperature of the object. The energy of thermal radiation is mainly concentrated within the wavelength range of 0.1–1000 μm, which basically coincides with the infrared wavelength range (0.7–1000 μm) (Tan 2006; Modest 2013; Siegel and Howell 2002). Except for some differences in terminology, the physical essence of the transmission process and the principles and methods of calculation are essentially the same. For the exhaust system of an aero-engine, since it operates in a high-temperature gas environment, accurately simulating the transmission of infrared radiation in a high-temperature gaseous medium is the key to the entire infrared radiation calculation of the aero-engine exhaust system (Huang and Liu 2013). This section mainly studies the calculation method of infrared radiation characteristics of the aero-engine exhaust system. Through the study of infrared radiation transmission methods in high-temperature gaseous media and the discrete transfer method for calculating the infrared characteristics of the target, an infrared radiation characteristics calculation program for the aero-engine exhaust system was developed based on the C++ programming language, providing support for the research work in this book.

3.2.1 Radiation Transfer in a Medium

In the engine exhaust system, infrared radiation propagates through a high-temperature gas medium, which exhibits both absorption and emission properties. When radiation travels along a ray path through such a medium, attenuation occurs due to the absorption and scattering effects of the medium. Simultaneously, radiation gain may also occur along the ray path due to scattering from surrounding regions and the medium's own emission.

3.2.1.1 Radiation Attenuation

The amount of radiation absorbed by the medium is proportional to the radiation intensity and the path length within the medium. Therefore, the radiation increment due to absorption can be expressed as:

$$\left(dL_\lambda\right)_{abs} = -\kappa_\lambda L_\lambda ds \tag{3.1}$$

where κ_λ is the spectral absorption coefficient.

In Eq. (3.1), the negative sign on the right-hand side denotes radiation attenuation.

Assuming the ray path length in the medium is s, integrating Eq. (3.1) over this path s gives:

$$L_\lambda(s) = L_\lambda(0)\exp\left(-\int_0^s \kappa_\lambda ds\right) = L_\lambda(0)e^{-t_\lambda} \tag{3.2}$$

where t_λ is the spectral absorption optical thickness and can be calculated by $t_\lambda = \int_0^s \kappa_\lambda ds$; $L_\lambda(0)$ is the initial spectral radiative intensity at the starting point of the path $s = 0$.

Then, the spectral transmittance of a medium of thickness s can be expressed as:

$$\tau_\lambda = \exp\left(-\int_0^s \kappa_\lambda ds\right) = e^{-t} \tag{3.3}$$

According to Kirchhoff's law, the spectral absorptivity is:

$$\alpha_\lambda = \frac{L_\eta(0) - L_\eta(s)}{L_\eta(0)} = 1 - e^{-t_\lambda} \tag{3.4}$$

Similar to absorption, the scattering effect of the medium also causes attenuation in the direction of radiative transfer. However, while absorption converts the attenuated radiation energy into the internal energy of the medium, scattering merely alters the direction of the attenuated energy, redistributing it into other directions.

The radiation increment due to scattering can be expressed as:

$$(dL_\lambda)_{sca} = -\sigma_{s\lambda} L_\lambda ds \tag{3.5}$$

where $\sigma_{s\lambda}$ is the spectral scattering coefficient. The negative sign in Eq. (3.5) also denotes radiation attenuation.

The phenomenon of radiation energy attenuation along the ray path caused by absorption and scattering in the medium is referred to as extinction. Thus, the extinction coefficient β_λ (also known as the spectral attenuation coefficient) is defined as (Pomraning 1973):

$$\beta_\lambda = \kappa_\eta + \sigma_{s\lambda} \tag{3.6}$$

The optical thickness based on the extinction phenomenon is defined as:

$$t_\lambda = \int_0^s \beta_\lambda ds = \int_0^s (\kappa_\lambda + \sigma_{s\lambda}) ds \tag{3.7}$$

Accordingly, the total change in radiation intensity during its transmission in the medium can be expressed as:

$$(dL_\lambda)_{ex} = -\beta_\lambda L_\lambda ds = -(\kappa_\lambda + \sigma_{s\lambda}) L_\lambda ds \tag{3.8}$$

In Eq. (3.8), the negative sign on the right-hand side indicates radiation attenuation.

3.2.1.2 Radiation Gain

When radiation propagates in a medium, due to the effects of absorption and scattering by the medium, part of the radiative energy in the propagation direction $\vec{s}$ will be lost. However, because the medium itself can also emit radiative energy, and radiative energy from other directions can also be scattered into the direction $\vec{s}$, the radiative energy in the direction $\vec{s}$ will also increase.

For emission gain, the energy gain emitted from a unit volume is proportional to the size of the volume element. Therefore, the energy radiated by the medium along a certain path is proportional to the path length and also to the local energy at that point in the medium. In the direction $\vec{s}$, the radiation from the medium is:

$$\left(dL_\lambda\right)_{em} = j_\lambda ds \tag{3.9}$$

where j_λ is the spectral emission coefficient.

According to local thermodynamic equilibrium, the emission intensity of the medium is equal to the blackbody radiation intensity, thus the following equality holds:

$$j_\lambda = \kappa_\lambda L_{b\lambda}, \quad \left(dL_\lambda\right)_{em} = \kappa_\lambda L_{b\lambda} ds \tag{3.10}$$

that is, under local thermodynamic equilibrium, the spectral emission coefficient is equal to the spectral absorption coefficient. By combining Eqs. (3.1) and (3.10), the radiative transfer equation for an absorbing-emitting medium (ignoring scattering) can be expressed as:

$$\frac{dL_\lambda}{ds} = \kappa_\lambda \left(L_{b\lambda} - L_\lambda\right) \tag{3.11}$$

where the first term on the right-hand side represents the radiation gain caused by emission from the medium, and the second term represents the radiation attenuation caused by absorption from the medium. For an isothermal gas medium with a thickness of s, the radiative transfer equation can be written as:

$$L_\eta\left(s\right) = L_\eta\left(0\right)e^{-t_\lambda} + L_{b\lambda}\left(1 - e^{-t_\lambda}\right) \tag{3.12}$$

When only the emission of the medium is considered, i.e. $L_\lambda(0) = 0$, the emissivity of the medium can be expressed as:

$$\varepsilon_\lambda = \frac{L_\lambda\left(s\right)}{L_{b\lambda}} = 1 - e^{-t_\lambda} \tag{3.13}$$

For scattering gain, the parameters used to describe the scattering direction mainly include directional scattering intensity and the scattering phase function. Let $\vec{s}_i$ denote the incident direction, and the spectral radiative intensity projected from that direction onto the scatterer be $L_\lambda(\vec{s}_i)$, part of which is scattered. This part of the spectral radiative intensity that is scattered is called the spectral scattered intensity and is denoted by $L_{s\lambda}(\vec{s}_i,4\pi)$, where 4π represents scattering into all spatial directions. The spectral directional scattering intensity $L_{s\lambda}(\vec{s}_i,\vec{s})$ is introduced and defined as follows:

$$L_{s\lambda}(\vec{s}_i,\vec{s})=\frac{L_{s\lambda}(\vec{s}_i,4\pi)}{d\Omega} \tag{3.14}$$

where $\vec{s}$ is the scattering direction and Ω is the solid angle of the scattering direction.

Obviously, the integral of the spectral directional scattering intensity over the 4π space is equal to the spectral scattered intensity, that is,

$$L_{s\lambda}(\vec{s}_i,4\pi)=\int_{\Omega=4\pi} L_{s\lambda}(\vec{s}_i,\vec{s})\,d\Omega \tag{3.15}$$

To describe the spatial distribution of the scattered energy, a scattering phase function (referred to as phase function) is introduced. It is defined as the ratio of the spectral directional scattering intensity to the average spectral directional scattering intensity uniformly distributed over the 4π scattering space, and can be expressed as:

$$\Phi_\lambda(\vec{s}_i,\vec{s})=\frac{L_{s\lambda}(\vec{s}_i,\vec{s})}{\frac{1}{4\pi}L_{s\lambda}(\vec{s}_i,4\pi)}=\frac{L_{s\lambda}(\vec{s}_i,\vec{s})}{\frac{1}{4\pi}\int_{\Omega=4\pi} L_{s\lambda}(\vec{s}_i,\vec{s})\,d\Omega} \tag{3.16}$$

Therefore, the radiation gain in the direction $\vec{s}$ caused by scattering from all directions in space is:

$$(dL_\lambda)_{sca}=\sigma_{s\lambda}L_{s\lambda}(\vec{s}_i,\vec{s})\,ds=\frac{\sigma_{s\lambda}}{4\pi}\int_{\Omega_i=4\pi} L_{s\lambda}(\vec{s})\Phi_\lambda(\vec{s}_i,\vec{s})\,d\Omega_i\,ds \tag{3.17}$$

3.2.1.3 Radiation Transfer Equation

By combining the radiative energy attenuation caused by absorption and scattering in the medium during radiation transmission, and the radiative energy gain caused by emission from the medium and scattering from other directions, and integrating Eqs. (3.1), (3.5), (3.10), and (3.17), the transfer equation for spectral radiative intensity in the medium can be expressed as:

$$\frac{dL_\lambda(s)}{ds}=\kappa_\lambda L_{b\lambda}(s)-\kappa_\lambda L_\lambda(s)-\sigma_{s\lambda}L_\lambda(s)+\frac{\sigma_{s\lambda}}{4\pi}\int_{\Omega_i=4\pi} L_\lambda(s,\vec{s}_i)\Phi_\lambda(\vec{s}_i,s)\,d\Omega_i \tag{3.18}$$

where s is the path length, $\vec{s}$ is the direction vector, $L_\lambda(s)$ is the spectral radiative intensity at position s and $L_{b\lambda}(s)$ is the blackbody spectral radiative intensity at each point along the radiation path.

For aero-engines, assuming complete combustion of the fuel in the combustion chamber and that the afterburner is not operating, the exhaust gas does not contain solid soot particles or liquid droplets. Under this assumption, the radiative attenuation and gain due to scattering in the medium can be neglected, i.e. $\sigma_{s\lambda} = 0$. Equation (3.18) can then be simplified into a one-dimensional linear differential equation:

$$\frac{dL_\lambda(s)}{ds} = \kappa_\lambda L_{b\lambda}(s) - \kappa_\lambda L_\lambda(s) \tag{3.19}$$

Combining the definition of spectral optical thickness in Eq. (3.2), and multiplying both sides of Eq. (3.19) by $\exp(t_\lambda)$, then dividing by κ_λ and rearranging the terms yields:

$$\frac{dL_\lambda(t_\lambda)}{dt_\lambda}\exp(t_\lambda) + L_\lambda(t_\lambda)\exp(t_\lambda) = L_{b\lambda}(s)\exp(t_\lambda) \tag{3.20}$$

The left-hand side of Eq. (3.20) can be written as:

$$\frac{dL_\lambda(t_\lambda)}{dt_\lambda}\exp(t_\lambda) + L_\lambda(t_\lambda)\exp(t_\lambda) = \frac{d}{dt_\lambda}\left[L_\lambda(t_\lambda)\exp(t_\lambda)\right] \tag{3.21}$$

By combining Eqs. (3.20) and (3.21), we obtain:

$$\frac{d}{dt_\lambda}\left[L_\lambda(t_\lambda)\exp(t_\lambda)\right] = L_{b\lambda}(s)\exp(t_\lambda) \tag{3.22}$$

Integrating Eq. (3.22) over the entire optical thickness $t_\lambda = 0 \sim t_\lambda(s)$ yields:

$$L_\lambda(t_\lambda) = L_\lambda(0)\exp(-t_\lambda) + \int_0^{t_\lambda} L_{b\lambda}(t'_\lambda)\exp\left[-(t_\lambda - t'_\lambda)\right]dt'_\lambda \tag{3.23}$$

This is the integral form of the radiative transfer Eq. (3.19) for spectral radiative intensity in an absorbing-emitting medium. Its physical meaning is that the spectral radiative intensity at location t_λ consists of two parts: the first part is the first term on the right-hand side of the equation, that is, the spectral radiative intensity $L_\lambda(0)$ at the starting point $t_\lambda = 0$, which, after being attenuated by the optical thickness t_λ, becomes the remaining spectral radiative intensity upon reaching point t_λ; the second part is the second term on the right-hand side of the equation, i.e. the spectral radiative intensity $L_{b\lambda}(t'_\lambda)$ emitted by the medium itself at point t'_λ $(0 \le t'_\lambda \le t_\lambda)$ along the transmission path, which, after being attenuated by the remaining optical thickness $t_\lambda - t'_\lambda$, becomes the remaining spectral radiative intensity upon reaching point t_λ; then, this is integrated over t'_λ from 0 to t_λ, as shown in Fig. 3.1.

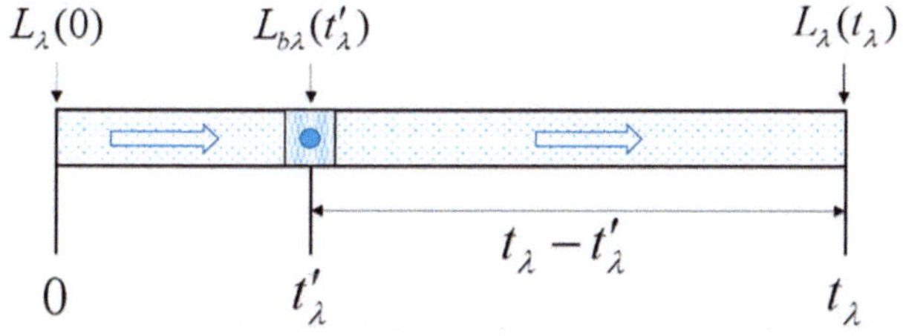

Fig. 3.1 Physical meaning of the integral form of the radiation transfer equation

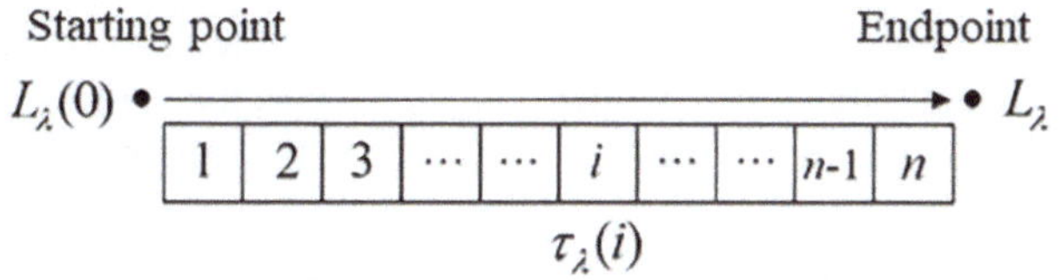

Fig. 3.2 Discretization of the radiation transfer equation

Rearranging Eq. (3.23) gives:

$$L_\lambda(t_\lambda) = L_\lambda(0)\exp(-t_\lambda) + \int_0^{t_\lambda} L_{b\lambda}(t'_\lambda)\, d\left\{\exp\left[-(t_\lambda - t'_\lambda)\right]\right\} \tag{3.24}$$

Substituting the spectral transmittance formula in Eq. (3.3) into the above equation gives:

$$L_\lambda(t_\lambda) = L_\lambda(0)\tau_\lambda(t_\lambda) + \int_0^{t_\lambda} L_{b\lambda}(t'_\lambda)\, d\tau_\lambda(t_\lambda - t'_\lambda) \tag{3.25}$$

Equation (3.25) is the integral form of the radiation transfer equation. In practical calculations, the radiation transfer equation is usually discretized along the direction of the radiation path and solved in a numerical summation form. As shown in Fig. 3.2, the radiation path is divided into n segments. Let the spectral transmittance of the i-th segment ($i = 0, 1, 2, \ldots, n$) be $\tau_\lambda(i)$, the initial spectral radiative intensity of the radiation transfer be $L_\lambda(0)$, and the spectral radiative intensity at the endpoint be L_λ. Then Eq. (3.25) can be discretized as:

$$L_\lambda = L_\lambda(0)\prod_{i=1}^{n}\tau_\lambda(i) + \sum_{i=1}^{n} L_{b\lambda}(i)\left[1 - \tau(i)\right]\prod_{j=i+1}^{n}\tau(j) \tag{3.26}$$

where, when$j = n + 1$, then $\tau_\lambda(j) = 1$; when the radiation path originates from a solid wall boundary, $L_\lambda(0)$ is the corresponding effective spectral radiative intensity of the wall; when the radiation path originates from a transparent boundary, $L_\lambda(0) = 0$.

Generally, the internal flow field of the aero-engine exhaust system is non-uniform. When calculating radiation transfer, the computational domain needs to be discretized into a finite number of spatially non-overlapping volume elements. Within each small element, flow field parameters such as the temperature, pressure, and composition of the gas can be approximated as uniform. Therefore, the above method can be used to calculate the infrared radiation transfer within the aero-engine exhaust system.

3.2.2 *Discrete Transfer Method*

The Discrete Transfer Method (DTM) was first proposed by Lockwood and Shah (1981) as a method for calculating radiation transfer problems in combustion chambers. Its main idea is to take the calculation boundary as the starting point of radiation transfer. The Discrete Transfer Method first requires discretizing the computational domain boundary into a finite number of micro boundary surfaces, as shown in Fig. 3.3. On this basis, a characteristic ray is formed between each micro boundary surface and the detector, and the infrared radiation of each micro boundary surface is transmitted to the detector along this characteristic ray. Finally, by superposing the infrared radiation of all micro boundary surfaces to the detector, the infrared radiation intensity of the target in the computational domain to the detector is obtained. The DTM transforms the complex three-dimensional spatial radiation transfer problem into a simple one-dimensional radiation transfer problem along the characteristic rays, simplifying the problem to be solved.

When calculating the infrared radiation characteristics of the target using the Discrete Transfer Method, it is first necessary to solve the problem of calculating the effective radiative intensity of each micro boundary surface. As shown in Fig. 3.4, according to the properties of each boundary, the boundary of the computational domain is divided into solid wall boundaries and transparent boundaries, in

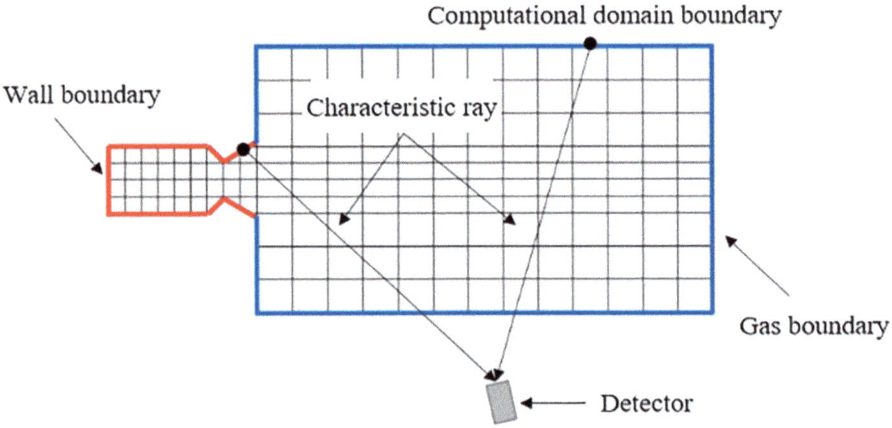

Fig. 3.3 Schematic diagram of the principle of the Discrete Transfer Method

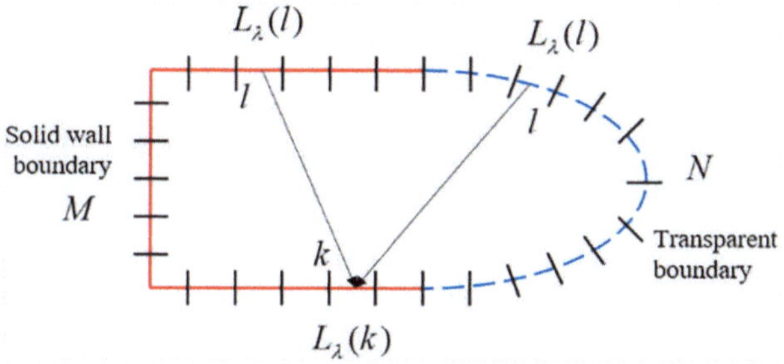

Fig. 3.4 Schematic diagram of the discretization of the computational domain boundary

which the solid wall boundary contains M micro surfaces, and the transparent boundary contains N micro surfaces. Taking micro surface k, assuming all other boundary micro surfaces are visible to k, the effective radiative intensity of micro surface k is:

$$L_{\lambda,k}(0) = \varepsilon_{\lambda,k} L_{\lambda b}(T_k) + \rho_{\lambda,k} L_{\lambda,k}^{in} \tag{3.27}$$

where $\varepsilon_{\lambda,\,k}$ is the emissivity of micro surface k; $\rho_{\lambda,\,k}$ is the reflectivity of micro surface k.

The first term on the right-hand side of Eq. (3.27) represents the spectral radiative intensity generated by the temperature of micro surface k itself, and the second term represents the reflection of all the spectral radiative intensity incident on micro surface k.

The spectral radiation irradiance from other micro surfaces to surface k is:

$$H_{\lambda,k}^{in} = \sum_{l=1}^{M+N-1} L_\lambda(l,k)\pi F_{k-l} \tag{3.28}$$

where F_{k-l} is the view factor from micro surface k to micro surface l; $L_\lambda(l,k)$ is the spectral radiative intensity transmitted from micro surface l to micro surface k.

In Eq. (3.28), $L_\lambda(l,k)$ can be calculated using Eq. (3.26). Then Eq. (3.28) can be transformed into:

$$H_{\lambda,k}^{in} = \sum_{l=1}^{M+N-1} \pi F_{k-l} \left\{ L_{\lambda,l}(0)\prod_{i=1}^{n}\tau_\lambda(i) + \sum_{i=1}^{n} L_{b\lambda}(i)\left[1-\tau(i)\right]\prod_{j=i+1}^{n}\tau(j) \right\} \tag{3.29}$$

where $L_{\lambda,\,l}(0)$ is the effective spectral radiative intensity of micro surface l.

Combining Eqs. (3.29) and (3.27) can be transformed into:

$$L_{\lambda,k}(0) = \varepsilon_{\lambda,k} L_{\lambda b}(T_k) + \rho_{\lambda,k} \frac{H_{\lambda,k}^{in}}{\pi} = \varepsilon_{\lambda,k} L_{\lambda b}(T_k)$$
$$+ \frac{\rho_{\lambda,k}}{\pi} \sum_{l=1}^{M+N-1} \pi F_{k-l} \left\{ L_{\lambda,l}(0) \prod_{i=1}^{n} \tau_\lambda(i) + \sum_{i=1}^{n} L_{b\lambda}(i)[1 - \tau(i)] \prod_{j=i+1}^{n} \tau(j) \right\} \tag{3.30}$$

For the exhaust system of an aero-engine, when the computational domain is sufficiently large, the transparent boundary is far away from the high-temperature core region of the gas, and the density inside and outside the boundary is basically equal, with the same refractive index. At this time, for the transparent boundary micro surfaces, $L_{\lambda,l}(0) = 0$ can be assumed. Therefore, the influence of the solid wall boundary micro surfaces and the transparent boundary micro surfaces on k in Eq. (3.30) can be separated, i.e.:

$$L_{\lambda,k}(0) = \varepsilon_{\lambda,k} L_{\lambda b}(T_k) + \frac{\rho_{\lambda,k}}{\pi} \sum_{l=1}^{M-1} \pi F_{k-l} \left[L_{\lambda,l}(0) \prod_{i=1}^{n} \tau_\lambda(i) \right]$$
$$+ \frac{\rho_{\lambda,k}}{\pi} \sum_{l=1}^{M+N-1} \pi F_{k-l} \left\{ \sum_{i=1}^{n} L_{b\lambda}(i)[1 - \tau(i)] \prod_{j=i+1}^{n} \tau(j) \right\} \tag{3.31}$$

From Eq. (3.31), it can be seen that to obtain the effective spectral radiative intensity of the solid wall boundary micro surfaces of the exhaust system, it is necessary to solve a large linear system of equations by combining the effective spectral radiative intensity equations of all solid wall boundary micro surfaces.

After obtaining the effective spectral radiative intensity of all solid wall boundary micro surfaces, the infrared radiation intensity of the target aero-engine exhaust system at the detection point can be calculated. When the detection point is far from the target aero-engine exhaust system, the infrared radiation intensity at the detection point is:

$$I_\lambda = \sum_{m=1}^{n} L_{\lambda,m} \Delta A_m \tag{3.32}$$

where $L_{\lambda,m}$ is the spectral radiative intensity transmitted from the visible boundary micro surface m to the detector at the detection point; ΔA_m is the projected area of micro surface m in the direction of the characteristic ray to the detector.

In this book, the infrared radiation of the aero-engine exhaust system is calculated using the Discrete Transfer Method based on the C++ programming language.

3.2.3 Numerical Simulation Method of Flow Field

The calculation of the infrared radiation characteristics of the serpentine nozzle is based on the flow field calculation. Through the flow field calculation, the temperature distribution of the serpentine nozzle wall surface, the temperature and pressure distribution of the combustion gas, as well as the mole fractions of gas components such as CO_2, H_2O, and CO are obtained. These flow field parameters are then imported into the infrared radiation characteristics calculation program for computation. This paper uses commercial CFD software to solve the three-dimensional Reynolds-averaged N–S equations, and applies the SST k–ω turbulence model to carry out the numerical calculation of the flow field characteristics of the serpentine nozzle.

Structured mesh generation for the geometric model of the serpentine nozzle is conducted using ICEM software. Figure 3.5 shows the full three-dimensional computational grid and boundary conditions. The near-wall mesh is refined, and the wall $y^+ < 5$. After mesh independence verification, when the number of computational grids reaches approximately 1.6 million, the static pressure distribution on the upper and lower walls of the symmetric plane of the serpentine nozzle no longer changes significantly with the increase of grid quantity.

A pressure inlet boundary condition is applied at the inlet of the serpentine nozzle, with a total temperature of $T^* = 1072$ K and a total pressure of $p^* = 1.54$ atm. The inlet and far field of the external flow field are both set as pressure far-field boundaries, with given $Ma_\infty = 0.05$, static temperature $T_0 = 300$ K, and static pressure $p_0 = 1$ atm. The outlet of the external flow field is set as a static pressure $p_b = 1$ atm. The wall surface is modeled as a no-slip adiabatic wall. This paper assumes that the fuel in the engine combustion chamber is fully burned, and the mole fractions of H_2O and CO_2 in the combustion gas at the nozzle inlet are set to x (H_2O) = 12.87% and x (CO_2) = 13.12%, respectively. The gas components at the inlet and far field of the external flow field are consistent with the ambient atmospheric environment.

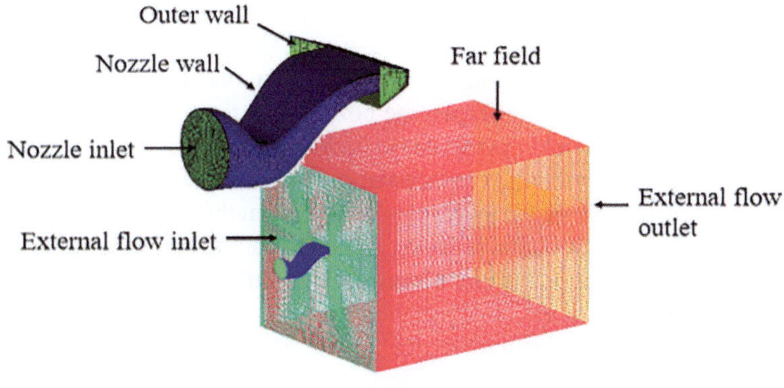

Fig. 3.5 Computational grid and boundary conditions

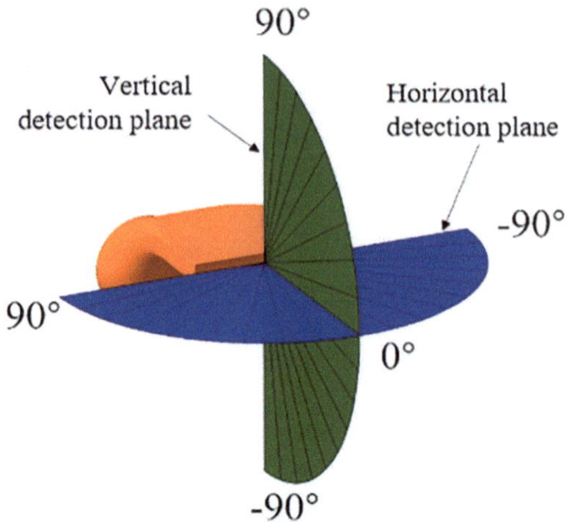

Fig. 3.6 Schematic diagram of detector arrangement

In the infrared radiation calculation of the serpentine nozzle, the nozzle inlet is simplified as a gray body wall, and its emissivity and that of the nozzle wall are both set to $\varepsilon = 0.9$ (Rao and Mahulikar 2005). Considering that in actual conditions the nozzle is installed inside the aircraft body and the outer wall of the nozzle is shielded by the aircraft skin, the radiation of the outer wall surface of the nozzle is not considered in the infrared radiation calculation. The infrared calculation wavelength range is 3 ~ 5 μm, the detection distance is set to 1000 m, and the attenuation of infrared radiation by the environment is not considered. Detection points are set on the horizontal detection plane and vertical detection plane, as shown in Fig. 3.6.

3.3 Study on the Influence of Geometric Parameters on the Infrared Radiation Characteristics of the Serpentine Nozzle

The infrared radiation characteristics of the serpentine nozzle are mainly determined by the nozzle wall surface, the temperature distribution of the combustion gas, and the concentration distribution of the gas components. Changes in the geometric design parameters of the serpentine nozzle alter the nozzle's geometric configuration and the flow field distribution, including temperature and component concentrations, thereby affecting the infrared radiation characteristics of the serpentine nozzle. Therefore, in order to develop a design method for serpentine nozzles with low infrared radiation characteristics, it is necessary to study the infrared radiation characteristics of the serpentine nozzle and the influence of key design parameters on its infrared radiation, analyze the characteristics and mechanisms of the infrared radiation, and provide a reference for the design of serpentine nozzles from the perspective of infrared suppression. In this section, the centerline variation pattern, exit aspect ratio, and the axial length ratio of the two bends are selected as key geometric parameters to investigate their influence on the infrared radiation characteristics of the serpentine nozzle.

3.3.1 Infrared Radiation Characteristics of the Serpentine Nozzle

To study the infrared radiation characteristics of the serpentine nozzle, the distribution of infrared radiation intensity is analyzed for a serpentine nozzle with a "steep-to-gradual" centerline variation in both bends. The geometric parameters of the nozzle are taken as the design-state values shown in Table 3.1. The temperature distribution of the combustion gas on the symmetry plane and the horizontal center plane at the nozzle exit is shown in Fig. 3.7. Due to the nozzle's exit aspect ratio being 7 and the exit shape being a flat quasi-rectangular form, the jet height in the vertical direction is significantly smaller than its width in the horizontal direction. This is the reason for the difference in infrared radiation intensity of the combustion gas between the horizontal and vertical detection planes.

Figure 3.8 presents the Mach number distribution and streamline direction on different cross sections and the symmetry plane of the serpentine nozzle. From the Mach number distribution inside the nozzle, it can be seen that as the inner surface of the nozzle contracts, the cross-sectional area of the nozzle gradually decreases

Table 3.1 Design-state values of geometric parameters of the serpentine nozzle

Geometric parameter	A_1/A_{in}	$\Delta Y_1/L_1$	L_1/L_2	W_e/H_e
Design-state value	0.6	0.28	1:2	7

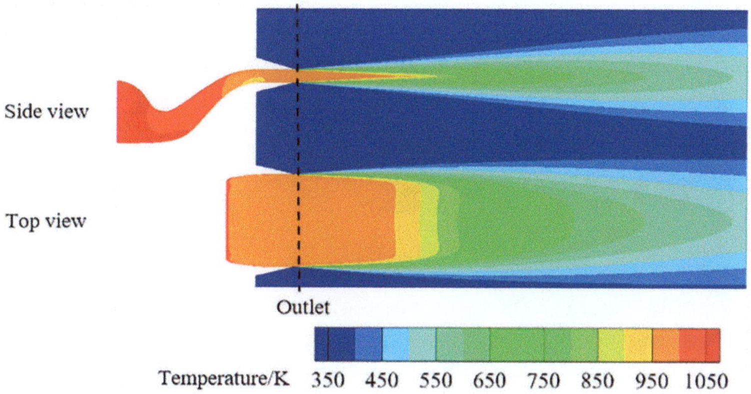

Fig. 3.7 Combustion gas temperature distribution of the serpentine nozzle

along the flow direction. As a result, the airflow inside the nozzle exhibits an overall acceleration trend along the flow direction. Near the nozzle inlet, the nozzle contour deflects downward. Due to the constraint of the wall structure, the upper part of the airflow near the inlet also deflects downward. Under the centripetal force exerted by the upper wall, the upper flow acceleration decreases. At cross-section 1, the velocity of the lower part of the flow is greater than that of the upper part, and accordingly, the pressure in the lower part is lower than that in the upper part. Therefore, the streamlines at cross-section 1 are directed downward. At the junction of the two bends, due to the upward deflection of the nozzle, the airflow also deflects upward under the constraint of the nozzle walls. Since the curvature of the upper wall is greater than that of the lower wall, the airflow near the upper wall accelerates more significantly under the action of centrifugal force, forming a local high-speed region near the upper wall. As shown in cross-section 3, the velocity of the upper part of the airflow is higher than that of the lower part, which leads to lower pressure in the upper part. Therefore, the streamlines at cross-section 3 are directed upward. In the middle section between the two bends, the acceleration of the upper and lower parts of the flow is constrained by the upper and lower nozzle walls. At cross-sections 2 and 4, the central part of the flow has the highest velocity and lowest pressure, thus the streamlines from the upper and lower parts are directed toward the center, forming a transverse flow. Similarly, at cross-section 5, the flow directions of the upper and lower parts are also directed toward the center, forming a lateral flow. As the transverse flow further develops along the nozzle, a transverse vortex is eventually formed at the nozzle exit, as shown in cross-section 6.

The infrared radiation intensity distribution of the serpentine nozzle on the horizontal and vertical detection planes is shown in Fig. 3.9. For the purpose of analysis and comparison, the figure presents the total infrared radiation intensity, wall infrared radiation intensity, and gas infrared radiation intensity of the serpentine nozzle. The infrared radiation intensity results of the serpentine nozzle in this chapter have been normalized.

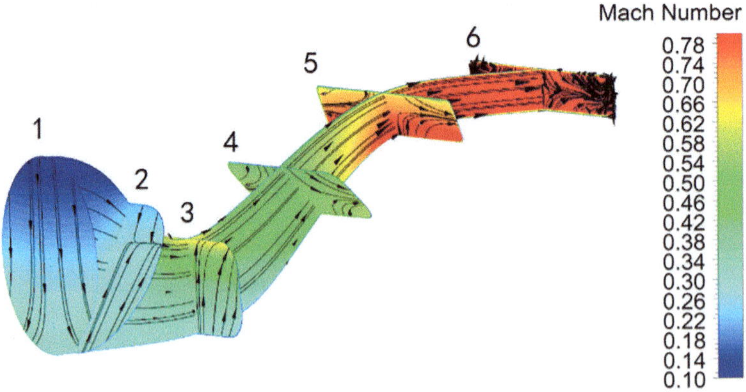

Fig. 3.8 Mach number and streamline direction distribution inside the serpentine nozzle

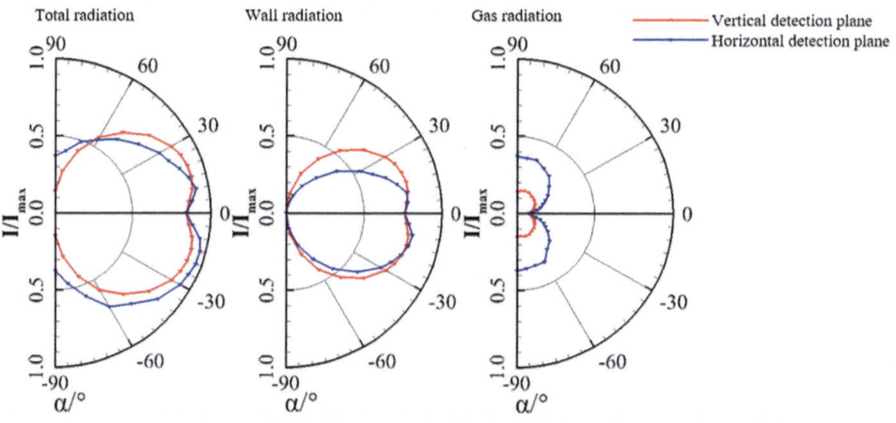

Fig. 3.9 Infrared radiation intensity distribution of the serpentine nozzle

As shown in Fig. 3.9, the gas infrared radiation intensity on both detection planes of the serpentine nozzle increases with the detection angle |α|. This is mainly caused by the variation in the projection angle of the gas to the detection point. As the detection angle |α| increases, the projected area of the gas in the detection direction gradually increases, so the area of the high-temperature region of the gas captured by the detector also becomes larger, and consequently, the infrared radiation intensity from the gas to the detection point becomes greater. Meanwhile, in the process of radiation transmission from the gas to the detector, the radiation from the gas farther from the detector needs to pass through the gas closer to the detector, and the radiation is attenuated due to the self-absorption of the gas. As the detection angle |α| increases, the path length of the gas that the radiation must travel through becomes shorter, thus reducing the self-absorption effect, which results in an increase in gas infrared radiation intensity.

For the serpentine nozzle, compared with the gas infrared radiation intensity on the vertical detection plane, that on the horizontal detection plane is significantly lower. As shown in Fig. 3.9, the average decrease reaches 14.2%. When |α| = 90°, the gas infrared radiation intensity on the horizontal detection plane decreases by as much as 61.1% compared to that on the vertical detection plane. This is caused by the distribution characteristics of the exhaust jet of the serpentine nozzle. As shown in Fig. 3.7, the width of the exhaust jet in the horizontal direction is greater than its height in the vertical direction. Therefore, under the same detection angle, the detector on the vertical detection plane can capture more of the high-temperature region of the exhaust jet.

The infrared radiation emitted from the wall must pass through the gas during its transmission to the detector. Due to the absorption effect of the gas on radiation, the radiation emitted from the wall is attenuated to some extent during transmission. Therefore, the infrared radiation intensity of the wall received by the detector is determined by two factors: the wall's own infrared radiation and the absorption effect of the gas. When the detection angle |α| < 10°, the wall infrared radiation

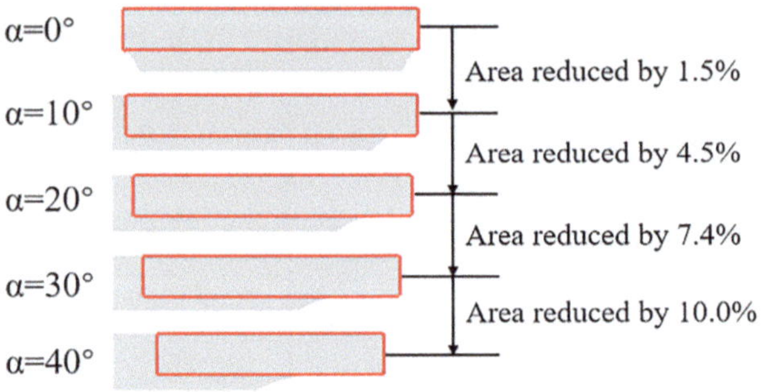

Fig. 3.10 Variation of visible wall area with detection angle

intensity of the serpentine nozzle increases with |α|, as shown in Fig. 3.9. This is because when the detection angle increases from 0° to 10°, the wall area visible to the detector decreases by only 1.5%, as shown in Fig. 3.10, and the change in the wall's own radiation is relatively small. At this time, the change in wall infrared radiation intensity is mainly caused by the absorption effect of the gas on the radiation from the wall. When the detection angle is small, the radiation emitted from the wall needs to travel a longer path through the gas to reach the detector, and the absorption effect of the gas is strong. As the detection angle increases, the gas path that the wall radiation must pass through gradually shortens, and the absorption effect of the gas is weakened accordingly. Therefore, when $|\alpha| < 10°$, the wall infrared radiation intensity of the serpentine nozzle increases with |α|. When $|\alpha| > 10°$, the wall radiation intensity begins to decrease gradually with the increase of detection angle |α|, as shown in Fig. 3.9. This is because as |α| further increases, the wall area visible to the detector begins to decrease significantly, as shown in Fig. 3.10, resulting in a noticeable change in the wall's own radiation. At this point, the change in the wall's own radiation becomes the main factor affecting the wall infrared radiation intensity. As the wall area visible to the detector decreases, the wall infrared radiation intensity gradually decreases. When $|\alpha| = 90°$, the wall radiation intensity decreases to 0.

As shown in Fig. 3.9, the wall infrared radiation intensity on the lower part of the vertical detection plane of the serpentine nozzle is close to that on the horizontal detection plane. However, the wall infrared radiation intensity on the upper part of the vertical detection plane is lower than that on the lower part. This is mainly caused by the temperature distribution difference between the upper and lower walls of the serpentine nozzle. Figure 3.11 shows the temperature distribution of the upper and lower walls of the serpentine nozzle. It can be observed that downstream of the serpentine nozzle, the overall temperature of the upper wall is higher than that of the lower wall. Therefore, the wall temperature detectable by the detector in the lower part of the vertical detection plane is higher than that in the upper part. Since wall radiation is strongly influenced by wall temperature, the wall infrared radiation intensity on the lower part of the vertical detection plane is higher than that on the upper part. The temperature distribution on the serpentine nozzle wall is related to

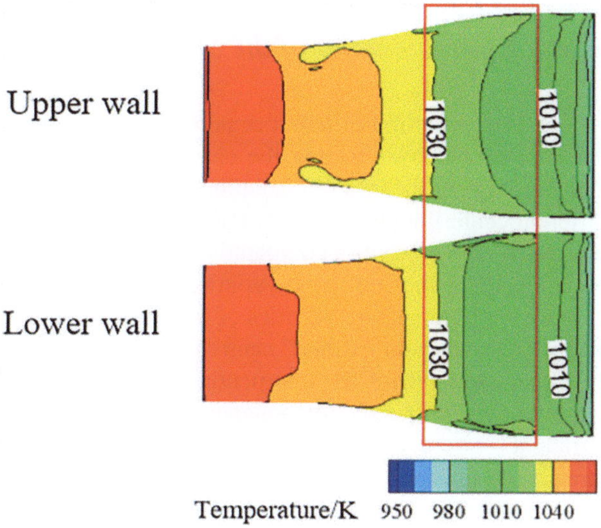

Fig. 3.11 Temperature distribution of upper and lower walls of the serpentine nozzle

the internal flow field characteristics. Downstream of the serpentine nozzle, the internal nozzle surface deflects downward. Due to the structural constraints of the nozzle wall, the airflow is also deflected downward. Since the lower wall of the nozzle is closer to the center of curvature compared to the upper wall, and the closer the airflow is to the center of curvature, the faster it accelerates (as shown in Fig. 3.8), the temperature of the upper part of the airflow is higher than that of the lower part downstream of the serpentine nozzle. As a result, the upper wall temperature is higher than the lower wall temperature.

Overall, the wall infrared radiation intensity of the serpentine nozzle is higher than its gas infrared radiation intensity, and the distribution of total infrared radiation intensity generally corresponds to that of the wall infrared radiation intensity. When the detection angle $|\alpha|$ is small, the visible wall area to the detector is large, and the wall infrared radiation intensity is high, exceeding that of the gas. As $|\alpha|$ increases, the visible wall area to the detector rapidly decreases, and the wall infrared radiation intensity also decreases accordingly, while the gas infrared radiation intensity increases with $|\alpha|$. Therefore, when $|\alpha|$ is large, the gas infrared radiation intensity exceeds that of the wall. When $|\alpha| = 90°$, the wall infrared radiation intensity of the serpentine nozzle is 0, and the total infrared radiation intensity equals the gas infrared radiation intensity.

In general, due to the flat exit geometry of the serpentine nozzle, the gas infrared radiation intensity on the vertical detection plane is higher than that on the horizontal detection plane. The higher temperature of the upper wall downstream of the serpentine nozzle results in a higher wall infrared radiation intensity on the lower part of the vertical detection plane compared to the upper part.

3.3.2 Influence of Centerline Variation Pattern on the Infrared Radiation Characteristics of the Serpentine Nozzle

For the serpentine nozzle, adopting different centerline variation patterns alters the geometry of the nozzle, thereby affecting the characteristics of the internal flow field. This section investigates the infrared radiation characteristics of the serpentine nozzle when different centerline variation patterns are applied to the first and second serpentine sections, and presents a comparative analysis.

To study the influence of the centerline variation pattern of the first serpentine section on the infrared radiation characteristics of the nozzle, the second serpentine section was fixed with a "steep-to-gradual" centerline variation. Based on this, the infrared radiation characteristics of three serpentine nozzles with different centerline variation patterns in the first section—namely "uniform," "gradual-to-steep," and "steep-to-gradual"—were analyzed. The geometrical configurations of the three nozzles are shown in Fig. 3.12. In the figure, the digits before and after the hyphen represent the centerline variation patterns of the first and second serpentine sections, respectively. The digits "1", "2", and "3" correspond to "uniform," "gradual-to-steep," and "steep-to-gradual" patterns, respectively. Figure 3.13 presents a comparison of the thrust coefficients and friction losses of the three nozzles. It can be seen that the aerodynamic performance differences among the nozzles are minimal, with thrust coefficient variations within 0.003 and maximum friction loss differences less than 0.05%. Among them, nozzle 1–3, which adopts a uniform variation pattern in the first section, has the highest thrust coefficient and the lowest friction loss.

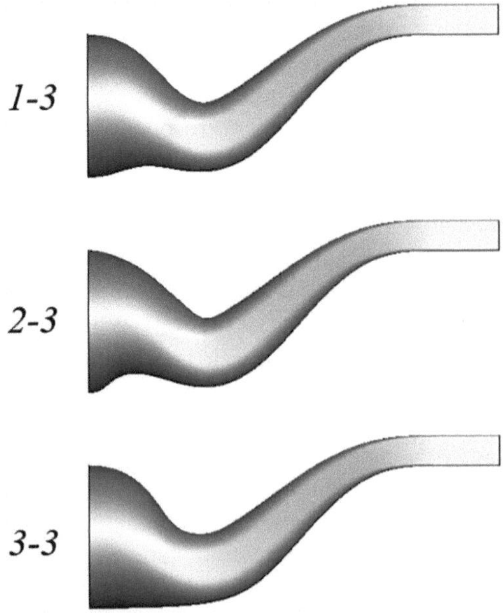

Fig. 3.12 Serpentine nozzle models with different centerline variation patterns in the first section

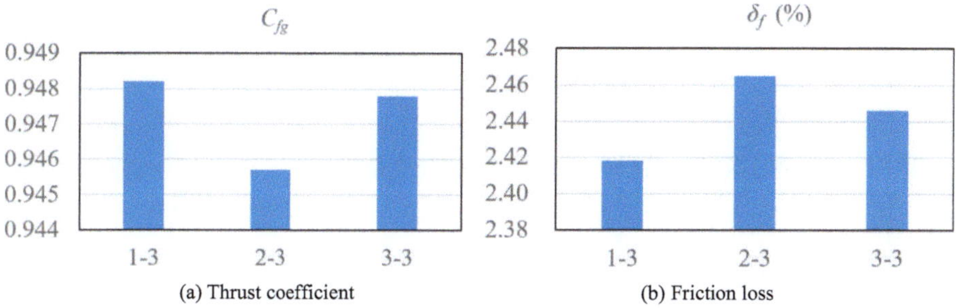

Fig. 3.13 Performance parameters of serpentine nozzles with different centerline variation patterns in the first section. (**a**) Thrust coefficient. (**b**) Friction loss

The infrared radiation intensity distributions of the three serpentine nozzles with different centerline variation patterns in the first serpentine section are shown in Fig. 3.14. It can be seen that the variation pattern of the centerline in the first serpentine section has essentially no effect on the gas-phase infrared radiation of the nozzle. This is because the jet plume lengths and the sizes of the high-temperature core regions among the nozzles are almost unaffected by the centerline variation pattern of the first serpentine section, as shown in Fig. 3.15.

On the horizontal detection plane, the wall infrared radiation intensity distributions of the nozzles are generally consistent. On the vertical detection plane, however, slight differences are observed in the wall infrared radiation intensities. Below the vertical detection plane, nozzle 1–3 has the lowest wall infrared radiation intensity, nozzle 3–3 is intermediate, and nozzle 2–3 has the highest. This is mainly caused by the differences in the temperature distributions on the downstream region of the upper wall surface of the nozzles. Compared to nozzle 1–3, nozzle 3–3 has a larger region on the upper wall downstream with temperatures above 1010 K; in comparison to nozzle 3–3, nozzle 2–3 has a larger region with temperatures above 1020 K, as shown in Fig. 3.16. Since wall infrared radiation is highly sensitive to wall temperature, the infrared radiation intensity below the vertical detection plane is lowest when the centerline of the first serpentine section adopts a “uniform” variation pattern (1–3), highest when it adopts a “gradual-to-steep” variation pattern (2–3), and intermediate when it adopts a “steep-to-gradual” variation pattern (3–3).

The variation in the total infrared radiation intensity of serpentine nozzles with different centerline patterns in the first serpentine section generally follows the same trend as the wall radiation intensity. The main differences in total infrared radiation intensity occur below the vertical detection plane. Below this plane, compared with the nozzle adopting the “gradual-to-steep” centerline variation (2–3), the nozzles using the “uniform” (1–3) and “steep-to-gradual” (3–3) variation patterns in the first serpentine section exhibit average reductions in total infrared radiation intensity of 6.3% and 3.7%, respectively.

When the centerline of the first serpentine section is fixed as “steep-to-gradual,” and the centerline of the second serpentine section is varied among “uniform,” “gradual-to-steep,” and “steep-to-gradual,” the geometries of the resulting nozzles

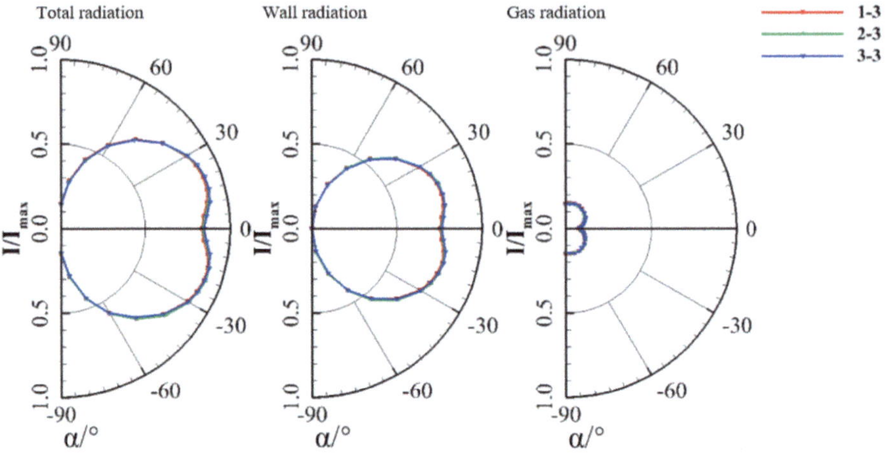

(a) Infrared radiation intensity distribution on the horizontal detection plane

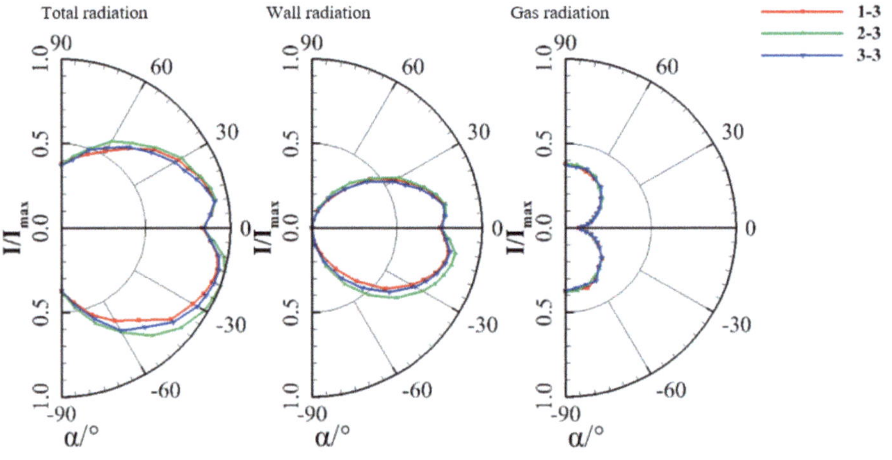

(b) Infrared radiation intensity distribution on the vertical detection plane

Fig. 3.14 Infrared radiation intensity distribution of serpentine nozzles with different centerline variation patterns in the first section. (**a**) Infrared radiation intensity distribution on the horizontal detection plane. (**b**) Infrared radiation intensity distribution on the vertical detection plane

are shown in Fig. 3.17. Figure 3.18 compares the thrust coefficients and friction losses of these nozzles. The difference in thrust coefficients is relatively significant, with a maximum deviation of 0.015, while the friction losses differ only slightly, with a maximum difference under 0.08%. The nozzle using the "steep-to-gradual" centerline variation in the second section achieves the highest thrust coefficient and the lowest friction loss, indicating the best aerodynamic performance.

The infrared radiation intensity distributions of three serpentine nozzles with different centerline variation patterns in the second serpentine section are shown in

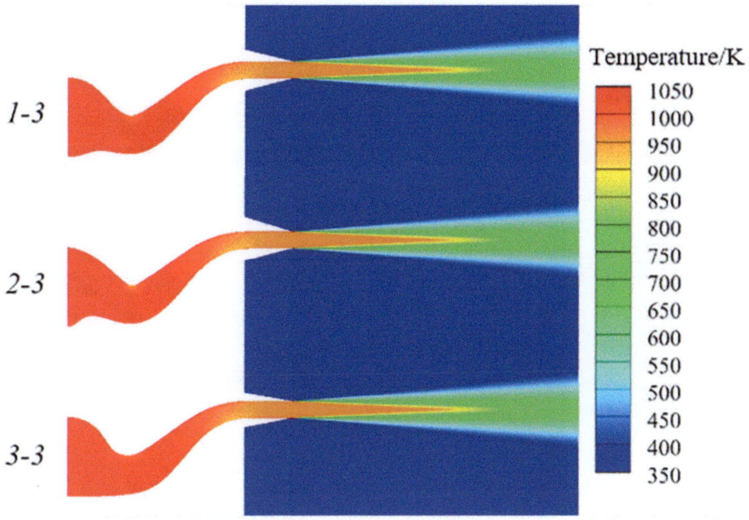

Fig. 3.15 Gas temperature distribution on the symmetry plane of serpentine nozzles with different centerline variation patterns in the first section

Fig. 3.19. Similar to the influence of the centerline variation pattern in the first serpentine section on the infrared radiation characteristics of the serpentine nozzles

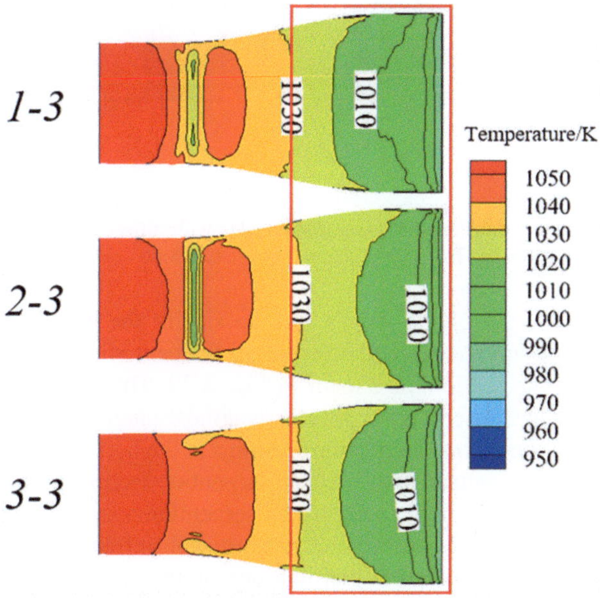

Fig. 3.16 Upper wall temperature distribution of serpentine nozzles with different centerline variation patterns in the first section

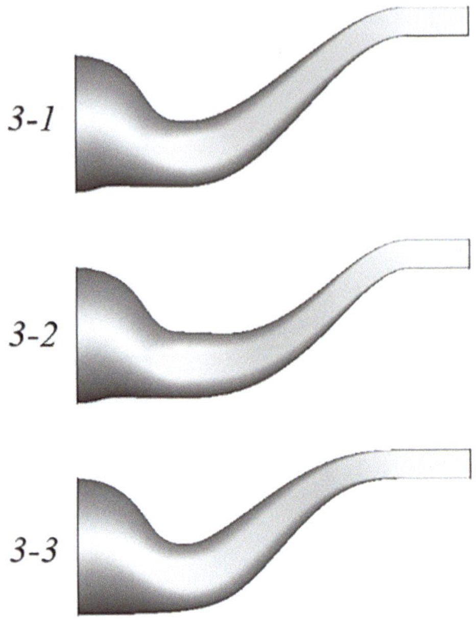

Fig. 3.17 Serpentine nozzle models with different centerline variation patterns in the second section

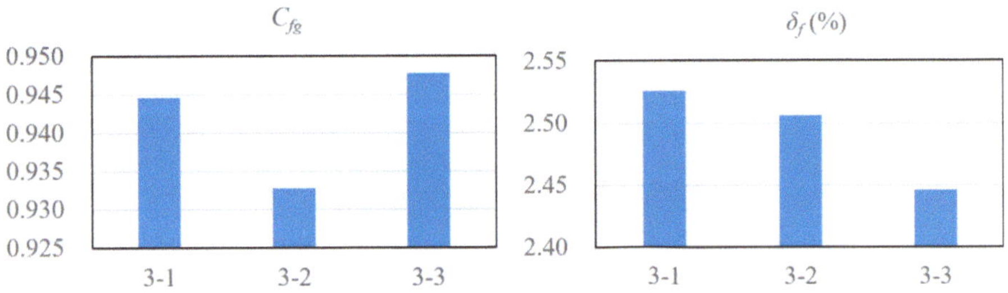

(a) Thrust coefficient (b) Friction loss

Fig. 3.18 Performance parameters of serpentine nozzles with different centerline variation patterns in the second section. (**a**) Thrust coefficient. (**b**) Friction loss

described above, the tail plume length and the size of the high-temperature core zone of each nozzle are basically unaffected by the centerline variation pattern of the second serpentine section, as shown in Fig. 3.20. Therefore, the gas infrared radiation intensity of the serpentine nozzles is almost unaffected by the centerline variation pattern of the second serpentine section. On the horizontal detection plane, the centerline variation pattern of the second serpentine section has almost no effect on the wall infrared radiation intensity of the serpentine nozzle. Below the vertical detection plane, nozzle 3–2 exhibits the greatest wall infrared radiation intensity. Figure 3.21 shows the temperature distributions on the upper wall surfaces of the nozzles. It can be seen that downstream of the serpentine nozzle, the region where

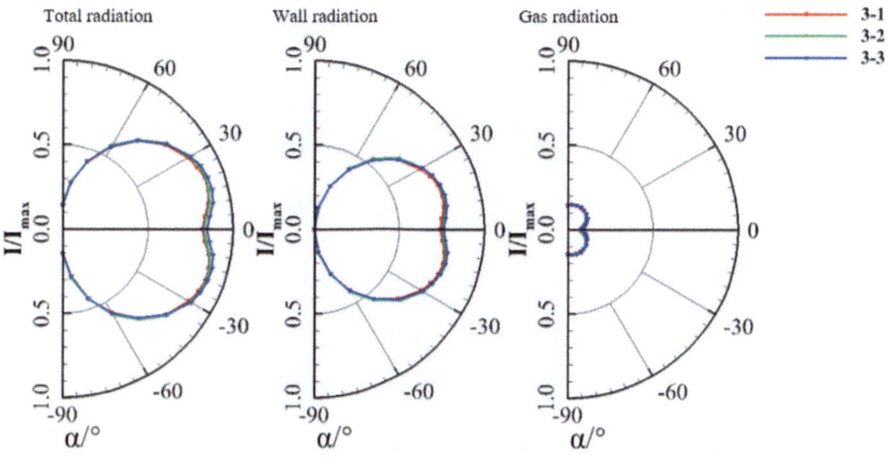

(a) Infrared radiation intensity distribution on the horizontal detection plane

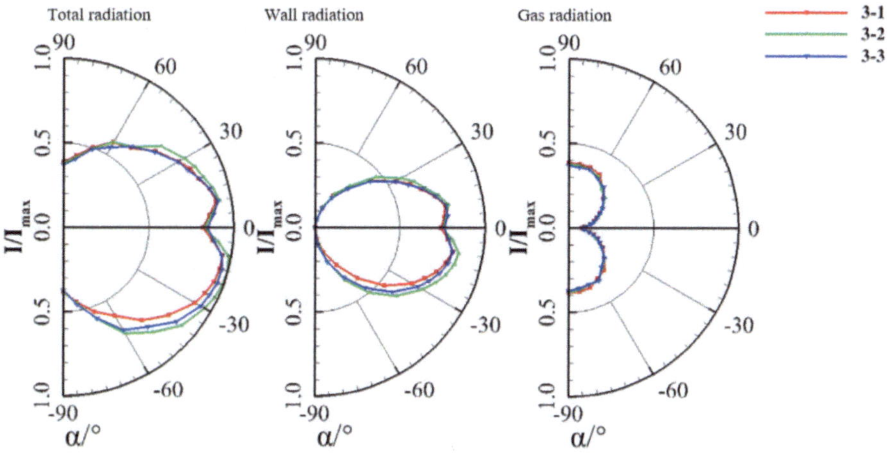

(b) Infrared radiation intensity distribution on the vertical detection plane

Fig. 3.19 Infrared radiation intensity distribution of serpentine nozzles with different centerline variation patterns in the second section. (**a**) Infrared radiation intensity distribution on the horizontal detection plane. (**b**) Infrared radiation intensity distribution on the vertical detection plane

the temperature of nozzle 3–2's upper wall surface exceeds 1020 K is the largest, and thus its corresponding wall radiation is also the greatest.

The variation of the total infrared radiation intensity of serpentine nozzles with different centerline variation patterns in the second serpentine section is basically consistent with the distribution of wall infrared radiation intensity. Below the vertical detection plane, the serpentine nozzle with a "gradual-to-steep" centerline variation pattern in the second serpentine section (3–2) has the greatest total infrared

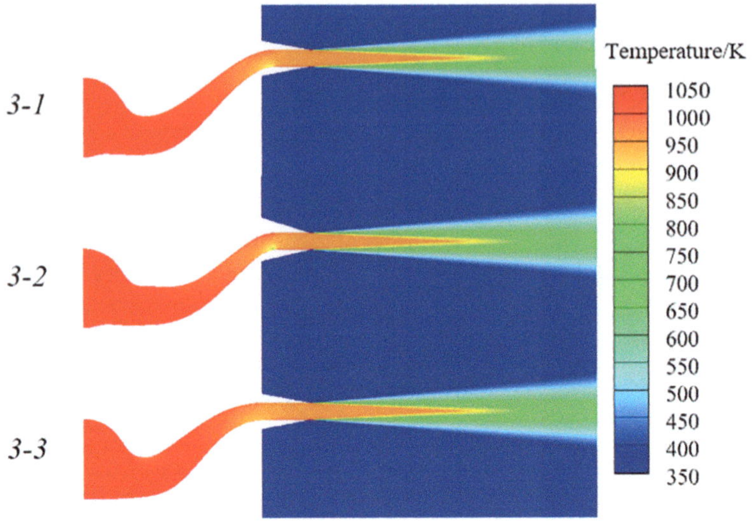

Fig. 3.20 Gas temperature distribution on the symmetry plane of serpentine nozzles with different centerline variation patterns in the second section

radiation intensity. Compared with the serpentine nozzles with a "uniform" centerline variation pattern (3–1) and a "steep-to-gradual" centerline variation pattern (3–3) in the second serpentine section, the total infrared radiation intensity of nozzle 3–2 is on average 5.9% and 2.2% higher, respectively.

Overall, the centerline variation pattern of the first serpentine section has little effect on the aerodynamic performance of the nozzle, while the centerline variation pattern of the second serpentine section has a significant impact on the thrust coefficient. The optimal aerodynamic performance is achieved when the centerline of the second serpentine section adopts the "steep-to-gradual" variation pattern. The gas infrared radiation intensity of serpentine nozzles with different centerline variation patterns is basically the same, but the distribution of wall infrared radiation intensity shows slight differences. The effects of the centerline variation patterns of both serpentine sections on the infrared radiation characteristics of the serpentine nozzle are consistent: when the centerline adopts the "uniform" variation pattern, the infrared radiation intensity is the lowest; when the centerline adopts the "gradual-to-steep" variation pattern, the infrared radiation intensity is the highest.

3.3.3 Influence of Exit Aspect Ratio on Infrared Radiation Characteristics of the Serpentine Nozzle

The variation of the exit aspect ratio of the serpentine nozzle affects the longitudinal offset and lateral curvature distribution of the second serpentine section, leading to different acceleration effects of the airflow at the bends of the serpentine nozzle and

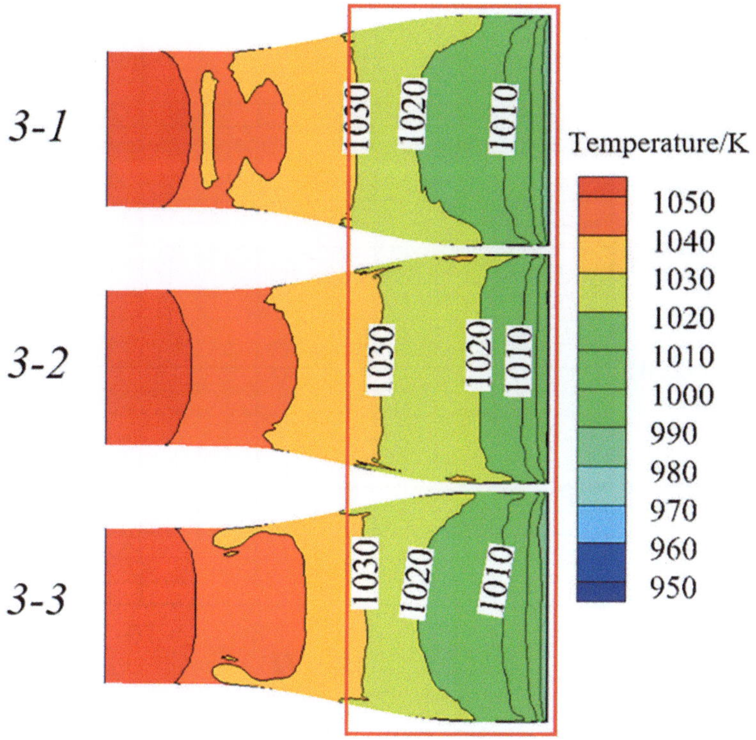

Fig. 3.21 Upper wall temperature distribution of serpentine nozzles with different centerline variation patterns in the second section

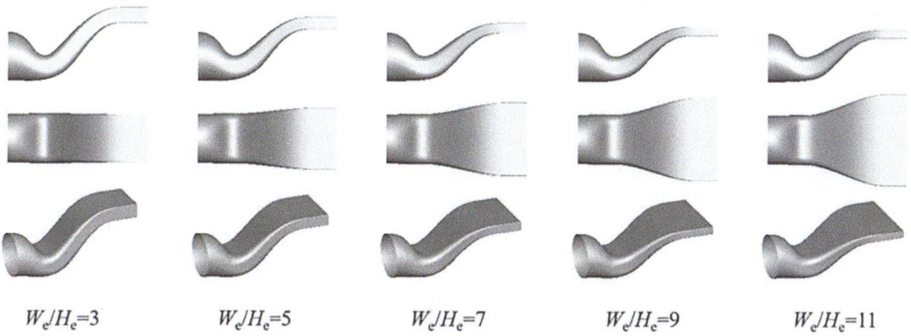

Fig. 3.22 Models of serpentine nozzles with different exit aspect ratios

thus altering the flow field characteristics. At the same time, a larger exit aspect ratio increases the mixing area between the jet plume and the ambient atmosphere, changing the temperature distribution of the jet plume and thereby influencing the infrared radiation characteristics of the serpentine nozzle. This section studies the infrared radiation characteristics of five serpentine nozzles with different exit aspect

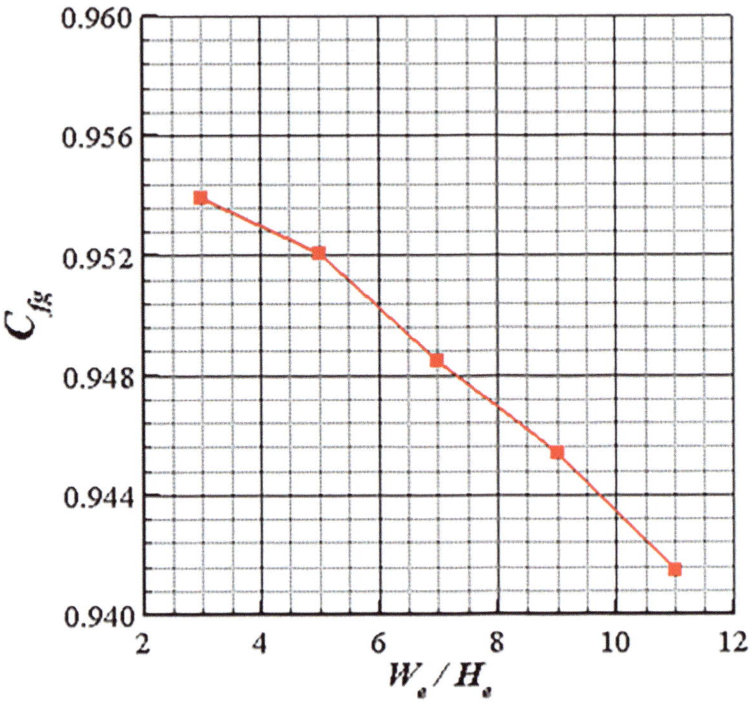

Fig. 3.23 Thrust coefficient of serpentine nozzles with different exit aspect ratios

ratios (W_e/H_e = 3, 5, 7, 9 and 11). Other geometric design parameters of the nozzles remain unchanged, and the geometric models are shown in Fig. 3.22. With the increase of the exit aspect ratio, the longitudinal offset of the second serpentine section gradually decreases, and the curvature radius at both bends also decreases.

Figure 3.23 shows the comparison of the thrust coefficients of each nozzle. It can be seen that with the increase of the exit aspect ratio, the thrust coefficient of the serpentine nozzle gradually decreases. Compared with the serpentine nozzle with an exit aspect ratio of $W_e/H_e = 3$, the thrust coefficient of the serpentine nozzle with an exit aspect ratio of $W_e/H_e = 11$ decreases by 0.012.

The distribution of infrared radiation intensity of serpentine nozzles with different exit aspect ratios is shown in Fig. 3.24. On the horizontal detection plane, the gas infrared radiation intensity of the serpentine nozzle gradually decreases with the increase of the exit aspect ratio. When the exit aspect ratio increases from $W_e/H_e = 3$ to $W_e/H_e = 11$, the average gas infrared radiation intensity on the horizontal detection plane decreases by 31.9%. This is because as the exit aspect ratio increases, the wetted perimeter at the exit becomes larger, increasing the mixing area between the jet plume and the ambient atmosphere, thus enhancing the mixing effect. As a result, the jet plume length decreases, and the length of the jet plume core zone correspondingly shortens, as shown in Fig. 3.25. At the same time, due to the increase in exit aspect ratio, the height of the nozzle exit decreases, leading to a reduction in the

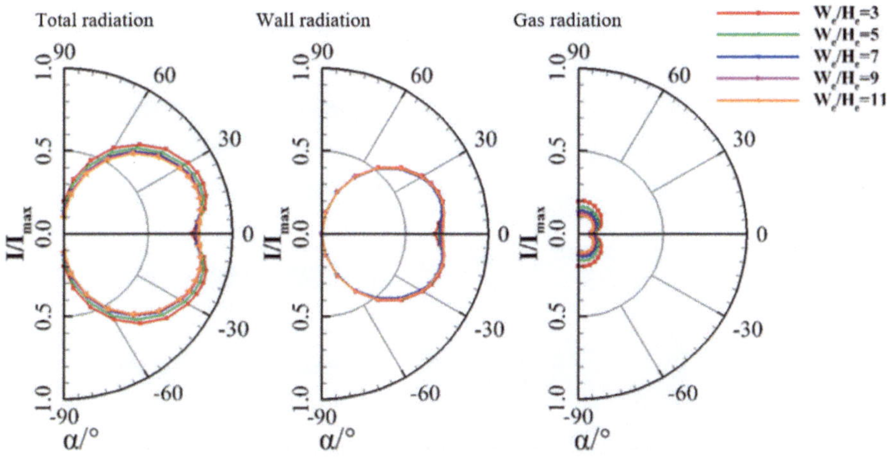

(a) Infrared radiation intensity distribution on the horizontal detection plane

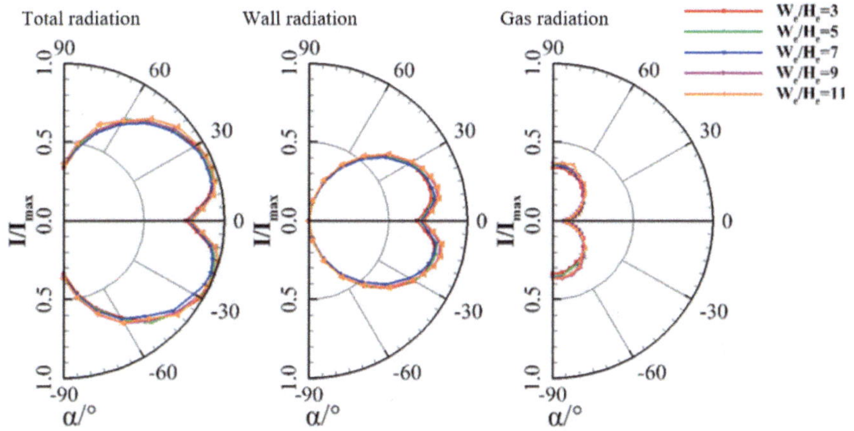

(b) Infrared radiation intensity distribution on the vertical detection plane

Fig. 3.24 Infrared radiation intensity distribution of serpentine nozzles with different exit aspect ratios. (**a**) Infrared radiation intensity distribution on the horizontal detection plane. (**b**) Infrared radiation intensity distribution on the vertical detection plane

height of the high-temperature core zone of the jet plume. Therefore, the high-temperature area detectable by the detector on the horizontal detection plane is correspondingly reduced. Thus, on the horizontal detection plane, the gas infrared radiation intensity gradually decreases with increasing exit aspect ratio. When the detection angle $|\alpha| = 90°$ on the horizontal detection plane, the gas infrared radiation intensity decreases gradually with the increase of the exit aspect ratio, and the decreasing rate gradually slows down, as shown in Fig. 3.26. Compared with the

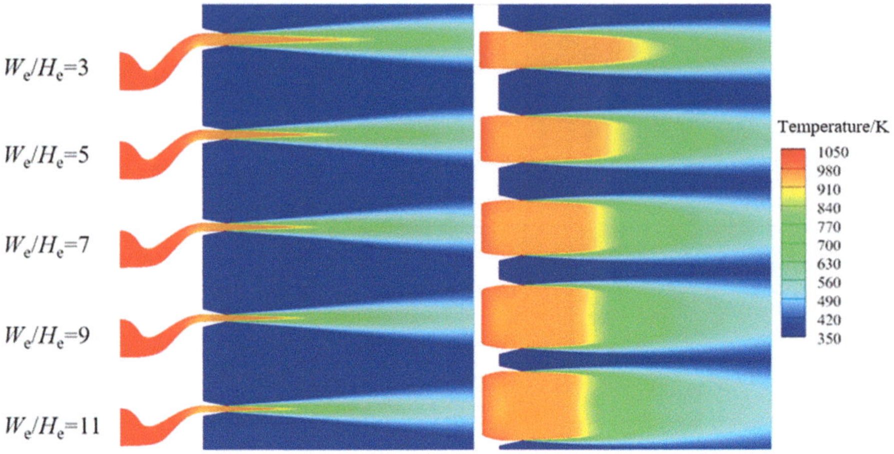

Fig. 3.25 Gas temperature distribution of serpentine nozzles with different exit aspect ratios

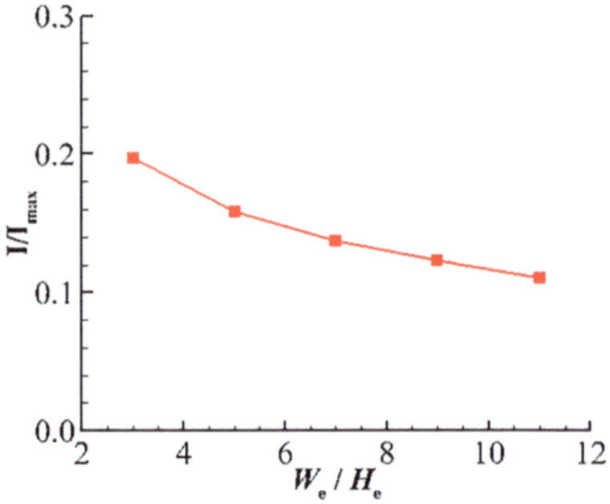

Fig. 3.26 Variation of gas infrared radiation intensity at detection angle |α| = 90° on the horizontal detection plane with exit aspect ratio

serpentine nozzle with $W_e/H_e = 3$, the gas infrared radiation intensity of the serpentine nozzle with $W_e/H_e = 11$ in the |α| = 90° direction decreases by 46.0%. On the vertical detection plane, the gas infrared radiation intensity does not show a similar regularity as on the horizontal detection plane. On one hand, the length of the high-temperature core zone of the jet plume gradually decreases with the increase of the exit aspect ratio, which helps reduce the gas infrared radiation intensity. On the other hand, with the increase of the exit aspect ratio, the width of the jet plume in the horizontal direction also increases, as shown in Fig. 3.25, thereby increasing the high-temperature area of the jet plume detected by the detector on the vertical detection plane and increasing the gas infrared radiation intensity. Under the influence of these two opposing effects, the gas infrared radiation intensity on the vertical detection plane varies among the nozzles, but the variation is not regular. The maximum average difference in gas infrared radiation intensity among the nozzles on the vertical detection plane is 3.1%.

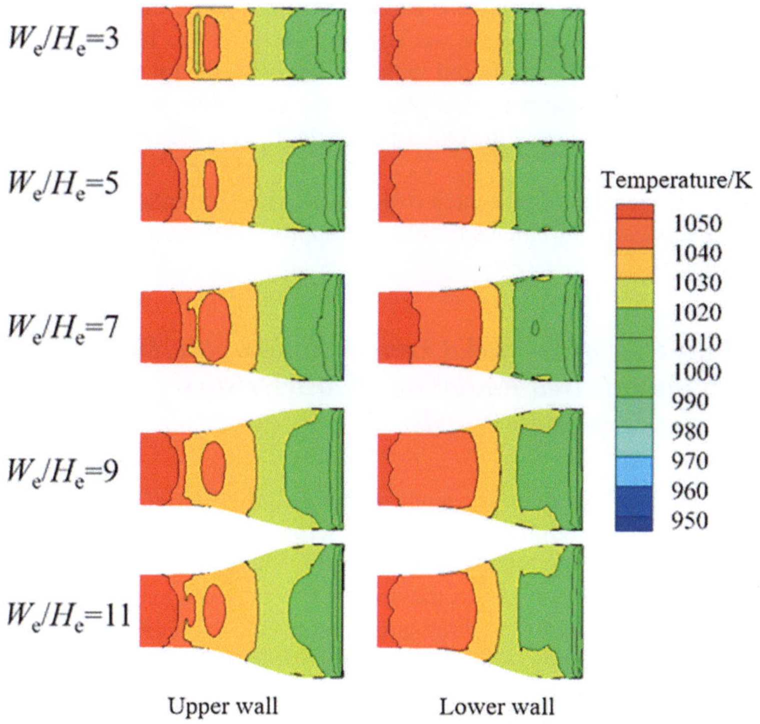

Fig. 3.27 Wall temperature distribution of serpentine nozzles with different exit aspect ratios

On the horizontal detection plane, the effect of the exit aspect ratio on wall infrared radiation intensity is small. The maximum average difference in wall infrared radiation intensity among the nozzles on the horizontal detection plane is 1.5%. On the vertical detection plane, the wall infrared radiation intensity increases with the increase of the exit aspect ratio. When the exit aspect ratio increases from $W_e/H_e = 3$ to $W_e/H_e = 11$, the average wall infrared radiation intensity on the vertical detection plane increases by 2.6%. This is because, as the exit aspect ratio increases, the temperature of the downstream wall surface of the serpentine nozzle shows an overall increasing trend, as shown in Fig. 3.27.

Affected by both gas and wall infrared radiation, the total infrared radiation intensity on the horizontal detection plane shows an overall decreasing trend with the increase of the exit aspect ratio. Compared with the serpentine nozzle with an exit aspect ratio of $W_e/H_e = 3$, the total infrared radiation intensity on the horizontal detection plane of the serpentine nozzle with $W_e/H_e = 11$ decreases by an average of 10.8%. The variation of total infrared radiation intensity on the vertical detection plane is small, with the maximum average difference between nozzles on the vertical detection plane being 3.4%.

Overall, increasing the exit aspect ratio leads to a decrease in the thrust coefficient of the serpentine nozzle. However, a larger exit aspect ratio can shorten the length of the high-temperature core zone of the jet plume, thereby reducing the infrared radiation intensity of the serpentine nozzle on the horizontal detection plane. Influenced by the gas infrared radiation intensity, the total infrared radiation intensity on the horizontal detection plane of the serpentine nozzle also gradually decreases with the increase of the exit aspect ratio. During the design process of the serpentine nozzle, the impact of the exit aspect ratio on both aerodynamic performance and infrared radiation characteristics should be comprehensively considered according to design requirements to select an appropriate exit aspect ratio.

3.3.4 Influence of the Axial Length Ratio of the Two Bends on the Infrared Radiation Characteristics of the Serpentine Nozzle

When the total axial length and straight section length of the serpentine nozzle are fixed, different axial length ratios of the two serpentine sections have a significant impact on the nozzle contour and cause substantial changes in the internal flow field characteristics. This section studies the influence of the axial length ratio of the two serpentine sections on the infrared radiation characteristics of the serpentine nozzle. The geometric configurations of the nozzles are shown in Fig. 3.28, where the axial length ratios L_1/L_2 of the two serpentine sections are 3:2, 2:2, and 2:4, respectively. Other geometric design parameters are consistent with those of the baseline serpentine nozzle. As the axial length ratio decreases, the axial length of the first serpentine section decreases, while that of the second serpentine section increases.

Figure 3.29 compares the thrust coefficients of nozzles with different axial length ratios. It can be seen that the differences in thrust coefficients among the nozzles are small, with a maximum difference less than 0.005. The serpentine nozzle with

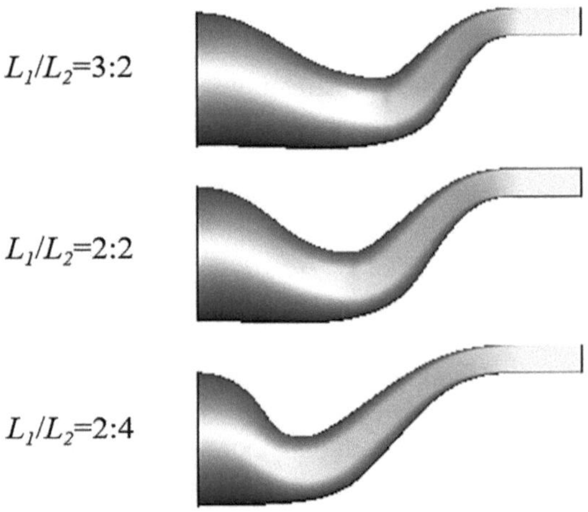

Fig. 3.28 Serpentine nozzle models with different axial length ratios of two bends

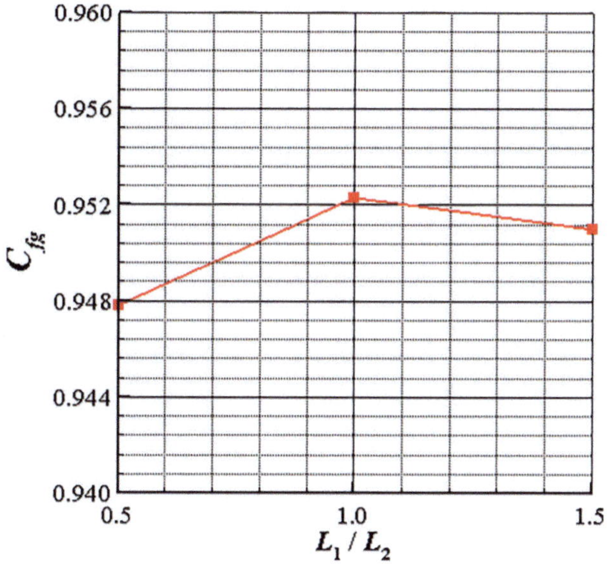

Fig. 3.29 Thrust coefficients of serpentine nozzles with different axial length ratios of two bends

L_1/L_2 = 2:4 has the smallest thrust coefficient, while that with L_1/L_2 = 2: 2 has the largest.

Figure 3.30 shows the infrared radiation intensity distributions of the serpentine nozzles with different axial length ratios. The axial length ratio has little effect on the gas infrared radiation intensity of the serpentine nozzle. The maximum average difference of gas infrared radiation intensity among nozzles on the horizontal and vertical detection planes is 1.5% and 1.6%, respectively. This is because the jet plume lengths of the nozzles with different axial length ratios are basically equal, as shown in Fig. 3.31. Since the gas infrared radiation intensity mainly depends on the size of the high-temperature core region, and the high-temperature core lengths of the jet plumes are basically equal among nozzles, the differences in gas infrared radiation intensity are small.

On the horizontal detection plane, the differences in wall infrared radiation intensity among the nozzles with different axial length ratios are also small, with a maximum average difference of 0.7%, as shown in Fig. 3.30a. On the lower part of the vertical detection plane, compared with the nozzles with axial length ratios of L_1/L_2 = 3:2 and L_1/L_2 = 2:2, the nozzle with L_1/L_2 = 2:4 shows an average increase in wall infrared radiation intensity by 7.6% and 13.8%, respectively, as shown in Fig. 3.30b. This is because the nozzle with L_1/L_2 = 2:4 has the longest axial length in the second serpentine section, resulting in the smoothest downstream contour variation of the serpentine nozzle, causing the internal flow direction changes to be gentlest, the smallest acceleration at the second bend, and the slowest flow velocity, as shown in Fig. 3.32. Therefore, at the second bend, the nozzle with L_1/L_2 = 2:4 has the highest gas temperature. The wall temperature distribution is related to the gas

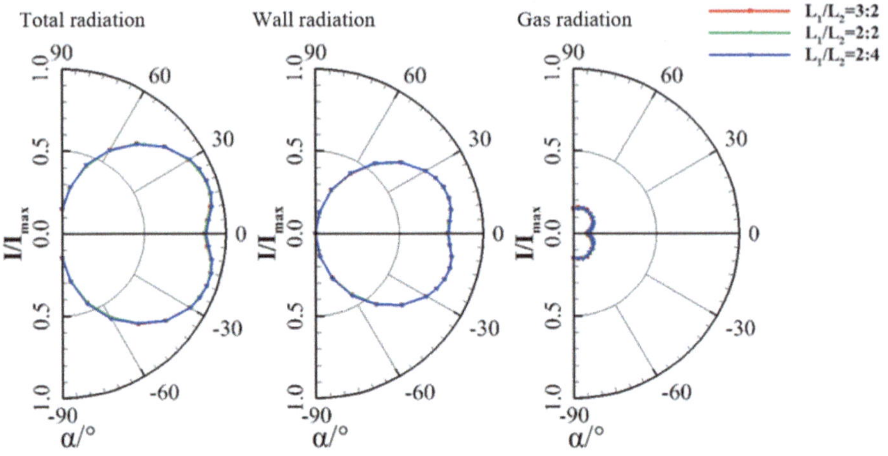

(a) Infrared radiation intensity distribution on the horizontal detection plane

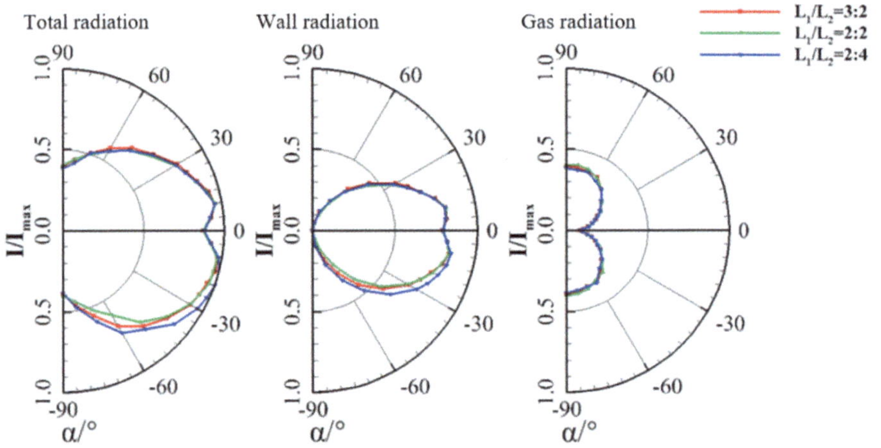

(b) Infrared radiation intensity distribution on the vertical detection plane

Fig. 3.30 Infrared radiation intensity distribution of serpentine nozzles with different axial length ratios of two bends. (**a**) Infrared radiation intensity distribution on the horizontal detection plane. (**b**) Infrared radiation intensity distribution on the vertical detection plane

temperature inside the nozzle; higher gas temperature near the wall leads to higher wall temperature. Figure 3.18 shows the temperature distribution on the upper wall of each nozzle. It can be seen that downstream, the upper wall temperature distribution of the nozzles with L_1/L_2 = 3:2 and L_1/L_2 = 2:2 are similar, while that of L_1/L_2 = 2:4 is higher. Thus, under the vertical detection plane, the wall temperature detected for the L_1/L_2 = 2:4 nozzle is higher than the other two, and correspondingly, the wall infrared radiation intensity is also greater (Fig. 3.33).

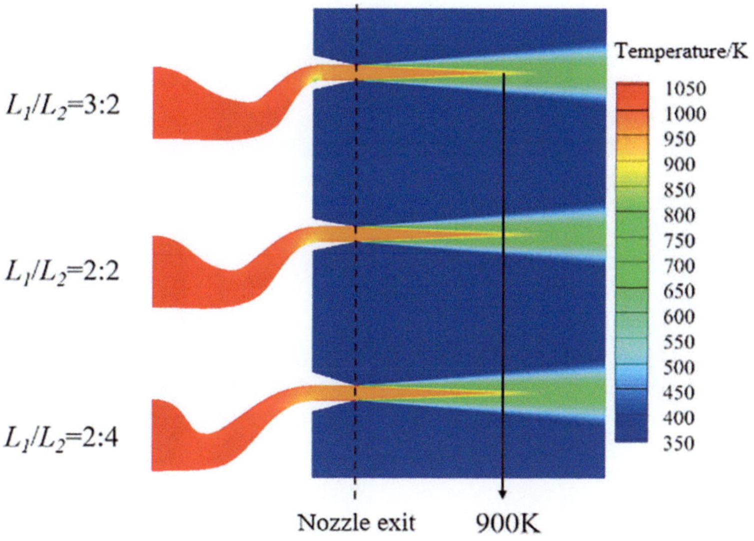

Fig. 3.31 Temperature distribution on the symmetry plane of serpentine nozzles with different axial length ratios of two bends

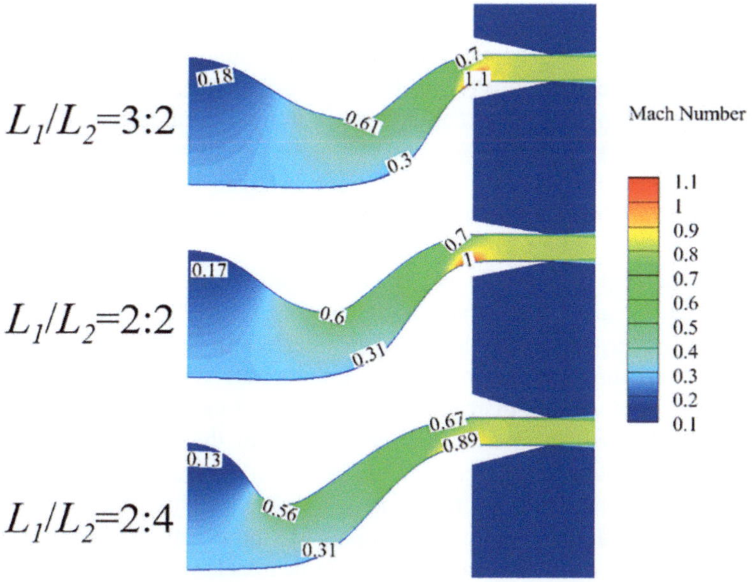

Fig. 3.32 Mach number distribution on the symmetry plane of serpentine nozzles with different axial length ratios of two bends

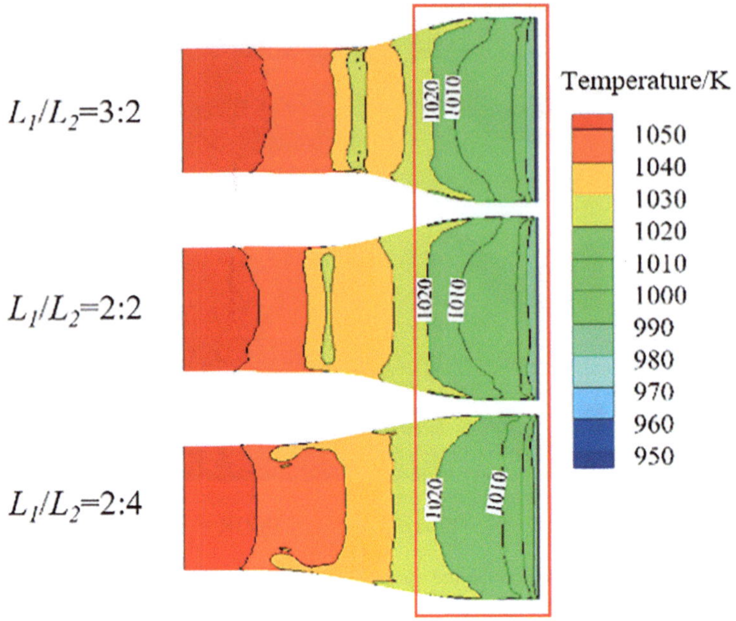

Fig. 3.33 Upper wall temperature distribution of serpentine nozzles with different axial length ratios of two bends

The variation trend of the total infrared radiation intensity of serpentine nozzles with different axial length ratios of two bends is basically consistent with that of the wall infrared radiation intensity. The difference in total infrared radiation intensity on the horizontal detection plane is small, while the variation in total infrared radiation intensity below the vertical detection plane is larger. Below the vertical detection plane, the total infrared radiation intensity of the serpentine nozzles with axial length ratios of $L_1/L_2 = 3{:}2$ and $L_1/L_2 = 2{:}2$ is on average 3.3% and 5.2% lower than that of the serpentine nozzle with an axial length ratio of $L_1/L_2 = 2{:}4$, respectively.

In summary, the axial length ratio of the two bends has little effect on the thrust coefficient of the serpentine nozzle. Since the nozzle wall detectable by the detector behind the serpentine nozzle is mainly the downstream wall section of the serpentine nozzle, the temperature distribution of the downstream wall is the main cause of changes in the wall infrared radiation intensity of the serpentine nozzle. The upper wall temperature downstream of the serpentine nozzle with an axial length ratio of $L_1/L_2 = 2{:}4$ is the highest, resulting in a higher wall infrared radiation intensity below the vertical detection plane compared to those with axial length ratios of $L_1/L_2 = 3{:}2$ and $L_1/L_2 = 2{:}2$. Overall, the serpentine nozzle with an axial length ratio of $L_1/L_2 = 2{:}2$ exhibits the lowest total infrared radiation intensity and the highest thrust coefficient.

3.4 Study on the Influence of Different Inlet Conditions on the Infrared Radiation Characteristics of Serpentine Nozzles

In real aero-engine environments, differences in engine configurations can alter the inlet boundary conditions of serpentine nozzles, thereby significantly affecting the composition and flow characteristics of the internal gas, and consequently changing the infrared radiation characteristics of the serpentine nozzles. Additionally, engine components such as the core/bypass ducts, struts, tail cones, and turbines can induce swirl and generate non-uniform nozzle inlet flows, influencing the mixing of gases inside the serpentine nozzle and altering the temperature and species concentration distributions, which in turn affect the nozzle's infrared radiation characteristics. This section investigates the influence of different inlet conditions on the infrared radiation characteristics of serpentine nozzles through numerical simulations. The study examines the differences in infrared radiation characteristics of serpentine nozzles used in engines with different configurations and the effects of engine swirl and varying aerodynamic parameters at the nozzle inlet.

3.4.1 Influence of Engine Configuration on the Infrared Radiation Characteristics of the Serpentine Nozzles

Turbojet and turbofan engines are two commonly used types of aero-engine configurations. Compared to turbojet engines, turbofan engines typically include an additional bypass duct. The high-temperature gas flowing through the core duct and the low-temperature air flowing through the bypass duct mix within the exhaust system and are finally discharged through the exhaust nozzle. Due to the mixing of low-temperature airflow, the temperature of the exhaust jet decreases significantly. Meanwhile, the low-temperature flow in the bypass duct acts as a thermal barrier between the nozzle wall and the high-temperature core flow, which greatly reduces the nozzle wall temperature. Since the infrared radiation intensity of a serpentine nozzle is primarily determined by the temperatures of the nozzle wall and jet plume, different engine configurations can significantly affect the infrared radiation characteristics of the nozzle. This section compares the infrared radiation characteristics of serpentine nozzles under turbojet and turbofan engine configurations.

In this study, the serpentine nozzle model used for the turbojet engine is the same as the one described in Sect. 3.3, whose infrared radiation characteristics have already been discussed in detail and will not be repeated here. The serpentine nozzle model for the turbofan engine is based on the model used in Sect. 3.3, with additional core and bypass inlet boundaries set at the nozzle inlet, as shown in Fig. 3.34. In addition, the turbofan serpentine nozzle model adopts a shielding design principle at the high-temperature core inlet, where the common tangent MN of the upper

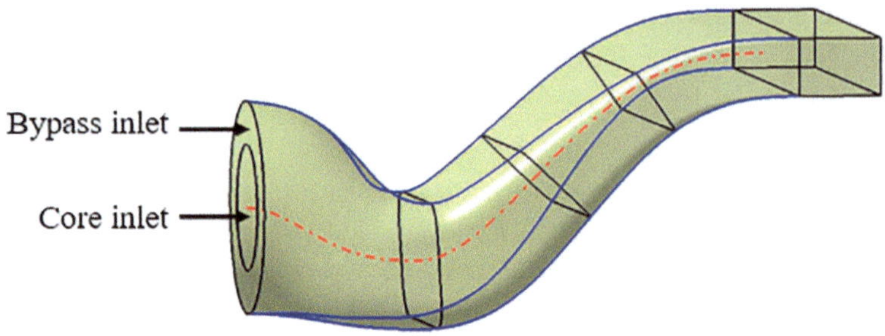

Fig. 3.34 Serpentine nozzle model for turbofan engine

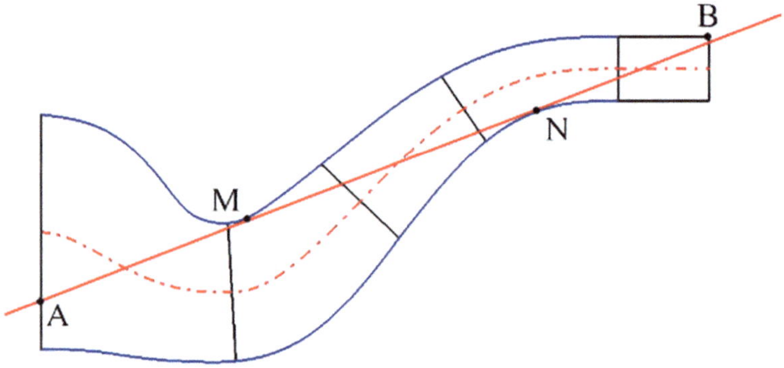

Fig. 3.35 Shielding criterion for serpentine nozzle used in turbofan engine

and lower longitudinal lines passes through the lowest point A of the core inlet, as shown in Fig. 3.35.

Both the core and bypass inlets are defined as pressure inlet boundaries. The total temperature and total pressure of the core inlet are 858.477 K and 1.866 atm, respectively; the total temperature and total pressure of the bypass inlet are 365.914 K and 1.934 atm, respectively. The ambient temperature and pressure are set to 300 K and 1 atm. The mole fractions of H_2O and CO_2 at the core inlet are x (H_2O) = 12.87% and x (CO_2) = 13.12%, respectively. The gas compositions at the bypass inlet and in the far field are consistent with the ambient atmosphere. The wall is modeled as an adiabatic no-slip surface with an emissivity of 0.9. The detector setup is the same as in Sect. 3.3.

The infrared radiation intensity distributions of serpentine nozzles used in the two engine configurations are shown in Fig. 3.36. For ease of comparison, the infrared radiation intensities of the two types of nozzles are normalized. As shown in Fig. 3.36, the infrared radiation intensity distributions of serpentine nozzles used in different engine configurations differ significantly. For the serpentine nozzle used in the turbojet engine, wall infrared radiation contributes more to the total infrared radiation; for the turbofan engine serpentine nozzle, gas infrared radiation is the dominant contributor. Figure 3.37 shows the proportions of wall and gas infrared

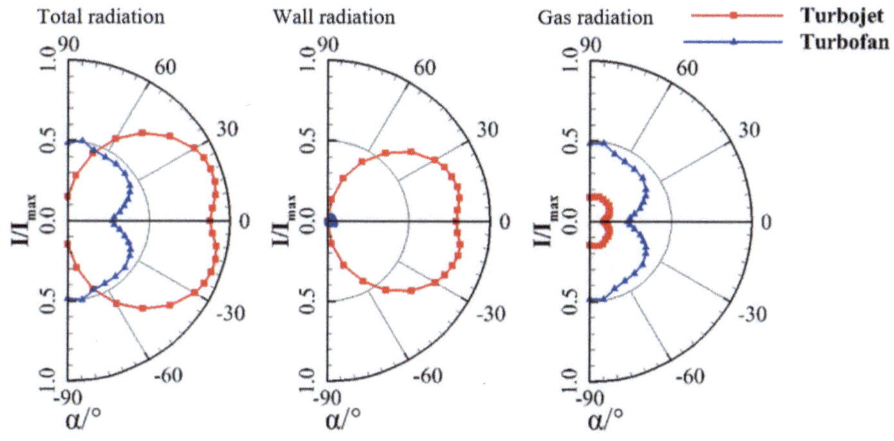

(a) Infrared radiation intensity distribution on the horizontal detection plane

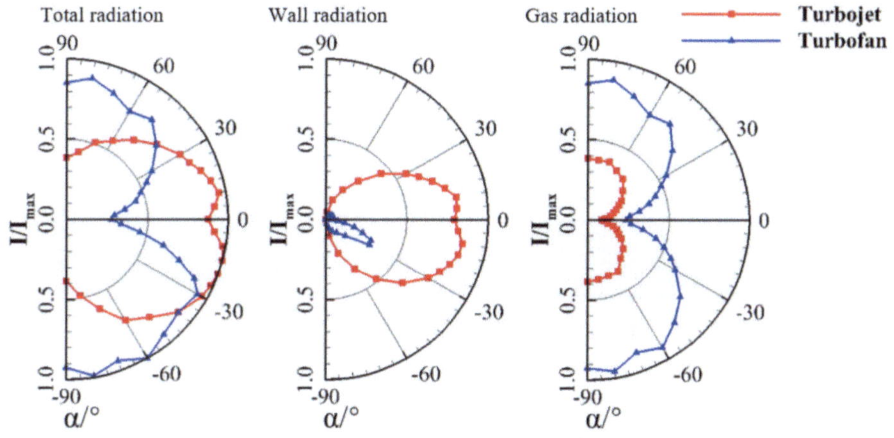

(b) Infrared radiation intensity distribution on the vertical detection plane

Fig. 3.36 Infrared radiation intensity distribution of serpentine nozzles with different engine configurations. (**a**) Infrared radiation intensity distribution on the horizontal detection plane. (**b**) Infrared radiation intensity distribution on the vertical detection plane

radiation in the total infrared radiation for the two types of nozzles. It can be seen that, for the turbojet serpentine nozzle, in most detection angle ranges, the proportion of wall infrared radiation exceeds that of gas infrared radiation. Only at large detection angles $|\alpha|$, where the detectable wall area is small, does gas infrared radiation surpass wall radiation. For the turbofan serpentine nozzle, the proportion of gas infrared radiation is significantly higher than that of wall infrared radiation across both the horizontal and vertical detection planes. On the horizontal detection plane, the gas infrared radiation ratio exceeds 0.85, while the minimum on the vertical detection plane is 0.65. This difference in infrared radiation intensity distributions

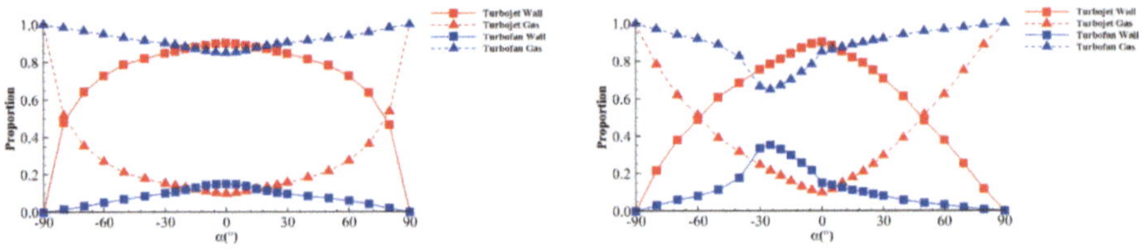

(a) Infrared radiation ratio on the horizontal detection plane (b) Infrared radiation ratio on the vertical detection plane

Fig. 3.37 Composition ratio of total infrared radiation for serpentine nozzles of different engine configurations. (**a**) Infrared radiation ratio on the horizontal detection plane. (**b**) Infrared radiation ratio on the vertical detection plane

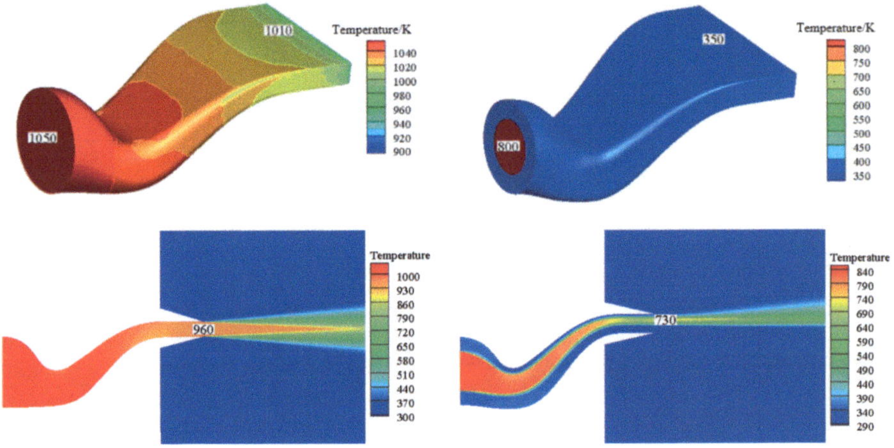

(a) Serpentine nozzle for turbojet engine (b) Serpentine nozzle for turbofan engine

Fig. 3.38 Temperature distribution on the wall and symmetry plane of serpentine nozzles with different engine configurations. (**a**) Serpentine nozzle for turbojet engine. (**b**) Serpentine nozzle for turbofan engine

of serpentine nozzles used in different engine configurations is mainly caused by the temperature difference between the nozzle wall and the core flow. Figure 3.38 presents the temperature distributions on the wall and symmetry plane for the two nozzle types. For the turbojet serpentine nozzle, the downstream wall temperature is higher than the core flow temperature at the nozzle exit, so when the detectable wall area is large, the wall radiation dominates. For the turbofan serpentine nozzle, the wall temperature is only about 350 K, while the core flow temperature at the exit reaches around 730 K, which is 380 K higher than the downstream wall temperature. Hence, the gas radiation dominates over wall radiation.

For the serpentine nozzle used in the turbofan engine, the wall infrared radiation intensity within the negative detection angle range on the vertical detection plane is greater than that in the positive angle range, as shown in Fig. 3.36b. This is mainly due to the difference in effective radiation between the upper and lower walls. Figure 3.39 shows the regions inside the nozzle that are visible to points A and B on

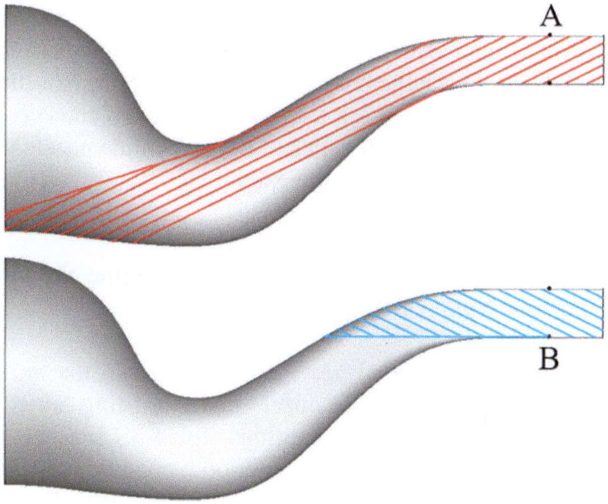

Fig. 3.39 Schematic diagram of incident radiation regions on the inner wall of the serpentine nozzle

the wall near the exit—i.e. regions from which radiation can directly reach points A and B. Points A and B are located on the upper and lower walls respectively, with the same axial position. It can be seen that most of the inner nozzle region (shaded in the top diagram) can directly radiate to point A on the upper wall, and part of the high-temperature core flow is also located within this region. For point B on the lower wall, only a small region near the nozzle exit (shaded in the bottom diagram) can radiate directly to it. Therefore, the incident radiative energy on the upper wall is greater than that on the lower wall, resulting in higher effective radiation on the upper wall. Consequently, the wall infrared radiation intensity is greater in the negative detection angle range on the vertical detection plane than in the positive range.

In summary, there are significant differences in the infrared radiation characteristics of serpentine nozzles used in different engine configurations. For the serpentine nozzle used in the turbojet engine, the total infrared radiation intensity is more influenced by the wall infrared radiation intensity. For the serpentine nozzle used in the turbofan engine, the wall temperature is much lower than the core flow temperature, making the total infrared radiation intensity more influenced by the gas infrared radiation intensity.

3.4.2 Influence of Inlet Swirl on the Infrared Radiation Characteristics of the Serpentine Nozzles

Inlet swirl alters the flow field characteristics inside the nozzle and the temperature distribution of the flow at the nozzle exit, and thus inevitably affects the infrared radiation characteristics of the serpentine nozzle. This section focuses on investigating the impact of inlet swirl on the infrared radiation characteristics of the serpentine nozzle. The inlet swirl angle is defined as the angle between the tangential component of the airflow at the nozzle inlet and the engine axis direction. In this study, inlet swirl is realized by setting the flow direction of the core airflow at the inlet. The infrared radiation characteristics of the serpentine nozzle under different swirl angles (0°, 5°, 10°, 15°, and 20°) are examined.

3.4.2.1 Infrared Radiation Characteristics of Serpentine Nozzle Under Swirling and Non-swirling Conditions

This subsection compares and analyzes the differences in the infrared radiation intensity distribution of the serpentine nozzle under swirling (swirl angle of 10°) and non-swirling (swirl angle of 0°) conditions. Table 3.2 presents the thrust and thrust coefficient of the serpentine nozzle under both conditions. It can be seen that the thrust differences between the two conditions are minimal. Compared with the non-swirling condition, the serpentine nozzle with swirl has a thrust decrease of only 0.72%, and the thrust coefficient differs by only 0.007.

The wall temperature distribution of the serpentine nozzle under swirling and non-swirling conditions is shown in Fig. 3.40. It can be observed that under both conditions, the temperature distributions on the upper and side walls of the serpentine nozzle are quite similar. However, the temperature distribution on the lower wall differs significantly. Under the swirling condition, a local hot spot with relatively high temperature appears downstream on the lower wall of the serpentine nozzle. This is mainly caused by the effect of inlet swirl on the internal flow characteristics of the nozzle. Figure 3.41 shows the gas temperature distributions at different cross-sections of the serpentine nozzle under swirling and non-swirling conditions. It can be seen that the temperature distribution of the gas in the downstream part of the serpentine nozzle differs significantly under the two conditions. Under the swirling condition, starting from cross-section C, the low-temperature bypass airflow separates the high-temperature core airflow into two parts. Meanwhile, the high-temperature core airflow moves downward and contacts the lower wall, as shown in Fig. 3.40, resulting in a hot spot on the lower wall.

To better understand the reasons behind the differences in gas temperature distribution inside the serpentine nozzle under swirling and non-swirling conditions, Fig. 3.42 shows the distribution of vorticity in the x-direction at different cross-sections of the serpentine nozzle under these two conditions. It can be seen that under the swirling condition, a strong positive vortex is formed slightly below the center of cross-section A. At cross-section C, influenced by the central vortex, the

Table 3.2 Comparison of thrust and thrust coefficient under swirling and non-swirling conditions

Swirl condition	Thrust/kN	Thrust coefficient
Non-swirling	118.623	0.955
Swirling	117.774	0.948

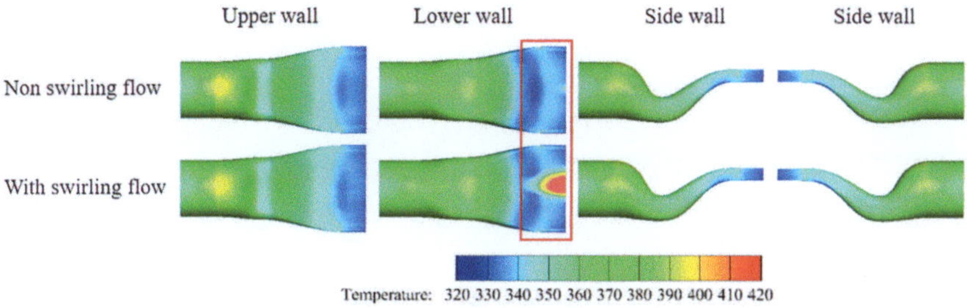

Fig. 3.40 Wall temperature distribution under swirling and non-swirling conditions

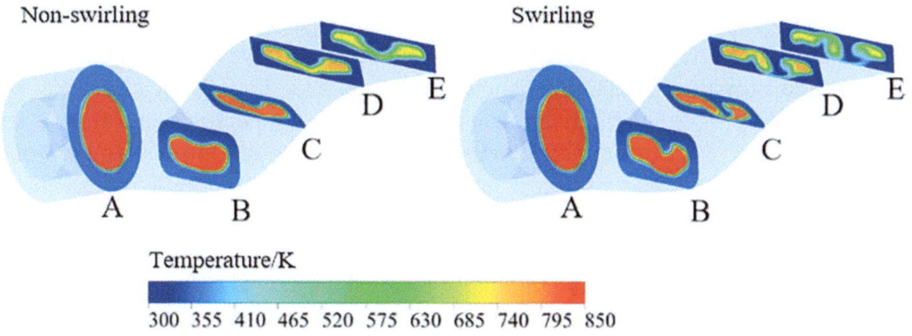

Fig. 3.41 Gas temperature distribution under swirling and non-swirling conditions

right-side airflow is drawn into the core flow, resulting in an increase in negative vorticity on the right side and a decrease in positive vorticity on the left side. At cross-section D, the negative vorticity on the right side further increases, forming a negative vortex that, together with the central positive vortex, creates a pair of counter-rotating vortices. This vortex pair enhances the mixing between the core and bypass streams and penetrates the core flow at cross-section D, dragging the high-temperature gas downward and causing it to impinge on the lower wall, as shown in Fig. 3.41. Due to the presence of swirl, both the wall temperature distribution and the gas temperature distribution inside the serpentine nozzle are altered, leading to differences in the infrared radiation intensity distribution under swirling and non-swirling conditions.

Figure 3.43 shows the infrared radiation intensity distribution of the serpentine nozzle under swirling and non-swirling conditions. Compared with the non-swirling case, the average total infrared radiation intensity of the serpentine nozzle decreases by 4.7% on the horizontal detection plane and by 14.2% on the vertical detection plane under swirling conditions. Under both swirl conditions, the wall infrared radiation intensity on the horizontal detection plane is significantly lower than the gas infrared radiation intensity. Therefore, on the horizontal detection plane, the

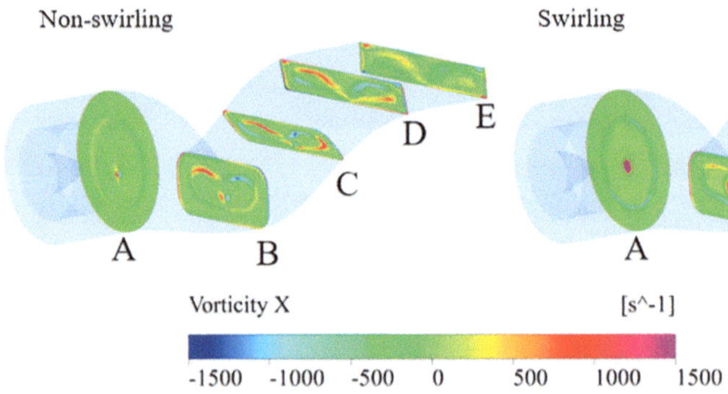

Fig. 3.42 Vorticity diagrams under swirling and non-swirling conditions

variation trend of the total infrared radiation intensity is consistent with that of the gas infrared radiation intensity, as shown in Fig. 3.43a. This is because the low-temperature bypass flow effectively insulates the high-temperature core flow from the nozzle wall, resulting in wall temperatures that are much lower than the temperature of the high-temperature core gas. Moreover, due to the adoption of a design principle that shields high-temperature components in the serpentine nozzle, only the low-temperature nozzle wall is visible at the detection point. As a result, the wall infrared radiation intensity is significantly lower than that of the gas. Figure 3.44 shows the proportion of wall and gas infrared radiation in the total infrared radiation on the horizontal detection plane under the two swirl conditions. It can be seen that throughout the detection range on the horizontal plane, the proportion of gas infrared radiation in the total radiation exceeds 0.84 under both conditions. Therefore, on the horizontal detection plane, the total infrared radiation of the serpentine nozzle is dominated by gas infrared radiation.

Compared with the infrared radiation intensity of the serpentine nozzle under non-swirling conditions, the gas-phase infrared radiation intensity under swirling conditions decreases by 5.5% on the horizontal detection plane and by 18.0% on the vertical detection plane, as shown in Fig. 3.43. This reduction is attributed to enhanced mixing between the core and bypass streams induced by swirl, which effectively reduces the size of the high-temperature gas core. Figure 3.45 illustrates the isosurfaces of gas temperature at 700 K under both swirling and non-swirling conditions. From the side and top views, it is evident that the length of the high-temperature core region is shorter under swirling conditions. As shown in Fig. 3.41, low-temperature bypass air is entrained into the high-temperature core flow under swirling conditions. Consequently, the temperature of the core flow in the middle region drops, and the volume of the high-temperature core is reduced, as indicated by the top and rear views in Fig. 3.45. Since gas-phase infrared radiation intensity is primarily determined by the size of the high-temperature core, the infrared radiation intensity is lower under swirling conditions.

To better understand the mechanism underlying the wall infrared radiation intensity distribution, the wall infrared radiation can be decomposed into two

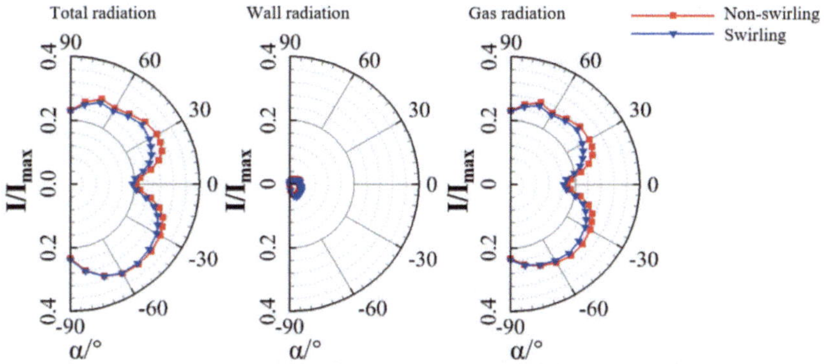

(a) Infrared radiation intensity distribution on the horizontal detection plane

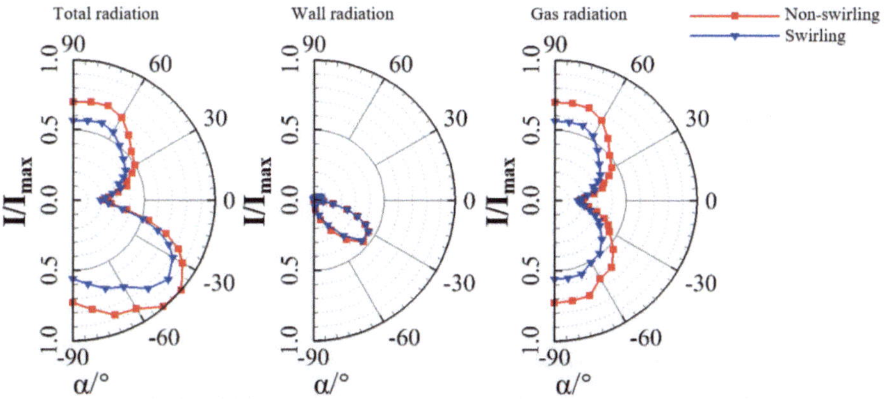

(b) Infrared radiation intensity distribution on the vertical detection plane

Fig. 3.43 Infrared radiation intensity distribution of serpentine nozzles under swirling and non-swirling conditions. (**a**) Infrared radiation intensity distribution on the horizontal detection plane. (**b**) Infrared radiation intensity distribution on the vertical detection plane

components: self-emission and reflected radiation. The self-emission originates from the wall itself, while the reflected radiation results from the reflection of all incident infrared radiation onto the wall. Figure 3.46 presents the variation of wall self-emission and reflected radiation under both swirling and non-swirling conditions. On the horizontal detection plane, the wall self-emission remains nearly unchanged under both conditions, as the sidewall temperature distributions are nearly identical (see Fig. 3.40). Thus, changes in the wall radiation intensity on the horizontal detection plane are mainly caused by variations in reflected radiation, as shown in Fig. 3.46a. The swirl alters the internal flow field of the serpentine nozzle, resulting in changes in both gas and wall temperatures, which in turn lead to changes in the intensity of incident radiation on the sidewalls and, consequently, the reflected radiation.

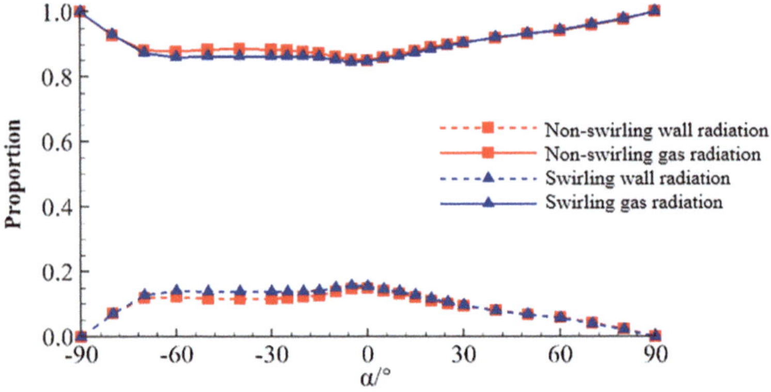

Fig. 3.44 Proportion of wall and gas radiation on the horizontal detection plane under swirling and non-swirling conditions

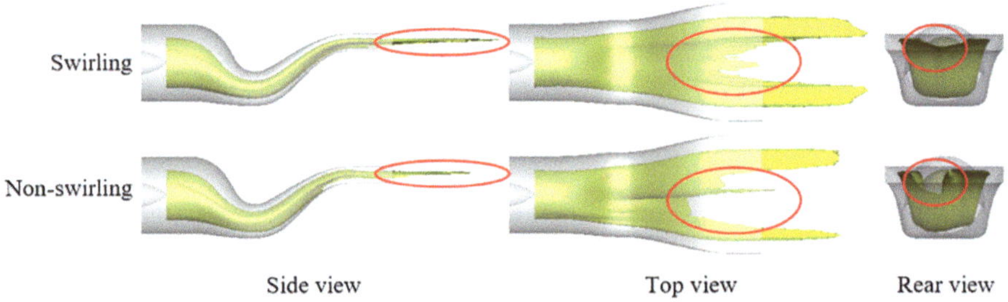

Fig. 3.45 Isosurface of gas temperature at 700 K under swirling and non-swirling conditions

The variation in total infrared radiation intensity of the serpentine nozzle depends on the changes in both wall and gas-phase radiation. Figure 3.47 shows the relative changes in infrared radiation intensity under swirling conditions compared to non-swirling conditions. It is evident that the relative change in gas-phase radiation is significantly greater than that in wall radiation. On the horizontal and vertical detection planes, the average relative changes in gas-phase infrared radiation intensity are 7.0 and 8.3 times greater than those of the wall, respectively. This is because the infrared radiation of serpentine nozzles used in turbofan engines is primarily dominated by gas-phase radiation. Therefore, compared to wall radiation, the variation in gas-phase radiation plays a more critical role. Across the entire horizontal detection plane and in the positive detection angle range of the vertical plane, the change in total infrared radiation intensity is mainly due to the change in gas-phase radiation intensity, as shown in Fig. 3.47. In the negative detection angle range on the vertical plane, the change in total infrared radiation intensity is jointly determined by both gas-phase and wall radiation variations.

Figure 3.48 presents the contour maps of total infrared radiation intensity in the rear hemisphere space under swirling and non-swirling conditions. It can be observed that the lower hemisphere exhibits stronger infrared radiation due to higher wall-reflected radiation at downward viewing angles. Additionally, a

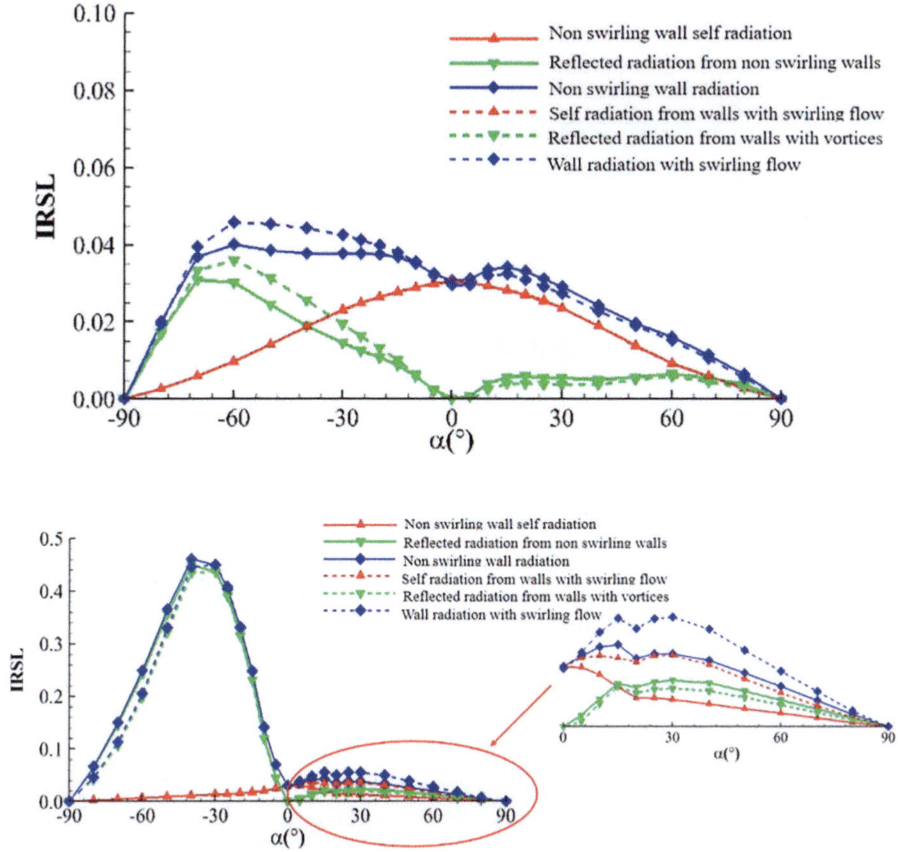

Fig. 3.46 Composition of wall radiation under swirling and non-swirling conditions. (**a**) Composition of wall radiation on the horizontal detection plane. (**b**) Composition of wall radiation on the vertical detection plane

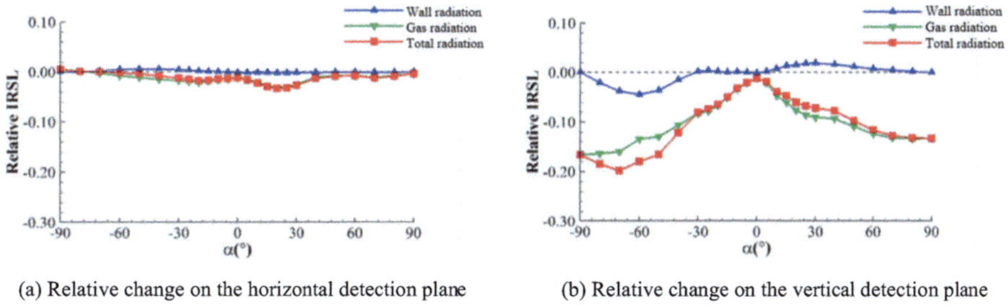

(a) Relative change on the horizontal detection plane

(b) Relative change on the vertical detection plane

Fig. 3.47 Relative changes in infrared radiation intensity under swirling conditions. (**a**) Relative change on the horizontal detection plane. (**b**) Relative change on the vertical detection plane

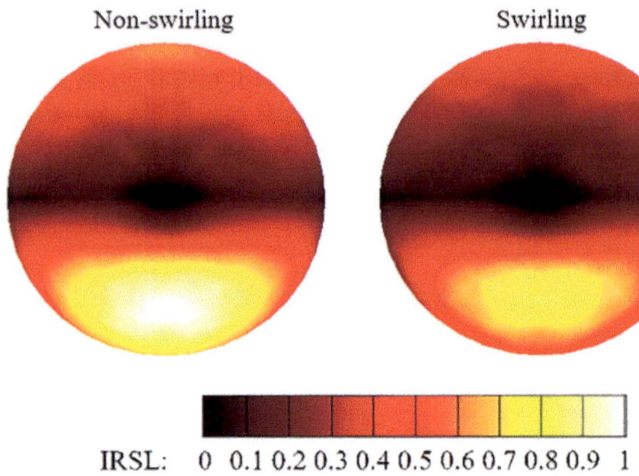

Fig. 3.48 Infrared radiation contour map of the serpentine nozzle rear hemisphere space under swirling and non-swirling conditions

localized high-radiation region appears in the lower hemisphere due to changes in wall infrared radiation intensity. Compared with the non-swirling condition, the infrared radiation intensity across the entire rear hemisphere is reduced under swirling conditions due to the diminished high-temperature core region caused by enhanced core-bypass mixing.

In conclusion, inlet swirl effectively suppresses the infrared radiation intensity of the serpentine nozzle with minimal thrust loss. Compared to the non-swirling case, the thrust under swirling conditions decreases by only 0.72%, while the infrared radiation intensity on the horizontal and vertical detection planes decreases by 4.7% and 14.2%, respectively. On the vertical detection plane, higher wall-reflected radiation in the negative detection angle range results in stronger wall radiation compared to the positive angle range. Under swirling conditions, a thermal hot spot appears downstream on the lower wall, causing a significant increase in wall infrared radiation in the positive angle range of the vertical plane. However, since gas-phase radiation dominates the total radiation and is much higher than wall radiation in this region, the effect of the hot spot on total infrared radiation is limited. Overall, the impact of inlet swirl on the infrared characteristics of the serpentine nozzle is primarily reflected in the suppression of gas-phase infrared radiation.

3.4.2.2 Influence of Swirl Angle on the Infrared Radiation Characteristics of the Serpentine Nozzle

This subsection compares and analyzes the differences in the infrared radiation intensity distribution of the serpentine nozzle under different swirl angles (5°, 10°, 15° and 20°). Table 3.3 presents the thrust and thrust coefficient of the serpentine nozzle under different swirl angle conditions. It can be seen that as the swirl angle increases from 5° to 20°, the thrust of the serpentine nozzle gradually decreases, and the thrust coefficient decreases from 0.953 to 0.936. Compared with the non-swirling condition, when the swirl angle is less than 10°, the thrust loss is less than 1%; when the swirl angle is 20°, the thrust loss is the largest, reaching 1.98%.

Table 3.3 Comparison of thrust and thrust coefficient under different swirl angles

Swirl angle/°	Thrust/kN	Thrust coefficient	Thrust loss (%)
5	118.340	0.953	0.24
10	117.774	0.948	0.72
15	117.063	0.942	1.32
20	116.274	0.936	1.98

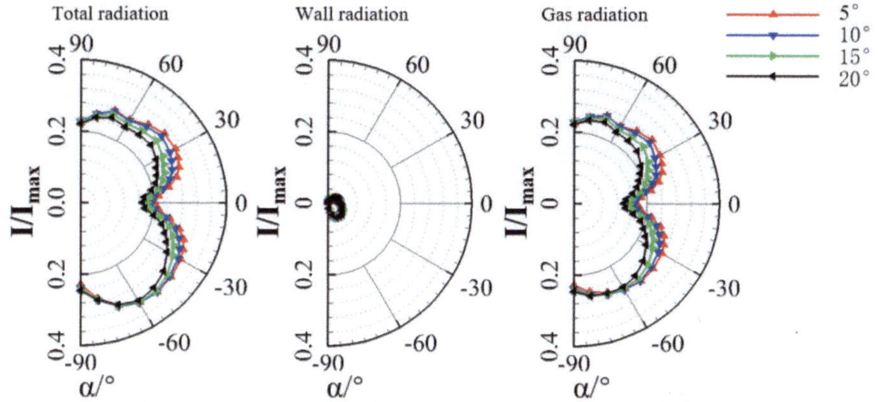

(a) Infrared radiation intensity distribution on the horizontal detection plane

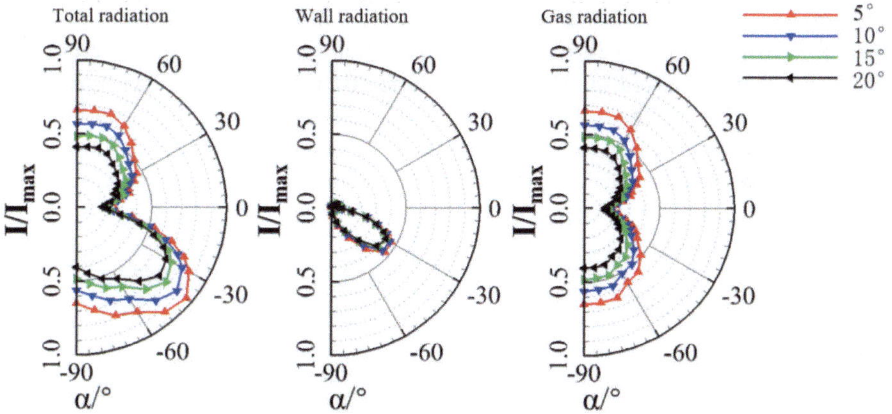

(b) Infrared radiation intensity distribution on the vertical detection plane

Fig. 3.49 Infrared radiation intensity distribution of serpentine nozzles under different swirl angles. (**a**) Infrared radiation intensity distribution on the horizontal detection plane. (**b**) Infrared radiation intensity distribution on the vertical detection plane

The infrared radiation intensity distribution of the serpentine nozzle under different swirl angles is shown in Fig. 3.49. It can be seen that the variation trend of the infrared radiation intensity of the serpentine nozzle under different swirl angles is the same. Over the entire horizontal detection plane and in the positive detection angle range on the vertical detection plane, the total infrared radiation intensity is mainly dominated by gas infrared radiation intensity. Meanwhile, on the vertical detection plane, the wall infrared radiation intensity at negative detection angles is significantly higher than that at positive detection angles. From Fig. 3.49, it can be seen that the larger the swirl angle, the more significant the suppression of the infrared radiation intensity of the serpentine nozzle. Compared with the serpentine nozzle under non-swirling conditions, as the swirl angle increases from 5° to 20°, the average reduction in total infrared radiation intensity on the horizontal detection plane increases from 2.3 to 13.8%. The average reduction in total infrared radiation intensity on the vertical detection plane increases from 5.8 to 31.2%.

As the swirl angle increases, the mixing between the high-temperature core flow and the low-temperature bypass flow is enhanced, resulting in a reduction in the high-temperature core zone of the gas, as shown in Fig. 3.50. Therefore, with the increase of the swirl angle, the gas infrared radiation intensity on the horizontal detection plane and the vertical detection plane decreases, as shown in Fig. 3.49. Compared with the infrared radiation intensity of the serpentine nozzle under non-swirling conditions, when the swirl angle increases from 5° to 20°, the average reduction in gas infrared radiation intensity on the horizontal detection plane increases from 2.9 to 16.4%. The average reduction in gas infrared radiation intensity on the vertical detection plane increases from 7.5 to 37.8%.

The variations in wall self-emission and wall reflection radiation of the serpentine nozzle under different swirl angles are shown in Fig. 3.51. It can be seen that on the horizontal detection plane, the differences in wall self-emission under different swirl angles are minimal. This is because the swirl angle has almost no effect on the temperature distribution of the serpentine nozzle sidewall, as shown in Fig. 3.52. Therefore, the differences in wall infrared radiation intensity under different swirl angles are mainly caused by wall reflection radiation, as shown in Fig. 3.51a. The change in swirl angle leads to variations in the mixing intensity between the core and bypass flows, resulting in changes in the temperature distribution of the gas and the upper and lower walls. Accordingly, the radiation intensity incident on the

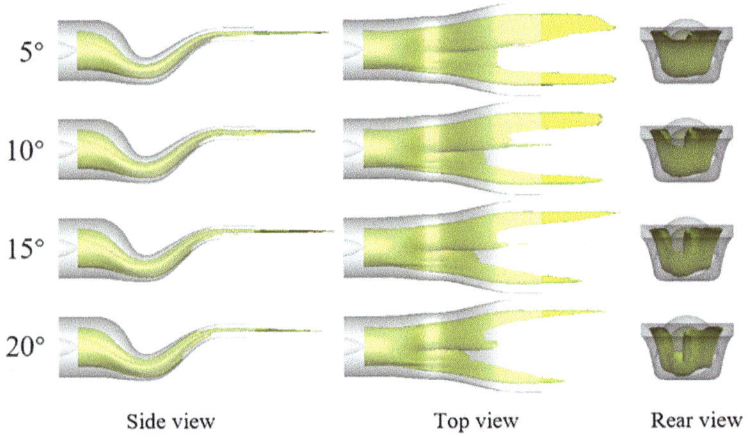

Fig. 3.50 Isosurface of gas temperature at 700 K under different swirl angles

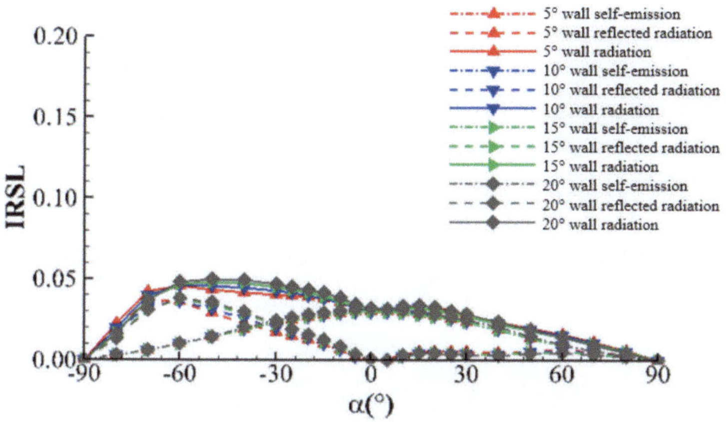

(a) Composition of wall radiation on the horizontal detection plane

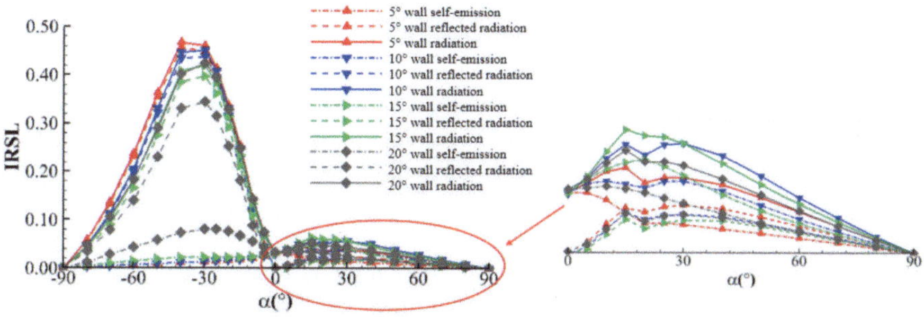

(b) Composition of wall radiation on the vertical detection plane

Fig. 3.51 Composition of wall radiation of serpentine nozzles under different swirl angles. (**a**) Composition of wall radiation on the horizontal detection plane. (**b**) Composition of wall radiation on the vertical detection plane

sidewall also changes. Since the gas radiation incident on the sidewall is relatively weak, the differences in wall reflection radiation on the horizontal detection plane are also small, as shown in Fig. 3.51a.

On the vertical detection plane, the wall self-emission at negative detection angles increases with the increase of swirl angle, as shown in Fig. 3.51b. This is due to the enhancement of the hot spot at the downstream section of the upper wall of the serpentine nozzle as the swirl angle increases, as shown in Fig. 3.52. At negative detection angles, wall reflection radiation gradually decreases with increasing swirl angle. This is because the mixing between core and bypass flows becomes stronger as the swirl angle increases, and the high-temperature core region of the gas gradually shrinks, as shown in Fig. 3.50. Since the wall infrared radiation at negative detection angles mainly comes from wall reflection radiation, the wall infrared radiation intensity also gradually decreases with increasing swirl angle. In the positive

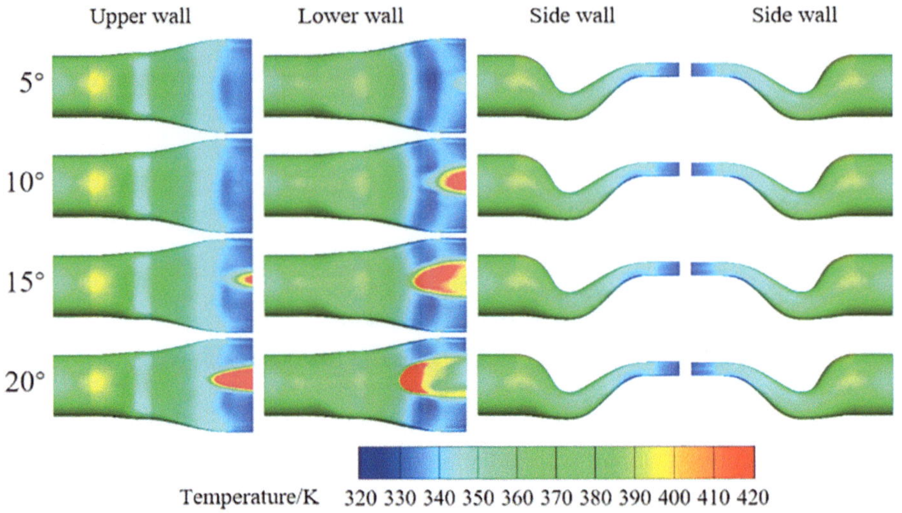

Fig. 3.52 Wall temperature distribution of serpentine nozzles under different swirl angles

detection angle range, when α > 30°, the wall self-emission intensity increases as the swirl angle increases from 5° to 10°. When the swirl angle continues to increase, the wall self-emission intensity begins to decrease. This is due to the changes in the intensity and position of the hot spot on the downstream section of the lower wall of the serpentine nozzle. Compared with the condition at a 5° swirl angle, the hot spot on the lower wall of the nozzle under a 10° swirl angle is stronger and larger in area, as shown in Fig. 3.52. As a result, the temperature of the wall (near the nozzle exit) detected is higher, and the corresponding wall self-emission infrared radiation intensity is also greater. As the swirl angle further increases from 10° to 20°, the position of the hot spot on the upper wall gradually moves forward, and the temperature of the detected wall gradually decreases, so the wall self-emission infrared radiation intensity also decreases accordingly. The variation in wall self-emission intensity in the positive detection angle range is the main reason for the variation in wall infrared radiation intensity. Therefore, the trend of wall infrared radiation intensity variation in the positive detection angle range is consistent with that of wall self-emission intensity.

Under different swirl angles, the contour plots of total infrared radiation intensity at different detection angles in the rear hemisphere space of the serpentine nozzle are shown in Fig. 3.53. The variation in total infrared radiation intensity of the serpentine nozzle with swirl angle can be intuitively observed from the figure. As the swirl angle increases, the high-temperature core region of the gas gradually shrinks; thus, the total infrared radiation intensity in the rear hemisphere space of the serpentine nozzle decreases gradually with increasing swirl angle. At the same time, in the lower part of the rear hemisphere space, both the size and intensity of the local high-infrared-radiation region also decrease gradually with increasing swirl angle.

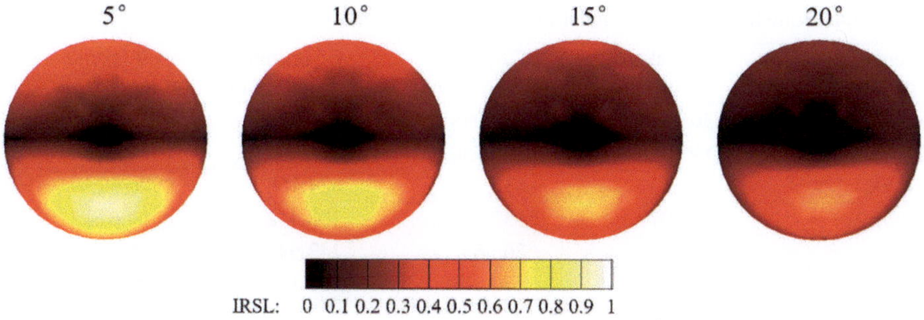

Fig. 3.53 Infrared radiation contour maps of the serpentine nozzle rear hemisphere space under different swirl angles

In conclusion, the swirl angle significantly affects the temperature distribution of both the wall and the gas, which in turn greatly alters the infrared radiation characteristics of the serpentine nozzle. As the swirl angle increases, the intensity and position of the hot spot on the lower wall of the serpentine nozzle change, causing the wall infrared radiation intensity in the positive detection angle range on the vertical detection plane to first increase and then decrease. In the negative detection angle range on the vertical detection plane, the wall reflection radiation decreases gradually, resulting in a gradual decrease in wall infrared radiation intensity. A larger swirl angle enhances the mixing between the core and bypass flows, thereby reducing the size of the high-temperature gas core region. Therefore, the gas infrared radiation intensity decreases gradually with increasing swirl angle. Under swirling conditions, the reduction in total infrared radiation intensity is mainly due to the decrease in gas infrared radiation intensity; hence, the total infrared radiation intensity of the serpentine nozzle also decreases with increasing swirl angle.

3.4.3 Influence of Different Inlet Aerodynamic Parameters on the Infrared Radiation Characteristics of the Serpentine Nozzle

Within the aircraft flight envelope, as flight conditions and engine operating states change, both the total temperature and total pressure at the inlet of the serpentine nozzle vary, which alters the temperature distribution of the nozzle wall and the gas, thereby causing changes in the infrared radiation intensity of the serpentine nozzle. This section mainly investigates the infrared radiation characteristics of the serpentine nozzle under four typical engine operating conditions. These four typical engine conditions include: take-off condition (C1, design point, $H = 0$ km, $Ma = 0$), climb condition 1 (C2, $H = 3$ km, $Ma = 0.4$), climb condition 2 (C3, $H = 7$ km, $Ma = 0.6$), and cruise condition (C4, $H = 11$ km, $Ma = 0.8$). Table 3.4 presents the total

Table 3.4 Boundary conditions under different engine operating conditions

Performance parameters	C1 (take-off condition)	C2 (climb condition 1)	C3 (climb condition 2)	C4 (cruise condition)
T_1^*/K	365.914	357.068	343.826	311.222
P_1^*/kPa	195.963	157.750	111.852	67.092
T_2^*/K	858.477	838.954	824.168	726.423
P_2^*/kPa	189.072	152.393	108.412	64.674
T_b/K	288.14	268.64	242.64	216.64
P_b/kPa	101.325	70.121	41.105	22.700

Table 3.5 Thrust coefficients of the serpentine nozzle under different engine operating conditions

Performance parameters	C1	C2	C3	C4
C_{fg}	0.955	0.959	0.962	0.961

temperature $T_1^*$and total pressure P_1^* at the bypass inlet, the total temperature T_2^* and total pressure P_2^* at the core inlet, as well as the ambient temperature T_b and pressure P_b under each typical operating condition.

The comparison of thrust coefficients of the serpentine nozzle under different operating conditions is shown in Table 3.5. It can be seen that the differences in thrust performance of the serpentine nozzle under various conditions are minimal, with the maximum difference being only 0.007. This indicates that the changes in flight conditions and operating states have little effect on the thrust performance of the serpentine nozzle.

Figure 3.54 shows the *Ma* number distribution on the symmetry plane of the serpentine nozzle under different engine conditions. It can be seen that the *Ma* number distribution inside the serpentine nozzle is basically consistent, and the differences in *Ma* number distribution under different conditions mainly appear in the downstream region of the nozzle exit. This is because the serpentine nozzle is in a critical/supercritical state under different conditions, and the *Ma* number at the nozzle exit reaches the critical value. According to the mass flow rate formula, the *Ma* number distribution at the same cross-section inside the serpentine nozzle is consistent under different conditions. As the engine condition changes from C1 to C4, the pressure ratio of the serpentine nozzle gradually increases, the under-expansion state of the nozzle intensifies, and the velocity core region of the exhaust jet gradually extends, as shown in Fig. 3.54.

Figures 3.55 and 3.56 show the wall temperature distribution of the serpentine nozzle and the gas temperature distribution at different cross-sections under different engine conditions, respectively. It can be seen that with the change of flight conditions and engine operating states, from condition C1 to condition C4, due to the gradual decrease of total temperature at the bypass and core inlets, the wall temperature of the serpentine nozzle gradually decreases, and the gas temperature at various cross-sections along the flow path also gradually decreases. Since the flow characteristics inside the serpentine nozzle are the same, the mixing effect between

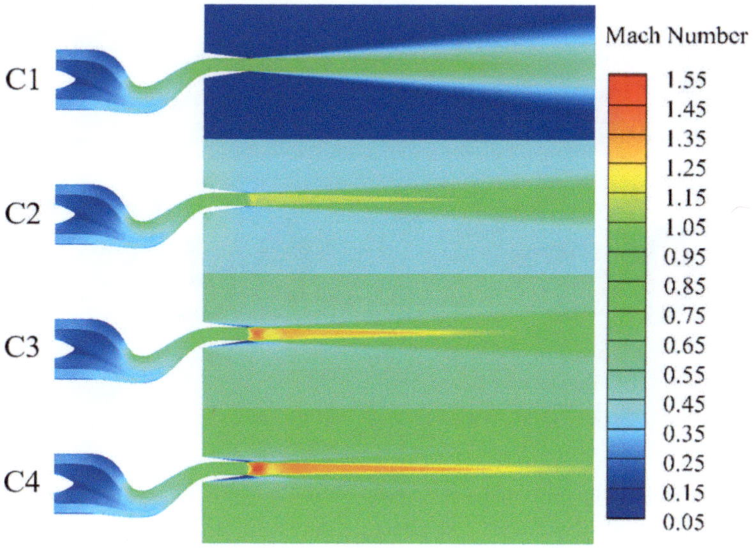

Fig. 3.54 Ma number distribution on the symmetry plane of serpentine nozzles under different engine conditions

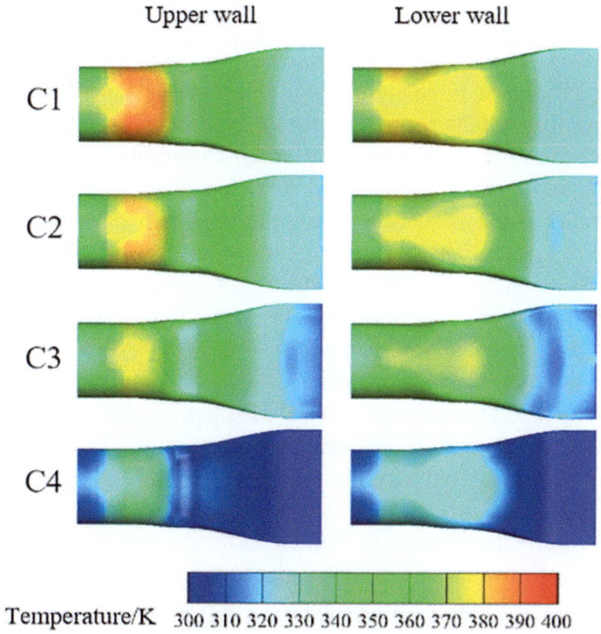

Fig. 3.55 Wall temperature distribution of serpentine nozzles under different engine conditions

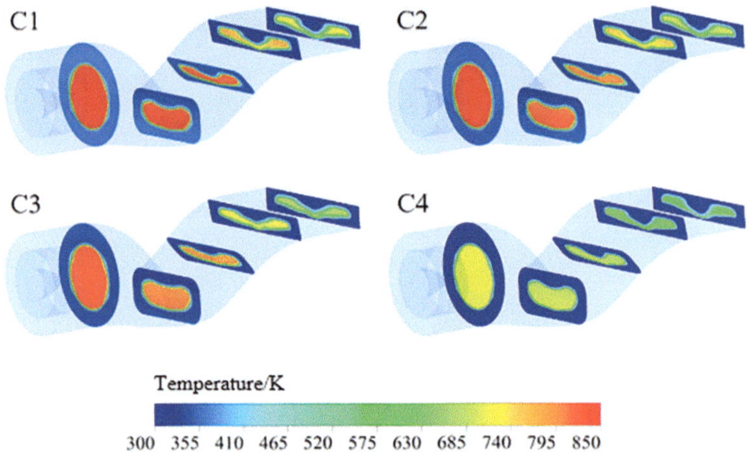

Fig. 3.56 Gas temperature distribution at different cross-sections of serpentine nozzles under different engine conditions

the bypass and core flows is consistent under different conditions, so the shape and area of the temperature distribution on the same cross-section inside the serpentine nozzle under different conditions basically remain unchanged, as shown in Fig. 3.56.

The infrared radiation intensity distribution of the serpentine nozzle under different engine operating conditions is shown in Fig. 3.57. It can be seen that, as the flight conditions and engine operating states change from condition C1 to condition C4, the total infrared radiation intensity gradually decreases due to the reduction in both wall infrared radiation intensity and gas infrared radiation intensity on the horizontal and vertical detection planes. From condition C1 to condition C4, the total temperature at the core and bypass inlets of the nozzle gradually decreases, and the gas temperature at various cross-sections along the flow path gradually drops, resulting in a gradual shrinkage of the high-temperature core region of the nozzle gas, as shown in Fig. 3.58. Therefore, from condition C1 to condition C4, the gas infrared radiation intensity of the serpentine nozzle gradually decreases. As for the wall infrared radiation, since the wall temperature of the serpentine nozzle gradually decreases from condition C1 to condition C4, as shown in Fig. 3.55, the wall self-radiation intensity also gradually decreases. Meanwhile, due to the shrinkage of the high-temperature core region of the gas, the wall reflection radiation also gradually decreases. Consequently, from condition C1 to condition C4, the wall infrared radiation intensity of the serpentine nozzle gradually decreases, as shown in Fig. 3.57.

In summary, with the change in flight conditions and engine operating states from condition C1 to condition C4, although the velocity core region of the exhaust jet gradually extends, the total temperature at the core and bypass inlets of the nozzle gradually decreases, causing the high-temperature core region of the serpentine nozzle gas to shrink, thereby reducing the gas infrared radiation intensity. Meanwhile, as the wall infrared radiation intensity of the serpentine nozzle also decreases, the total infrared radiation intensity of the serpentine nozzle gradually decreases from condition C1 to condition C4.

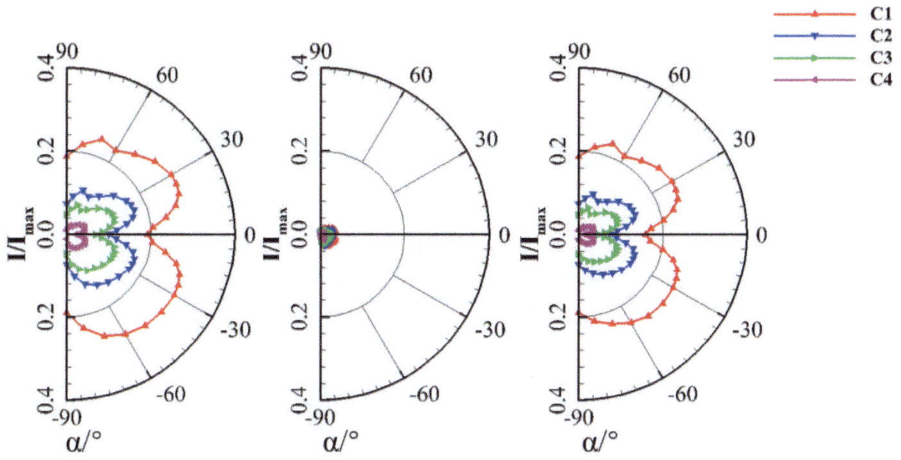

(a) Infrared radiation intensity distribution on the horizontal detection plane

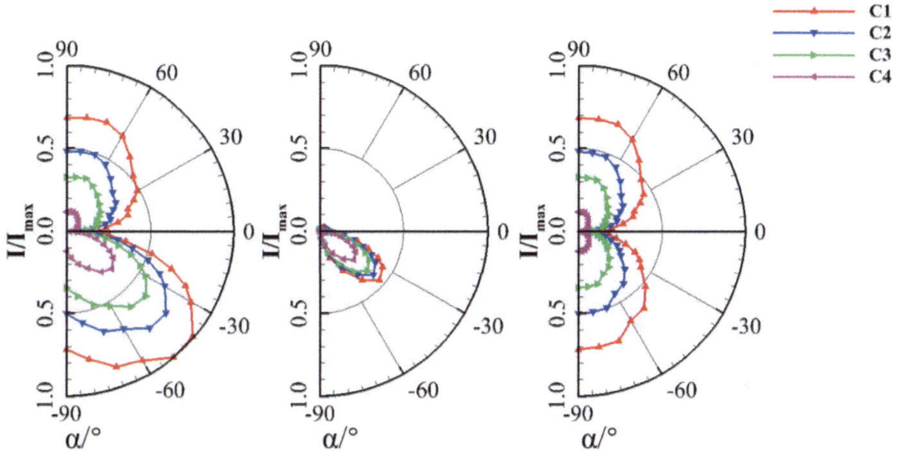

(b) Infrared radiation intensity distribution on the vertical detection plane

Fig. 3.57 Infrared radiation intensity distribution of serpentine nozzles under different engine operating conditions. (**a**) Infrared radiation intensity distribution on the horizontal detection plane. (**b**) Infrared radiation intensity distribution on the vertical detection plane

3.5 Study on Low Infrared Radiation Design Criteria for Serpentine Nozzles

One of the main objectives of designing a serpentine nozzle is to shield the high-temperature components of the exhaust system in order to reduce the rearward infrared radiation intensity of the exhaust system (Cheng et al. 2017, 2019). Currently, to simplify the design process, a fully shielded design criterion is

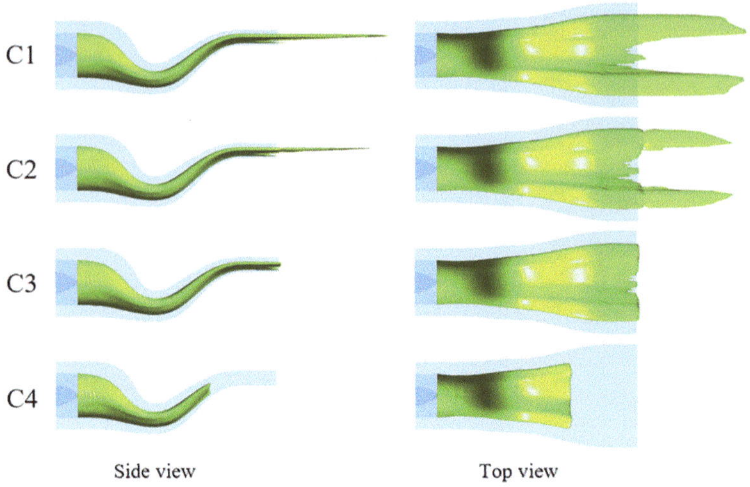

Fig. 3.58 Isosurface of gas temperature at 670 K under different engine operating conditions

commonly adopted, which means that the geometric configuration of the serpentine nozzle is used to shield the high-temperature components of the exhaust system, making them invisible from all detection directions behind the nozzle. Serpentine nozzles designed based on the fully shielded criterion usually have large outlet offsets, which can negatively affect the aerodynamic performance of the nozzle. Therefore, the fully shielded design criterion cannot fully balance the infrared stealth performance and aerodynamic performance of serpentine nozzles. In addition, the infrared radiation characteristics of serpentine nozzles used for engines with different configurations vary significantly, which also limits the applicability of the fully shielded design criterion. This section mainly investigates the influence of shielding ratio on the infrared radiation characteristics of serpentine nozzles used in turbojet and turbofan engines. From the perspective of infrared stealth, low infrared radiation design criteria for serpentine nozzles are proposed to provide a reference for nozzle design.

3.5.1 Influence of Shielding Ratio on the Infrared Radiation Characteristics of Serpentine Nozzles for Turbojet Engines

This section mainly investigates the influence of shielding ratio on the infrared radiation characteristics of serpentine nozzles used for turbojet engines. The definition of shielding ratio is detailed in Sect. 2.4.3 and will not be repeated here. For single serpentine nozzles and double serpentine nozzles, serpentine nozzles with shielding ratios of $\Delta Y = 0$, 0.25, 0.5, 0.75 and 1 were selected to study the effect of shielding ratio on the infrared radiation characteristics of the two types of serpentine nozzles.

3.5.1.1 Influence of Shielding Ratio on the Infrared Radiation Characteristics of Single Serpentine Nozzles

The geometric models of single serpentine nozzles with different shielding ratios are shown in Fig. 3.59. It can be seen that with the increase of shielding ratio ΔY, the visible inlet area of the nozzle gradually decreases. When the shielding ratio $\Delta Y = 0$, the single serpentine nozzle becomes an ordinary two-dimensional nozzle. The thrust coefficients of each nozzle are shown in Table 3.6. It can be seen that the thrust coefficient of the single serpentine nozzle gradually decreases with the increase of shielding ratio.

The infrared radiation intensity distributions of single serpentine nozzles with different shielding ratios are shown in Fig. 3.60. It can be seen that on the horizontal detection plane, when the detection angle $|\alpha| < 70°$, the variation trend of total infrared radiation intensity is consistent with that of wall infrared radiation, as shown in Fig. 3.60a. On the vertical detection plane, when the detection angle $|\alpha| < 50°$, the variation trend of total infrared radiation intensity is also consistent with that of wall infrared radiation, as shown in Fig. 3.60b. When the detection angle $|\alpha|$ is small, the wall infrared radiation intensity is much higher than the gas infrared radiation intensity, and the total infrared radiation intensity mainly originates from wall radiation. As the detection angle $|\alpha|$ increases, the visible area of the nozzle wall gradually decreases, the influence of wall radiation on the total infrared radiation weakens,

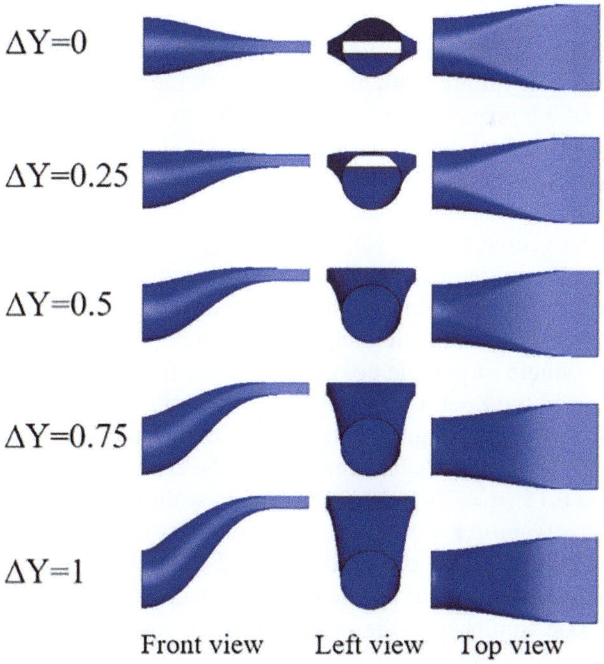

Fig. 3.59 Geometric models of single serpentine nozzles with different shielding ratios

Table 3.6 Thrust coefficients of single serpentine nozzles with different ΔY

ΔY	0	0.25	0.5	0.75	1
Thrust coefficient	0.9583	0.9579	0.9578	0.9571	0.9563

and the proportion of gas radiation in the total infrared radiation gradually increases. When $|\alpha| = 90°$, the visible area of the nozzle wall is 0, and the total infrared radiation comes entirely from the gas radiation. At this point, the total infrared radiation intensity equals the gas infrared radiation intensity.

The primary purpose of serpentine nozzles is to shield high-temperature components (simplified as the nozzle inlet). For single serpentine nozzles with shielding ratios of $\Delta Y = 0$ and $\Delta Y = 1$, on the horizontal detection plane, when $|\alpha|$ is small, the gas infrared radiation intensity of the nozzle with $\Delta Y = 0$ is greater than that of the nozzle with $\Delta Y = 1$, as shown in Fig. 3.60a. Figures 3.61 and 3.62 show the gas temperature distribution and the concentration distribution of CO_2 and H_2O, respectively, for each nozzle. The mixing between the exhaust gas and the ambient air determines the distribution of gas temperature and component concentrations. Since the nozzles have the same exit aspect ratio, the distribution of exhaust gas temperature and component concentrations downstream of the nozzle exit are basically the same. When the detection angle $|\alpha|$ is small, the gas upstream of the nozzle exit (i.e. inside the nozzle) is visible to the detector, but the visible gas volume decreases with the increase of shielding ratio. Because the gas temperature distribution and component concentrations outside the nozzle are essentially the same, the difference in gas infrared radiation intensity at small detection angles $|\alpha|$ mainly comes from the visible gas inside the nozzle. The larger the visible gas volume inside the nozzle, the greater the infrared radiation intensity of the gas.

Tables 3.7 and 3.8 present the visible area ratios at different detection points on the horizontal and vertical detection planes of single serpentine nozzles with varying shielding ratios, respectively. The visible area ratio is defined as the ratio of the visible inlet area of the nozzle to the total visible wall area (including the visible inlet area and the visible nozzle wall area) at a given detection point. Due to the symmetry of the serpentine nozzle, Table 3.7 only provides the visible area ratios within the range of 0° ~ 90° on the horizontal detection plane. On the horizontal detection plane, when the detection angle α is in the range of 0° ~ 20°, the visible area ratio of the single serpentine nozzle with $\Delta Y = 0$ is significantly greater than that of the nozzle with $\Delta Y = 1$. Correspondingly, the wall infrared radiation intensity of the $\Delta Y = 0$ nozzle is also significantly greater than that of the $\Delta Y = 1$ nozzle, as shown in Fig. 3.60a. As the visible area ratio of the $\Delta Y = 0$ nozzle decreases from 0.67 ($\alpha = 0°$) to 0.15 ($\alpha = 20°$), the difference in wall infrared radiation intensity between the $\Delta Y = 0$ and $\Delta Y = 1$ nozzles also gradually diminishes. On the vertical detection plane, within the detection angle range of −5° to 10°, the wall infrared radiation intensity of the $\Delta Y = 0$ nozzle is greater than that of the $\Delta Y = 1$ nozzle, as shown in Fig. 3.60b. Similarly, as the visible area ratio of the $\Delta Y = 0$ nozzle decreases from 0.67 ($\alpha = 0°$) to 0.13 ($\alpha = 10°$), the difference in wall infrared radiation intensity between the $\Delta Y = 0$ and $\Delta Y = 1$ nozzles also gradually decreases. Since the

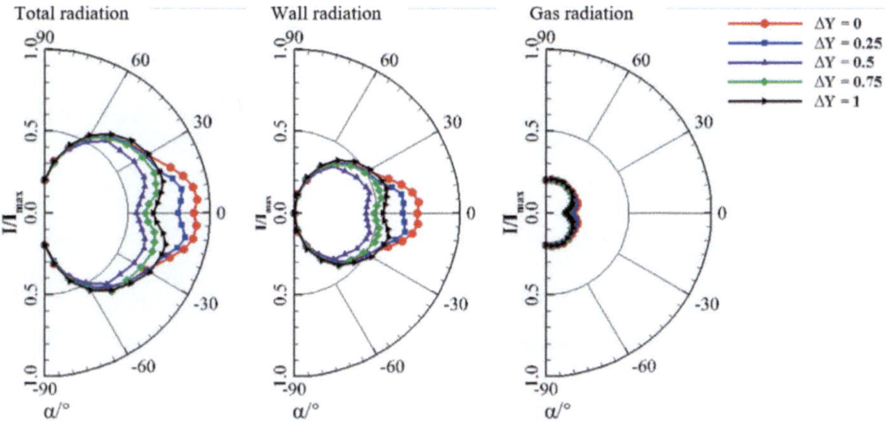

(a) Infrared radiation intensity distribution on the horizontal detection plane 布

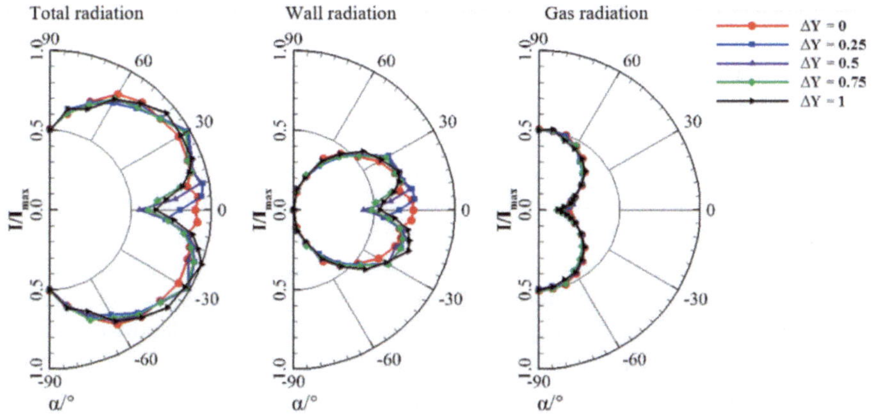

(b) Infrared radiation intensity distribution on the vertical detection plane

Fig. 3.60 Infrared radiation intensity distribution of single serpentine nozzles with different shielding ratios. (**a**) Infrared radiation intensity distribution on the horizontal detection plane布. (**b**) Infrared radiation intensity distribution on the vertical detection plane

temperature and component concentration distributions of the gas outside the nozzle are essentially the same, the difference in gas infrared radiation intensity among serpentine nozzles with different shielding ratios is very small. Thus, the attenuation effect of the gas on wall infrared radiation can be considered equivalent. Consequently, the temperature distribution of the visible nozzle wall is the main factor affecting the variation in wall infrared radiation intensity for each nozzle. Because the inlet temperature of the nozzle is higher than that of the wall surface, the visible area ratio has a significant impact on the wall infrared radiation intensity. Within a certain range, the greater the visible area ratio, the greater the wall infrared radiation intensity.

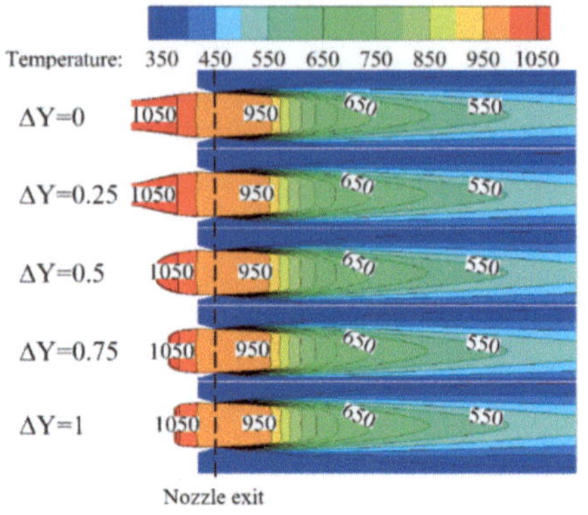

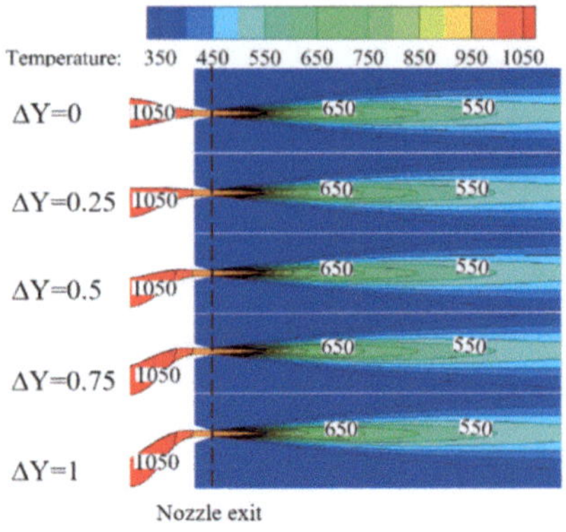

Fig. 3.61 Gas temperature distribution of single serpentine nozzles with different shielding ratios. (**a**) Horizontal plane. (**b**) Vertical plane

On the vertical detection plane, when the detection angles are α = 10° and α = −10°, the visible area ratio of the ΔY = 0 nozzle is 0.13, as shown in Table 3.8. However, at α = 10°, the wall infrared radiation intensity of the ΔY = 0 nozzle is greater than that of the ΔY = 1 nozzle, whereas at α = −10°, it is lower. This is due to the differences in temperature distribution between the upper and lower walls of the ΔY = 1 nozzle. Figure 3.63 shows the temperature distribution of the visible nozzle walls at α = 10° and α = −10° on the vertical detection plane for both ΔY = 0 and ΔY = 1 nozzles. Due to the geometrical symmetry of the ΔY = 0 nozzle, the temperature distributions of its upper and lower walls are nearly identical, resulting in nearly equal wall infrared radiation intensities at α = 10° and α = −10°, as shown in Fig. 3.60b. For the ΔY = 1 nozzle, when α = 10°, both the upper and lower walls are visible, whereas at α = −10°, only the upper wall is visible, as shown in Fig. 3.63. At α = 10°, most of the visible wall surface of the ΔY = 1 nozzle is below 920 K, while at α = −10°, most of the visible wall surface is above 920 K. Therefore, the wall infrared radiation intensity of the ΔY = 1 nozzle is higher at α = −10° than at α = 10°, as shown in Fig. 3.60b.

The wall infrared radiation intensity of the serpentine nozzle is mainly determined by the temperature distribution of the visible wall. According to the comparison of the infrared radiation intensity distribution between single serpentine nozzles with ΔY = 0 and ΔY = 1, it can be seen that when the visible area ratio is greater than 0.15, the visible nozzle inlet has a significant influence on the wall infrared radiation intensity. When the visible area ratio is less than 0.15, as the visible area ratio decreases, the influence of the visible nozzle inlet on wall infrared radiation becomes small. At this time, the wall infrared radiation intensity is mainly determined by the temperature distribution of the nozzle wall.

On the horizontal detection plane, when $|\alpha| < 30°$, due to the variation of the visible gas volume inside the nozzle, the gas infrared radiation intensity of single serpentine nozzles with different shielding ratios differs slightly, as shown in Fig. 3.60a.

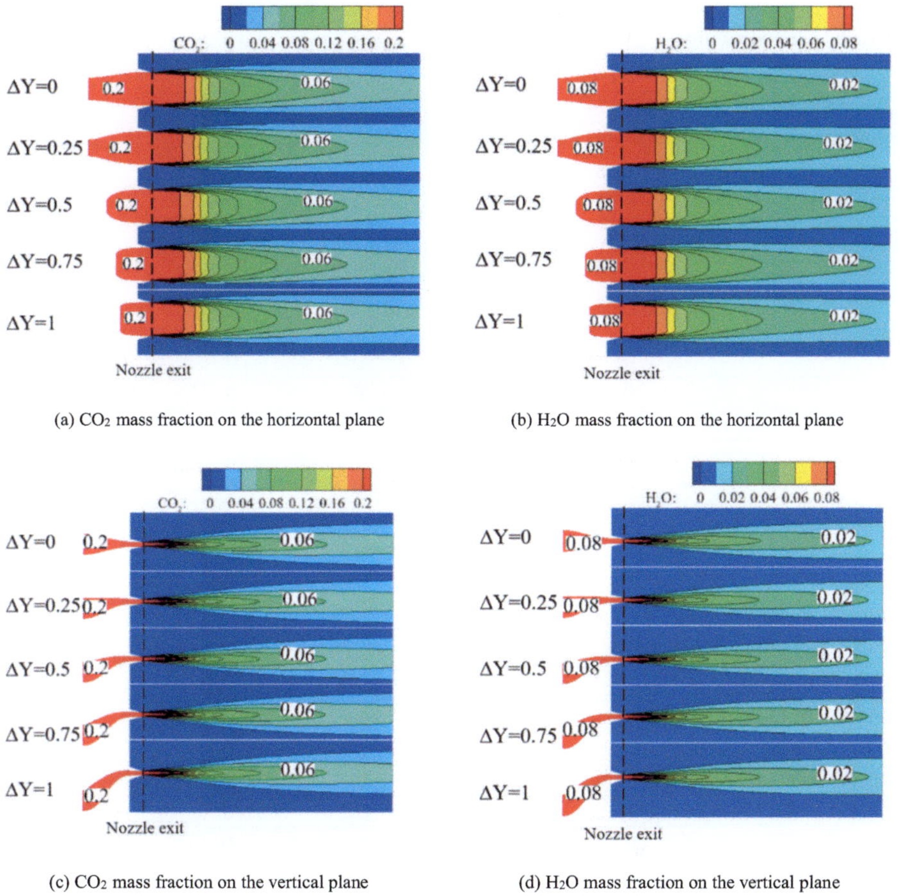

Fig. 3.62 CO_2 and H_2O concentration distributions of single serpentine nozzles with different shielding ratios. (**a**) CO_2 mass fraction on the horizontal plane. (**b**) H_2O mass fraction on the horizontal plane. (**c**) CO_2 mass fraction on the vertical plane. (**d**) H_2O mass fraction on the vertical plane

When $|\alpha| > 30°$, due to the similar gas temperature distribution and component concentration distribution outside the nozzles, the difference in gas infrared radiation intensity between single serpentine nozzles with different shielding ratios is small. On the horizontal detection plane, the wall infrared radiation intensity of the single serpentine nozzle with $\Delta Y = 0$ in the range of $-20°$ to $20°$ and the single serpentine nozzle with $\Delta Y = 0.25$ in the range of $-15°$ to $15°$ is significantly higher than that of the single serpentine nozzles with $\Delta Y = 0.5$, 0.75 and 1. This is because the visible area ratio of the single serpentine nozzle with $\Delta Y = 0$ in the range of $-20°$ to $20°$ and that of the single serpentine nozzle with $\Delta Y = 0.25$ in the range of $-15°$ to $15°$ are both greater than 0.15, as shown in Table 3.7. On the horizontal detection plane, when the shielding ratio increases from 0 to 0.5, the total infrared radiation intensity gradually decreases. As the shielding ratio further increases from 0.5 to 1,

Table 3.7 Visible area ratios on the horizontal detection plane for single serpentine nozzles with different shielding ratios

	ΔY				
Detection angle/°	0	0.25	0.5	0.75	1
0	0.67	0.52	0	0	0
5	0.66	0.50	0	0	0
10	0.51	0.43	0	0	0
15	0.33	0.26	0	0	0
20	0.15	0.06	0	0	0
30	0	0	0	0	0
40	0	0	0	0	0
50	0	0	0	0	0
60	0	0	0	0	0
70	0	0	0	0	0
80	0	0	0	0	0
90	0	0	0	0	0

Table 3.8 Visible area ratios on the vertical detection plane for single serpentine nozzles with different shielding ratios

	ΔY				
Detection angle/°	0	0.25	0.5	0.75	1
90	0	0	0	0	0
80	0	0	0	0	0
70	0	0	0	0	0
60	0	0	0	0	0
50	0	0	0	0	0
40	0	0	0	0	0
30	0	0	0	0	0
20	0	0	0	0	0
15	0	0.01	0.07	0.07	0
10	0.13	0.31	0.29	0	0
5	0.46	0.52	0.22	0	0
0	0.67	0.51	0	0	0
−5	0.46	0.02	0	0	0
−10	0.13	0	0	0	0
−15	0	0	0	0	0
−20	0	0	0	0	0
−30	0	0	0	0	0
−40	0	0	0	0	0
−50	0	0	0	0	0
−60	0	0	0	0	0
−70	0	0	0	0	0
−80	0	0	0	0	0
−90	0	0	0	0	0

the total infrared radiation intensity gradually increases. This is caused by the change in wall temperature distribution. On the vertical detection plane, the visible area ratio of the single serpentine nozzle with $\Delta Y = 0$ in the range of −5° to 5°, the single serpentine nozzle with $\Delta Y = 0.25$ in the range of 0°–10°, and the single serpentine nozzle with $\Delta Y = 0.5$ in the range of 5°–10° are all greater than 0.15, as shown in Table 3.8. Therefore, on the vertical detection plane, the wall infrared radiation intensity of the single serpentine nozzle with $\Delta Y = 0$ in the range of −5° to 5°, that of the single serpentine nozzle with $\Delta Y = 0.25$ in the range of 0°–10°, and that of the single serpentine nozzle with $\Delta Y = 0.5$ in the range of 5°–10° are all greater than that of the single serpentine nozzles with $\Delta Y = 0.75$ and 1, as shown in Fig. 3.60b.

From the above results, it can be seen that on the horizontal detection plane, the single serpentine nozzle with $\Delta Y = 0.5$ has the lowest total infrared radiation intensity. On the vertical detection plane, when $|\alpha|$ is large, the differences in total infrared radiation intensity among the nozzles are small. However, at small $|\alpha|$, the single serpentine nozzle with $\Delta Y = 0.75$ has the lowest total infrared radiation intensity. Considering that the total infrared radiation intensity of the serpentine nozzle on the vertical detection plane is greater than that on the horizontal detection plane, the single serpentine nozzle with $\Delta Y = 0.75$ has better infrared stealth performance than the other nozzles.

In summary, a larger shielding ratio does not necessarily improve the infrared stealth performance of the serpentine nozzle. When the visible area ratio is sufficiently large, the visible high-temperature nozzle inlet has a significant influence on the wall infrared radiation intensity of the single serpentine nozzle. For single serpentine nozzles with $\Delta Y = 0.75$ and 1, the visible area ratio at all detection angles is less than 0.15, so the influence of the visible high-temperature nozzle inlet on the wall infrared radiation intensity is limited. When $\Delta Y > 0.75$, affected by the wall temperature distribution, the total infrared radiation intensity of the single serpentine nozzle with $\Delta Y = 1$ is greater than that of the one with $\Delta Y = 0.75$. Therefore, when $\Delta Y > 0.75$, a larger ΔY cannot improve the infrared stealth performance of the single serpentine nozzle.

3.5.1.2 Influence of Shielding Ratio on the Infrared Radiation Characteristics of Double Serpentine Nozzles

The geometric models of double serpentine nozzles with different shielding ratios are shown in Fig. 3.64. It can be seen that as the shielding ratio decreases, the visible nozzle inlet area gradually increases. When the shielding ratio $\Delta Y = 0$, the visible nozzle inlet area is the largest. In this section, the inlet diameter, axial length, outlet area, and outlet aspect ratio of the double serpentine nozzles are kept consistent with those of the single serpentine nozzle model. The thrust coefficients of each nozzle are listed in Table 3.9. It can be seen that the thrust coefficient of the double serpentine nozzle gradually decreases with the increase of the shielding ratio.

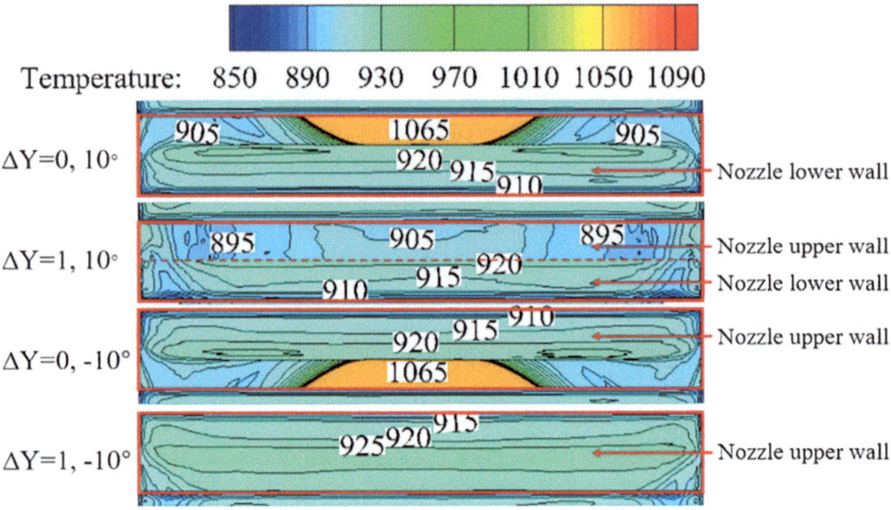

Fig. 3.63 Temperature distribution of visible nozzle walls for ΔY = 0 and ΔY = 1 nozzles at 10° and −10° detection angles on the vertical detection plane

Table 3.9 Thrust coefficients of double serpentine nozzles with different ΔY

ΔY	0	0.25	0.5	0.75	1
Thrust coefficient	0.9481	0.9478	0.9468	0.9459	0.9457

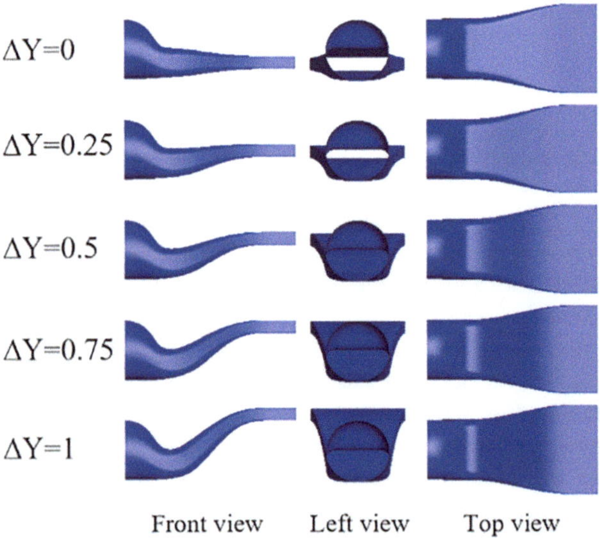

Fig. 3.64 Geometric models of double serpentine nozzles with different shielding ratios

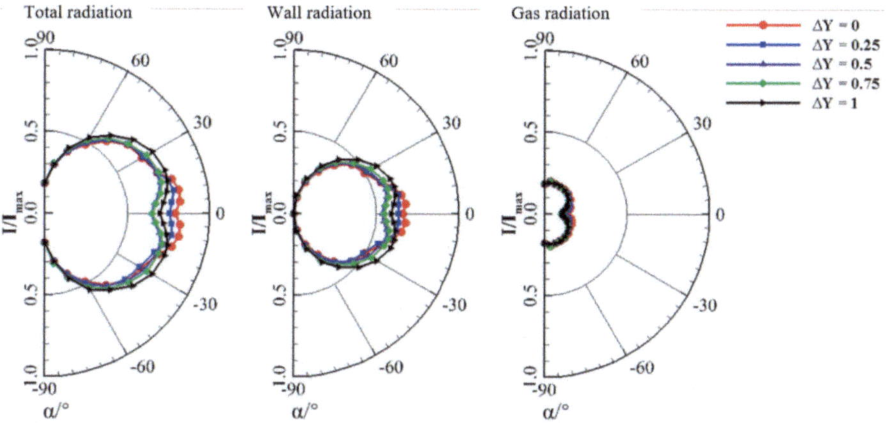

(a) Infrared radiation intensity distribution on the horizontal detection plane

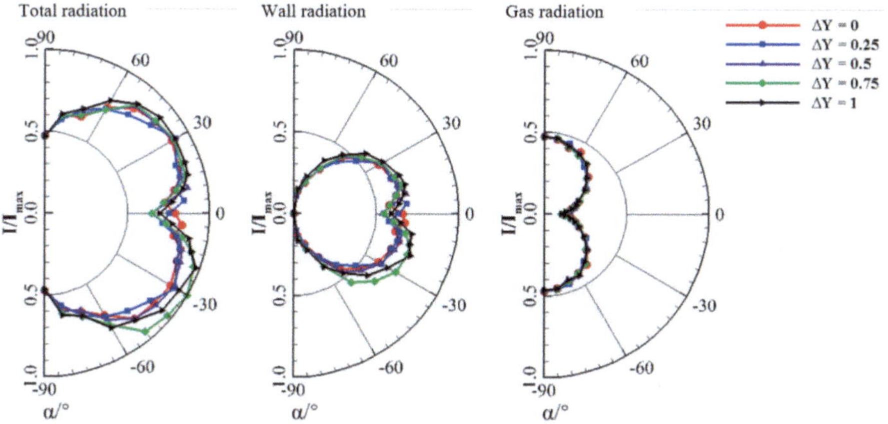

(b) Infrared radiation intensity distribution on the vertical detection plane

Fig. 3.65 Infrared radiation intensity distribution of double serpentine nozzles with different shielding ratios. (**a**) Infrared radiation intensity distribution on the horizontal detection plane. (**b**) Infrared radiation intensity distribution on the vertical detection plane

The distribution of infrared radiation intensity of double serpentine nozzles with different shielding ratios is shown in Fig. 3.65. It can be seen that the distribution pattern of the infrared radiation intensity of double serpentine nozzles is similar to that of single serpentine nozzles. On the horizontal detection plane, the total infrared radiation intensity of the double serpentine nozzle with $\Delta Y = 0.5$ is the lowest. When $\Delta Y < 0.5$, the total infrared radiation intensity decreases with the increase of ΔY. When $\Delta Y > 0.5$, the total infrared radiation intensity increases with the increase of ΔY.

Tables 3.10 and 3.11 respectively show the visible area ratios at different detection points on the horizontal and vertical detection plane for double serpentine nozzles. On the horizontal detection plane, for double serpentine nozzles with $\Delta Y = 0$

Table 3.10 Visible area ratios on the horizontal detection plane of double serpentine nozzles with different shielding ratios

Detection angle/°	ΔY				
	0	0.25	0.5	0.75	1
0	0.62	0.44	0	0	0
5	0.59	0.42	0	0	0
10	0.47	0.35	0	0	0
15	0.30	0.24	0	0	0
20	0.11	0.10	0	0	0
30	0	0	0	0	0
40	0	0	0	0	0
50	0	0	0	0	0
60	0	0	0	0	0
70	0	0	0	0	0
80	0	0	0	0	0
90	0	0	0	0	0

Table 3.11 Visible area ratios on the vertical detection plane of double serpentine nozzles with different shielding ratios

Detection angle/°	ΔY				
	0	0.25	0.5	0.75	1
90	0	0	0	0	0
80	0	0	0	0	0
70	0	0	0	0	0
60	0	0	0	0	0
50	0	0	0	0	0
40	0	0	0	0	0
30	0	0	0	0	0
20	0	0	0	0	0
15	0	0	0.02	0.08	0
10	0	0.12	0.27	0	0
5	0.23	0.45	0.17	0	0
0	0.62	0.44	0	0	0
−5	0.44	0	0	0	0
−10	0	0	0	0	0
−15	0	0	0	0	0
−20	0	0	0	0	0
−30	0	0	0	0	0
−40	0	0	0	0	0
−50	0	0	0	0	0
−60	0	0	0	0	0
−70	0	0	0	0	0
−80	0	0	0	0	0
−90	0	0	0	0	0

and $\Delta Y = 0.25$, the visible detection angle range of the nozzle inlet is from −20° to 20°, which is the same as that of the single serpentine nozzle. On the vertical detection plane, for double serpentine nozzles with $\Delta Y = 0$ and $\Delta Y = 0.25$, the visible angle range of the nozzle inlet is smaller than that of the single serpentine nozzle. Meanwhile, at the same detection angle, the visible area ratio of the double serpentine nozzle is smaller than that of the single serpentine nozzle.

For double serpentine nozzles with different shielding ratios, on the horizontal detection plane, the visible area ratios of the nozzles with $\Delta Y = 0$ and $\Delta Y = 0.25$ are greater than 0.35 in the range of 0°–10°, as shown in Table 3.10. Correspondingly, the wall infrared radiation intensities of the nozzles with $\Delta Y = 0$ and $\Delta Y = 0.25$ are higher than those with $\Delta Y = 0.5$, 0.75 and 1 in the 0°–10° range. On the horizontal detection plane, at $\alpha = 15°$, the visible area ratios of the nozzles with $\Delta Y = 0$ and $\Delta Y = 1$ are 0.30 and 0 respectively; however, the wall infrared radiation intensity of the nozzle with $\Delta Y = 0$ is lower than that of the nozzle with $\Delta Y = 1$. When $|\alpha| > 10°$, the influence of the nozzle wall on the wall infrared radiation intensity gradually increases, while the influence of the nozzle inlet gradually decreases. When $|\alpha| > 10°$, the wall infrared radiation intensity increases gradually with the increase of ΔY. When $\Delta Y = 1$, the wall infrared radiation intensity of the double serpentine nozzle is the greatest, as shown in Fig. 3.65a. This is caused by the change in the wall temperature distribution with ΔY. Figure 3.66 shows the temperature distribution of the visible wall surfaces of the nozzles at $\alpha = 30°$ on the horizontal detection plane. It can be seen that the temperature of the visible wall surfaces generally

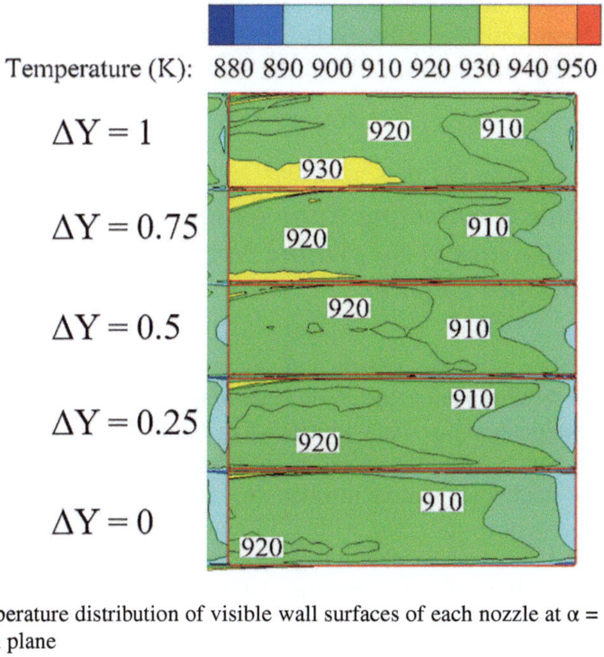

Fig. 3.66 Temperature distribution of visible wall surfaces of each nozzle at $\alpha = 30°$ on the horizontal detection plane

increases gradually with the increase of the shielding ratio. The temperature of the lower left corner of the visible wall surface of the nozzle with $\Delta Y = 1$ is higher than 930 K, while the temperatures of the visible wall surfaces of the other nozzles are basically below 930 K. Therefore, the wall infrared radiation intensity of the nozzle with $\Delta Y = 1$ is the greatest.

On the vertical detection plane, the visible nozzle inlet has a significant impact on the infrared radiation intensity of the double serpentine nozzle with $\Delta Y = 0$ in the detection angle range of $-5°$ to $0°$ and the double serpentine nozzle with $\Delta Y = 0.25$ in the detection angle range of $0°$–$5°$. The visible area ratios of the double serpentine nozzle with $\Delta Y = 0$ in the range of $-5°$ to $0°$ and with $\Delta Y = 0.25$ in the range of $0°$ to $5°$ are both greater than 0.35, as shown in Table 3.11. Correspondingly, the wall infrared radiation intensities of the double serpentine nozzle with $\Delta Y = 0$ in the range of $-5°$ to $0°$ and with $\Delta Y = 0.25$ in the range of $0°$–$5°$ are greater than those of $\Delta Y = 0.5$, 0.75 and 1.

In summary, when the visible area ratio is greater than 0.35, the visible nozzle inlet has a considerable effect on the infrared radiation intensity of the double serpentine nozzle. On the horizontal detection plane, the visible area ratios of the double serpentine nozzles with $\Delta Y = 0.5$ and $\Delta Y = 0.75$ are 0, and their total infrared radiation intensities are lower than those with $\Delta Y = 0$, 0.25 and 1. On the vertical detection plane, the visible area ratios of the double serpentine nozzles with $\Delta Y = 0.5$, 0.75 and 1 are less than 0.35; however, the total infrared radiation intensities of the double serpentine nozzles with $\Delta Y = 0.75$ and $\Delta Y = 1$ in the range of $-10°$ to $-60°$ are higher than that of the double serpentine nozzle with $\Delta Y = 0.5$. Considering the infrared radiation intensities on both horizontal and vertical detection planes, the double serpentine nozzle with $\Delta Y = 0.5$ has the best infrared stealth performance. Meanwhile, to effectively suppress the infrared radiation of the exhaust system, the visible area ratio of the double serpentine nozzle should be controlled below 0.35.

3.5.2 *Influence of Shielding Ratio on the Infrared Radiation Characteristics of Serpentine Nozzles for Turbofan Engines*

This section mainly studies the influence of shielding ratio on the infrared radiation characteristics of serpentine nozzles used in turbofan engines. For double serpentine nozzles, double serpentine nozzles with shielding ratios $\Delta Y = 0$, 0.25, 0.5, 0.75 and 1 were selected to investigate the effect of shielding ratio on the infrared radiation characteristics of double serpentine nozzles used in turbofan engines. The geometric models of serpentine nozzles with different shielding ratios are shown in Fig. 3.67. The inlet diameter, axial length, outlet area, and outlet aspect ratio of each nozzle are the same. It can be seen that the visible area of the core inlet varies with ΔY. When $\Delta Y < 0.25$, the core inlet is clearly visible. The thrust coefficients of each

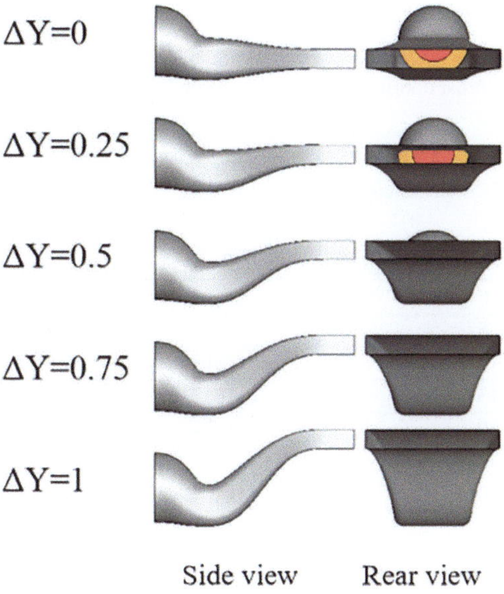

Fig. 3.67 Geometric models of serpentine nozzles with different shielding ratios

nozzle are shown in Table 3.12. It can be seen that the thrust coefficient of the serpentine nozzle gradually decreases with the increase of the shielding ratio.

Tables 3.13 and 3.14 respectively present the visible area ratios of the serpentine nozzle on the horizontal detection plane and the vertical detection plane, defined as the ratio of the visible core inlet area to the total visible wall area (including the visible core inlet area, bypass inlet area, and visible nozzle wall area). Due to the geometric symmetry of the serpentine nozzle, Table 3.13 only provides the visible area ratios within the 0°–90° range on the horizontal detection plane. As ΔY varies, the mutual shading relationships between the nozzle walls also change. Because the outlet offset of the $\Delta Y = 0$ serpentine nozzle is smaller than that of the $\Delta Y = 0.25$ serpentine nozzle, directly behind the serpentine nozzle, only the lower part of the core inlet is visible for $\Delta Y = 0$, while the middle part of the core inlet is visible for $\Delta Y = 0.25$, as shown in Fig. 3.67. Since the outlet heights of each nozzle are the same and the core inlet is circular, the closer to the center, the larger the visible core inlet area. Therefore, the $\Delta Y = 0.25$ serpentine nozzle has the largest visible area ratio at a 0° detection angle. Moreover, at the same detection points within the 0°–20° range on the horizontal detection plane, the visible area ratio of the $\Delta Y = 0.25$ serpentine nozzle is greater than that of the $\Delta Y = 0$ serpentine nozzle, as shown in Table 3.13. As ΔY increases, the nozzle outlet moves gradually upward, and the shading effect of the serpentine nozzle on the nozzle inlet becomes more significant. Therefore, on the vertical detection plane, the detection angle range with a visible area ratio greater than 0 gradually decreases and shifts upward as ΔY increases, as shown in Table 3.14.

The distribution of infrared radiation intensity of serpentine nozzles with different shielding ratios is shown in Fig. 3.68. It can be seen that on the horizontal

Table 3.12 Thrust coefficients of serpentine nozzles with different ΔY

ΔY	0	0.25	0.5	0.75	1
Thrust coefficient	0.9624	0.9604	0.0.9502	0.0.9401	0.9241

Table 3.13 Visible area ratios of serpentine nozzles on the horizontal detection plane with different shielding ratios

	ΔY				
Detection angle/°	0	0.25	0.5	0.75	1
0	0.13	0.21	0	0	0
5	0.13	0.21	0	0	0
10	0.13	0.21	0	0	0
15	0.12	0.18	0	0	0
20	0.04	0.08	0	0	0
30	0	0	0	0	0
40	0	0	0	0	0
50	0	0	0	0	0
60	0	0	0	0	0
70	0	0	0	0	0
80	0	0	0	0	0
90	0	0	0	0	0

Table 3.14 Visible area ratios of serpentine nozzles on the vertical detection plane with different shielding ratios

	ΔY				
Detection angle/°	0	0.25	0.5	0.75	1
90	0	0	0	0	0
80	0	0	0	0	0
70	0	0	0	0	0
60	0	0	0	0	0
50	0	0	0	0	0
40	0	0	0	0	0
30	0	0	0	0	0
20	0	0	0	0	0
15	0	0	0	0.06	0
10	0	0	0.10	0	0
5	0	0.12	0.09	0	0
0	0.13	0.21	0	0	0
−5	0.23	0.02	0	0	0
−10	0.07	0	0	0	0
−15	0	0	0	0	0
−20	0	0	0	0	0
−30	0	0	0	0	0
−40	0	0	0	0	0
−50	0	0	0	0	0

Detection angle/°	ΔY				
	0	0.25	0.5	0.75	1
−60	0	0	0	0	0
−70	0	0	0	0	0
−80	0	0	0	0	0
−90	0	0	0	0	0

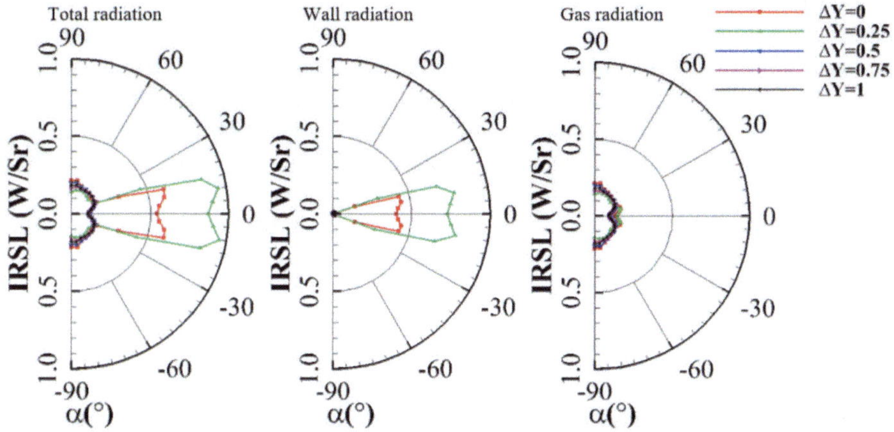

(a) Infrared radiation intensity distribution on the horizontal detection plane

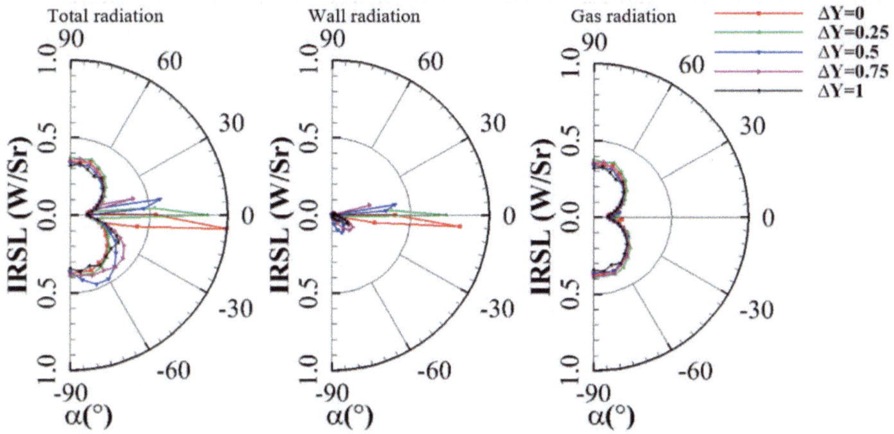

(b) Infrared radiation intensity distribution on the vertical detection plane

Fig. 3.68 Infrared radiation intensity distribution of serpentine nozzles with different shielding ratios. (**a**) Infrared radiation intensity distribution on the horizontal detection plane. (**b**) Infrared radiation intensity distribution on the vertical detection plane

detection plane, when $\Delta Y > 0.25$, the differences in infrared radiation intensity among serpentine nozzles are small. However, on the vertical detection plane, the shielding ratio has a significant effect on the infrared radiation intensity distribution of serpentine nozzles.

On the horizontal detection plane, for serpentine nozzles with $\Delta Y = 0$ and $\Delta Y = 0.25$, the visible area ratio is greater than zero within the detection angle range of −20° to 20°, as shown in Table 3.13, meaning that the high-temperature core inlet is visible in this angle range. Correspondingly, the wall infrared radiation intensity of the $\Delta Y = 0$ and $\Delta Y = 0.25$ serpentine nozzles within the −20° to 20° range is significantly higher than that of the $\Delta Y = 0.5$, 0.75 and 1 serpentine nozzles. Because the high-temperature core inlet is visible within this range for $\Delta Y = 0$ and $\Delta Y = 0.25$, the wall infrared radiation intensity is significantly higher than the gas infrared radiation intensity, and thus the total infrared radiation intensity variation matches that of the wall infrared radiation intensity. When $|\alpha| > 20°$, the core inlet is not visible, and the gas infrared radiation intensity is higher than the wall infrared radiation intensity. The total infrared radiation intensity variation of the serpentine nozzles corresponds to that of the gas infrared radiation intensity, as shown in Fig. 3.68a. Due to the introduction of the low-temperature bypass flow, the nozzle wall temperature approaches that of the outer bypass inlet, and there is a large temperature difference between the core inlet, outer bypass inlet, and the nozzle wall. Therefore, when the core inlet is visible, the high-temperature core inlet greatly influences the wall infrared radiation intensity. Within the −20° to 20° detection angle range on the horizontal detection plane, the visible area ratios of the $\Delta Y = 0$ and $\Delta Y = 0.25$ serpentine nozzles are greater than 0.04, while those of the $\Delta Y = 0.5$, 0.75 and 1 serpentine nozzles are zero. Affected by the high-temperature core inlet, the wall infrared radiation intensity of the $\Delta Y = 0$ and $\Delta Y = 0.25$ serpentine nozzles is higher than that of the $\Delta Y = 0.5$, 0.75 and 1 serpentine nozzles. The larger the visible area ratio, the higher the corresponding wall infrared radiation intensity. Within the −20° to 20° range, the visible area ratio of the $\Delta Y = 0.25$ serpentine nozzle is much larger than that of the $\Delta Y = 0$ serpentine nozzle, as shown in Table 3.13, thus the wall infrared radiation intensity of the $\Delta Y = 0.25$ serpentine nozzle is also significantly higher than that of the $\Delta Y = 0$ nozzle, as shown in Fig. 3.68a. When $|\alpha|$ increases from 15° to 25°, the visible area ratio of the $\Delta Y = 0$ serpentine nozzle decreases from 0.12 to 0, and that of the $\Delta Y = 0.25$ serpentine nozzle decreases from 0.18 to 0. Correspondingly, the wall infrared radiation intensities of the $\Delta Y = 0$ and $\Delta Y = 0.25$ serpentine nozzles also rapidly decrease.

On the vertical detection plane, as ΔY increases, the maximum visible area ratio of each nozzle gradually decreases from 0.23 for the $\Delta Y = 0$ serpentine nozzle to 0.06 for the $\Delta Y = 0.75$ serpentine nozzle. The angles corresponding to the maximum visible area ratio also gradually shift upward from −5° for the $\Delta Y = 0$ serpentine nozzle to 15° for the $\Delta Y = 0.75$ serpentine nozzle, as shown in Table 3.14. Corresponding to the maximum visible area ratio, except for the $\Delta Y = 1$ serpentine nozzle, the wall infrared radiation intensity distribution of each nozzle exhibits an extremum. Moreover, the angles at which these extrema occur also shift upward from −5° for the $\Delta Y = 0$ serpentine nozzle to 15° for the $\Delta Y = 0.75$ serpentine nozzle, as shown in Fig. 3.68b. Because the larger the visible area ratio, the higher the corresponding wall infrared radiation intensity, the extremum of wall infrared radiation intensity gradually decreases with increasing ΔY. For the $\Delta Y = 1$ serpentine nozzle, the visible area ratio is zero at all detection angles, meaning the high-temperature core inlet is not visible at any detection point; thus, the wall infrared radiation intensity distribution of the $\Delta Y = 1$ serpentine nozzle does not exhibit a similar extremum.

On the vertical detection plane, the visible area ratios of the $\Delta Y = 0.25$ serpentine nozzle at detection angles of −5° and 5° are 0.02 and 0.12, respectively. However, at the −5° detection angle, the effect of the wall infrared radiation intensity of the $\Delta Y = 0.25$ serpentine nozzle on the total infrared radiation intensity is not obvious, as shown in Fig. 3.68b. The minimum visible area ratio of the $\Delta Y = 0$ serpentine nozzle within the −10° to 0° detection angle range is 0.07; the minimum visible area ratio of the $\Delta Y = 0.5$ serpentine nozzle within the 5°–10° detection angle range is 0.09; and the visible area ratio of the $\Delta Y = 0.75$ serpentine nozzle at the 15° detection angle is 0.06. The wall infrared radiation intensities of the respective nozzles at these detection angles have significant impacts on the total infrared radiation intensity, as shown in Fig. 3.68b. On the horizontal detection plane, when the visible area ratio is greater than 0.04, the wall infrared radiation intensity significantly influences the total infrared radiation intensity. Combining the infrared radiation intensity distribution characteristics on both the horizontal and vertical detection planes, it can be concluded that when the visible area ratio is greater than 0.04, the visible high-temperature core inlet has a significant impact on both the wall infrared radiation intensity and the total infrared radiation intensity.

The mixing between the serpentine nozzle jet plume and the ambient atmosphere is greatly influenced by the outlet aspect ratio. Since all nozzles have the same outlet aspect ratio, the mixing effects between the jet plumes and the ambient atmosphere differ little among the nozzles. Figure 3.69 shows the axial velocity distribution

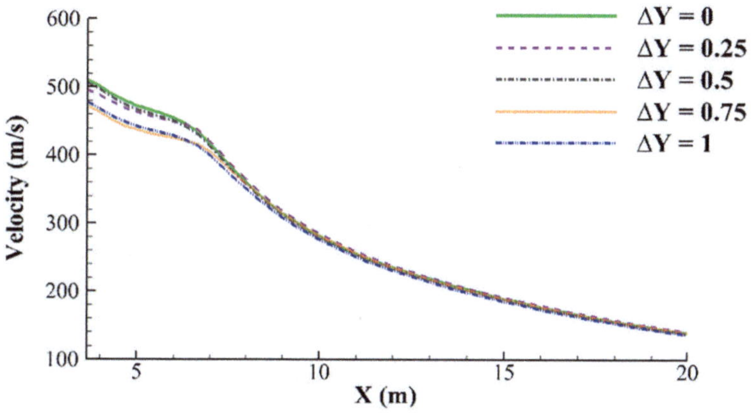

Fig. 3.69 Axial velocity distribution at the center of the jet plume of serpentine nozzles with different shielding ratios

along the axis at the center of the jet plume for serpentine nozzles with different shielding ratios. It can be seen that the axial velocity trends at the jet plume centers are the same for all nozzles. Before 7 m, there are slight differences in the axial velocities at the jet plume centers, but after 7 m, the axial velocities are basically the same. Therefore, the lengths of the high-temperature core flows of all nozzles are basically identical, and the differences in the corresponding gas infrared radiation intensities are small. Since the total infrared radiation intensity equals the sum of the wall infrared radiation intensity and the gas infrared radiation intensity, the characteristics of the wall infrared radiation intensity distribution also appear in the total infrared radiation intensity distribution.

In summary, the shielding ratio has a significant impact on the infrared radiation intensity distribution of serpentine nozzles used in aero-engines. On the horizontal detection plane within the −20° to 20° detection angle range, the visible area ratios of the $\Delta Y = 0.5$, 0.75 and 1 serpentine nozzles are all zero, and their corresponding total infrared radiation intensities are significantly lower than those of the $\Delta Y = 0$ and $\Delta Y = 0.25$ serpentine nozzles. On the vertical detection plane, when $\Delta Y < 1$, the high-temperature core inlet is visible at some detection angles, corresponding to the appearance of extrema in the wall infrared radiation intensity distribution. Overall, the visible area ratio of serpentine nozzles used in aero-engines should be controlled below 0.04.

3.6 Study on Multi-objective Aerodynamic/Infrared Optimization of the Serpentine Nozzle

In the design process of serpentine nozzles, the aerodynamic performance design objectives and the infrared stealth performance design objectives are often conflicting. To achieve optimal aerodynamic performance, the design of the serpentine nozzle needs to minimize the change in the flow direction inside the nozzle, that is, to reduce the nozzle offset and the variation of the nozzle contour as much as possible, thereby reducing the local losses along the airflow path. However, serpentine nozzles with small offsets have limited shielding effect on the high-temperature components at the front end, resulting in a smaller suppression effect on the infrared radiation intensity of the exhaust system. Serpentine nozzles with larger offsets can effectively shield the high-temperature components at the front end, thus better suppressing the infrared radiation intensity of the exhaust system, but their negative impact on aerodynamic performance is also more pronounced. To address this issue, this section mainly studies a multi-objective optimization method for serpentine nozzle aerodynamics and infrared characteristics. Based on the radial basis neural network technology, an approximate model between the serpentine nozzle design parameters and the aerodynamic and infrared design objectives is established. On this basis, through an improved multi-objective particle swarm optimization algorithm, a multi-objective aerodynamic/infrared optimization design of the serpentine nozzle is carried out with the goals of high aerodynamic performance and low infrared radiation intensity.

3.6.1 Experimental Design of Serpentine Nozzle Optimization

Experimental design methods are a branch of natural science research methodologies (He et al. 2012). In multi-objective optimization design, experimental design is an important component of approximate modeling techniques. For complex systems, analysis and computation based on precise models require a large amount of time and effort. Therefore, in experimental design, it is desired to minimize the number of sample points to reduce the workload of precise model analysis and computation. At the same time, to obtain a reliable approximate model, the number of sample points must satisfy the needs of approximate modeling. How to obtain an approximate model that meets the requirements with fewer sample points, and realistically reflect the variation trend of design objectives with design parameters using the least number of trials, requires reasonable arrangement through experimental design methods. In this study, the aerodynamic performance and infrared radiation characteristics of the serpentine nozzle are both obtained through computer simulation analysis. Since there is no known model for reference in the analysis and computation of aerodynamic and infrared characteristics of serpentine nozzles, the "space-filling" experimental design method (Simpson et al. 1998; Palmer 1998) is adopted, which treats each region of the design space equally, and employs uniform design methods to arrange sample points. Uniform experimental design was first constructed by the Chinese scholar Fang Kaitai using number theory methods. Uniform design requires arrangement of experiments via normalized tables. Usually, $U_n\,(m^k)$ denotes a uniform design table, where U is the symbol of the uniform design table; n is the number of trials; m is the number of levels of factors; k is the maximum number of factors arranged. Uniform design tables are usually accompanied by usage tables, which need to be used together.

Considering comprehensively the influence of geometric design parameters of the serpentine nozzle on its aerodynamic performance and infrared radiation characteristics, one geometric parameter is selected in each of the three dimensions of the serpentine nozzle as an experimental design factor. Among them, the axial design parameter is the ratio of the two bend axial lengths L_1/L_2; the transverse design parameter is the outlet aspect ratio W_e/H_e; the vertical design parameter is the shielding ratio ΔY. The parameter value ranges are shown in Table 3.15. The aerodynamic optimization objective and infrared optimization objective of the serpentine nozzle are the thrust coefficient C_{fg} and the infrared radiation intensity IRSL, respectively. The infrared radiation intensity IRSL is taken as the average value at azimuth angles 0°, 5°, 10°, 15°, 20°, 30°, 40°, 60° and 90°.

The $U_{21}^*\left(21^7\right)$ uniform design table is adopted to conduct a 3-factor, 21-level experimental design. The $U_{21}^*\left(21^7\right)$ uniform design table and its usage table are shown in Tables 3.16 and 3.17, respectively. According to the usage table of $U_{21}^*\left(21^7\right)$, the first, third, and fourth columns of the $U_{21}^*\left(21^7\right)$ uniform design table are selected for the experimental design.

Finally, the parameter values of the 21 experimental samples of the serpentine nozzle are shown in Table 3.18. For each of these 21 serpentine nozzle experimental

Table 3.15 Design parameter value ranges

L_1/L_2	0.5 ~ 1.5
We/He	3 ~ 11
ΔY	0 ~ 1

Table 3.16 $U_{21}^{*}(21^{7})$ Uniform design table

	Column no.						
Test no.	1	2	3	4	5	6	7
1	1	5	7	9	13	17	19
2	2	10	14	18	4	12	16
3	3	15	21	5	17	7	13
4	4	20	6	14	8	2	10
5	5	3	13	1	21	19	7
6	6	8	20	10	12	14	4
7	7	13	5	19	3	9	1
8	8	18	12	6	16	4	20
9	9	1	19	15	7	21	17
10	10	6	4	2	20	16	14
11	11	11	11	11	11	11	11
12	12	16	18	20	2	6	8
13	13	21	3	7	15	1	5
14	14	4	10	16	6	18	2
15	15	9	17	3	19	13	21
16	16	14	2	12	10	8	18
17	17	19	9	21	1	3	15
18	18	2	16	8	14	20	12
19	19	7	1	17	5	15	9
20	20	12	8	4	18	10	6
21	21	17	15	13	9	5	3

Table 3.17 Usage table of $U_{21}^{*}(21^{7})$

Number of factors	Column no.					D
2	1	5				0.0697
3	1	3	4			0.1121
4	1	2	3	5		0.1381
5	1	4	5	6	7	0.1759

Table 3.18 Serpentine nozzle experimental samples and aerodynamic/infrared response values

Samples	L_1/L_2	W_e/H_e	ΔY	C_{fg}	IRSL
N1	0.5	5.4	0.4	0.9585	0.7631
N2	0.55	8.2	0.85	0.9522	0.7661
N3	0.6	11	0.2	0.9498	0.7561
N4	0.65	5	0.65	0.9591	0.7710
N5	0.7	7.8	0	0.9565	0.7549
N6	0.75	10.6	0.45	0.9510	0.7562
N7	0.8	4.6	0.9	0.9547	0.7710
N8	0.85	7.4	0.25	0.9579	0.7540
N9	0.9	10.2	0.7	0.9516	0.7576
N10	0.95	4.2	0.05	0.9645	0.7630
N11	1	7	0.5	0.9586	0.7610
N12	1.05	9.8	0.95	0.9507	0.7633
N13	1.1	3.8	0.3	0.9654	0.7640
N14	1.15	6.6	0.75	0.9579	0.7638
N15	1.2	9.4	0.1	0.9556	0.7561
N16	1.25	3.4	0.55	0.9649	0.7726
N17	1.3	6.2	1	0.9381	0.7670
N18	1.35	9	0.35	0.9563	0.7465
N19	1.4	3	0.8	0.9488	0.7833
N20	1.45	5.8	0.15	0.9636	0.7612
N21	1.5	8.6	0.6	0.9561	0.7606

samples, modeling, meshing, and numerical simulation calculations are performed to obtain the aerodynamic performance response value C_{fg} and the infrared response value IRSL, as shown in Table 3.18. The infrared response value IRSL is normalized.

3.6.2 *Influence of Serpentine Nozzle Design Parameters on Optimization Objectives*

When using the uniform design method for experimental design, due to the lack of neat comparability of the experimental results, general analysis of variance cannot be applied. Therefore, multiple regression analysis is usually adopted to analyze the results to obtain the influence relationship between multiple factors and the target response value. In the analysis of the results of the uniform design method, cubic terms and the interaction effects among more than two factors are generally not considered. Thus, the response function of multiple regression can be expressed as:

$$\hat{y} = b_0 + \sum_{j=1}^{m} b_j x_j + \sum_{j,k=1}^{m} b_{jk} x_j x_k + \sum_{j=1}^{m} b_{jj} x_j^2 \tag{3.33}$$

where M is the total number of factors; x_j^2 is the quadratic term of factor j.

In Eq. (3.33), $x_j x_k$ reflects the interaction between factor j and factor k. When the interaction among factors is not considered, Eq. (3.33) can be further simplified as:

$$\hat{y} = b_0 + \sum_{j=1}^{m} b_j x_j + \sum_{j=1}^{m} b_{jj} x_j^2 \tag{3.34}$$

where the regression coefficients b_0, b_j, and b_{jj} (j = 1, 2, ..., m) can be solved by referring to relevant textbooks and literature.

In the multi-objective optimization study of aerodynamic/infrared performance of the serpentine nozzle, without considering the mutual influence among the factors, the relationship between the target response and the factors can be described by the following equation:

$$\hat{y} = b_0 + b_1 x_1 + b_2 x_2 + b_3 x_3 + b_4 x_1^2 + b_5 x_2^2 + b_6 x_3^2 \tag{3.35}$$

where $\hat{y}$ represents the target response C_{fg} or IRSL; x_1, x_2 and x_3 respectively represent the factors L_1/L_2, W_e/H_e, and ΔY. Assuming $x_4 = x_1^2$, $x_5 = x_2^2$ and $x_6 = x_3^2$, then Eq. (3.35) can be linearized as:

$$\hat{y} = b_0 + b_1 x_1 + b_2 x_2 + b_3 x_3 + b_4 x_4 + b_5 x_5 + b_6 x_6 \tag{3.36}$$

Using the uniform design experimental data in Table 3.16, statistical analysis software can conveniently be used to obtain the variance analysis results of the regression equations for different objective functions (C_{fg} and IRSL), as shown in Table 3.19. In the table, the F value is the result of variance analysis, representing the test of the overall regression equation, and Sig. is the significance value corresponding to the F value. It can be seen that for both objective functions C_{fg} and IRSL, the Sig. values are less than 0.05. Therefore, it can be considered that the established regression equations are effective and can reflect the relationship between the factors and the responses to a certain extent.

Since the absolute value of the regression coefficient b_j is related to the unit of the corresponding factor, in order to compare the degree of influence of each factor on the response value, it is necessary to standardize the units of the factors to obtain the standardized regression coefficients b_j'. The larger the absolute value of b_j', the

Table 3.19 Analysis of variance table for regression equations

Objective function	Source of variation	Sum of squares	Degrees of freedom	Variance	F	Sig.
C_{fg}	Regression	0.000586	6	0.000098	5.681721	0.003550
	Residual	0.000241	14	0.000017		
	Statistics	0.000826	20			
IRSL	Regression	0.001112	6	0.000185	17.449604	0.000009
	Residual	0.000149	14	0.000011		
	Statistics	0.001261	20			

greater the influence of the corresponding factor on the response. Table 3.20 lists the regression coefficients b_j and the standardized regression coefficients $b_j^{'}$ of the regression equations for the two objective functions C_{fg} and IRSL. It can be seen that for the objective function C_{fg}, we have $\left|b_1^{'}\right| > \left|b_3^{'}\right| > \left|b_2^{'}\right|$, therefore, the priority order of the three factors is $x_1 > x_3 > x_2$, i.e. under the same standard, the axial length ratio of the two bends L_1/L_2 has the greatest impact on the thrust coefficient, while the outlet aspect ratio W_e/H_e has the least impact. For the objective function IRSL, we have $\left|b_2^{'}\right| > \left|b_1^{'}\right| > \left|b_3^{'}\right|$, therefore, the priority order of the three factors is $x_2 > x_1 > x_3$, i.e. under the same standard, the outlet aspect ratio W_e/H_e has the greatest impact on the infrared radiation intensity, while the shielding ratio ΔY has the least impact.

3.6.3 Aerodynamic/Infrared Multi-objective Optimization of the Serpentine Nozzle

The aerodynamic/infrared multi-objective optimization process of the serpentine nozzle mainly consists of two steps: surrogate modeling and multi-objective optimization. In the process of constructing surrogate models for the aerodynamic performance objective C_{fg} and the infrared objective IRSL using the radial basis function neural network technique, the expansion coefficient δ was optimized within the range of 0.01–2, with a step size of 0.01. Figure 3.70 shows the variation of the average error of sample objective responses with δ during the optimization process. It can be seen that for the objective C_{fg}, the average error of the samples is minimized when $\delta = 1.15$. For the objective IRSL, the minimum average error occurs at $\delta = 0.23$. Therefore, when building the surrogate models, δ is set to 1.15 for C_{fg} and 0.23 for IRSL, respectively.

After constructing the surrogate models between the optimization objectives and the design parameters of the serpentine nozzle, an improved particle swarm multi-objective optimization algorithm was employed to carry out the aerodynamic/

Table 3.20 Regression coefficients and standardized regression coefficients of the regression equations

$\hat{y}$ represents C_{fg}			$\hat{y}$ represents IRSL		
Model	Regression coefficient b_j	Standardized regression coefficient $b_j^{'}$	Model	Regression coefficient b_j	Standardized regression coefficient $b_j^{'}$
b_0	0.942616		b_0	0.808966	
b_1	0.030312	1.462904	b_1	−0.031860	−1.244858
b_2	0.001745	0.673766	b_2	−0.008725	−2.727161
b_3	0.019911	0.960938	b_3	0.000259	0.010103
b_4	−0.015254	−1.485718	b_4	0.015096	1.190338
b_5	−0.000213	−1.163169	b_5	0.000465	2.060324
b_6	−0.032133	−1.606232	b_6	0.012659	0.512314

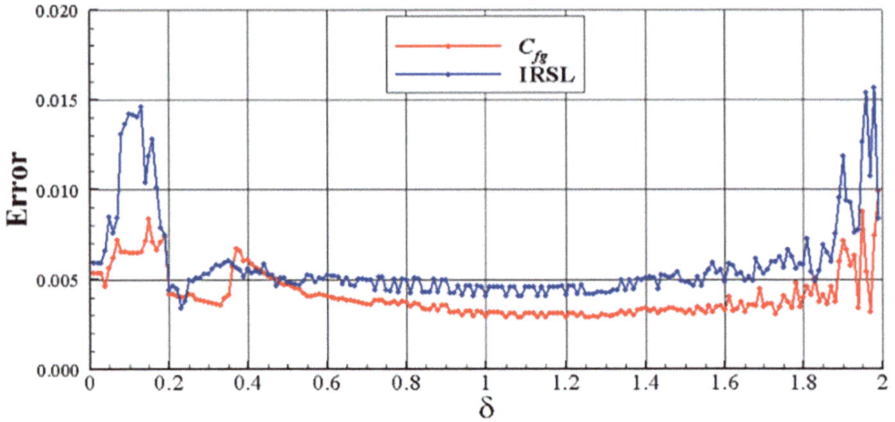

Fig. 3.70 Optimization process of expansion constant δ

infrared multi-objective optimization of the serpentine nozzle. During the optimization process, the population size of the particle swarm was set to 800, the external archive size to 200, and the maximum number of iterations to 5000. Figures 3.71 and 3.72 show the distributions of the samples and the optimized Pareto solutions in the design space and the objective space, respectively. As shown in Fig. 3.71, the three optimization parameters of the serpentine nozzle converge toward a certain region. Table 3.21 lists the range of design parameters for the obtained Pareto solutions. From Fig. 3.72, it can be seen that the Pareto solutions move in the direction of increasing C_{fg} and decreasing IRSL, which aligns with the optimization goals of high aerodynamic performance and low infrared radiation intensity for the serpentine nozzle.

To verify the reliability of the optimization results, two Pareto solutions were selected from the Pareto solution set for validation. The design parameters of the selected Pareto solutions are shown in Table 3.22. Based on the design parameters of the Pareto solutions to be validated, modeling, meshing, and numerical simulation were carried out for the serpentine nozzles, and the objective function values of the two validation points were obtained, as shown in Table 3.23. Figure 3.62 shows the position of the two validation points relative to the experimental samples and the Pareto solution set in the objective space. It can be seen that, due to the error of the surrogate model, there is a slight deviation between the validation points and the Pareto solution set, but both validation points are distributed near the Pareto solution set and show good agreement.

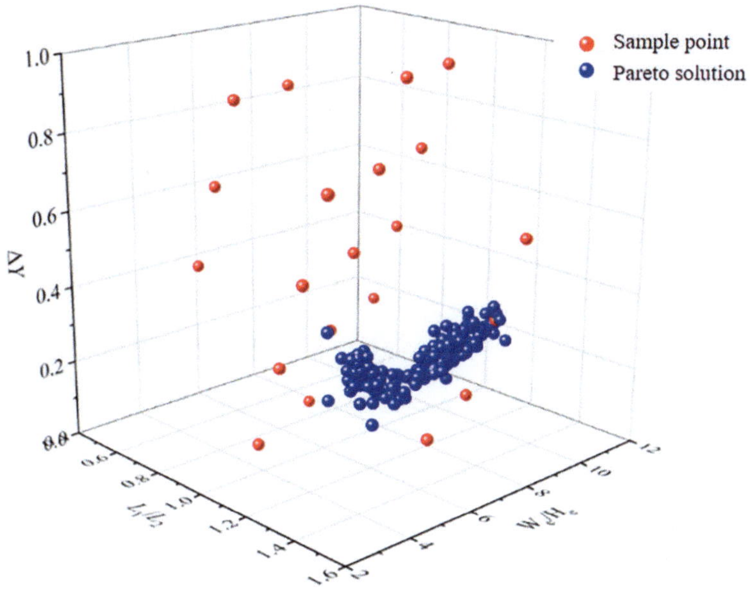

Fig. 3.71 Distribution of particles in the design space

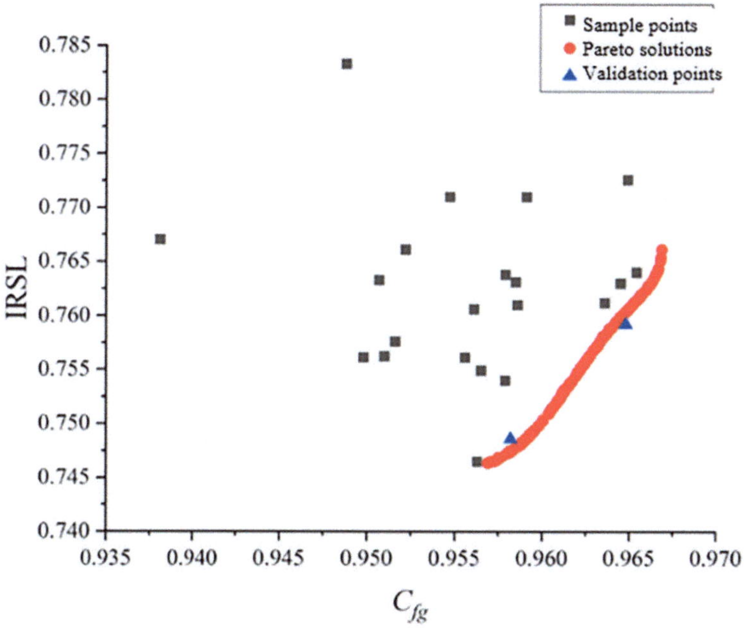

Fig. 3.72 Distribution of particles in the objective space

Table 3.21 Design parameter ranges of pareto solutions

L_1/L_2	1.22 ~ 1.44
We/*He*	4.11 ~ 9.17
ΔY	0.21 ~ 0.42

Table 3.22 Design parameters of Pareto solutions for validation

Pareto solutions	L_1/L_2	W_e/H_e	ΔY
V1	1.33	8.11	0.33
V2	1.37	4.80	0.26

Table 3.23 Objective function values of validation points

Validation points	C_{fg}	IRSL
V1	0.9582	0.7486
V2	0.9648	0.7592

Chapter 4
Design Methodology and Application of Three-Bearing Swivel Nozzle

4.1 Introduction

Short/vertical takeoff and landing (S/VTOL) aircraft combine the advantages of helicopters and conventional fighter jets. The aircraft realizes S/VTOL through the lift generated by the propulsion system. Therefore, the propulsion system for S/VTOL aircraft has always been the focus of research in various developed aviation countries (Anderson 1983; Ransone 2002). The three-bearing swivel nozzle has emerged as a key research focus for S/VTOL thrust vectoring nozzles due to its structural simplicity, lightweight design, and high degree of freedom characteristics (Suo et al. 2014). Comprising three mutually rotatable cylindrical sections, the nozzle achieves thrust vectoring by adjusting the rotational angles and combinations of these sections (Liu et al. 2015).

During near-ground vertical takeoff/landing, S/VTOL aircraft experience interference between the high-speed exhaust jet and ground surfaces. The jet's entrainment effect reduces pressure on the aircraft's lower surface, causing lift loss. Concurrently, engine intakes ingest high-temperature exhaust gases, inducing temperature distortion in inlet airflow and compromising engine stability (Smith et al. 1991; Barata 1993; Cabrita et al. 2002; Saddington et al. 2009).

This chapter investigates the design methodology and application of three-bearing swivel nozzles. The rationality of numerical simulation methods is first validated through experimentally obtained nozzle wall static pressure distributions. Aerodynamic contour design is performed using numerical simulations, examining both the nozzle's aerodynamic performance and thrust vectoring implementation. Based on a hybrid propulsion system configuration, lift losses under dual-jet near-ground conditions are quantified. Thermal model testing further verifies the integrated compatibility between the three-bearing swivel nozzle and aero-engines.

J. Shi et al., *Special Purpose Nozzles*,
https://doi.org/10.1007/978-981-95-2939-1_4

4.2 Aerodynamic Contour Design Methodology for Three-Bearing Swivel Nozzle

4.2.1 Geometric Characteristics Analysis of Three-Bearing Swivel Nozzle

The three-bearing swivel nozzle is primarily composed of an engine transition adapter, three cylindrical nozzle sections, three-bearing assemblies, sealing mechanisms, an actuation system, and a control system. Adjacent cylindrical sections are connected via bearings to enable relative rotation (Kentfield 1967). To achieve thrust vectoring capability and high aerodynamic performance, the contour design of the three cylindrical sections must meet specific profile requirements, while their relative rotations adhere to predefined kinematic laws, as illustrated in Fig. 4.1. The structural configuration and operating principles of the three-bearing swivel nozzle differ fundamentally from conventional convergent/divergent nozzles, 2D vectoring nozzles, and axisymmetric vectoring nozzles, resulting in unique contour design characteristics.

Figure 4.2 provides axial cross-sectional views of each section: Section 0 denotes the interface between the engine adapter and the first cylindrical section, while sections 12 and 23 represent interconnection interfaces between the cylindrical sections. To minimize frictional resistance during rotation, all interconnection sections (0, 12, 23) utilize standardized bearings with circular cross-sections. The exit section 3 of the N4 convergent nozzle retains a circular cross-section through

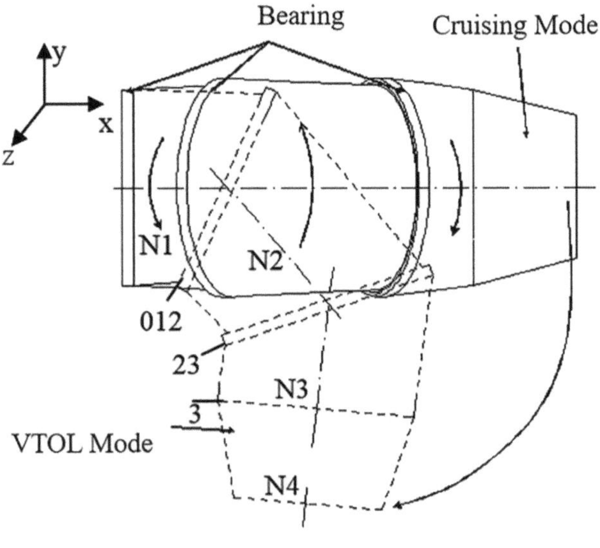

Fig. 4.1 Working principle diagram of three-bearing swivel nozzle

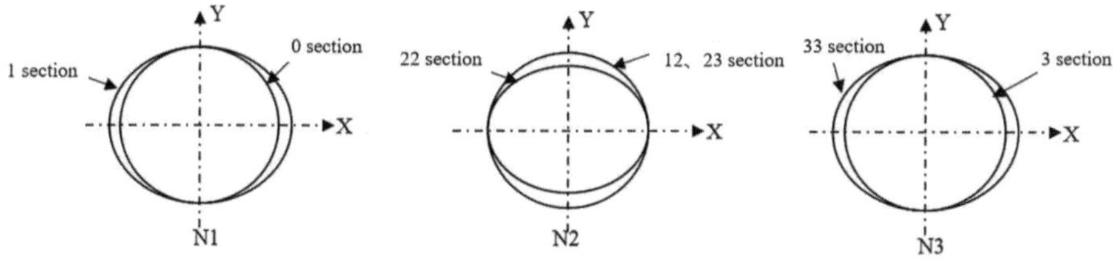

Fig. 4.2 Axial cross-sectional views of each cylindrical section

proportional scaling of a turbojet engine exhaust reference. Geometric analysis reveals the following configuration:

- The N1 segment integrates a circular-to-elliptical transition and an elliptical section.
- The N2 segment constitutes a pure elliptical cylinder.
- The N3 segment combines an elliptical section with an elliptical-to-circular transition.

Critical reference planes include Sect. 11 (terminal plane of the N1 transition segment), section 33 (initial plane of the N3 transition segment), and section 22 (cross-section of the N2 segment). The simplified cross-sectional schematic is depicted in Fig. 4.2.

4.2.2 *Contour Design of Three-Bearing Swivel Nozzle*

Based on the preceding analysis, the contour design of the three-bearing swivel nozzle can be divided into three components: the constant-diameter section, transition section, and convergent section. The constant-diameter section serves for connection and flow rectification. Its design methodology involves equally dividing the section length into N segments (N being a positive integer), with the y-coordinates of discrete points remaining unchanged. All elliptical sections in this study adopt this constant-diameter design.

The transition section ensures smooth contour transitions to minimize aerodynamic losses. To effectively mitigate secondary flow losses caused by uneven pressure gradients along the inner wall, the transition contour is designed using Eqs. (2.1), (2.2), and (2.3) from Chap. 2, following the theoretical framework of Lee et al. and operational requirements. For the circular-to-elliptical transition design, a moderate variation strategy is implemented to balance area change and profile width adjustment, thereby reducing transition losses. The design procedure involves:

(1) Profile area calculation. Assuming the inlet area of the transition section is A_{in} and the outlet area is A_{ex}, then $\Delta A = A_{ex} - A_{in}$. A controlled area variation law is applied to determine the profile area at each control point.

$$A_i = A_{in} + \Delta A\left[3\left(\frac{x_i}{L}\right)^2 - 2\left(\frac{x_i}{L}\right)^3\right] \tag{4.1}$$

(2) Profile width calculation. For the transition section, let the inlet width be W_{in} and the outlet width W_{ex}, yielding a width difference $\Delta W = W_{ex} - W_{in}$. A moderate-width variation law is applied to determine the profile width at each control point.

$$W_i = W_{in} + \Delta W\left[3\left(\frac{x_i}{L}\right)^2 - 2\left(\frac{x_i}{L}\right)^3\right] \tag{4.2}$$

(3) Surface generation. Given the area and width at each control point, since the transition surfaces in this book are all ellipses, assuming the height (the length of the long or short half axis of the ellipse) is H_i, and calculate the height of the ellipse based on the ellipse area formula.

$$A_i = \pi \times W_{in} \times \frac{H_i}{4} \tag{4.3}$$

The parameter equation of the ellipse can be known, and the coordinates of the characteristic points can be obtained in the spatial coordinate system, discretizing the ellipse to generate the profile curve. Connect the corresponding feature points to generate longitudinal lines, and use the profile lines and longitudinal lines to generate internal profiles using multi-section surfaces in CATIA software to obtain the design of transition profiles. The length of the transition section is determined by other design elements.

The convergent profile is designed using the Witoszynski formula to define wall contours:

$$\left(\frac{r_e}{r}\right)^2 = 1 - \left(1 - \frac{1}{C}\right)\frac{\left[1-(x/l)^2\right]^2}{\left[1+\frac{1}{3}(x/l)^2\right]^3} \tag{4.4}$$

where C denotes the contraction ratio $C = (r_0/r_e)^2$, with parameters illustrated in Fig. 4.3. Here, l (selected as $l>r_0$) is adjustable within a broad range, making this profile suitable for connecting ducts of differing dimensions in subsonic wind tunnels. In the current design, empirical values $r_0 = 2r_e$ and $l = 2r_0/3$ are adopted to optimize flow quality.

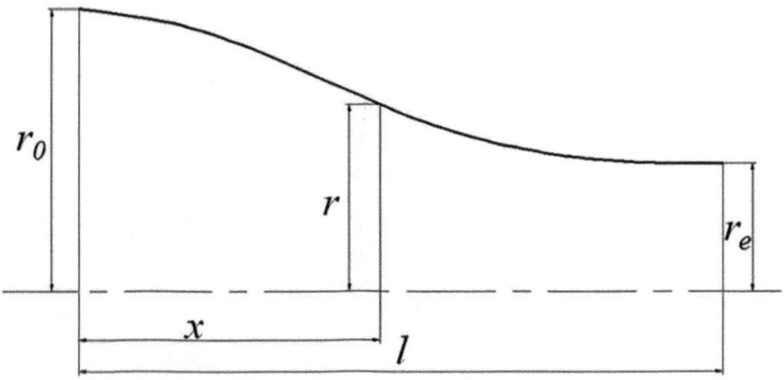

Fig. 4.3 Schematic diagram of contraction section surface design

4.3 Numerical Validation of Three-Bearing Swivel Nozzle Simulation Methodology

Under thrust vectoring conditions, the airflow characteristics within the three-bearing swivel nozzle resemble those in curved ducts, with potential secondary flow formation and pressure differentials between the inner and outer walls. This section validates the numerical simulation methodology by analyzing flow features in both non-vectoring and 95° deflection states. Experimental static pressure distributions along the inner/outer walls are obtained across multiple pressure ratio conditions and compared with numerical results to verify simulation accuracy. The experimental bench installation of the nozzle at 95° deflection is shown in Fig. 4.4.

4.3.1 Experimental Bench

The experimental study is conducted using the low-pressure ratio experimental bench at Nanjing University of Aeronautics and Astronautics. Wall static pressure measurements provide comprehensive experimental data to validate numerical simulations, aerodynamic contour design, and performance analysis.

The low-pressure ratio wind tunnel experimental bench (Fig. 4.5) comprises: high/low-pressure air sources, main flow valve bank, transition section, three-component balance, total pressure probe, test chamber, nozzle model and ejector duct.

Auxiliary systems include pressure scanning valves, schlieren apparatus, and balance signal acquisition. The air supply delivers inflow at up to 6atm absolute pressure, while the vacuum chamber operates below 0.03atm. A 48-channel pressure scanning valve system measures wall pressure distributions:

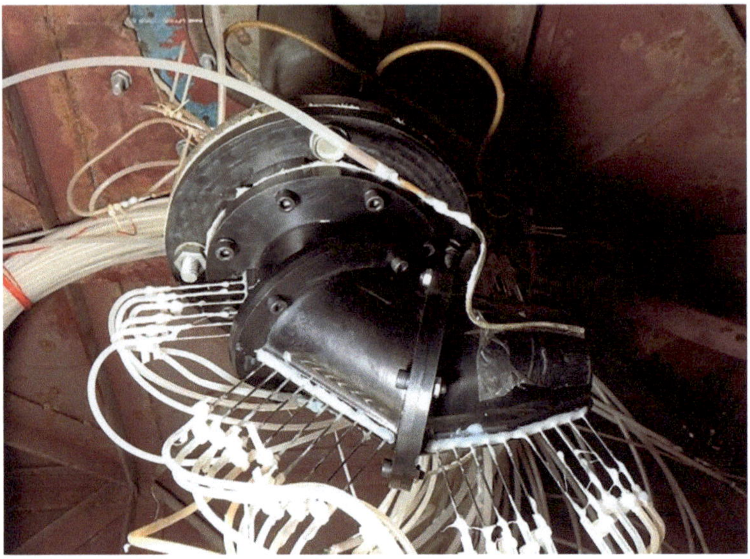

Fig. 4.4 Experimental bench installation view of the nozzle at 95° deflection

Fig. 4.5 The low-pressure ratio wind tunnel experimental bench

- Modules A1–A8: 60 PSI range
- Modules A1–A8: 60 PSI range
- Modules A1–A8: 60 PSI range

4.3.2 Experimental Bench

Figures 4.6 and 4.7 show the 3D views of the nozzle test article in non-vectoring and 95° deflection configurations, respectively. The test article comprises three independent cylindrical sections fixed by screws, forming either non-vectoring or 95° deflection profiles. The nozzle features an inlet-to-outlet area ratio of 2.23, with 32 measurement points distributed along the wide and narrow edges of the upper and lower walls. For cross-referencing experimental results, the upper wall measurement points in the non-vectoring state are categorized as Groups A1 and A2, while the lower wall points are grouped as B1, B2, and B3.

Two experimental groups are conducted based on nozzle deflection angles to measure pressure distributions on the upper and lower walls under multiple pressure ratio conditions. Specific test configurations are listed in Table 4.1.

According to the experimental plan for measuring the static pressure distribution of the three-bearing deflection nozzle, the nozzle pressure ratio is changed and the pressure measurement of the two states of the nozzle is continuously measured. The experimental steps are as follows:

(1) Inspect flatness of pressure taps, ventilation, and airtightness of static pressure tubes;
(2) Record ambient atmospheric pressure and temperature;
(3) Activate vacuum pumps and verify rig airtightness until chamber pressure meets test requirements;
(4) Sequentially measure critical aerodynamic parameters: inflow total pressure, wall static pressure distribution, and nozzle pressure ratio;

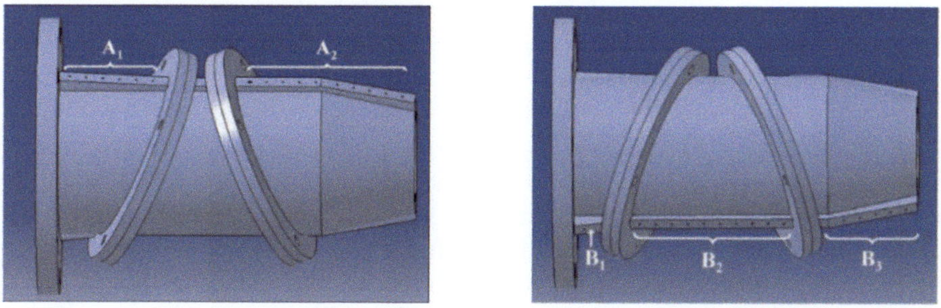

(a) Upper wall measurement layout

(b) Lower wall measurement layout

Fig. 4.6 3D view of the nozzle test article in non-vectoring state. (**a**) Upper wall measurement layout. (**b**) Lower wall measurement layout

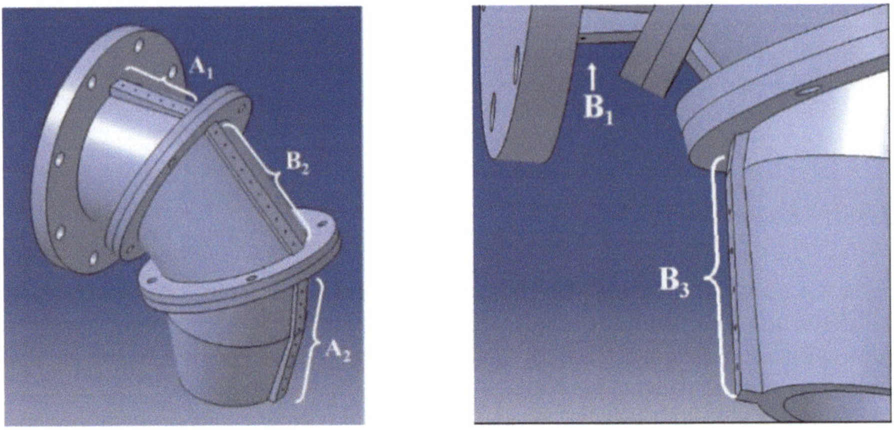

(a) Upper wall measurement layout (b) Lower wall measurement layout

Fig. 4.7 3D view of the nozzle test article at 95° deflection. (**a**) Upper wall measurement layout. (**b**) Lower wall measurement layout

Table 4.1 Test configurations for pressure measurement of nozzle wall

Experimental grouping	Experimental nozzle pressure ratio
Non-vector state of the nozzle	1.3, 1.4, 1.5, 1.6, 2, 2.5, 3, 4
95° deflection state of the nozzle	1.4, 1.5, 1.6

(5) Repeat measurements across varying mainstream pressure ratios until all target pressure ratio conditions are acquired;
(6) Deactivate vacuum pumps and valves after completing all tests.

4.3.3 Numerical Validation of Flow Characteristics in Three-Bearing Swivel Nozzle

4.3.3.1 Cold-Flow Experimental and Numerical Validation Results for Non-vectoring State

Three-dimensional CFD simulations were conducted for the non-vectoring nozzle flow field. A structured O-grid with wall refinement was employed, ensuring wall $y^+ = 1$–3. The governing equations utilized the 3D conservative Navier-Stokes formulation, with second-order upwind schemes for convective terms, turbulent kinetic energy, and dissipation rate, while central differencing was applied to viscous terms. Turbulence models evaluated included: Spalart-Allmaras (S-A) one-equation model, Standard k-ε model (enhanced wall treatment) and RNG k-ε model (enhanced wall treatment).

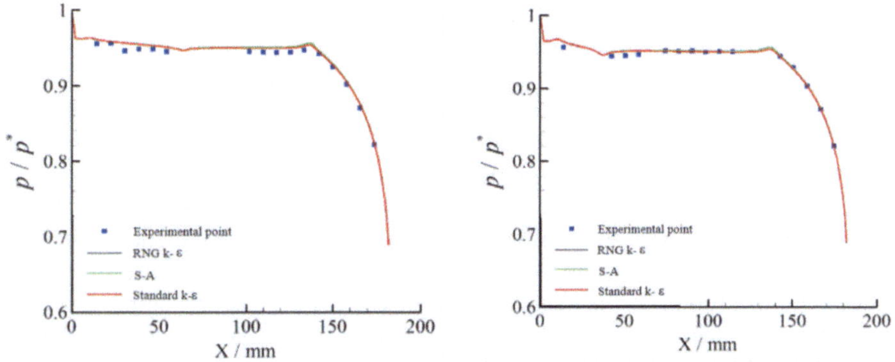

(a) Upper wall pressure distribution

(b) Lower wall pressure distribution

Fig. 4.8 Comparison of experimental and numerical simulation results of pressure distribution under NPR = 1.5. (**a**) Upper wall pressure distribution. (**b**) Lower wall pressure distribution

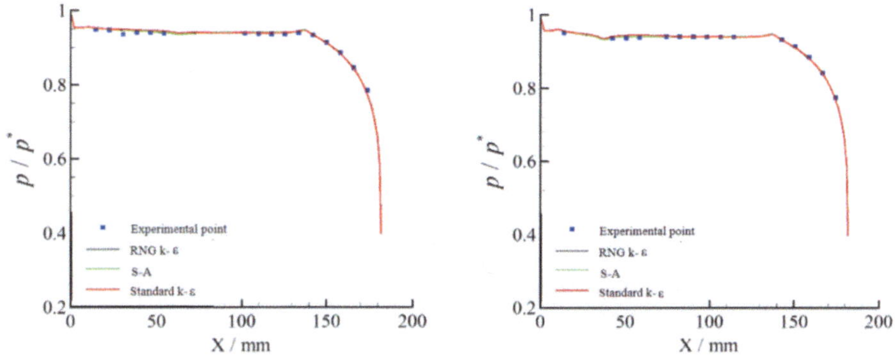

(a) Upper wall pressure distribution

(b) Lower wall pressure distribution

Fig. 4.9 Comparison of experimental and numerical simulation results of pressure distribution under NPR = 4. (**a**) Upper wall pressure distribution. (**b**) Lower wall pressure distribution

Figures 4.8 and 4.9 present dimensionless static pressure distributions (normalized by inlet total pressure) on the upper/lower walls at pressure ratios of 1.5 and 4, respectively. All turbulence models yielded similar predictions that closely matched experimental data. Upstream of the convergent section, flow behavior resembled straight duct flow with minimal pressure variation. Accelerated flow in the convergent section induced pressure reduction near the wall, with high agreement between simulations and experiments. Thus, all models demonstrated comparable accuracy for straight and convergent wall pressure predictions under non-vectoring conditions.

4.3.3.2 Cold-Flow Experimental and Numerical Validation Results for 95° Deflection State

For the 95° deflection case, turbulence models assessed were: S-A one-equation model, Standard k-ε model (enhanced wall treatment), RNG k-ε model (enhanced wall treatment), Realizable k-ε model (enhanced wall treatment) and SST k-ω model with compressibility effects (two-layer wall treatment). Figure 4.10 shows dimensionless pressure distributions at a pressure ratio of 1.5. All models exhibited nearly identical upper wall pressure predictions aligned with experiments. On the lower wall of section 3, the standard and Realizable k-ε models showed closest agreement with test data, while RNG k-ε produced the largest deviation, followed by SST and S-A models. The predictive accuracy for section 3 lower wall pressures reflects each model's capability to resolve vortical and separated flows influenced by upstream flow from section 2.

Figures 4.11 and 4.12 compare Standard k-ε predictions with experimental results at pressure ratios 1.4 and 1.6. It can be seen from the figure that the prediction results of the Standard k-ε model for the wall pressure distribution of nozzles have a high degree of consistency with the experimental results. Therefore, this book selects the Standard k-ε turbulence model to conduct a study on the aerodynamic performance of three-bearing swivel nozzles.

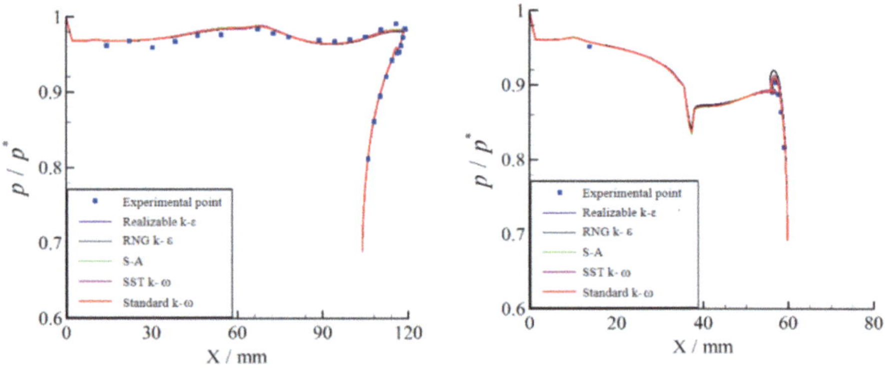

(a) Upper wall pressure distribution

(b) Lower wall pressure distribution

Fig. 4.10 Comparison of experimental and numerical simulation results of pressure distribution under NPR = 1.5. (**a**) Upper wall pressure distribution. (**b**) Lower wall pressure distribution

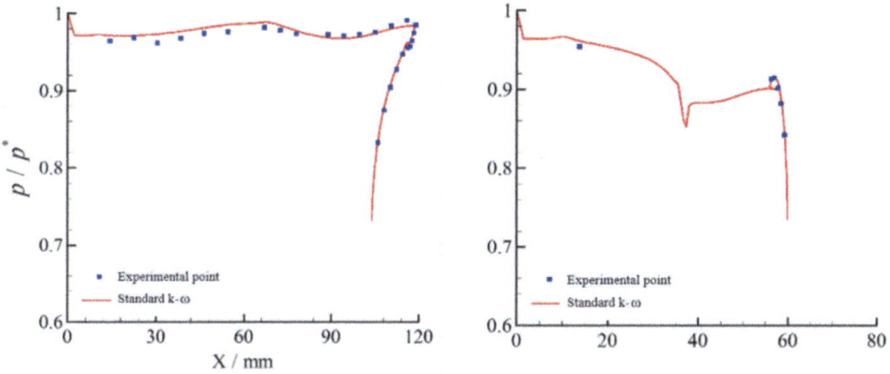

(a) Upper wall pressure distribution (b) Lower wall pressure distribution

Fig. 4.11 Comparison of experimental and numerical simulation results of pressure distribution under NPR = 1.4. (**a**) Upper wall pressure distribution. (**b**) Lower wall pressure distribution

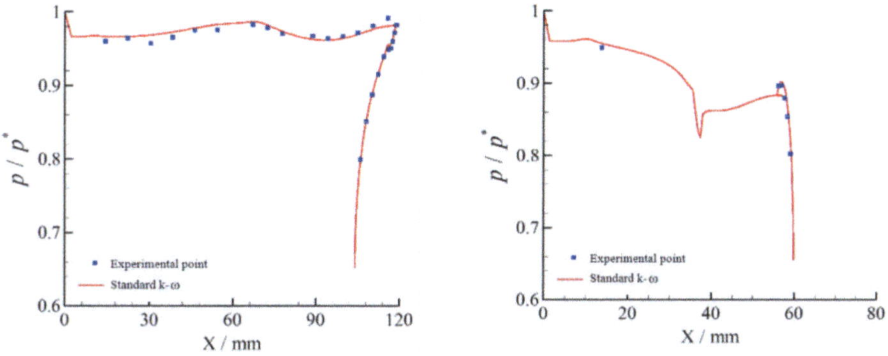

(a) Upper wall pressure distribution (b) Lower wall pressure distribution

Fig. 4.12 Comparison of experimental and numerical simulation results of pressure distribution under NPR = 1.6. (**a**) Upper wall pressure distribution. (**b**) Lower wall pressure distribution

4.4 Design of Three-Bearing Swivel Nozzle

4.4.1 Aerodynamic Contour Design for Non-vectoring State

4.4.1.1 Geometric Model and Area Variation Law

The 3D solid model of the three-bearing swivel nozzle includes three mutually rotatable cylindrical sections and a fixed section connected to the engine, with detailed geometric dimensions shown in Fig. 4.13. The fixed section features an annular inlet and a hemispherical tail cone at its center. The nozzle inlet total pressure ($p^* = 155817$ Pa) and total temperature ($T^* = 1072$ K) are defined based on the

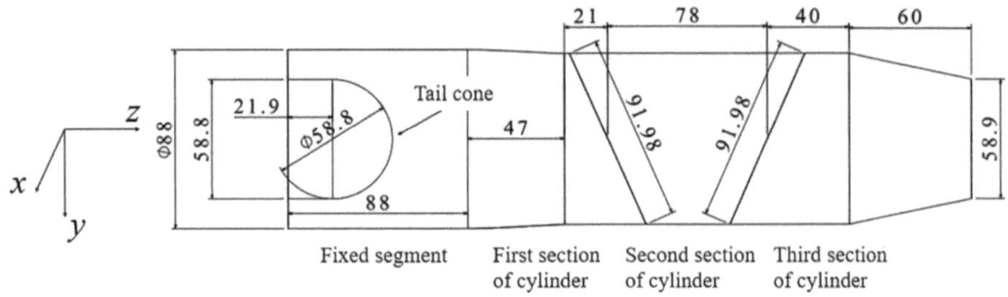

Fig. 4.13 Geometric model of nozzle

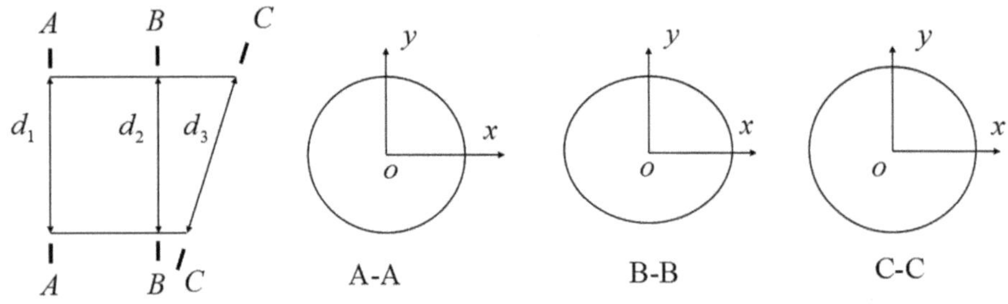

Fig. 4.14 Schematic diagram of the cylinder body shape of the nozzle transition section

turbine outlet conditions of a turbojet engine. Flow field characteristics are obtained by solving the Navier-Stokes equations using the standard k-ε turbulence model.

As analyzed in the geometric characteristics, the first cylindrical section incorporates a circular-to-elliptical transition contour. Three area variation laws for the transition section are proposed, and corresponding geometric models under non-vectoring conditions are developed using 3D modeling software.

Figure 4.14 illustrates the transition section contour, in which cross-section A to cross-section B represents transitional profiles from circular to elliptical sections, cross-section B to cross-section C corresponds to elliptical cylindrical bodies, and circular section C denotes the beveled surfaces of these elliptical cylindrical bodies. d_1 is the diameter of cross section A, d_2 is the elliptical minor axis of cross section B, and d_3 is the diameter of cross section C. The area variation rules of the three transition section cylinders are:

1. $A_1 > A_2$, and $d_1 = d_3$;
2. $A_1 = A_2$;
3. $A_1 < A_2$, and $d_1 = d_2$.

4.4.1.2 Aerodynamic Contour Optimization

Numerical simulation results for nozzle performance under three area variation laws are listed in Table 4.2. When $A_1 > A_2$, the total pressure recovery coefficient, flow coefficient, and thrust coefficient at the nozzle outlet are higher than the performance parameters under the other two types of area variation laws. The aerodynamic performance of the nozzle under this area variation law is optimal. The nozzle thrust coefficient under $A_1 = A_2$ condition is 0.9% lower than that under $A_1 > A_2$ condition, and the nozzle thrust coefficient under $A_1 < A_2$ condition is 1.8% lower than that under $A_1 > A_2$ condition. By analyzing the flow field characteristics inside the nozzle, the influence mechanism of the transition section area variation on the aerodynamic performance of the nozzle can be obtained.

Figure 4.15 shows the total pressure and streamline distribution of the airflow on the symmetrical plane and D-section of the nozzle under three different area variation patterns. From the streamline distribution in the figure, it can be seen that the flow trends of the airflow under the three types of area variation laws are generally the same. After passing through the tail cone, due to the expansion of the flow area, a portion of the airflow generates a recirculation zone behind the tail cone, and the total pressure of the airflow in the recirculation zone decreases. Another portion of the airflow bypasses the recirculation zone and mixes after the recirculation zone, and the total pressure of the airflow in the central area of the nozzle gradually increases. Comparing the total pressure distribution of the airflow on the symmetrical plane and D-section under three different area change patterns, it can be seen that the recirculation zone behind the tail cone is the smallest when $A1 > A2$, and the recirculation zone behind the tail cone is larger under the conditions of $A1 = A2$ and $A1 < A2$. This is because after the airflow passes through the tail cone, the expansion of the transition section surface causes a part of the airflow outside the central area to flow along the expansion wall and another part to flow towards the central area. The flow of the airflow along the expansion wall weakens the restriction on the recirculation zone, resulting in an increase in the recirculation zone and flow loss inside the nozzle. Therefore, the cross-sectional area of the transition section will affect the recirculation zone after the tail cone, thereby affecting the flow loss of the airflow inside the nozzle.

Figure 4.16 shows the total pressure distribution of the nozzle outlet airflow. As shown in the figure, the total pressure distribution results under the three types of area variation laws correspond to the total pressure recovery coefficient results in Table 4.2. The total pressure distribution gradient at the nozzle outlet is the largest under the condition of $A_1 > A_2$. As the cross-sectional area of the transition section increases, the total pressure at the nozzle outlet gradually decreases and tends to be uniform. This indicates that as the cross-sectional area of the transition section increases, the range of the recirculation zone inside the nozzle increases, and the annular inlet airflow mixes more evenly inside the nozzle, resulting in an increase in energy loss.

Table 4.2 Numerical simulation results of nozzle aerodynamic performance under three types of area variation laws

	σ	C_d	C_{fg}
$A_1 > A_2$	0.9751	0.9154	0.9677
$A_1 = A_2$	0.9745	0.9053	0.959
$A_1 < A_2$	0.9737	0.8952	0.9502

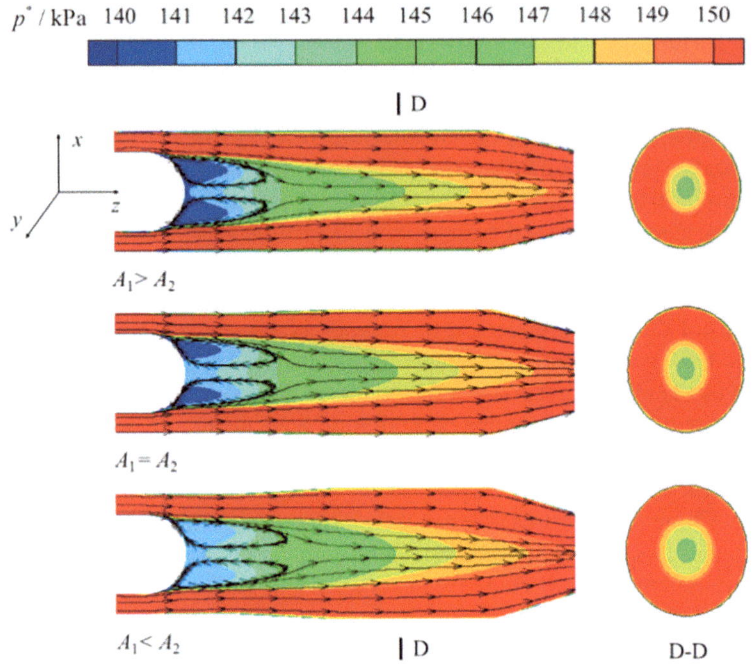

Fig. 4.15 Total pressure distribution of nozzle symmetry plane and D-section airflow

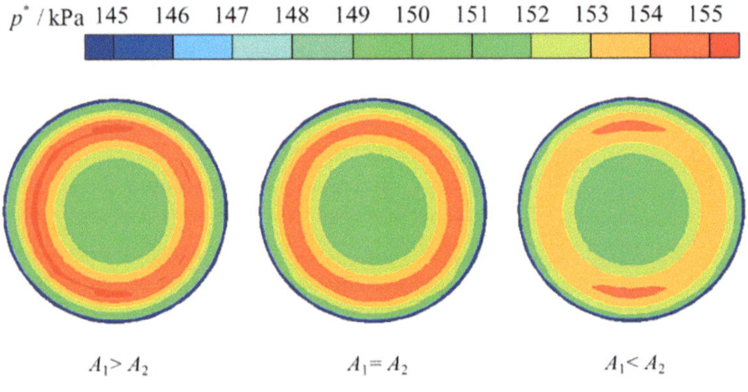

Fig. 4.16 Total pressure distribution of airflow at the nozzle outlet section

4.4.2 Aerodynamic Contour Design for Thrust-Vectoring State of Three-Bearing Swivel Nozzle

During vertical takeoff/landing, the nozzle deflects to 90° to provide the required lift, making this state critical for aerodynamic contour design.

Figure 4.17 is a schematic diagram of the profile of the nozzle at a 90° angle. l_1, l_2, and l_3 in the figure are the axis lengths of the three sections of the nozzle, and *d* is the diameter of the inlet section of the nozzle and the oblique circle section of each section of the nozzle. From the figure, it can be seen that the axis length of each section of the nozzle and the diameter of the oblique circle section are the key parameters determining the geometric profile of the nozzle at a 90° angle. Therefore, this book analyzes the influence of the axis length of each section of the nozzle on the aerodynamic performance of the nozzle at a 90° angle by changing the axis length of the three sections of the nozzle, establishes an approximate performance model of the nozzle at a 90° angle, optimizes the maximum thrust coefficient of the nozzle, and the corresponding axis length of the three sections of the nozzle. During the research process, the length of the three cylinder axis lines was dimensionless, and the dimensionless lengths of the three cylinder axis lines were $L_1 = l_1/d$, $L_2 = l_2/d$ and $L_3 = l_3/d$, respectively.

4.4.2.1 Impact of First Section Length on Aerodynamic Performance

Table 4.3 compares the performance parameters of the nozzle when $L_1 = 0.3$ and $L_1 = 1$. It can be seen from the table that the thrust coefficient of the nozzle increases by 0.84% and the flow coefficient increases by 0.3% when $L_1 = 1$ compared to $L_1 = 0.3$. This indicates that under the conditions of $L_2 = 0.7$ and $L_3 = 0.22$, L_1 has a certain impact on the aerodynamic performance of the nozzle.

Figure 4.18 shows the streamline and turbulent kinetic energy distribution of the nozzle profile under $L_1 = 0.3$ and $L_1 = 1$ profiles. From the figure, it can be seen that when $L_1 = 0.3$, due to the shorter length of the first section of the cylinder, the airflow immediately enters the second section of the cylinder after bypassing the tail cone. The flow direction of the airflow changes with the shape of the second section of the cylinder, thereby limiting the full development of the recirculation zone and reducing its range. From the streamline distribution, it can be seen that after the

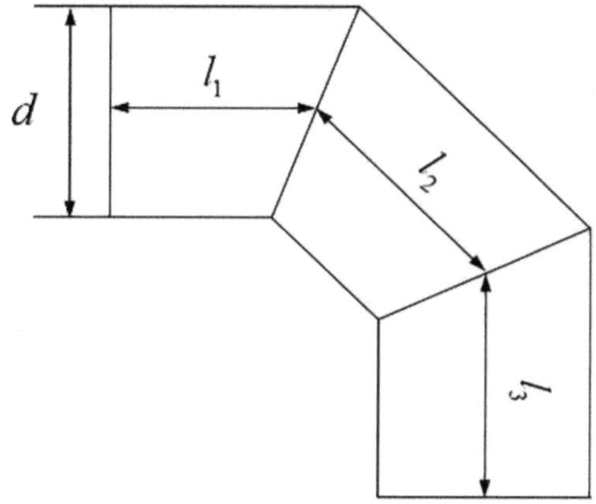

Fig. 4.17 Schematic diagram of nozzle under 90° deflection angle

Table 4.3 Performance parameters of nozzle for $L_1 = 0.3$ and $L_1 = 1$ ($L_2 = 0.7$, $L_3 = 0.22$)

	C_{fg}	C_d
$L_1 = 0.3$	0.9318	0.8948
$L_1 = 1$	0.9396	0.8978

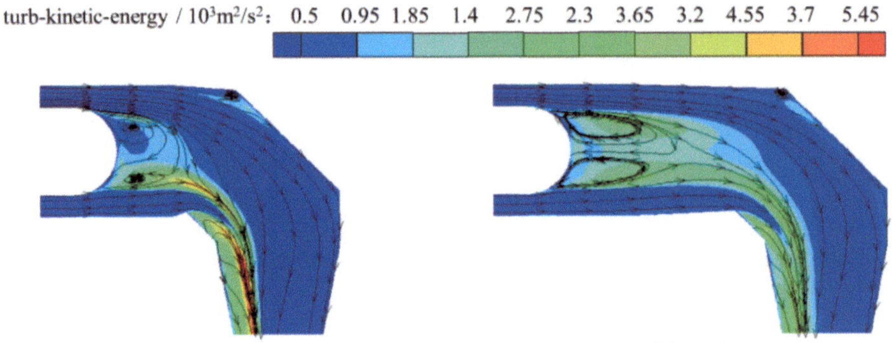

(a) L_1=0.3 (b) L_1=1

Fig. 4.18 Streamline and turbulent energy distribution of nozzle profile for $L_1 = 0.3$ and $L_1 = 1$ ($L_2 = 0.7$, $L_3 = 0.22$). (**a**) $L_1 = 0.3$. (**b**) $L_1 = 1$

airflow bypasses the recirculation zone, the velocity direction of the airflow outside the recirculation zone changes, and the flow direction of the airflow changes from horizontal to nearly vertical. Due to the large vertical velocity component of the airflow in this area, when mixed with the airflow flowing horizontally inside the recirculation zone, the turbulent kinetic energy in the mixing zone increases, accompanied by a certain amount of energy loss. In addition, due to the shorter length of the first section of the cylinder, the turning angle of the first and second nozzle cylinders at the connection of the inner wall surface increases, resulting in a significant separation of the airflow passing through the turning angle on the inner wall surface of the second section of the cylinder, and an increase in energy loss of the airflow. When $L_1 = 1$, due to the longer length of the first section of the cylinder, the recirculation zone behind the tail cone is fully developed, and the range of the recirculation zone increases. Compared with the $L_1 = 0.3$ profile, the energy loss of the airflow in the recirculation zone increases. From the streamline distribution, it can be seen that after bypassing the recirculation zone, the airflow mixes around the axis of the first section of the cylinder. Due to the horizontal flow direction of the airflow, there is no significant energy loss after the airflow is mixed. In addition, compared with the $L_1 = 0.3$ profile, due to the reduced turning angle of the first and second nozzle bodies at the connection of the inner wall surface, the degree of separation of the inner wall surface of the second nozzle body is weakened, and the energy loss is relatively reduced. From the comparison of nozzle flow coefficients, it can be seen that the nozzle flow coefficient increases by 0.3% when $L_1 = 1$ compared to $L_1 = 0.3$. This is because as L_1 increases, the degree of airflow separation between the second- and third- cylinder walls decreases, and the effective flow area of the nozzle is less affected by wall airflow separation, resulting in a relative increase in nozzle flow rate.

4.4.2.2 Impact of Second Section Length on Aerodynamic Performance

Table 4.4 compares the performance parameters of the nozzle when $L_2 = 0.41$ and $L_2 = 1$. It can be seen from the table that the thrust coefficient of the nozzle increases by 2.54% and the flow coefficient increases by 2.8% when $L_2 = 1$ compared to $L_2 = 0.41$, indicating that L_2 has a certain impact on the aerodynamic performance of the nozzle under the conditions of $L_1 = 1$ and $L_3 = 0.61$.

Figure 4.19 shows the streamline and turbulent energy distribution of the nozzle profile under $L_2 = 0.41$ and $L_2 = 1$ profiles. It can be seen from the figure that when $L_2 = 0.41$, due to the small length of the inner wall surface of the second section of the cylinder, the air flow close to the inner wall of the nozzle flows out of the first section of the cylinder and directly enters the third section of the cylinder. And the turning angle of the air flow is close to 90°, which leads to a large degree of separation of the air flow on the inner wall of the third section of the cylinder, and the turbulent energy of the air flow in this area increases, accompanied by a large energy loss. When $L_2 = 1$, due to the large length of the inner wall of the second section of the cylinder, the airflow close to the inner wall of the nozzle flows out of the first section of the cylinder and enters the second section of the cylinder. The turning angle of the airflow is close to 45°, and the airflow is separated on the wall surface of the second section of the cylinder. Compared with $L_2 = 0.41$, the range of the separation zone and the turbulent energy are reduced, and the energy loss of the airflow in this area is reduced. The separation zone of the inner wall of the second

Table 4.4 Performance parameters of nozzle for $L_2 = 0.41$ and $L_2 = 1$ ($L_1 = 1$, $L_3 = 0.61$)

	C_{fg}	C_d
$L_2 = 0.41$	0.9112	0.877
$L_2 = 1$	0.9343	0.8979

Fig. 4.19 Streamlines and turbulent kinetic energy distribution of nozzle profiles for $L_2 = 0.41$ and $L_2 = 1$ ($L_1 = 1$, $L_3 = 0.61$). **(a)** $L_2 = 0.41$. **(b)** $L_2 = 1$

section extends to the inner wall of the third segment, and the turbulent energy of the airflow in the separation zone of the third section of the cylinder is reduced, and the energy loss of the airflow in this area is reduced. From the above analysis, it can be seen that the length of the second section of the cylinder directly affects the range and strength of the separation zone of the wall surface of the second and third stages of the cylinder. Thereby affecting the energy loss of the air flow in the nozzle, so it can be seen that the length of the second section of the cylinder has a great influence on the aerodynamic performance of the nozzle. According to the comparison of the flow coefficient of the nozzle, the flow coefficient of the nozzle for $L_2 = 1$ is increased by 2.8% compared with $L_2 = 0.41$. This is because when the length of the second cylinder is smaller, the separation zone on the inner wall of the third cylinder increases, and the airflow velocity at the nozzle outlet near the inner wall decreases, resulting in a decrease in the flow rate of the nozzle.

4.4.2.3 Analysis of the Effect of Cylinder Length on the Aerodynamic Performance of the Nozzle in the Third Segment

Table 4.5 shows the comparison of nozzle performance parameters at $L_3 = 0.22$ and $L_3 = 1$. It can be seen from the table that the thrust coefficient of the nozzle at $L_3 = 0.22$ increases by 1.53% and the flow coefficient increases by 0.74% compared with $L_3 = 1$, indicating that L_3 has some influence on the aerodynamic performance of the nozzle under the conditions of $L_1 = 1$, $L_2 = 0.7$.

Figure 4.20 shows the streamlines and turbulent kinetic energy distributions of the nozzle profiles for $L_3 = 0.22$ and $L_3 = 1$ profiles. As can be seen from the figures, when $L_3 = 1$, due to the larger length of the wall inside the third section of the barrel, the separation zone of the second section of the barrel is fully developed in the inner wall of the third section of the barrel, the turbulent kinetic energy of the separation zone increases and vortexes appear, and the energy loss of the airflow in this region increases. When $L_3 = 0.22$, the length of the inner wall of the third section of the cylinder is smaller, the scope of the separation zone and the turbulent kinetic energy of the inner wall decrease, and the energy loss of the airflow in the region decreases. From the comparison of nozzle flow coefficients, it can be seen that the nozzle flow coefficient at $L_3 = 0.22$ is increased by 0.74% compared to that at $L_3 = 1$. The change in the flow coefficient is also caused by the separation zone of the airflow at the inner wall surface of the nozzle.

Table 4.5 Flow coefficients and thrust coefficients for Nozzles at $L_3 = 0.22$ and $L_3 = 1$ ($L_1 = 1$, $L_2 = 0.7$)

	C_{fg}	C_d
$L_3 = 0.22$	0.9396	0.8978
$L_3 = 1$	0.9254	0.8904

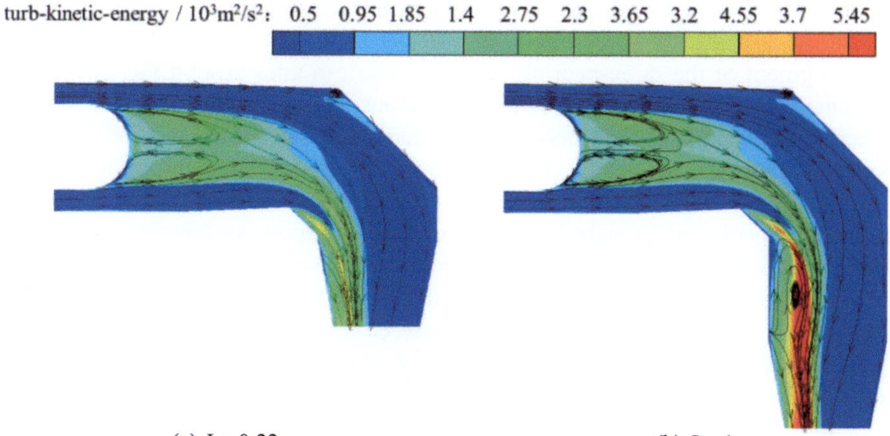

(a) L =0.22 (b) L =1

Fig. 4.20 Streamlines and turbulent kinetic energy distributions of nozzle profiles at $L_3 = 0.22$ and $L_3 = 1$ ($L_1 = 1$, $L_2 = 0.7$). (**a**) $L_3 = 0.22$. (**b**) $L_3 = 1$

4.4.3 Three-Bearing Deflection Nozzle Law of Motion Design Methodology

As can be seen from the working principle diagrams of Figure 4.1 three-bearing deflection nozzle, the three-bearing deflection nozzle realizes the nozzle deflection through the mutual rotation of the three sections of barrels, and the adjacent two sections of barrels are connected by bearings to reduce the friction of the rotation of each section of barrels. During the nozzle deflection process, the first and the third cylinder rotate in the same direction, the second cylinder rotates in the opposite direction relative to the first and the third cylinders, and the axes of the three cylinders are always kept in the same pitch plane, i.e., only the pitch force is generated during the nozzle deflection process, and the lateral force is not generated (Kentfield 1967). Regarding the advantages of the three-bearing deflector nozzle over other mechanical vector nozzles, the Curtiss Wright Company of the United States made the following description:

(1) During deflection, the vector thrust remains in the same pitch plane;
(2) The vector angle of the nozzle can be continuously varied during deflection;
(3) The design of the nozzle deflection mechanism does not affect the aerodynamic performance of the nozzle in the non-vectorized state;

In this book, the unique advantages of the above-mentioned three-bearing deflection nozzle are also taken as the design objectives of the nozzle to carry out the design of the deflection law of the three-bearing deflection nozzle.

4.4.3.1 Overview of Coordinate Transformation Methods

Each segment of the three-bearing deflector nozzle rotates around the axis of the diagonally tangent circular section of the previous segment, so each segment can be regarded as an independent coordinate system, and the coordinates of the mass points of each segment can be converted to the same coordinate system by the following coordinate transformation method (Han 2004).

There are two sets of right-handed Cartesian coordinate systems $x_iy_iz_i$ and $x_jy_jz_j$ with a common origin, the i system is called the old system and the j system is called the new system, as shown in Fig. 4.21. Let there be a point in space P (with vectorial diameter r), and the coordinates about these two sets of coordinate systems are (x_i, y_i, z_i) and (x_j, y_j, z_j), respectively, then the matrix $[C_{ij}]$ is introduced so that $(r)_i = [C_{ij}](r)_j$. Where $(r)_i$ and $(r)_j$ are the columns of the same point P in the old coordinate system i and the new coordinate system$j$, i.e., $(r)_i = (x_i, y_i, z_i)^T$ and $(r)_j = (x_j, y_j, z_j)^T$, respectively, and the matrix $[C_{ij}]$ can be expressed by Eq. (4.5).

$$\left[C_{ij}\right] = \begin{bmatrix} \cos\alpha_1 & \cos\alpha_2 & \cos\alpha_3 \\ \cos\beta_1 & \cos\beta_2 & \cos\beta_3 \\ \cos\gamma_1 & \cos\gamma_2 & \cos\gamma_3 \end{bmatrix} \tag{4.5}$$

In the equation, the lower-right corner label indicates that the square matrix is a coordinate transformation matrix transformed from the coordinate system j to the coordinate system i , where α_1, β_1, γ_1, α_2, β_2, γ_2, α_3, β_3, γ_3 are the direction angles of the x_j, y_j and z_j axes about $x_iy_iz_i$ respectively, and $[C_{ij}]$ is often called the direction cosine matrix.

To rotate around the z axis as an example of the coordinate transformation, there are two common origin coordinate system, for the coordinate system $x_iy_iz_i$, the coordinate axis direction of the coordinate system $x_jy_jz_j$ can be considered to be

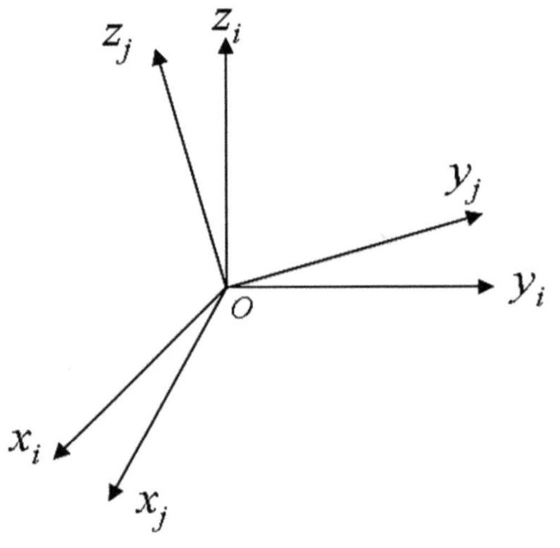

Fig. 4.21 Transformed coordinate system

around the z axis rotated by an angle θ, θ positive and negative according to the right-hand rule, then around the z axis to turn the θ angle of the direction of the cosine matrix for the

$$\left[C_{ij}^{(\theta)}\right]=\begin{bmatrix}\cos\theta & -\sin\theta & 0\\ \sin\theta & \cos\theta & 0\\ 0 & 0 & 1\end{bmatrix} \quad (4.6)$$

The upper right corner of the matrix, (θ), indicates that the coordinate system j is obtained by rotating the coordinate system by an angle of θ around the z axis. Similarly, the directional cosine matrices for rotating α by an angle around the x-axis and β by an angle around the$y$-axis are

$$\left[C_{ij}^{(\alpha)}\right]=\begin{bmatrix}1 & 0 & 0\\ 0 & \cos\alpha & -\sin\alpha\\ 0 & \sin\alpha & \cos\alpha\end{bmatrix} \quad (4.7)$$

respond in singing

$$\left[C_{ij}^{(\beta)}\right]=\begin{bmatrix}\cos\beta & 0 & \sin\beta\\ 0 & 1 & 0\\ -\sin\beta & 0 & \cos\beta\end{bmatrix} \quad (4.8)$$

4.4.3.2 Modeling of Nozzle Deflection Patterns

From the design objective of the three-bearing deflection nozzle, it can be seen that the nozzle deflection process should ensure that the axes of the three sections of the nozzle cylinder are in the same pitch plane. Based on this design objective, the nozzle deflection law model is established to obtain the rotation angle law of the three-segment cylinder and the relationship between the nozzle deflection angle and the rotation angle of each segment of the cylinder.

Figure 4.22 shows the nozzle deflection law research model, the coordinate system xyz is the absolute coordinate system, the coordinate system $x_0y_0z_0$, $x_1y_1z_1$, $x_2y_2z_2$, $x_3y_3z_3$, $x_4y_4z_4$ and $x_5y_5z_5$ are the rotational coordinate system connected with each cylinder and established in the inlet and outlet section of each section of the cylinder. Assuming the rotation angle of the first section of the cylinder around the x-axis is θ_1, the rotation angle of the second section of the cylinder around the x_1-axis is θ_2, and the rotation angle of the third section of the cylinder around the x_3-axis is θ_3. Points A B, and C are the mass points fixed on the first, second, and third sections of the cylinder, respectively. To simplify the model, the nozzle inlet radius of the y_0z_0 cross-section is normalized. The coordinate of A is $r_{A0} = (0,\ 0,\ 1)$, the coordinate

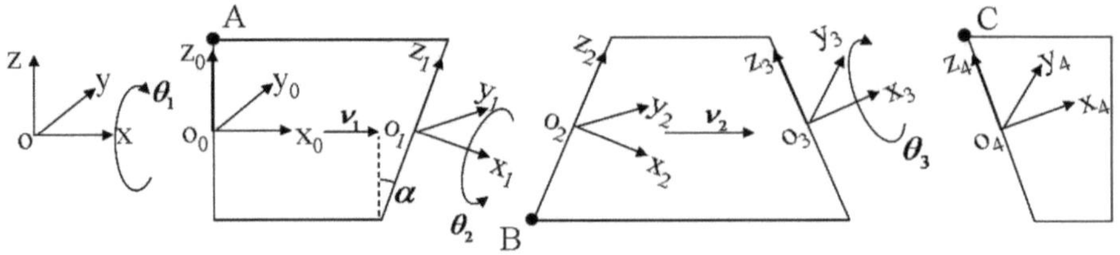

Fig. 4.22 Model of nozzle deflection law

of B is $r_{B2} = (0, \quad 0, \quad -1/\cos\alpha)$, and the coordinate of C is $r_{C4} = (0, \quad 0, \quad 1/\cos\alpha)$. The cosine matrix of the rotation angle $-\theta_1$ of the coordinate system $x_0y_0z_0$ around the absolute coordinate system $oxyz$ can be represented by Eq. (4.9).

$$C_{ab0} = \begin{bmatrix} 1 & 0 & 0 \\ 0 & \cos(-\theta_1) & -\sin(-\theta_1) \\ 0 & \sin(-\theta_1) & \cos(-\theta_1) \end{bmatrix} \tag{4.9}$$

The cosine matrix of the angle of rotation α of the coordinate system $x_1y_1z_1$ around the y_0-axis of the coordinate system $x_0y_0z_0$ is

$$C_{01} = \begin{bmatrix} \cos(\alpha) & 0 & \sin(\alpha) \\ 0 & 1 & 0 \\ -\sin(\alpha) & 0 & \cos(\alpha) \end{bmatrix} \tag{4.10}$$

The cosine matrix of the angle of rotation θ_2 of the coordinate system$x_2y_2z_2$ around the x_1-axis of the coordinate system $x_1y_1z_1$ is

$$C_{12} = \begin{bmatrix} 1 & 0 & 0 \\ 0 & \cos(\theta_2) & -\sin(\theta_2) \\ 0 & \sin(\theta_2) & \cos(\theta_2) \end{bmatrix} \tag{4.11}$$

The cosine matrix of the angle of rotation -2α of the coordinate system $x_3y_3z_3$ around the y_2 axis of the coordinate system $x_2y_2z_2$ is

$$C_{23} = \begin{bmatrix} \cos(-2\alpha) & 0 & \sin(-2\alpha) \\ 0 & 1 & 0 \\ -\sin(-2\alpha) & 0 & \cos(-2\alpha) \end{bmatrix} \tag{4.12}$$

The cosine matrix of the angle of rotation $-\theta_3$ of the coordinate system $x_4y_4z_4$ around the x_3 axis of the coordinate system $x_3y_3z_3$ is

$$C_{34} = \begin{bmatrix} 1 & 0 & 0 \\ 0 & \cos(-\theta_3) & -\sin(-\theta_3) \\ 0 & \sin(-\theta_3) & \cos(-\theta_3) \end{bmatrix} \tag{4.13}$$

Through the above coordinate transformation, it can be concluded that the coordinates of the points *A*, *B*, *C* in the absolute coordinate system *xyz* after the three segments of the cylinder are rotated around the axes x, x_1, x_3 by $-\theta_1$, θ_2, $-\theta_3$ angles respectively are

$$r_A = C_{ab0} r_{A0} \tag{4.14}$$

$$r_B = C_{ab0} C_{01} C_{12} r_{B2} \tag{4.15}$$

$$r_C = C_{ab0} C_{01} C_{12} C_{23} C_{34} r_{C4} \tag{4.16}$$

In order to achieve the design objective of keeping the axes of the three nozzle barrel segments in the same pitch plane during nozzle deflection, points *A*, *B*, and *C* should satisfy the following relationship equation (Woodward 1987):

$$|z_A| = |z_B| = |z_C| \tag{4.17}$$

Substituting Eqs. (4.14)–(4.16) into Eq. (4.17) yields the correlation equation betweenθ_1, θ_2, θ_3 as

$$\tan\theta_1 = \tan\left(\frac{\theta_2}{2}\right)\cos\alpha \tag{4.18}$$

$$\theta_2 = \theta_3 \tag{4.19}$$

From Equations (4.18) and (4.19), the correlation equation between the nozzle deflection angleβ and θ_1, θ_2 and θ_3 can be further derived. In Fig. 4.23, let v_{01} be the direction vector of the first cylinder axis in the coordinate system $x_0y_0z_0$, v_{02} be the direction vector of the second cylinder axis in the coordinate system $x_2y_2z_2$, where $v_{01} = (1, \ 0, \ 0)$, $v_{02} = (\cos\alpha, \ 0, \ \sin\alpha)$. After the three cylinder segments are rotated around the axes of x, x_1 and x_3 by $-\theta_1$, θ_2 and $-\theta_3$ angles respectively, v_{01} and v_{02} can be expressed in the absolute coordinate system as follows:

$$v_1 = (1, \ 0, \ 0) \tag{4.20}$$

$$v_2 = C_{ab0} C_{01} C_{12} v_{02} \tag{4.21}$$

From the geometric characterization of the nozzle in Sect. 4.2.1, the angle between v_1 andv_2 is $\beta/2$, i.e.

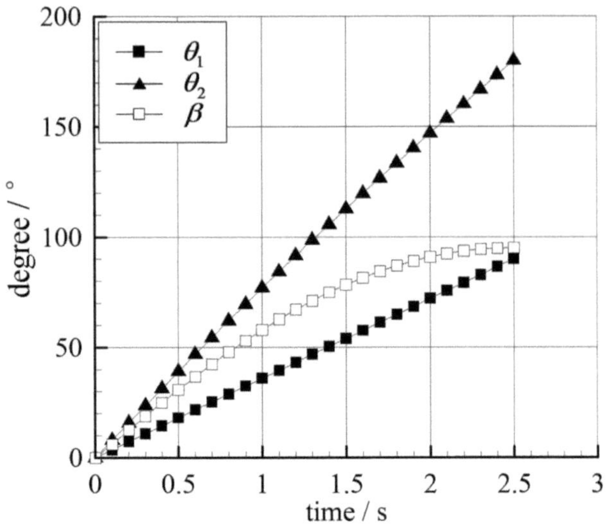

Fig. 4.23 Nonlinear deflection control law

$$\cos\left(\frac{\beta}{2}\right) = \frac{v_1 \cdot v_2}{|v_1||v_2|} \tag{4.22}$$

Substituting Eqs. (4.20) and (4.21) into Eq. (4.22) yields

$$\cos\left(\frac{\beta}{2}\right) = \cos^2\alpha + \sin^2\alpha\cos\theta_2 \tag{4.23}$$

Equation (4.23) can be reduced to

$$\beta = 2\arccos\left(\cos^2\alpha + \sin^2\alpha\cos\theta_2\right) \tag{4.24}$$

Equation (4.24) is the relationship between the nozzle deflection angle and the cylinder rotation angle.

From equations (4.18) and (4.19), it can be seen that the relative rotational angular velocity between the first and second cylinders needs to change nonlinearly with time, and the angular velocity of the third cylinder is the same as that of the second cylinder.

4.4.3.3 Preferred Nozzle Deflection Pattern

According to the relationship between the nozzle deflection angle β and the rotation angle of each section of the cylinder θ_1, θ_2, θ_3, two typical nozzle deflection control laws can be obtained, i.e., the nonlinear deflection control law and the linear deflection control law.

The nonlinear deflection angle control law refers to the vector angle control law that β changes nonlinearly with time. Set nozzle can realize the maximum deflection angle of 95°, nozzle from 0° deflection to 95° with 2.5s, in the nonlinear deflection angle control law, the first section of the cylinder around the$x$ axis of uniform rotation, then by the Equations (4.18) and (4.24) can be derived from the change rule of β, θ_1, θ_2 with the time, such as Fig. 4.23 is shown. As can be seen from the figure, according to the control law, the second and third section of the barrel rotational angular velocity and the first section of the barrel rotational angular velocity close to 2 times the relationship, and the nozzle deflection process, the first section of the barrel rotational angular velocity is a constant, the second and third section of the barrel rotational angular velocity is close to a constant, the nozzle deflection angle with the time nonlinear change, and the angular velocity of the deflection is gradually decreasing, and finally tends to be close to 0, that is, the nozzle can be a smooth and continuous from 0° to 95° deflection.

Linear deflection angle control law refers to β vector angle control law that changes linearly with time. Still set nozzle can realize the maximum deflection angle of 95°, nozzle deflection from 0° to 95° with 2.5s, in the linear deflection angle control law, the nozzle to maintain a uniform angular velocity deflection, that is, $d\beta/dt$ is a constant, then by the Eqs. (4.18) and (4.24) can be derived from the change rule of β, θ_1, θ_2 with the time, such as Fig. 4.24 is shown. As can be seen from the figure, according to the control law, the first, second and third sections of the cylinder rotational angular velocity are nonlinear with time. And in the late stage of the nozzle deflection, the first, second and third sections of the cylinder rotational angular acceleration increase dramatically, which increases the difficulty of the design of the control system of the nozzle.

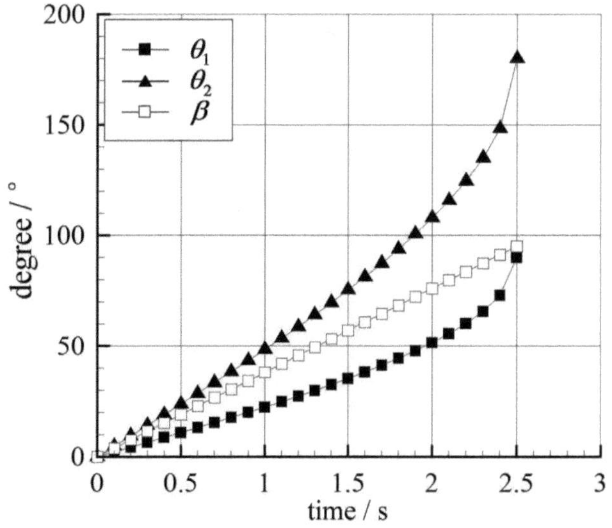

Fig. 4.24 Linear deflection control law

From the above analysis, it can be seen that in the nonlinear deflection control law, the rotational angular velocity of the first cylinder is constant, and the rotational angle of the second and third cylinders varies nonlinearly with time, whereas in the linear deflection control law, the rotational angular velocities of the three cylinders vary nonlinearly with time, so that adopting the nonlinear deflection control law reduces the nonlinear control variables and reduces the difficulty of the design of the nozzle control system.

4.4.3.4 Compound Rigid Body Motion

Simple rigid body motion can be divided into translational and rotational (Liu 2010). From the working principle and deflection law of the three-bearing deflection nozzle, it can be seen that the first section of the cylinder of the three-bearing deflection nozzle makes a simple uniform fixed-axis rotation, and the second and third sections of the cylinder make a variable-speed fixed-axis rotation relative to the previous section of the cylinder, and due to the constant change of the axial direction and position of the previous section of the cylinder, the motion of the second section of the cylinder can be described as a rotation of the rigid body around an instantaneous axis, and the motion of the third section of the cylinder can be described as a compound motion combining the rigid body's rotation around an instantaneous axis and The motion of the third cylinder can be described as a composite motion combining rotation around an instantaneous axis and translation. In this section, we firstly introduce the modeling method of the composite motion of the rigid body, and on the basis of which we establish the motion law model of each segment of the cylinder, and obtain the motion law of each segment of the cylinder in the absolute coordinate system.

According to the literature (Wang 1988), let the system be composed of n rigid bodies, rigid body 1 rotates on a fixed axis, and rigid body n rotates on a fixed axis relative to rigid body (n − 1). Let $\boldsymbol{\omega}_{1a}$, $\boldsymbol{\omega}_{2a}$, …, $\boldsymbol{\omega}_{na}$ be the absolute angular velocity of each rigid body, $\boldsymbol{\omega}_2$, $\boldsymbol{\omega}_3$, …, $\boldsymbol{\omega}_n$ be the relative angular velocity of rigid body n relative to rigid body (n − 1), O_1, O_2, …, O_n be any point on the line where the relative angular velocity vector $\boldsymbol{\omega}_2$, $\boldsymbol{\omega}_3$, …, $\boldsymbol{\omega}_n$ is located, $\mathbf{v}_{O_1}, \mathbf{v}_{O_2}, \supset, \mathbf{v}_{O_n}$ be the absolute velocity of point O_1, O_2, …, O_n. The absolute motion of rigid body 2 is the composite motion of rotation and translation, and the absolute angular velocity $\boldsymbol{\omega}_{2a}$ and linear velocity $\mathbf{v}_{O_2}$ of rigid body 2 can be expressed as follows

$$\omega_{2a} = \omega_{1a} + \omega_2 = \omega_1 + \omega_2 \tag{4.25}$$

$$\mathbf{v}_{O_2} = \omega_1 \times \mathbf{O}_1\mathbf{O}_2 \tag{4.26}$$

Considering the rigid body 2 as a dynamic coordinate system, and the rigid body 3 rotates with respect to the rigid body 2 in translation and fixed axis, the absolute angular velocity $\boldsymbol{\omega}_{3a}$ and linear velocity $\mathbf{v}_{O_3}$ of the rigid body 3 can be expressed as

$$\omega_{3a} = \omega_{2a} + \omega_3 = \omega_1 + \omega_2 + \omega_3 \tag{4.27}$$

$$\begin{aligned} \mathbf{v}_{O_3} &= \mathbf{v}_{O_3} + \omega_{2a} \times \mathbf{O}_2\mathbf{O}_3 \\ &= \omega_1 \times \mathbf{O}_1\mathbf{O}_2 + (\omega_1 + \omega_2) \times \mathbf{O}_2\mathbf{O}_3 \\ &= \omega_1 \times \mathbf{O}_1\mathbf{O}_3 + \omega_2 \times \mathbf{O}_2\mathbf{O}_3 \end{aligned} \tag{4.28}$$

In order to synthesize, it can be deduced that the rigid body n makes a compound motion of translation and rotation, and the absolute angular velocity $\boldsymbol{\omega}_{na}$ and linear velocity $\mathbf{v}_{O_n}$ of the rigid body n can be expressed as:

$$\mathbf{v}_{O_n} = \omega_1 \times \mathbf{O}_1\mathbf{O}_n + \omega_2 \times \mathbf{O}_2\mathbf{O}_n + \ldots + \omega_{n-1} \times \mathbf{O}_{n-1}\mathbf{O}_n = \sum_{i=1}^{n-1} \omega_i \times \mathbf{O}_i\mathbf{O}_n \tag{4.29}$$

$$\omega_{na} = \omega_1 + \omega_2 + \ldots + \omega_n = \sum_{i=1}^{n} \omega_i \tag{4.30}$$

4.4.3.5 Modeling of the Laws of Motion of the Cylinder Segments

The law of motion model of each section of the cylinder of the three-bearing deflection nozzle can be obtained by the simplified nozzle deflection law model, as shown in Figure 4.25. In the model, the *xyz* coordinate system is an absolute coordinate system, which does not move with the nozzle, the $x'y'z'$ coordinate system is a coordinate system fixed on the second section of the cylinder, the $y'z'$ cross section coincides with the inclined section of the second section of the nozzle, and the distance between the centers of the two inclined sections of the second section of the cylinder is l. Wherein, the coordinate origin of the *xyz* coordinate system coincides with the coordinate origin of the $x'y'z'$ coordinate system and is located at the geometric centers of the bearing mounting cross sections of the first and second nozzle segments. By analyzing the composite motion of the nozzle, it can be seen that the first section

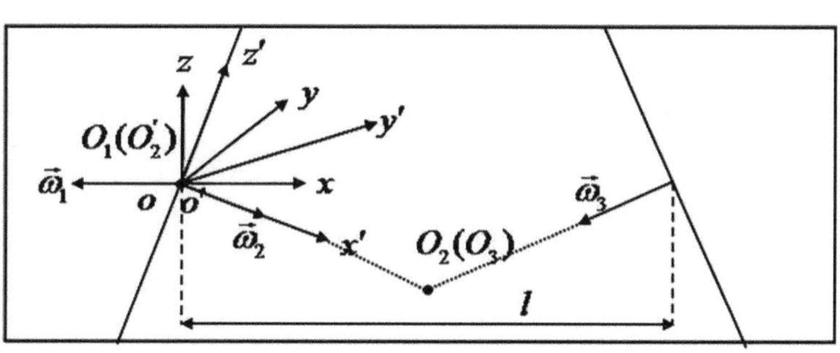

Fig. 4.25 Three-bearing deflection nozzle barrel motion law model

of the cylinder in the three-bearing deflection nozzle rotates around the x axis with fixed axis, and the angular velocity vector is $\boldsymbol{\omega}_1$, O_1 is the fixed point on $\boldsymbol{\omega}_1$ and it coincides with the origin of the $x'y'z'$ coordinate system. The second section of the cylinder makes the fixed-point motion of the rigid body around the origin of the coordinates in the absolute coordinate system, and the relative motion is the relative rotation of the first section of the cylinder around the x' axis with fixed axis, and the angular velocity vector is $\boldsymbol{\omega}_2$, O_2, and O_2' are all the fixed points on $\boldsymbol{\omega}_2$, the point of O_2' coincides with the point of O_1, and the point of O_2 is the normal line of two inclined cross sections of the second section of the cylinder. The third cylinder in the absolute coordinate system for the rigid body around a point of fixed-point motion and rigid body translation combined with the composite motion. The third section of the cylinder is relative to the second section of the cylinder for the fixed-axis rotation. The angular velocity vector is $\boldsymbol{\omega}_3$, O_3 is a fixed point on $\boldsymbol{\omega}_3$.

From the above analysis and the modeling method of rigid-body composite motion, it can be seen that the absolute angular velocity of the first segment of the cylinder is

$$\boldsymbol{\omega}_{1a} = \boldsymbol{\omega}_1 = \left(-\omega_1, 0, 0\right) \tag{4.31}$$

The absolute linear velocity of the point O_1 is

$$\mathbf{v}_{O_1} = 0 \tag{4.32}$$

The absolute angular velocity of the second section of the cylinder is

$$\boldsymbol{\omega}_{2a} = \boldsymbol{\omega}_1 + \boldsymbol{\omega}_2 \tag{4.33}$$

The absolute linear velocity of the point O_2' is

$$\mathbf{v}_{O_2'} = \boldsymbol{\omega}_1 \times \mathbf{O}_1\mathbf{O}_2' = 0 \tag{4.34}$$

The absolute angular velocity of the third section of the cylinder is

$$\boldsymbol{\omega}_{3a} = \boldsymbol{\omega}_1 + \boldsymbol{\omega}_2 + \boldsymbol{\omega}_3 \tag{4.35}$$

The absolute linear velocity of the point O_3 is

$$\mathbf{v}_{O_3} = \boldsymbol{\omega}_1 \times \mathbf{O}_1\mathbf{O}_3 + \boldsymbol{\omega}_2 \times \mathbf{O}_2\mathbf{O}_3 = \boldsymbol{\omega}_1 \times \mathbf{O}_1\mathbf{O}_3 = \begin{pmatrix} 0 \\ -\frac{l}{2}\omega_1 \tan\alpha\cos\theta_1 \\ \frac{l}{2}\omega_1 \tan\alpha\sin\theta_1 \end{pmatrix} \tag{4.36}$$

From the three-bearing deflection nozzle deflection law study model, $\boldsymbol{\omega}_2$ and $\boldsymbol{\omega}_3$ in the coordinate system xyz are

$$\boldsymbol{\omega}_2 = \begin{pmatrix} 1 & 0 & 0 \\ 0 & \cos\theta_1 & \sin\theta_1 \\ 0 & -\sin\theta_1 & \cos\theta_1 \end{pmatrix} \begin{pmatrix} \omega_2\cos\alpha \\ 0 \\ -\omega_2\sin\alpha \end{pmatrix} = \begin{pmatrix} \omega_2\cos\alpha \\ -\omega_2\sin\theta_1\sin\alpha \\ -\omega_2\cos\theta_1\sin\alpha \end{pmatrix} \tag{4.37}$$

$$\boldsymbol{\omega}_3 = \begin{pmatrix} 1 & 0 & 0 \\ 0 & \cos\theta_1 & \sin\theta_1 \\ 0 & -\sin\theta_1 & \cos\theta_1 \end{pmatrix} \begin{pmatrix} \cos\alpha & 0 & \sin\alpha \\ 0 & 1 & 0 \\ -\sin\alpha & 0 & \cos\alpha \end{pmatrix} \begin{pmatrix} 1 & 0 & 0 \\ 0 & \cos\theta_2 & -\sin\theta_2 \\ 0 & \sin\theta_2 & \cos\theta_2 \end{pmatrix} \begin{pmatrix} -\omega_3\cos 2\alpha \\ 0 \\ -\omega_3\sin 2\alpha \end{pmatrix}$$
$$= \begin{pmatrix} -\omega_2\cos\alpha\cos 2\alpha - \omega_2\sin\alpha\sin 2\alpha\cos\theta_2 \\ \omega_2\sin 2\alpha\sin\theta_2\cos\theta_1 - \omega_2\sin 2\alpha\cos\alpha\cos\theta_2\sin\theta_1 + \omega_2\sin\alpha\cos 2\alpha\sin\theta_1 \\ -\omega_2\sin 2\alpha\sin\theta_2\sin\theta_1 - \omega_2\sin 2\alpha\cos\alpha\cos\theta_2\cos\theta_1 + \omega_2\sin\alpha\cos 2\alpha\cos\theta_1 \end{pmatrix} \tag{4.38}$$

Substituting Eqs. (4.31) and (4.37) into (3.33) yields

$$\boldsymbol{\omega}_{2a} = \begin{pmatrix} \omega_2\cos\alpha - \omega_1 \\ -\omega_2\sin\theta_1\sin\alpha \\ -\omega_2\cos\theta_1\sin\alpha \end{pmatrix} \tag{4.39}$$

Substituting Eqs. (4.31), (4.37), and (4.38) into (4.35) yields

$$\boldsymbol{\omega}_{3a} = \begin{pmatrix} -\omega_2\cos\alpha\cos 2\alpha - \omega_2\sin\alpha\sin 2\alpha\cos\theta_2 + \omega_2\cos\alpha - \omega_1 \\ \omega_2\sin 2\alpha\sin\theta_2\cos\theta_1 - \omega_2\sin 2\alpha\cos\alpha\cos\theta_2\sin\theta_1 \\ + \omega_2\sin\alpha\cos 2\alpha\sin\theta_1 - \omega_2\sin\theta_1\sin\alpha \\ -\omega_2\sin 2\alpha\sin\theta_2\sin\theta_1 - \omega_2\sin 2\alpha\cos\alpha\cos\theta_2\cos\theta_1 \\ + \omega_2\sin\alpha\cos 2\alpha\cos\theta_1 - \omega_2\cos\theta_1\sin\alpha \end{pmatrix} \tag{4.40}$$

4.4.4 Three-Bearing Deflection Nozzle Drive Torque Design Methodology

During the deflection of the three-bearing deflection nozzle, the drive system needs to provide the torque required for the rotation of each section of the barrel, and too large a drive torque will lead to excessive energy consumption of the drive system and irrational allocation of engine power; too small a drive torque will lead to the failure of the nozzle deflection, which directly affects the safety and maneuverability of the aircraft. Therefore, the size of the driving torque has a certain influence on the volume, quality and design complexity of the nozzle actuation system (Wang et al. 2014).

The three-bearing deflection nozzle moment calculation model needs to consider that the nozzle deflection axis direction is different from the rotation axis direction of each section of the cylinder, and the rotation axis direction of each section of the cylinder changes with time. Figure 4.26 shows the three-bearing deflection nozzle deflection direction and force diagram. From the figure, it can be seen that during the nozzle deflection process, the rotation axes of each section of the cylinder are in different directions from the deflection axis of the nozzle, and the direction of the rotation axes of the second and third sections of the cylinder changes with time. And from the analysis in Sect. 4.4.3, it can be seen that the second and third sections of the cylinder rotate around the instantaneous axes in the absolute coordinates. In addition, it can be seen from the figure that the driving torque of each section of the cylinder needs to overcome the nozzle gravity, the circumferential friction of the rotating joint surface between each section of the nozzle and the torque on the rotational axis of each section of the cylinder generated by the aerodynamic force on the nozzle when the gas passes through the nozzle. The related literature on the calculation model of driving moment of three-bearing deflection nozzle at home and abroad has not been disclosed, and only the modeling research on the driving moment of oscillating nozzle for rocket engine has been carried out in China (Yang 1985, Yang 1989, Yang 1993; You et al. 2001), and its method of moment modeling is of some significance in guiding the modeling of the driving moment of the three-bearing deflection nozzle. In this section, the three segments of the cylinder of the three-bearing deflection nozzle are taken as the research objects to establish the driving moment calculation model of each segment of the cylinder.

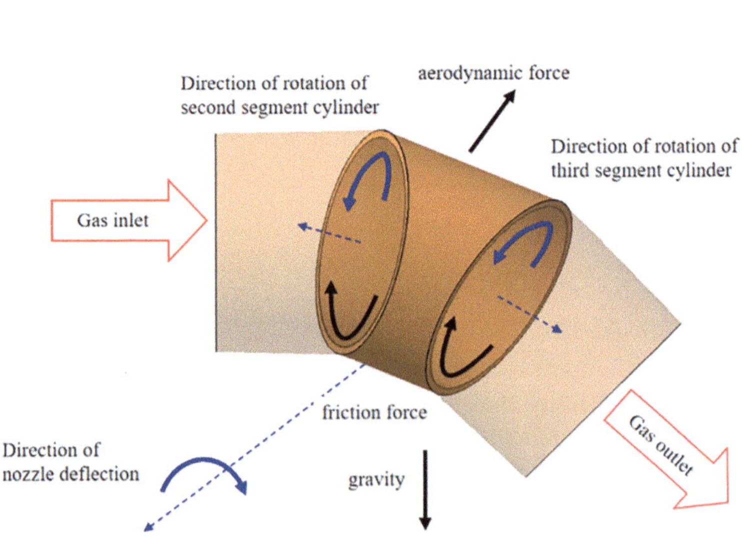

Fig. 4.26 Schematic diagram of deflection direction and force on three-bearing deflection nozzle

4.4.4.1 Basic Assumptions

In this book, based on the theory of classical mechanics, the driving moment calculation model of each segment of the three-bearing deflection nozzle's cylinder is established and the driving moment of each segment of the cylinder is calculated. In the model, the force of the three-bearing deflection nozzle is simplified, and the centrifugal inertia force of the three-segment cylinder in the non-inertial system is neglected.

Three-bearing deflection nozzle in the deflection process, each section of the cylinder rotation required to drive the torque to overcome the nozzle gravity, each section of the cylinder between the bearing mounting cross-section of the circumferential friction and the gas through the nozzle when the wall of the nozzle generated by aerodynamic force on the axis of rotation of each section of the cylinder torque. Due to the viscous effect of the gas flow, the gas through the nozzle wall produces a tangential force, the tangential force on the cylinder rotation axis of each section of the moment is very small and negligible. Since the axes of the cylinder segments are always kept in the same pitch plane when the nozzle is deflected, the aerodynamic force of the gas on the nozzle can be approximated and simplified to the two-dimensional pitch plane. In summary, the following reasonable assumptions are made for the three-bearing deflected nozzle driving moment calculation model:

(1) The driving torque required to rotate each section of the barrel of the three-bearing deflector nozzle is only the torque required to overcome the gravity of the nozzle, the circumferential friction of the bearing mounting cross-section, and the aerodynamic force on the wall of the nozzle when the gas passes through the nozzle to the rotational axis of each section of the barrel;
(2) The physical parameters of the gas in each cross section are uniformly distributed;
(3) The gas flow rate at the inlet and outlet sections of each cylinder is equal and does not vary with time;
(4) The aerodynamic force of the gas on the nozzle wall as it passes through each section of the cylinder is in the pitch plane;
(5) Changes in the position and velocity of the control body, cross-section are simplified to changes in the position and velocity of its geometric center;
(6) Deactivate vacuum pumps and valves after completing all tests.

Based on the deflection law model of the three-bearing deflection nozzle in Sect. 4.4.3, to simplify the amount of calculation, the points O and O_0 of the coordinate system xyz, $x_0y_0z_0$ are overlapped with the point O_1 of the coordinate system $x_1y_1z_1$, as shown in Fig. 4.27.

Let the rotation axis vector of the second cylinder relative to the first cylinder be $\mathbf{n}_2$, $\mathbf{n}_2$ coincides with the $\mathbf{x}_1$ axis in the coordinate system $x_1y_1z_1$, and the rotation axis vector of the third cylinder relative to the second cylinder be $\mathbf{n}_3$, $\mathbf{n}_3$ coincides with the $\mathbf{x}_3$ axis in the coordinate system $x_3y_3z_3$. In order to simplify the derivation of the formula, through the simplified model of the nozzle deflection law shown in

Figure 4.27, it can be derived that the rotation axes of the second and the third section of the nozzle $\mathbf{n}_2$, $\mathbf{n}_3$ in the direction of the absolute coordinate system xyz, respectively:

$$(\mathbf{n}_2)_o = C_{oo_0} C_{o_0 o_1} (\mathbf{n}_2)_{o_1} \tag{4.41}$$

$$(\mathbf{n}_3)_o = C_{oo_0} C_{o_0 o_1} C_{o_1 o_2} C_{o_2 o_3} (\mathbf{n}_3)_{o_3} \tag{4.42}$$

where C_{oo_0}, $C_{o_0 o_1}$, $C_{o_1 o_2}$, $C_{o_2 o_3}$ are the cosine matrices of the coordinate transformations of the coordinate systems $x_0y_0z_0$ and xyz, $x_1y_1z_1$ and $x_0y_0z_0$, $x_2y_2z_2$ and $x_1y_1z_1$, $x_3y_3z_3$ and $x_2y_2z_2$, respectively.

4.4.4.2 Calculation Model of the Driving Moment of the First Section of the Cylinder

Taking the first section of the cylinder as the research object, the driving moment calculation model of the first section of the cylinder is established, as shown in Fig. 4.28. In the process of nozzle deflection from 0° to the maximum deflection angle, the driving moment needs to overcome the circumferential friction of the first cylinder rotation, the gravity of the first, second and third cylinder and the aerodynamic force on the nozzle wall generated by the gas passing through the first, second and third cylinders to the moment of the rotation axis of the first cylinder. Taking an instantaneous nozzle as a whole as the object of study, let the moment sum at the rotation axis of the first cylinder be $\sum M_1$, which can be obtained from the momentum moment theorem

$$\Sigma M_1 = \left(J_{1n_1} + J_{2n_1} + J_{3n_1}\right)\frac{d\omega_1}{dt} \tag{4.43}$$

$$\Sigma M_1 = M_{d_1} + M_{f_1} + M_{G_1 n_1} + M_{G_2 n_1} + M_{G_3 n_1} + M_{j_1 n_1} + M_{j_2 n_1} + M_{j_3 n_1} \tag{4.44}$$

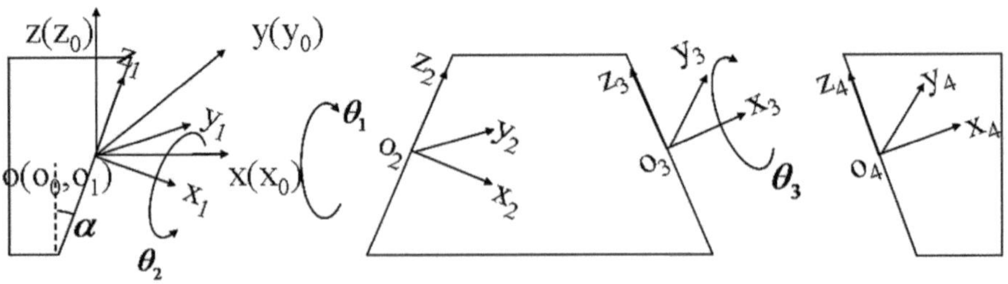

Fig. 4.27 Simplified model of nozzle deflection law

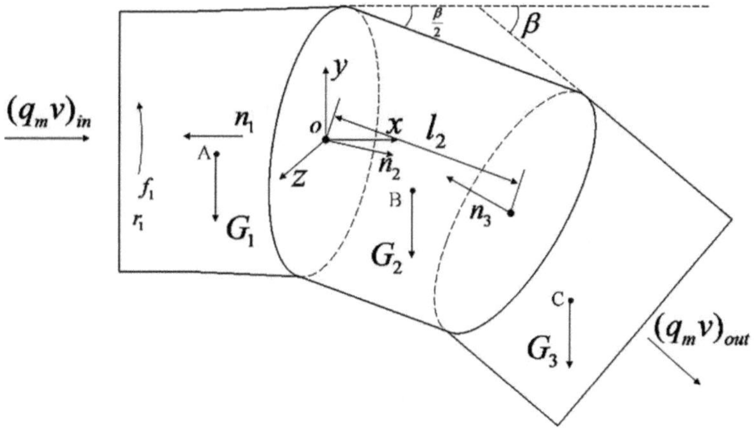

Fig. 4.28 Calculation model of driving moment of the first section of the cylinder

In the Eq. (4.43), ω_1 is the scalar of the rotational angular velocity of the first cylinder, J_{1n_1}, J_{2n_1}, J_{3n_1} are the rotational moment of inertia of the first, second and third cylinders on the rotational axis of the first cylinder, M_{d_1} is the driving moment of the first cylinder, M_{f_1} is the friction moment of the bearing rotation on the rotational axis of the first cylinder, $M_{G_1n_1}$, $M_{G_2n_1}$, $M_{G_3n_1}$ are the gravity moments of the first, second and third cylinders on the rotational axis of the first cylinder. $M_{j_1n_1}$, $M_{j_2n_1}$, and $M_{j_3n_1}$ are the aerodynamic moments of the first, second and third cylinders on the rotation axis of the first cylinder.

Since the direction of the rotation axis of the first cylindern_1 always coincides with the x axis in the coordinate system xyz, and the rotation direction of the first cylinder remains unchanged during the deflection of the nozzle, only the moment of the x axis is required. According to Assumption 4, the gas flow through the sections of the cylinder, the wall of the nozzle aerodynamic force in the pitch plane, so the three sections of the wall of the nozzle by the aerodynamic force on the n_1 axis of the moment of $M_{j_1n_1}$, $M_{j_2n_1}$, $M_{j_3n_1}$ for 0. According to the obtained deflection law of the nozzle, the angular velocity of the first section of the cylinder rotating around its axis is constant. By substituting equation (4.44) into equation (4.43), it can be obtained that

$$M_{d_1} + M_{f_1} + M_{G_1n_1} + M_{G_2n_1} + M_{G_3n_1} = 0 \tag{4.45}$$

Let the gravity force on the first, second and third sections of the cylinder be $\mathbf{G}_1$, $\mathbf{G}_2$ and$\mathbf{G}_3$ respectively, and the points of gravity force are point A, B and C respectively, as shown in Figure 4.28. The coordinates of the three points in the fixed coordinate system of the cylinder are $(A)_{o_0} = (x_A, y_A, z_A)$, $(B)_{o_2} = (x_B, y_B, z_B)$ and $(C)_{o_4} = (x_C, y_C, z_C)$, and the coordinates of the three points A, B and C in the absolute coordinate system xyz can be obtained through the coordinate transformation as

$(A)_o = C_{oo_0}(A)_{o_0}$, $(B)_o = C_{oo_0}C_{o_0o_1}C_{o_1o_2}(B)_{o_2}$ and $(C)_o = C_{oo_0}C_{o_0o_1}C_{o_1o_2}C_{o_2o_3}C_{o_3o_4}(C)_{o_4}$, and the vector forms of $\mathbf{G}_1$, $\mathbf{G}_2$ and $\mathbf{G}_3$ can be expressed as $(0, -|G_1|, 0)$, $(0, -|G_2|, 0)$ and $(0, -|G_3|, 0)$ in the absolute coordinate system, respectively. The direction of the rotation axis of the first section of the cylinder $\mathbf{n}_1 = (-1, 0, 0)$, the coordinates of the starting point of $\mathbf{n}_1$ coincide with the point O, then the moments of $\mathbf{G}_1$, $\mathbf{G}_2$, $\mathbf{G}_3$ to $\mathbf{n}_1$ can be expressed as follows

$$M_{G_1n_1} = (\mathbf{OA} \times \mathbf{G}_1) \cdot \mathbf{n}_1 \tag{4.46}$$

$$M_{G_2n_1} = (\mathbf{OB} \times \mathbf{G}_2) \cdot \mathbf{n}_1 \tag{4.47}$$

$$M_{G_3n_1} = (\mathbf{OC} \times \mathbf{G}_3) \cdot \mathbf{n}_1 \tag{4.48}$$

additionally

$$M_{f_1} = (\mathbf{f}_1 \times \mathbf{r}_1) \cdot \mathbf{n}_1 \tag{4.49}$$

In the equation, $\mathbf{f}_1$ is the friction force when the bearing rotates, $\mathbf{f}_1 = \mu\mathbf{P}$, μ are the friction coefficients, which are affected by the bearing type, load, rotational speed, lubrication method, etc., and $\mathbf{P}$ is the bearing load. Substitute Eqs. (4.46)–(4.49) into Eq. (4.45) to get the driving torque of the first cylinder M_{d_1}.

4.4.4.3 Model for Calculating the Driving Moment of the Second Section of the Cylinder

Taking the second segment cylinder as the research object, the moment calculation model of the second segment cylinder is established, as shown in Fig. 4.29. In the process of nozzle deflection from 0° to the maximum vector angle, the driving moment needs to overcome the circumferential friction force when the second segment cylinder is rotating, the gravity force of the second and third segment cylinders and the aerodynamic force generated by the gas through the second and third segment cylinders to the wall to the second segment cylinder rotating axis of the moment. Taking the second and third nozzle as a whole at a certain instant as the research object, let the moment sum at the rotation axis of the second cylinder be $\sum M_2$, which can be obtained by the momentum moment theorem

$$\Sigma M_2 = (J_{2n_2} + J_{3n_2})\frac{d\omega_2}{dt} \tag{4.50}$$

$$\Sigma M_2 = M_{d_2} + M_{f_2} + M_{G_2n_2} + M_{G_3n_2} + M_{j_2n_2} + M_{j_3n_2} \tag{4.51}$$

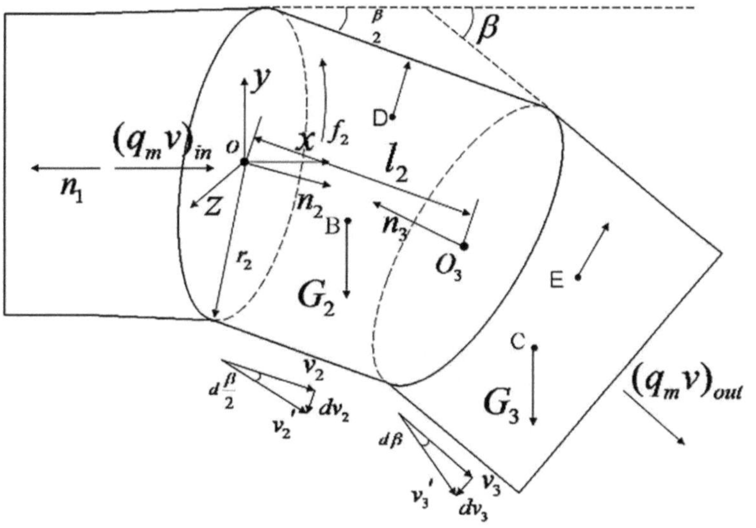

Fig. 4.29 Calculation model of driving moment of the second section of the cylinder

In the formula, ω_2 is the rotational angular velocity scalar of the second section of the cylinder, J_{2n_2} and J_{3n_2} are the rotational inertia of the second and third sections of the cylinder on the rotational axis of the second section of the cylinder, M_{d_2} is the driving moment of the second section of the cylinder, M_{f_2} is the friction moment of the bearing rotation friction on the rotational axis of the second section of the cylinder, $M_{G_2n_2}$ and $M_{G_3n_2}$ are the gravity moment on the rotational axis of the second section of the cylinder, respectively, and is the pneumatic force on the rotational axis of the second section of the cylinder. $M_{j_2n_2}$ and $M_{j_3n_2}$ are the pneumatic moments of the rotation axis of the second and third cylinders by the pneumatic force, respectively.

The vector direction of the rotation axis of the second cylinder relative to the first cylinder in the absolute coordinate system is $(\mathbf{n}_2)_o = C_{oo_0} C_{o_0o_1} (\mathbf{n}_2)_{o_1}$, and the starting point of $\mathbf{n}_2$ is set to coincide with the origin of the coordinate system $x_2y_2z_2$. The vector direction of the rotation axis of the third cylinder with respect to the second cylinder in the absolute coordinate system is $(\mathbf{n}_3)_o = C_{oo_0} C_{o_0o_1} C_{o_1o_2} C_{o_2o_3} (\mathbf{n}_3)_{o_3}$, then the moments of $\mathbf{G}_2$ and $\mathbf{G}_3$ with respect to $\mathbf{n}_2$ can be found as

$$M_{G_3n_2} = (\mathbf{O}_2\mathbf{E} \times \mathbf{G}_3) \cdot \mathbf{n}_2 \tag{4.52}$$

$$M_{G_2n_2} = (\mathbf{O}_2\mathbf{B} \times \mathbf{G}_2) \cdot \mathbf{n}_2 \tag{4.53}$$

Let the magnitude of the inlet and outlet airflow velocity of the second section of the cylinder be $\mathbf{v}$, the velocity of the second section of the cylinder be $\mathbf{u}$, $\mathbf{A}_1$ is the inlet cross-section of the second section of the cylinder, and $\mathbf{A}_2$ is the outlet cross-section of the second section of the cylinder, then according to the one-dimensional Reynolds Transport Theorem (Wang 2006), we can get

$$\frac{d\phi_s}{dt} = \frac{d\phi_{cv}}{dt} + \int_{cs} \eta\rho(\mathbf{v}-\mathbf{u})\mathrm{d}\mathbf{A}_2 - \int_{cs} \eta\rho(\mathbf{v}-\mathbf{u})\mathrm{d}\mathbf{A}_1 \tag{4.54}$$

By simplifying the above equation according to Assumption 2, the momentum equation for the passage of gas through the nozzle can be expressed as

$$\Sigma\mathbf{F} = \frac{d(m\mathbf{v})_s}{dt} = \frac{d(m_2\mathbf{v}_2)_{cv}}{dt} + \mathbf{v}_2\rho_2(\mathbf{v}_2-\mathbf{u}_2)\mathbf{A}_2 - \mathbf{v}_1\rho_1(\mathbf{v}_1-\mathbf{u}_1)\mathbf{A}_1 \tag{4.55}$$

The control body in the equation can be a motion control body, and $d(m_2\mathbf{v}_2)_{cv}/dt$ is the derivative of the gas momentum in the motion control body and the spatial position of the motion control body with respect to time. According to Assumption 3, the mass of the gas flow in the second cylinder does not change with time, and the geometric center of the second cylinder is set to be the point N_2, then the momentum equation of the gas passing through the second cylinder can be expressed as follows

$$\begin{aligned} &\mathbf{p}_1A_1 - \mathbf{p}_2A_2 + \mathbf{F}_{n2} \\ &= m_2\frac{\partial\mathbf{v}_2}{\partial t} + m_2\frac{\partial N_2(x,y,z,t)}{\partial t} + \mathbf{v}_2\rho_2(\mathbf{v}_2-\mathbf{u}_2)\mathbf{A}_2 - \mathbf{v}_1\rho_1(\mathbf{v}_1-\mathbf{u}_1)\mathbf{A}_1 \end{aligned} \tag{4.56}$$

where $\mathbf{p}_1$,$\mathbf{p}_2$ are perpendicular to the second section of the cylinder body inlet and outlet cross-section, the direction points to the cylinder body inside the static pressure, $\mathbf{F}_{n2}$ for the second section of the cylinder body wall surface force on the gas, set the gas on the cylinder body of the gas force for $\mathbf{F}_{j2} = -\mathbf{F}_{n2}$, substituting into the Eq. (4.56) can be obtained

$$\begin{aligned} \mathbf{F}_{j2} = &-m_2\frac{\partial\mathbf{v}_2}{\partial t} - m_2\frac{\partial N_2(x,y,z,t)}{\partial t} - \mathbf{v}_2\rho_2(\mathbf{v}_2-\mathbf{u}_2)\mathbf{A}_2 \\ &+ \mathbf{v}_1\rho_1(\mathbf{v}_1-\mathbf{u}_1)\mathbf{A}_1 + \mathbf{p}_1A_1 - \mathbf{p}_2A_2 \end{aligned} \tag{4.57}$$

The direction of gas flow in the second cylinder varies with the movement of the second cylinder over time, so the

$$\frac{\partial\mathbf{v}_2}{\partial t} = \left(-\frac{|\mathbf{v}_2|}{2}\frac{d(\beta)}{dt}\sin\left(\frac{\beta}{2}\right), -\frac{|\mathbf{v}_2|}{2}\frac{d(\beta)}{dt}\cos\left(\frac{\beta}{2}\right), 0\right) \tag{4.58}$$

where $\mathbf{v}_2$ is the gas velocity in the second cylinder, which can be expressed as $\mathbf{v}_2 = (|\mathbf{v}_2|, 0, 0)$ in the relative coordinate system and $\mathbf{v}_2 = (|\mathbf{v}_2| \cos(\beta/2), -|\mathbf{v}_2| \sin(\beta/2), 0)$ in the absolute coordinate system.

Since $\mathbf{p}_1$ is perpendicular to the inlet cross-section of the second cylinder, the rotation axis of the second cylinder is also perpendicular to the inlet cross-section of the second cylinder, so $\mathbf{p}_1$ is parallel to the rotational axis, and the torque on the rotational axis is 0. $\mathbf{p}_2$ is perpendicular to the outlet cross-section of the second cylinder, and according to the numerical simulation results of the flow field of the nozzle in the literature, the hydrostatic pressure distribution of the outlet cross-section is approximately symmetric with respect to the pitch plane, and $\mathbf{p}_2$ is approximately simplified in the pitch plane, so the torque on the rotational axis is simplified to 0. Equation (4.57) can be simplified to 0. $\mathbf{p}_2$ can be simplified to the rotational axis. Equation (4.57) can be simplified as follows

$$\mathbf{F}_{j2} = -m_2 \frac{\partial \mathbf{v}_2}{\partial t} - m_2 \frac{\partial N_2(x,y,z,t)}{\partial t} - \mathbf{v}_2 \rho_2 (\mathbf{v}_2 - \mathbf{u}_2) \mathbf{A}_2 + \mathbf{v}_1 \rho_1 (\mathbf{v}_1 - \mathbf{u}_1) \mathbf{A}_1 \tag{4.59}$$

When the nozzle inlet and outlet boundary conditions are given, the gas parameters in Eq. (4.59) such as the gas mass in the nozzle m and the nozzle cross-sectional velocity $\mathbf{v}$ can be derived by CFD numerical methods.

Let the point of action of $\mathbf{F}_{j2}$ on the second cylinder segment D have the coordinates $(D)_{o_2} = (x_D, y_D, z_D)$ in the coordinate system $x_2y_2z_2$, which is fixed to the second cylinder segment, and $D = C_{oo_0} C_{o_0o_1} C_{o_1o_2} (D)_{o_2}$ in the absolute coordinate system xyz, and let the point of action of $\mathbf{F}_{j3}$ on the third cylinder segment E have the coordinates $(E)_{o_4} = (x_E, y_E, z_E)$ in the coordinate system $x_4y_4z_4$, which is fixed to the third cylinder segment, and $E = C_{oo_0} C_{o_0o_1} C_{o_1o_2} C_{o_2o_3} C_{o_3o_4} (E)_{o_4}$ in the absolute coordinate system xyz, then we can see that the moments of $\mathbf{F}_{j2}$, $\mathbf{F}_{j3}$ on $\mathbf{n}_2$ are

$$M_{j_2n_2} = (\mathbf{O}_2\mathbf{D} \times \mathbf{F}_{j2}) \cdot \mathbf{n}_2 \tag{4.60}$$

$$M_{j_3n_2} = (\mathbf{O}_2\mathbf{E} \times \mathbf{F}_{j2}) \cdot \mathbf{n}_2 \tag{4.61}$$

additionally

$$M_{f_2} = (\mathbf{f}_2 \times \mathbf{r}_2) \cdot \mathbf{n}_2 \tag{4.62}$$

Substituting Equations (4.53), (4.54) and (4.60)-(4.62) into Equation (4.52) yields the driving torque for the second section of the cylinder M_{d_2}.

4.4.4.4 Calculation Model of the Driving Moment of the Third Section of the Cylinder

Taking the third section of the cylinder as the research object, the moment calculation model of the third section of the cylinder is established, as shown in Fig. 4.30. During the deflection of the nozzle from 0° to the maximum deflection angle, the sum of moments at the rotation axis of the third section of the cylinder is $\sum M_3$, which can be obtained by the momentum moment theorem

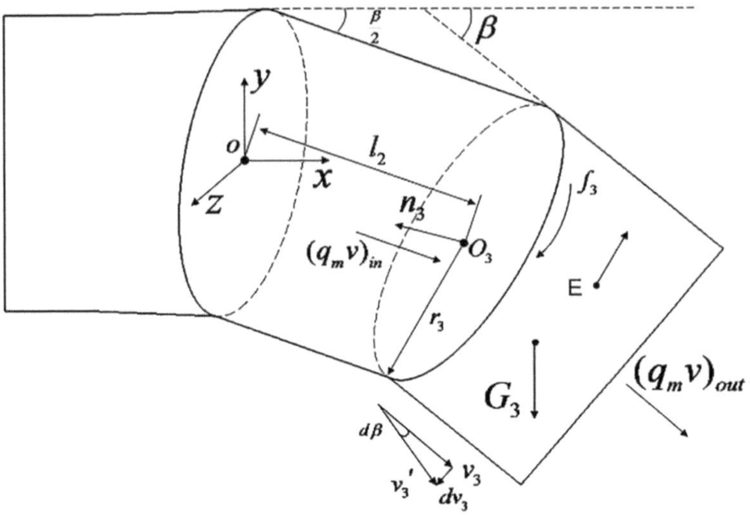

Fig. 4.30 Calculation model of the driving moment of the third section of the cylinder

$$\Sigma M_3 = J_{3n_3} \frac{d\omega_3}{dt} \tag{4.63}$$

$$\Sigma M_3 = M_{d_3} + M_{f_3} + M_{G_3 n_3} + M_{j_3 n_3} \tag{4.64}$$

In Eq. (4.63), ω_3 is the rotational angular velocity scalar of the third cylinder, J_{3n_3} is the rotational moment of inertia of the third cylinder on the rotating shaft of the third cylinder, M_{d_3} is the driving moment on the rotating shaft of the third cylinder, M_{f_3} is the friction moment on the rotating shaft of the third cylinder when the bearing rotates, and $M_{G_3 n_3}$ is the gravitational moment of the gravity force of the third cylinder on the rotating shaft of the third nozzle. $M_{j_3 n_3}$ is the aerodynamic moment of the aerodynamic force exerted on the nozzle rotation axis of the third section of the cartridge.

The vector direction of the axis of rotation of the third cylinder relative to the second cylinder in the absolute coordinate system is $(\mathbf{n}_3)_o = C_{oo_0} C_{o_0 o_1} C_{o_1 o_2} C_{o_2 o_3} C_{o_3 o_4} (\mathbf{n}_3)_{o_4}$, let the starting point of $\mathbf{n}_3$ coincide with the coordinate origin O_3 of the coordinate system $x_3 y_3 z_3$, and the coordinates of the point O_3 in the absolute coordinate system are $O_3 = C_{oo_0} C_{o_0 o_1} C_{o_1 o_2} (l_2 \cos\alpha, 0, l_2 \sin\alpha)$, then the moment of $\mathbf{G}_3$ to $\mathbf{n}_3$ can be known as

$$M_{G_3 n_3} = (\mathbf{O}_3\mathbf{E} \times \mathbf{G}_3) \cdot \mathbf{n}_3 \tag{4.65}$$

Set the third section of the cylinder inlet gas velocity for $\mathbf{v}_2$, outlet velocity for $\mathbf{v}_3$, from the Eq. (4.59) can be derived from the gas through the third section of the cylinder on the wall of the aerodynamic force $\mathbf{F}_{j3}$ for

$$\mathbf{F}_{j3} = -m_3 \frac{\partial \mathbf{v}_3}{\partial t} - m_3 \frac{\partial N_3(x,y,z,t)}{\partial t} - \mathbf{v}_3 \rho_3 (\mathbf{v}_3 - \mathbf{u}_3) \mathbf{A}_3 + \mathbf{v}_2 \rho_2 (\mathbf{v}_2 - \mathbf{u}_2) \mathbf{A}_2 \tag{4.66}$$

From the direction of gas flow in the third section of the cylinder is the same as the direction of motion of the third section of the cylinder and changes with time can be obtained by

$$\frac{d\mathbf{v}_3}{dt} = \left(-|\mathbf{v}_3| \frac{d\beta}{dt} \sin(\beta), -|\mathbf{v}_3| \frac{d\beta}{dt} \cos\beta, 0 \right) \tag{4.67}$$

where $\mathbf{v}_3$ is the gas velocity in the third cylinder, which can be expressed as $\mathbf{v}_3 = (|\mathbf{v}_3|, 0, 0)$ in the relative coordinate system and $\mathbf{v}_3 = (|\mathbf{v}_3| \cos(\beta/2), -|\mathbf{v}_3| \sin(\beta/2), 0)$ in the absolute coordinate system.

The point E is the action point of $\mathbf{F}_{j3}$ on the third section of the cylinder, let the coordinates of the point$E$ in the coordinate system $x_4y_4z_4$, which is solidly connected with the third section of the cylinder, be $(E)_{o_4} = (x_E, y_E, z_E)$, and the coordinates of $E = C_{oo_0} C_{o_0o_1} C_{o_1o_2} C_{o_2o_3} C_{o_3o_4} (E)_{O_4}$ in the absolute coordinate system xyz, then the moments of $\mathbf{F}_{j3}$ on $\mathbf{n}_3$ is

$$M_{j_3n_3} = (\mathbf{O}_4\mathbf{E} \times \mathbf{F}_{j_3}) \cdot \mathbf{n}_3 \tag{4.68}$$

additionally

$$M_{f_3} = \mathbf{f}_3 \mathbf{r}_3 \tag{4.69}$$

Substituting Equations (4.36), (4.68), and (4.69) into Eq. (4.64) yields the driving torque for the third cylinder segment M_{d_3}.

4.4.4.5 Driving Moment Analysis of the Nozzle Cylinder

From the driving moment calculation model of the three segments of the cylinder, it can be seen that when the friction force is certain, the moment of the friction force on the axis of rotation of the cylinder is a constant. Assuming that the aerodynamic force and gravitational force on each segment of the cylinder act on the geometric center of each segment of the cylinder, the driving moment of the first segment of the cylinder only needs to overcome the moment of friction on the rotational axis during the process of deflection of the nozzle from 0° to the maximum deflection angle. Taking the physical model of a three-bearing deflection nozzle assembled with a turbojet engine developed by this group as an example, as shown in Fig. 4.31, the driving moments required to overcome the aerodynamic force and gravity during the rotation of the cylinder in the second and third segments are calculated, and

Fig. 4.31 Physical model of three-bearing deflection nozzle

the influence of the geometric parameters of the nozzle on the driving moments is analyzed.

From the driving moment calculation model of each section of the cylinder, we can get the change rule of the aerodynamic force suffered by the second and third sections of the cylinder and the moment of gravity on the rotation axis of the second and third sections of the cylinder with the nozzle deflection angle. The maximum deflection angle of the nozzle is 95°, the tilt angle of the tilted section of each cylinder is 23.75°, the centerline length of the second cylinder is 54mm, and the centerline length of the third cylinder is 30mm. Under the non-vectorized state of the nozzle, the size of the gas momentum at the inlet section of the second cylinder is $79.5kg \cdot m/s^2$, the size of the gas momentum at the outlet section of the second cylinder is $82kg \cdot m/s^2$, and the size of the gas momentum at the outlet section of the third cylinder is , and for simple calculation, the gas momentum at the outlet section of the third cylinder is $161kg \cdot m/s^2$. For the sake of simplicity of calculation, the change of gas velocity in each cross-section is ignored when the nozzle is deflected. The gravity of the second section of the nozzle is 7.13N, and the gravity of the third section of the nozzle is 3.73N.

Figs. 4.32 and 4.33 show the change rule of the moments of aerodynamic force and gravity force on the rotation axis of the second and third sections of the cylinder with the deflection angle of the nozzle, respectively. Comparing the two figures, it can be seen that the change rule of gravity and aerodynamic force on the rotation axis of the second cylinder with the nozzle deflection angle is more or less the same as that of the third cylinder. The moment of gravity on the nozzle axis of rotation with the increase of the nozzle deflection angle and decrease, the second and third section of the cylinder by the aerodynamic force on their respective rotary axis of the moment with the increase of the nozzle deflection angle is the first increase and then decrease, the second and third section of the cylinder by the aerodynamic force of their respective rotary axis of the maximum moment appear in the position of the

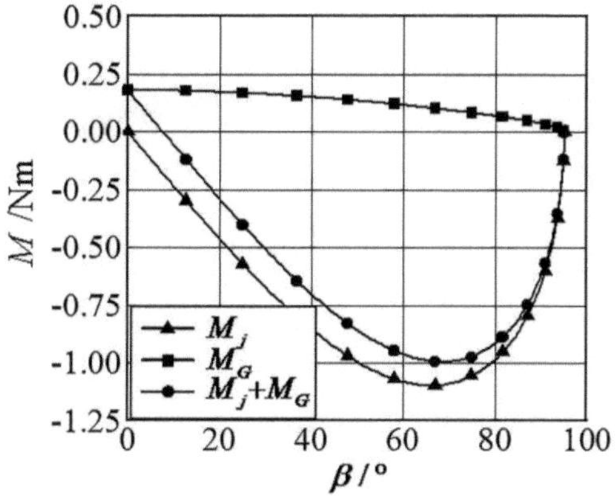

Fig. 4.32 Variation of driving moment with deflection angle for the second section of the cylinder

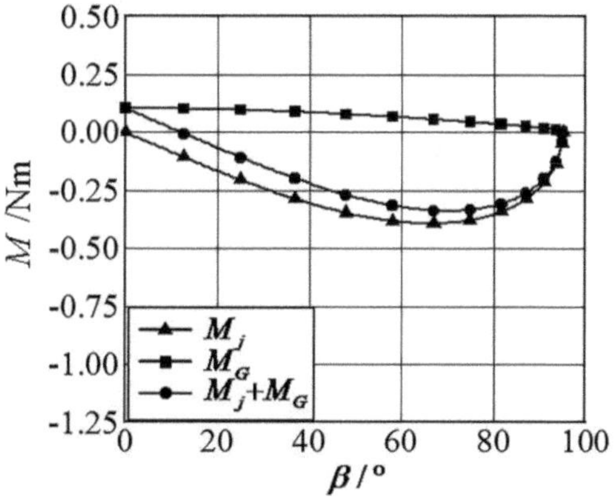

Fig. 4.33 Variation of driving moment with deflection angle for the third section of the cylinder. (**a**) The variation of M_G with β. (**b**) The variation of M_j with β. (c) Variation of the combined moments of M_G and M_j with β

nozzle deflection to about 70°. The moment effect of aerodynamic force on the cylinder rotation axis is opposite to the moment effect of gravity on the cylinder rotation axis. When the nozzle is deflected to an angle, the combined aerodynamic and gravitational moments on the axis of rotation of the third cylinder are less than the combined aerodynamic and gravitational moments on the axis of rotation of the second cylinder.

When the maximum deflection angle achievable by the nozzle is changed, the inclination angle of the inclined cross section of the barrel of each section of the nozzle is also changed, and the change of the geometric parameters of the nozzle will have a certain effect on the driving torque of the second and third sections of the barrel.

Figures 4.34 and 4.35 show the variation rules of the aerodynamic force and the moment of gravity on the rotational axis of the second and third sections of the cylinder with the nozzle deflection angle when the maximum deflection angle achievable by the nozzle is 85°, 90° and 95°, respectively. From Figures 4.34a, b and 4.35a, b can be seen, gravity and aerodynamic force on the second and third section of the cylinder rotary axis of the moment with the nozzle deflection angle of

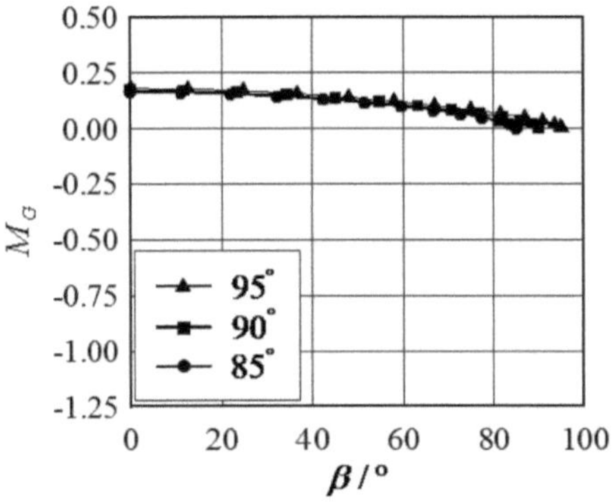

(a) The variation of M_G with β

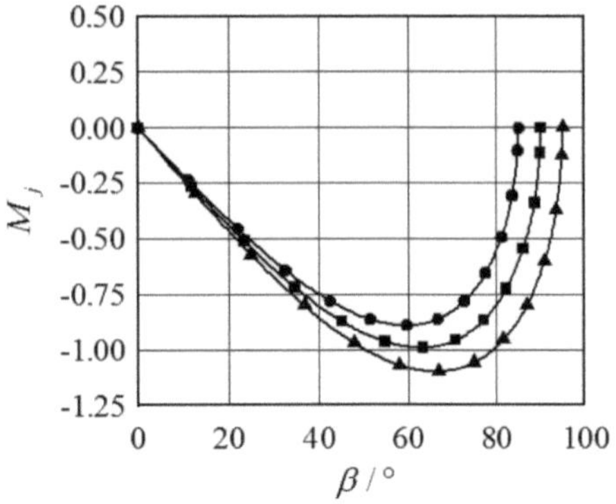

(b) The variation of M_j with β

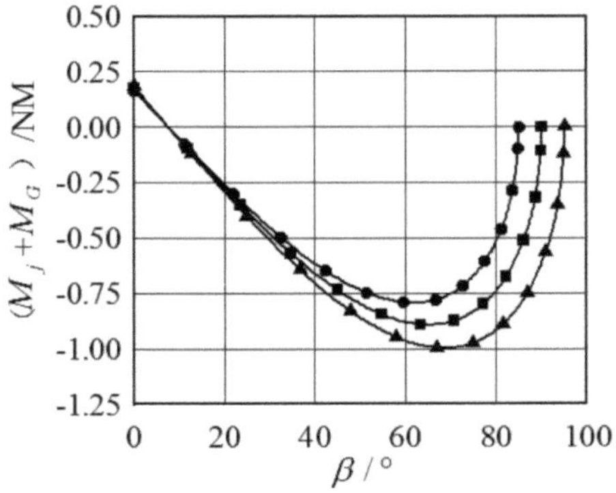

(c) Variation of the combined moments of M_G and M_j with β

Fig. 4.34 M_j and M_G variation of combined moments with β for the second section of the cylinder at β_{max}=85°, 90° and 95°. **(a)** $L_3 = 0.22$. **(b)** $L_3 = 1$

the rule of law is generally the same, and the gravity moment with the nozzle maximum deflection angle increases basically unchanged. When the maximum deflection angle of the nozzle can be realized increases, the moment of the pneumatic force on the rotary axis also increases, the maximum moment of the pneumatic force on the rotary axis of the second and third sections of the cylinder increases with the increase of the maximum deflection angle of the nozzle, and the maximum moment of the pneumatic force on the rotary axis corresponds to the nozzle deflection angle of about 75% of the maximum deflection angle. From Figs. 4.34c and 4.35c can be seen, aerodynamic force and gravity on the cylinder rotation axis of the combined torque with the nozzle deflection angle of the rule of change and aerodynamic force on the cylinder rotation axis of the rule of change of the rule of change of the rule of law is generally the same.

In summary, when the rotational friction of the bearing is ignored, the driving moment of the second cylinder is larger than that of the third cylinder, and the

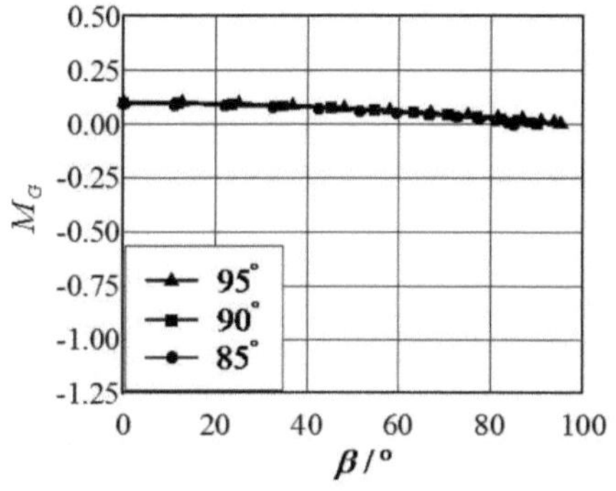

(a) The variation of M_G with β

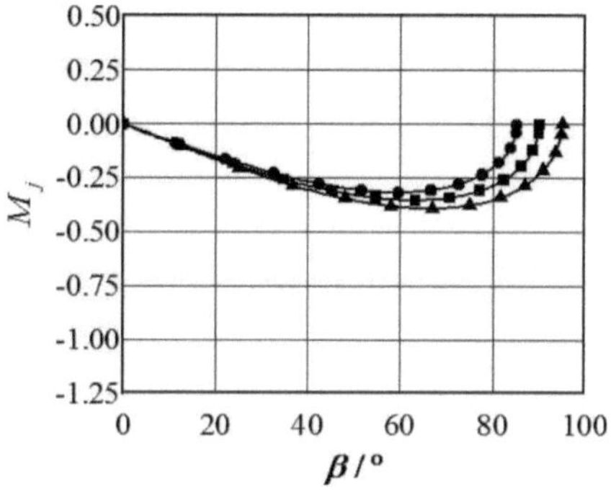

(b) The variation of M_j with β

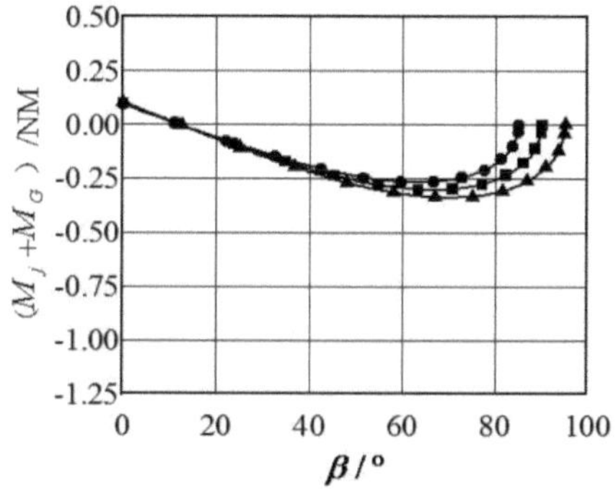

(c) Variation of the combined moments of M_G and M_j with β

Fig. 4.35 M_j and M_G variation of combined moments with β for the third section of the cylinder at β_{max}=85°, 90° and 95°. (**a**) The variation of M_G with β. (**b**) The variation of M_j with β. (**c**) Variation of the combined moments of M_G and M_j with β

moments of aerodynamic force and gravity on the rotating axis increase with the increase of the deflection angle, which increases and then decreases. In the engineering application of the three-bearing deflecting nozzle, the driving moments of the second and third cylinders should be adjusted dynamically, and the sizes of the driving moments of the second and third cylinders should be allocated reasonably. The geometric parameters of the nozzle also have a certain impact on the torque of the rotating shafts of the second and third sections of the cylinder. When the maximum deflection angle that the nozzle can achieve increases, the angle of the inclined section of each section of the cylinder changes, resulting in an increase in the driving torque of the second and third sections of the cylinder. The maximum torque of aerodynamic force on the rotating shaft of the cylinder corresponds to the nozzle deflection angle, which increases with the increase of the maximum deflection angle, and the deflection angle corresponding to the maximum driving torque is approximately 75% of the maximum vector angle.

4.5 Constant Aerodynamic Characteristics of a Three-Bearing Deflection Nozzle

In this section, a 3D CFD numerical method is used to obtain the flow field characterization and aerodynamic performance of the nozzle at a constant state. Figure 4.36 shows the geometrical model used in the numerical simulation process, the key geometrical dimensions of the nozzle have been marked in the figure, and the key cross-sections that will be mentioned in the flow field characterization are also marked in the figure.

The cross-sections 1, 2, and 4 are the inlet and outlet cross-sections of the first and the second cylinder sections, respectively, and the cross-sections 3 are located in between the cross-sections 2 and 4, which are perpendicular to the axial direction of the nozzle, and the cross-sections 5 are located in the inlet of the constricted section of the nozzle.

4.5.1 *Three-Bearing Deflection Nozzle Flow Characteristics*

Figure 4.37 shows the total pressure and streamline distribution of the nozzle profile, 1–5 cross-section and outlet cross-section at a 10° deflection angle. From the streamline and total pressure distribution of the nozzle profile, it can be seen that after the airflow passes through the tail cone, due to the expansion of the circulation area, a return area is generated after the tail cone, the total pressure of the airflow in the return area decreases, and a large energy loss occurs, the airflow outside the tail cone bypasses the return area, and mixes with the center area after the return area, and the total pressure of the airflow in the center area of the nozzle rises gradually in the axial direction.

Cross-section 1 is the inlet section of the first section of the cylinder, from the streamline and total pressure distribution of cross-section 1, it can be seen that the total pressure distribution of the airflow in the cross-section of the gradient is larger,

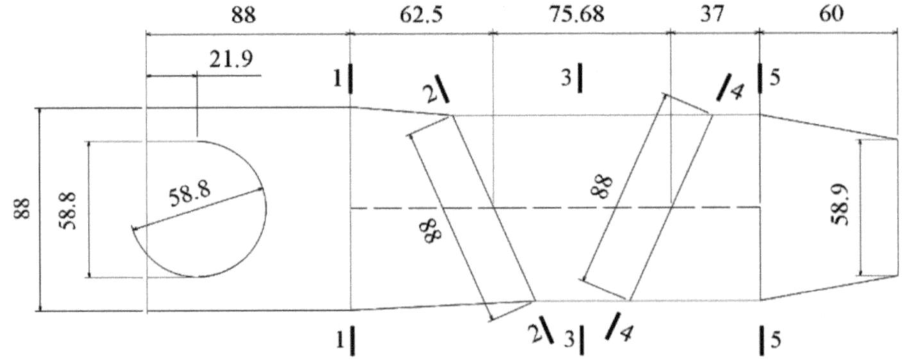

Fig. 4.36 Geometric model of a three-bearing deflector nozzle

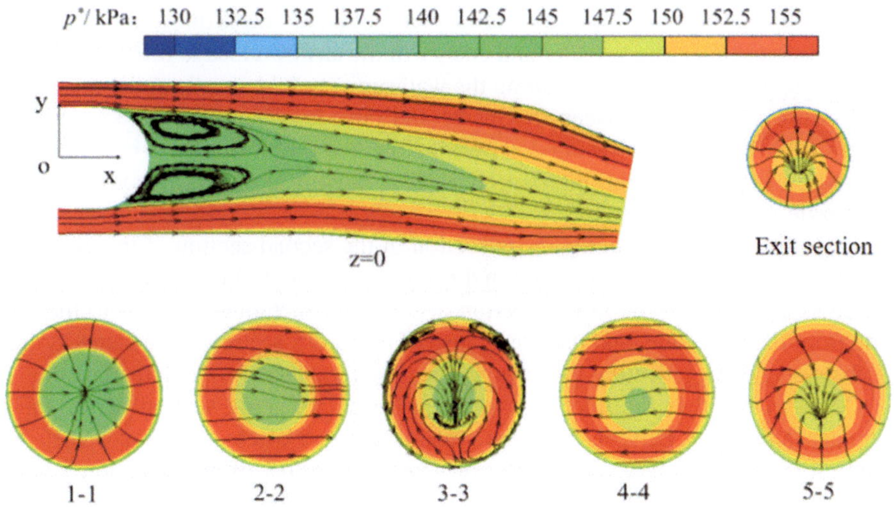

Fig. 4.37 Total pressure and streamline distribution of airflow in each section at 10° nozzle offset

which is due to the airflow around the return area after the cross-section is not sufficiently mixed caused by the flow direction of cross-section 1 for the airflow direction of the cross-section of the projection, the cross-section of the direction of the streamline by the wall of the nozzle pointing to the center of the position, this is due to the flow of the airflow along the mainstream direction of flow at the same time also flow to the center of the nozzle area, the flow trend can also be seen from the streamline distribution of the nozzle cross-section. This is because the airflow flows along the main flow direction and also flows to the center region of the nozzle, and the flow trend can also be seen from the distribution of the flow lines of the nozzle profile.

Section 2 for the second section of the cylinder inlet cross-section, cross-section 2 of the flow direction for the air flow direction in the cross-section of the projection, the cross-section of the flow direction from the left to the right, this is due to the cross-section of the 2 and the nozzle axis direction is not perpendicular to the existence of a certain angle of the cross-section, the cross-section of the flow direction reflects the direction of the flow of the air flow for the flow of the axial direction of flow along the nozzle.

Section 3 is perpendicular to the axis of the second section of the cylinder, and the streamline distribution of section 3 shows that the airflow is subjected to centripetal force under the action of wall bending, and the pressure near the upper wall of the nozzle is higher than that near the lower wall, and the airflow forms a pair of vortexes in the center region of section 3 under the action of pressure difference between the upper and lower walls, and the airflow velocity is larger in the regions at both sides of the nozzle and flows toward the upper wall of the nozzle, and the airflow in the center region of the nozzle is smaller and flows downward under the action of internal and external pressure difference. The airflow in the center region of the nozzle has a smaller velocity and flows downward under the effect of the pressure difference between the inside and outside of the nozzle. A small range of vortices appears near the upper wall surface of section 3. This is due to the influence of the viscous effect, the velocity and energy of the airflow close to the wall surface is lower, and it flows downward along the wall surface of the nozzle under the effect of pressure difference between the inside and outside of the nozzle, as shown in Figure 4.38. The airflow flowing downward close to the wall surface is in a different direction of flow from the high-speed airflow flowing upward on both sides of the nozzle.

Cross-section 4 is the outlet cross-section of the second section of the cylinder, and the flow direction of cross-section 4 is the projection of the airflow direction in the cross-section, and the flow direction of the cross-section is directed to the left side from the right side, and the flow cause of this cross-section is the same as that of cross-section 2, reflecting the airflow direction as axial flow along the nozzle.

Cross-section 5 is the inlet cross-section of the shrinking section of the third cylinder, and it can be seen from the streamline trend of cross-section 5 that the streamline direction of this cross-section is similar to that of cross-section 1. The difference is that the airflow intersection points and the low-pressure area of cross-section 5 are shifted downward, and the counter-vortex of cross-section 3 and a small range of vortices near the upper wall surface disappear.

Nozzle outlet cross-section of the flow direction is similar to the cross-section 5, in the centripetal force under the action of the upper wall of the nozzle airflow pressure is higher, the airflow from the nozzle inlet to the outlet of the flow process is constantly mixing, so the nozzle outlet airflow of the total pressure distribution is more uniform.

In summary, the nozzle 10° deflection angle state, the airflow in the tail cone after the formation of the return area, in the centripetal force, the airflow in the second section of the center of the cylinder region formed a pair of vortex turbulence, the airflow in the pair of vortex turbulence in the mixing and accompanied by the dissipation of energy, the cross-section of the airflow of the total pressure is gradually reduced and tends to be uniform.

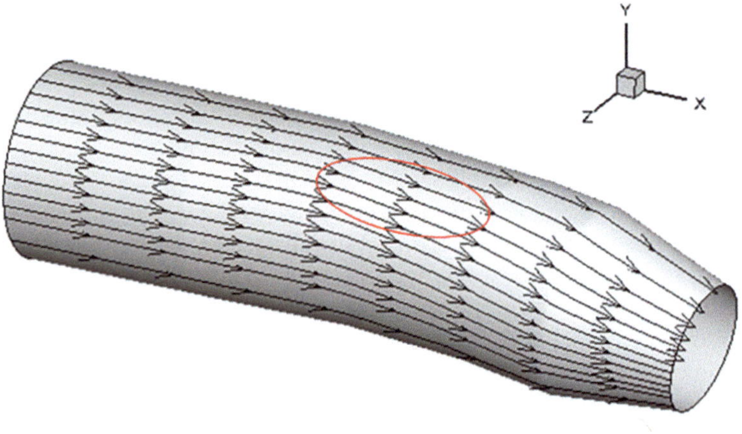

Fig. 4.38 Streamline distribution of near-wall airflow at 10° nozzle offset

Figure 4.39 shows the total pressure and streamline distributions of the nozzle profile, sections 1–5 and outlet section at 20° declination. As can be seen from the figures, the streamlines and total pressure distributions in each section of the nozzle are similar to those of the 10° declination condition. Compared with the 10° declination condition, the vortex in the center region of section 3 is more obvious, which is due to the fact that as the nozzle declination increases, the centripetal force on the airflow increases, and the partial velocity of the airflow in the two regions of the nozzle increases towards the upper wall, which leads to the increase of the vortex intensity in the center region of section 3. Compared with the 10° deflection state, the vortex range near the upper wall of section 3 increases and the position is shifted downward, which is due to the fact that the increase of the nozzle deflection angle leads to the increase of the pressure difference between the inside and the outside of the nozzle, and the airflow close to the wall increases the velocity of the flow downward along the wall of the nozzle under the effect of the pressure difference between the inside and the outside of the nozzle, as shown in Fig. 4.40. Therefore, the vortex range near the upper wall of section 3 increases and the position is shifted downward.

Figure 4.41 shows the total pressure and streamline distributions of the nozzle profile, sections 1–5 and outlet section at 30° declination. As can be seen from the figures, the streamlines and total pressure distributions in each section of the nozzle are similar to those of the 20° declination condition. Compared with the 20° off-angle state of the nozzle, the convective vortex in the center region of cross-section 3 is more obvious, and the vortex near the upper wall surface has increased in range and shifted downward, and the cause and change of the vortex on the upper wall surface is the same as that of the 20° off-angle state, which is caused by the downward flow of the airflow close to the wall along the two sides of the wall (Fig. 4.42).

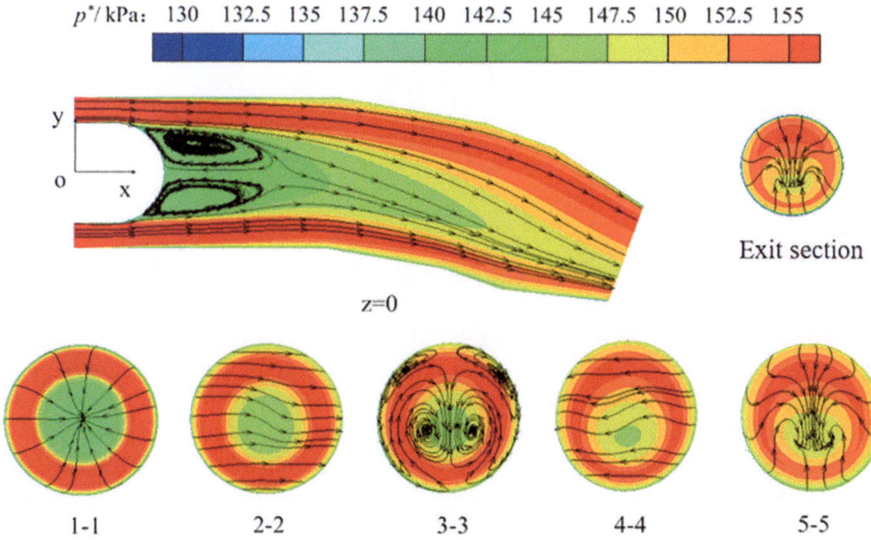

Fig. 4.39 Total pressure and streamline distribution of airflow in each section at 20° nozzle offset

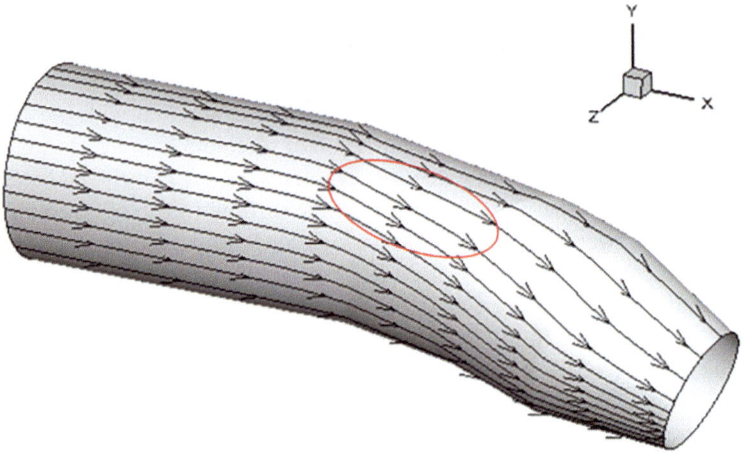

Fig. 4.40 Streamline distribution of near-wall airflow at 20° deflection angle

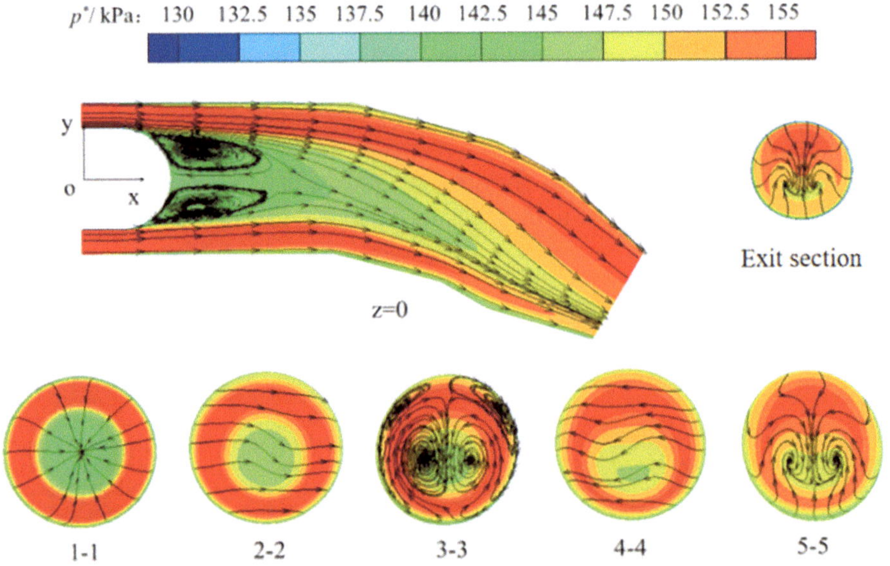

Fig. 4.41 Total pressure and streamline distribution of airflow in each section at 30° nozzle offset

Figure 4.43 shows the total pressure and streamline distribution of the nozzle profile, sections 1–5 and outlet section at 60° declination. From the streamline and total pressure distribution of the nozzle profile, it can be seen that the airflow passes through the inflection point of the wall of the first and second sections of the cylinder, separated from the inner wall of the nozzle and accompanied by energy loss, the total pressure of the airflow near the wall of the second and third sections of the cylinder decreases.

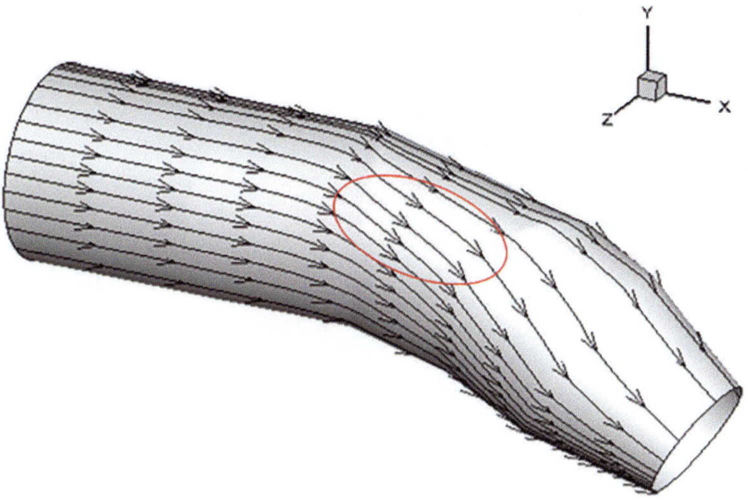

Fig. 4.42 Streamline distribution of near-wall airflow at 30° nozzle deflection angle

From the total pressure distribution of section 2, it can be seen that the total pressure of the airflow near the position of the upper wall surface of the nozzle is low, which is due to the upper wall surface of section 2 for the first and second section of the cylinder wall surface of the inflection point, where the airflow by the influence of the wall bends to produce the return area, the energy loss of the airflow increases, and the total pressure is reduced.

Figure 4.44 shows the streamline distribution close to the wall surface for the nozzle at 60° declination. It can be seen from the figures that the velocity component of the downward flow of the air stream increases due to the increase in differential pressure force compared to the 30° declination condition, which leads to the downward movement and increase in the extent of the vortex near the upper wall of section 3 in Fig. 4.43.

From the streamline distribution of section 3, it can be seen that vortices appear near the lower wall surface of section 3 compared to the 10°, 20° and 30° off-angle states. As can be seen from, the airflow close to the lower wall surface of the nozzle produces a backflow at the wall inflection points in the first and second sections of the cylinder, and the airflow close to the wall surface flows sideways, opposite to the direction of the convective vortex flow in the center region thereby forming a vortex near the lower wall surface of the cross-section 3.

From the total pressure distribution of sections 3, 4, and 5, it can be seen that the low-pressure region occurs near the lower wall surface of sections 3, 4, and 5, which is due to the separation of the airflow from the wall by the effect of wall bending, and the airflow generates an energy loss in the separation region, and the total pressure of the airflow decreases.

Figure 4.45 shows the total pressure and streamline distribution of the nozzle profile, 1–5 sections and outlet section at 90° declination. From the streamline and

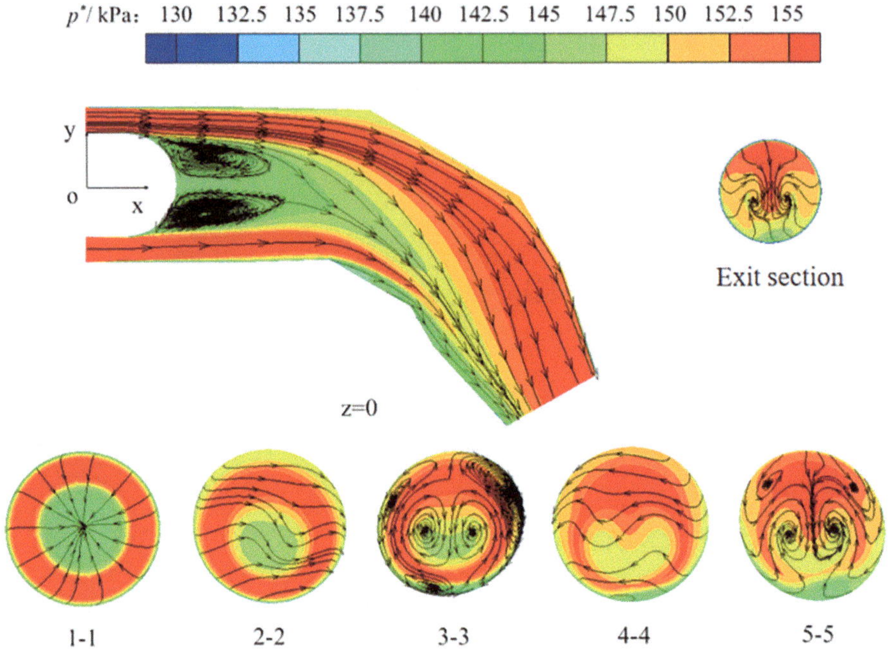

Fig. 4.43 Total pressure and streamline distribution of airflow by cross section at 60° nozzle offset

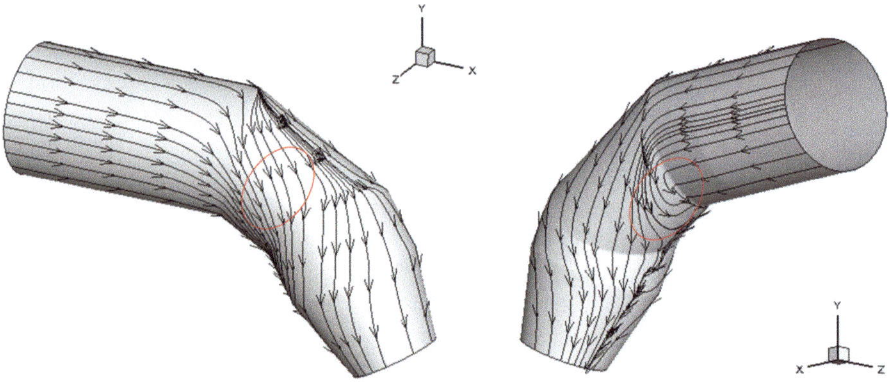

Fig. 4.44 Streamline distribution of near-wall airflow at 60° nozzle offset

total pressure distribution of the nozzle profile, it can be seen that, compared with the state of 60 ° declination angle, the airflow forms a vortex at the inflection point of the outer wall surface of the cylinder in the first and second segments, and the energy loss increases, and the scope of the separation zone formed on the inner wall surface of the cylinder in the second and third segments increases, the energy loss of the airflow increases, and the total pressure of the airflow in the separation zone decreases.

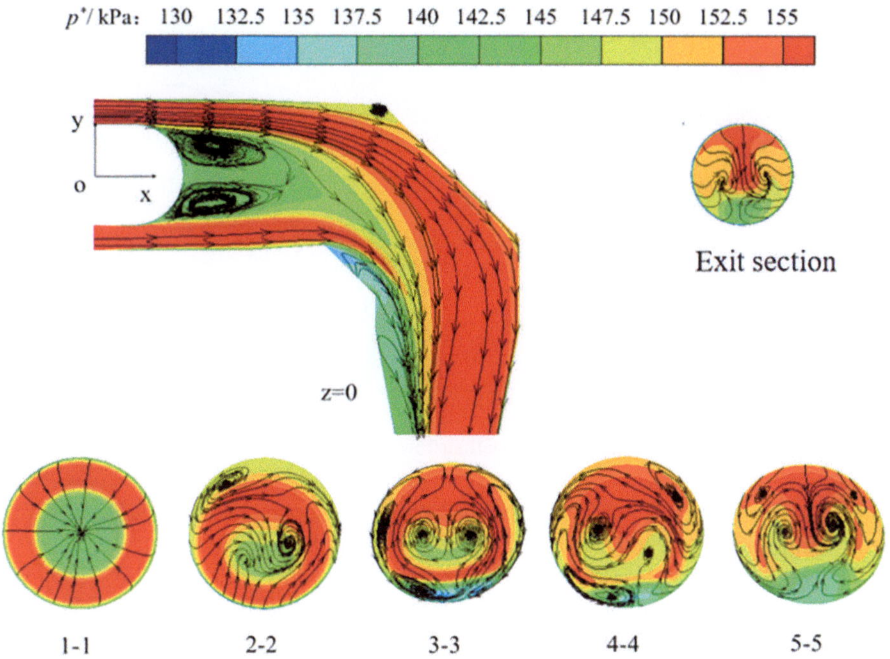

Fig. 4.45 Total pressure and streamline distribution of airflow by cross section at 90° nozzle offset

The vortex appears in the left position of the upper wall of section 2, which is due to the increase of centripetal force on the airflow under the 90° deviation of the nozzle, and it can be seen from the flow line of the section that there is a tendency for the airflow to flow downward at the end of the first section of the cylinder, so the airflow is already subjected to the centripetal force at the end of the first section of the cylinder, and the airflow at the end of the first section of the cylinder close to the upper wall is flowing downward with the effect of the pressure difference (Fig. 4.46), and the air flow close to the upper wall has a difference of velocity direction with the main stream, thus forming a vortex. The airflow close to the upper wall and the main flow has a difference in velocity direction, thus forming a vortex. A small counter-vortex also occurs in the center of section 2 due to the pressure difference.

The streamline trend of cross-section 3 is similar to the 60° off-angle condition, and from the streamline distribution, it can be seen that the airflow close to the upper wall surface flows downward along the two sides of the wall, and the vortex of cross-section 3 close to the two sides of the wall surface moves downward and increases in range due to the increase in the velocity component of the downward flow of the airflow (Fig. 4.46).

A vortex appears at the rightward position of the upper wall surface of the cross-section 4, which has the same formation mechanism as the vortex at the leftward position of the upper wall surface of the cross-section 2. Compared with the 60° off-angle state, there still exists a pair of vortices in the center region of the nozzle and a vortex near the lower wall surface of the nozzle in cross-section 4, and the pressure near the lower wall surface of cross sections 3, 4, and 5 is further reduced,

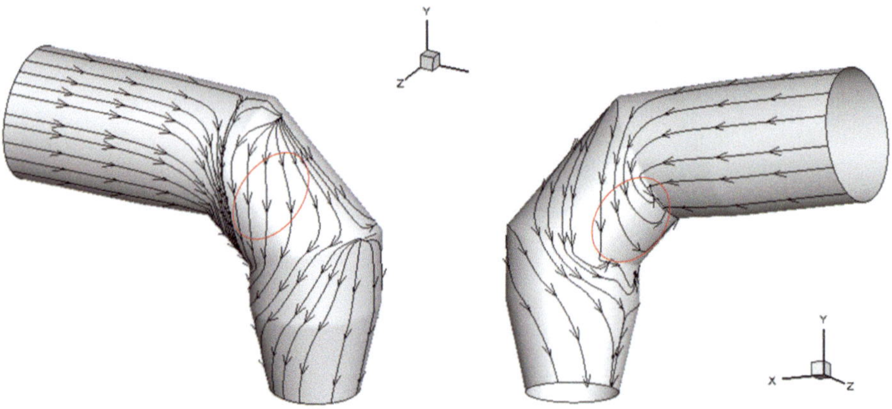

Fig. 4.46 Streamline distribution of near-wall airflow at 90° nozzle offset

and the intensity of the vortex of the airflow in the separation zone increases, the energy loss increases, and the total pressure of the airflow decreases.

By analyzing the flow field characteristics under the nozzle vector state, the energy loss mechanism of the airflow in the nozzle under the constant state is obtained. In small vector angle, the flow of the airflow is relatively simple, the airflow after the tail cone produces reflux area, the wall deflection has less influence on the airflow, the airflow under the action of centripetal force in the center of the formation of a weak vortex, nozzle on both sides of the wall, due to the close to the wall of the airflow and the mainstream flow direction of the formation of a small range of vortices, the energy dissipation of the airflow mainly occurs in the tail cone after the reflux area, and the center of the region of the vortex. In the large vector angle, the flow of the airflow is relatively complex, the intensity of vortex on both sides of the wall of the nozzle increases, the airflow in the two sections of the cylinder wall at the inflection point of the two sections of the wall to produce reflux zone, the formation of the separation zone at the inflection point of the inner wall, the energy dissipation of the airflow mainly occurs in the reflux zone after the tail cone, the center of the region of the vortex, the vortex close to both sides of the wall, the reflux zone of the inflection point of the wall and the separation zone of the inflection point of the wall, and the strength and energy loss of vortices vary with the direction of the main flow of the nozzle and the direction of the main flow. intensity and energy loss increase with the increase of the nozzle deflection angle.

4.5.2 Three-Bearing Deflection Nozzle Aerodynamic Performance

Figures 4.47 and 4.50 show the aerodynamic performance of the three-bearing deflector nozzle in the states of 0°, 30°, 60° and 90° deflection angle obtained by the CFD numerical method. Figure 4.48 shows the variation rule of the total pressure

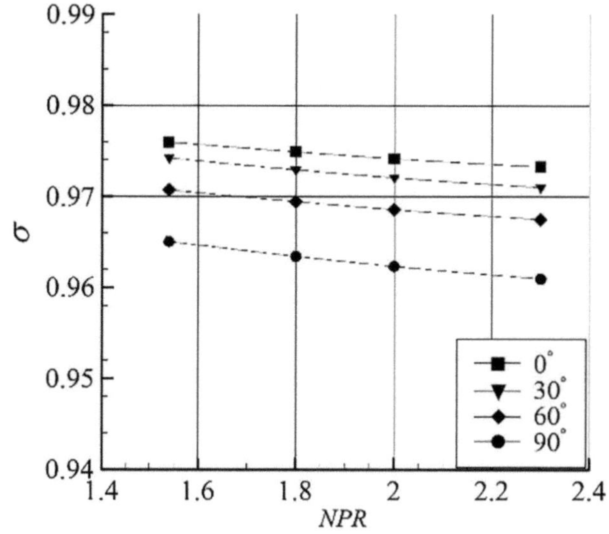

Fig. 4.47 The changes in σ follow the patterns of β and *NPR*

recovery coefficient at the nozzle outlet with the deflection angle and the drop pressure ratio, from which it can be seen that the total pressure recovery coefficient decreases with the increase of the deflection angle under the condition of the same drop pressure ratio, which corresponds to the increase of the flow loss of the airflow inside the nozzle with the increase of the nozzle deflection angle in the flow field characterization. When the deflection angle is certain, the total pressure recovery coefficient decreases with the increase of drop pressure ratio, which is due to the fact that the velocity of the airflow inside the nozzle increases with the increase of drop pressure ratio, and the increase of the velocity of the airflow leads to the increase of the flow loss, which leads to the decrease of the total pressure recovery coefficient.Figure 4.47 shows the flow coefficient of the nozzle with the change rule of deflection angle and drop pressure ratio. Figure 4.50 shows the thrust coefficient of the nozzle with the change rule of deflection angle and drop pressure ratio. As can be seen from the figure, under the condition of the same drop pressure ratio, the flow coefficient and thrust coefficient of the nozzle decreases with the increase of the deflection angle, and it can be seen from the characterization of the flow field inside the nozzle that the nozzle deflection makes the airflow form a separation in the inner wall, which leads to the reduction of the velocity of the airflow in the inner wall surface of the outlet of the nozzle, the flow rate of the nozzle and the thrust decreases.Figure 4.50 shows the change rule of nozzle thrust vector angle with deflection angle, it can be seen from the figure, the thrust vector angle and deflection angle of the linear relationship between and slightly greater than the deflection angle, this is due to the nozzle deflection angle increases, there is a certain amount of airflow separation of the nozzle inner wall surface, compared to the air velocity near the outer wall surface, the air velocity near the inner wall surface is small, the constriction of the nozzle exit surface leads to the nozzle outlet near the outer wall

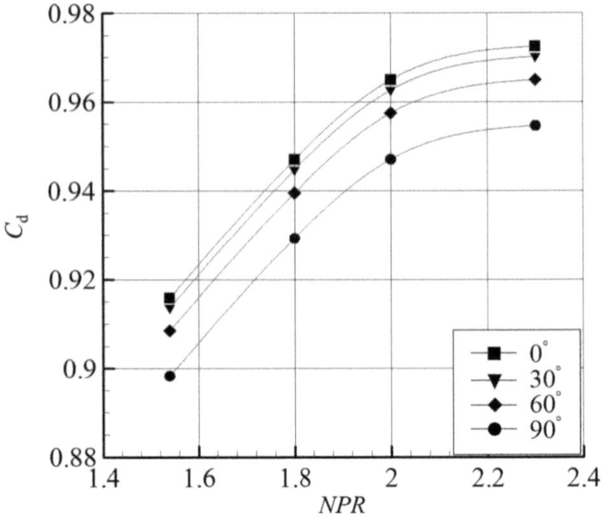

Fig. 4.48 The changes in C_d follow the patterns of β and *NPR*

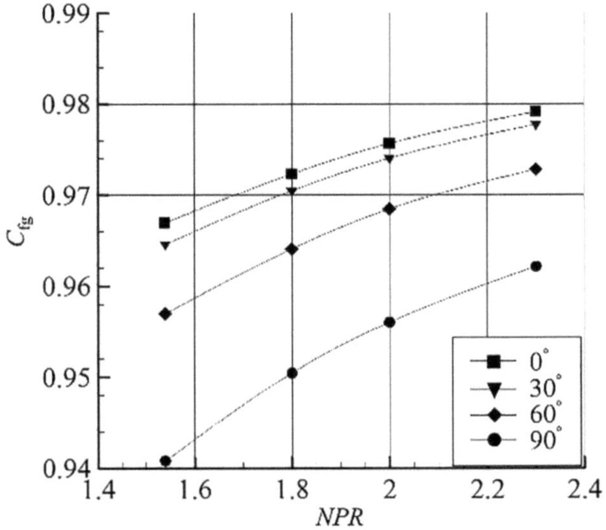

Fig. 4.49 The changes in C_{fg} follow the patterns of β and *NPR*

surface of the airflow deflection angle increases, coupled with the outer wall surface air velocity is larger than the inner wall surface, resulting in the nozzle thrust vector angle is slightly larger than the actual angle. The constricted shape of the nozzle exit causes the deflection angle of the airflow near the outer wall of the nozzle exit to increase, which, together with the fact that the airflow velocity near the outer wall is larger than that near the inner wall, results in the thrust vector angle of the nozzle being slightly larger than the actual deflection angle.

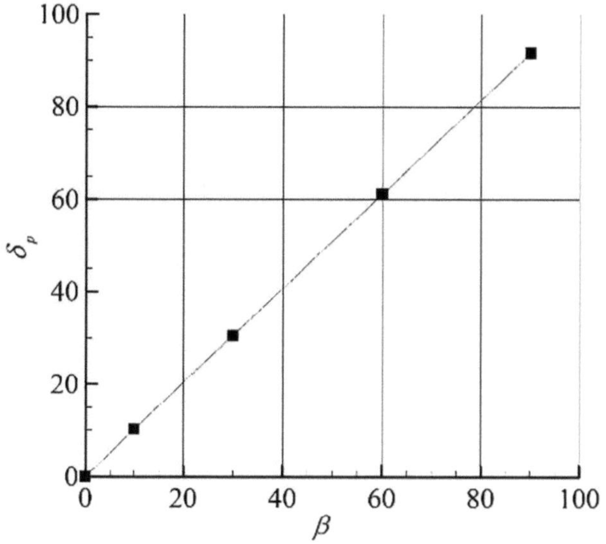

Fig. 4.50 The changes in δ_p follow the patterns of β and *NPR*

4.6 Analysis of Ground Effects of a Three-Bearing Deflection Nozzle

With the emergence of thrust vectoring technology and the improvement of engine technology level, the United States, Russia's S/VTOL fighters are used in front of the lift engine/lift fan + lift/cruise engine power layout, the main advantage of this power layout is to reduce the cross-sectional area of the propulsion system is conducive to the realization of the aircraft to achieve supersonic flight, and as far as possible to reduce the dead weight of the aircraft in flight. The power layout of the propulsion system is mainly distributed in the fuselage before and after the position of the power unit generated by the two high-speed jets to obtain upward lift, the aircraft in the near-ground vertical takeoff and landing, the propulsion system discharged by the high-speed jets and the ground to interfere with the jet's volume suction effect makes the aircraft under the lower surface pressure is reduced, resulting in a loss of lift, in addition, the engine intake will be inhaled in the intake tract to propel the system to discharge a high temperature gas, thus causing the airflow in the intake tract, which will result in a loss of lift. In addition, the inlet of the engine will inhale the high-temperature gas from the propulsion system, which will cause the temperature distortion of the airflow in the inlet, affecting the stable operation of the engine. The lift loss caused by the interference between the high temperature and high-speed jet produced by the propulsion system and the ground and the temperature distortion in the inlet are two important research directions in the ground effect of S/VTOL aircraft. In this section, for the power layout of front-mounted lift fan and rear-mounted lift/cruise engine vector nozzle, we carry out the study of the interference effect between the dual jets and the ground, and analyze the influence of aerodynamic and geometric parameters on the interference effect between the dual jets and the ground.

4.6.1 Computational Model and Parameter Definitions

Numerical modeling of the interference effect between the two jets and the ground is established using a three-dimensional CFD numerical method, and the computational domain is shown in Fig. 4.51. In the computational domain, the diameter of the jet outlet is *d*, the center distance between two jets is *s*, the height of the flat plate from the ground is *h*, the velocity of the incoming jet is *u*, the drop pressure ratio of the cold jet is *NPR*, the velocity is *v*, the momentum ratio of the cold and hot jets is *R*, and the computational domain also includes the models of the external environment, the flat plate and the ground, etc. When the velocity u ≠ 0, the inlet (A) and the ground are modeled as the inlet of the computational domain. When the incoming velocity u≠0, the far-field boundary conditions are used for the inlet (A) and upper surface (B) of the computational domain, when the incoming velocity u = 0, the pressure exit boundary conditions are used for the inlet (A) and upper surface (B) of the computational domain, the pressure exit boundary conditions are also used for the outlet (D) and the side surfaces (C) of the computational domain, and the wall boundary conditions are used for the ground (E) and the upper and lower surfaces of the flat plate in the computational domain. Since the model is symmetric with respect to the x-z plane, the x-z plane adopts the symmetry plane boundary condition. Mesh encryption is carried out near the upper and lower wall surfaces of the ground and the flat plate, respectively, to ensure that the y+ value of the mesh does not exceed 5.

From Fig. 4.51, it can be seen that the cold and hot jets below the flat plate form a fountain area between the flat plate and the ground, and the airflow in the fountain area produces upward lift on the flat plate. Meanwhile, it can be seen from the single-jet ground interference model that the ambient airflow forms a low-pressure

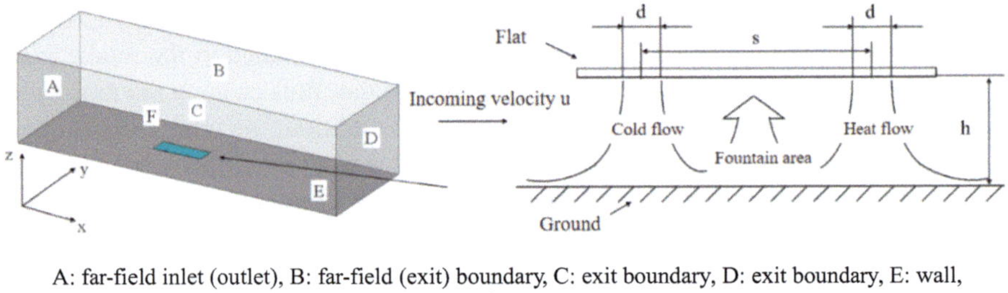

A: far-field inlet (outlet), B: far-field (exit) boundary, C: exit boundary, D: exit boundary, E: wall, F: symmetry plane

Fig. 4.51 Calculation domain of double jet and ground interference effect. (**a**) far-field inlet (outlet), (**b**) far-field (exit) boundary, (**c**) exit boundary, (**d**) exit boundary, (**e**) wall, (**f**) symmetry plane

area on the lower surface of the flat plate under the suction effect of the jets, and produces downward suction on the flat plate, therefore, the loss of lift in the near-ground condition of the dual-jet is the result of the joint action of the fountain effect and the downward suction effect. Therefore, the loss of lift under the near-surface conditions of the double jet is the result of the fountain effect and the downward suction effect together. According to the force on the lower surface of the plate, the lift loss coefficient of the dual-jet flow is defined as

$$C_{\mathrm{f}} = \frac{\left| \int_{A_{\mathrm{t}}} p_{\mathrm{t}} dA + \int_{A_{\mathrm{u}}} p_{\mathrm{u}} dA \right|}{F_{\mathrm{j1}} + F_{\mathrm{j2}}} \tag{4.70}$$

In the Equation:

F_{j1}—Lift generated by the front nozzle;
F_{j2}—Lift generated by the rear nozzle;
$\int_{A_{\mathrm{t}}} p_{\mathrm{t}} dA$—The combined force applied to the upper table of the flat plate;
$\int_{A_{\mathrm{u}}} p_{\mathrm{u}} dA$—The combined force exerted on the table below the flat plate.

Define the lift force generated by the airflow in the fountain area as F_{f} and the downward suction force generated by the airflow in the coiling area as F_{i}. The combined force on the lower surface of the flat plate can be expressed as

$$\int_{A_{\mathrm{u}}} p_{\mathrm{u}} dA = F_{\mathrm{f}} + F_{\mathrm{i}} \tag{4.71}$$

Since the pressure change on the upper surface of the flat plate is small, the mechanism of lift loss caused by the interference between the twin jets and the ground is analyzed only by the change of the pressure on the lower surface of the flat plate. When there is a certain velocity of the incoming flow, the incoming flow may have a certain effect on the fountain effect and the downward suction effect, in addition to the jet drop pressure ratio, before and after the jet flow ratio, the height of the flat plate from the ground, the jet spacing and other parameters will affect the interference effect between the twin jets and the ground. Therefore, this paper establishes a numerical model of the interference effect between the dual jets and the ground, and studies the influence of each geometric parameter (nozzle spacing s, flat plate height h) and aerodynamic parameter (incoming velocity u, cold flow pressure ratio NPR, hot and cold jet volume ratio R) on the interference effect between the dual jets and the ground, and to make the results of the study of a certain degree of general significance, respectively, the spacing of the jets, the height of the flat plate, and the incoming velocity of the dimensionless processing as: $S = s/d$, $H = h/d$, $U = u/v_j$.

Table 4.6 Forces on the lower surface of the plate at $H = 3$ and $H = 11$ (S = 7.5, $NPR = 2.25$, R = 1, U = 0)

	$H = 3$	$H = 11$
F_f/N	147.8	41.63
F_i/N	−561.5	−73.9
$F_f + F_i$/N	−413.7	−32.27
C_f	0.2356	0.0184

(a) H = 3; (b) H = 11

4.6.2 Analysis of the Effect of Plate Height on Lift Loss

Table 4.6 shows the force on the lower surface of the plate when $H = 3$ and $H = 11$. From the table, it can be seen that the fountain force on the lower surface of the flat plate is smaller than the lower suction force, and the direction of the combined force on the flat plate is the same as the direction of the lower suction force. When *H is* increased from 3 to 11, the fountain force and the lower suction force are both reduced, among which the fountain force is reduced by 71.8% and the lower suction force by 86.8%, the combined force on the lower surface of the flat plate is reduced, and the coefficient of the loss of lift force is reduced.

Figure 4.52 shows the static pressure distribution on the lower surface of the plate at $H = 3$ and $H = 11$. From the figure, it can be seen that the high-pressure area on the lower surface of the flat plate at both heights appears between the two jets, and the low-pressure area appears around the jets. Comparing the hydrostatic pressure distribution on the lower surface of the flat plate at $H = 3$ and $H = 11$, it can be seen that the pressure in the high-pressure area decreases, the fountain force decreases, and the pressure in the low-pressure area near the jets rises, and the downward suction decreases, when H is increased.

Figure 4.53 shows the Mach number and streamline distribution on the symmetry plane. From the figure, it can be seen that, when the height of the flat plate increases, the velocity of the jet near the ground decreases with the increase of the height of the flat plate, the velocity of the fountain area formed by the convergence of the two air streams decreases, and the momentum of the fountain area air stream decreases, when the fountain area air stream flows upward, the velocity decreases with the height, and the momentum of the air stream gradually decreases with the height, and the fountain force decreases with the height. Therefore, the fountain force decreases with the increase of the plate height. From the flow distribution in the figure can be seen, when H is small, the fountain area airflow and the ground between the angle of inclination is small, flat plate near the lower surface of the fountain area airflow along the lower surface of the flat plate to both sides of the direction of the jet flow, the fountain area airflow in the vicinity of the jet by the jet's volume suction in the jet and fountain area between the formation of low-pressure zone. When H is larger, the tilt angle between the fountain area airflow and the ground is larger, which is due to the environmental airflow around the hot jet heat pressure increases, resulting in the fountain area airflow tilted in the direction of the cold flow.

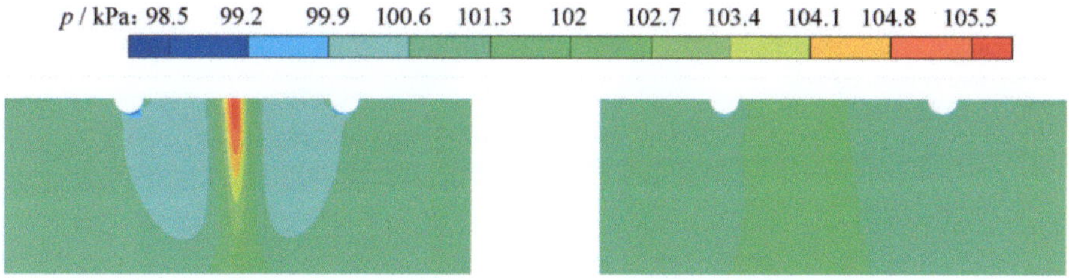

(a) H=3

(b) H=11

Fig. 4.52 Hydrostatic pressure distribution on the lower surface of a flat plate at $H = 3$ and $H = 11$ (S = 7.5, *NPR* = 2.25, R = 1, U = 0). (**a**) H = 3. (**b**) H = 11

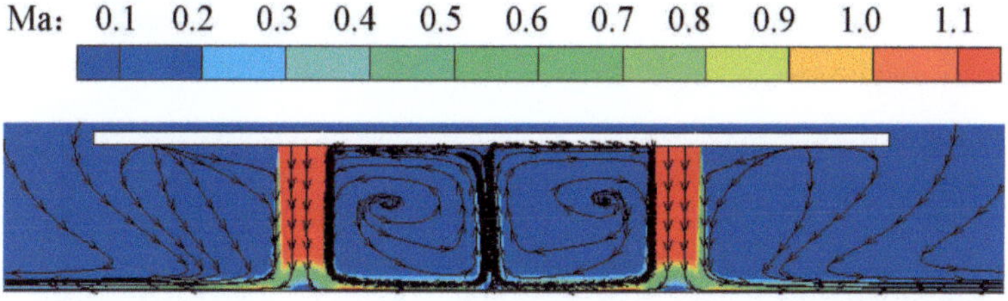

(a) *H*=3

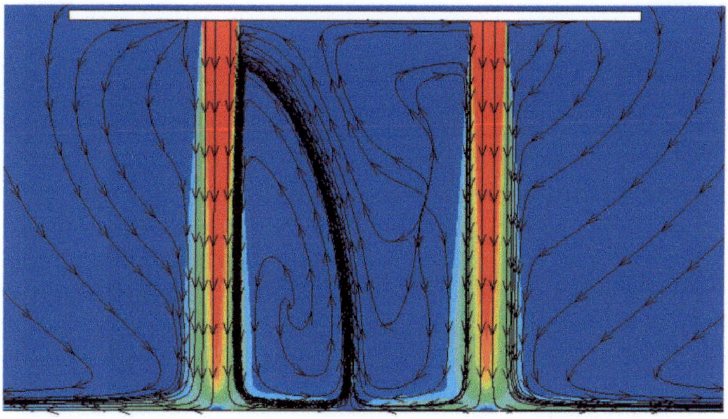

(b) *H*=11

Fig. 4.53 Mach number and streamline distribution of the symmetry plane at $H = 3$ and $H = 11$ (S = 7.5, *NPR* = 2.25, R = 1, U = 0). (**a**) $H = 3$. (**b**) $H = 11$

4.6.3 Analysis of the Effect of Jet Spacing on Lift Loss

Table 4.7 shows the force on the lower surface of the plate when $S = 5$ and $S = 10$. From the table, it can be seen that the fountain force on the lower surface of the flat plate is smaller than the lower suction force, when *S is* increased from 5 to 10, the

Table 4.7 Forces on the lower surface of the flat plate at S = 5 and S = 10 ($H = 7$, $NPR = 2.25$, R = 1, U = 0)

	$S = 5$	$S = 10$
F_f/N	108.49	99.27
F_i/N	−167	−164
$F_f + F_i$/N	−58.51	−64.73
C_f	0.033	0.037

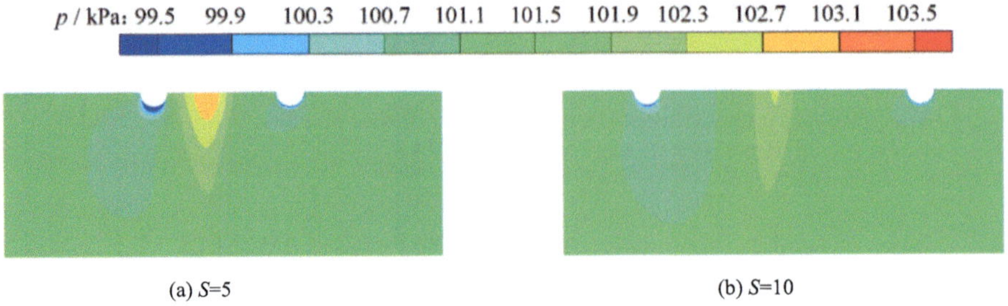

Fig. 4.54 Hydrostatic pressure distribution on the lower surface of a flat plate at $S = 5$ and $S = 10$ ($H = 7$, $NPR = 2.25$, R = 1, U = 0). (**a**) $S = 5$. (**b**) $S = 10$

fountain force and the lower suction force are reduced, in which the fountain force is reduced by 8.5%, the lower suction force is reduced by 1.8%, the combined force on the lower surface of the flat plate is increased, and the coefficient of loss of lift force is increased.

Figure 4.54 shows the static pressure distribution on the lower surface of the flat plate at $S = 5$ and $S = 10$. Comparing the static pressure distribution on the lower surface of the flat plate with $S = 5$ and $S = 10$, it can be seen that when S is larger, the pressure and range of the fountain area decreases, the fountain force decreases, and the low-pressure area near the front jet appears in the area between the jet and the fountain area, and when S is smaller, the pressure and range of the fountain area increases, and the fountain force increases, and the low-pressure area near the front jet appears in the outside area of the jet, because when S is smaller, the location of the fountain area of the flat plate is close to the front jet, and due to the small spacing between the jets, the airflow in the fountain area is between the two jets. Surface of the fountain area location near the front jet, due to the jet spacing is small, the fountain area airflow between the two jets does not have enough space to form a low-pressure area under the effect of jet suction, so the fountain area airflow in the lower surface of the flat plate bypassing the front jet to form a low-pressure area under the effect of jet suction on the outside of the jet stream, when the S is larger, the two jets have enough space between the two jets, fountain area airflow in the area between the two jets to form a low-pressure area under the effect of jet suction. The low-pressure region is under the effect of jet suction. By comparing the static pressure distribution on the lower surface of the flat plate with $S = 5$ and $S = 10$, it can be concluded that there is interference between the fountain zone airflow on the lower surface of the flat plate and the jet, and the fountain zone airflow forms a low-pressure zone around the jet under the effect of jet coiling and sucking, which

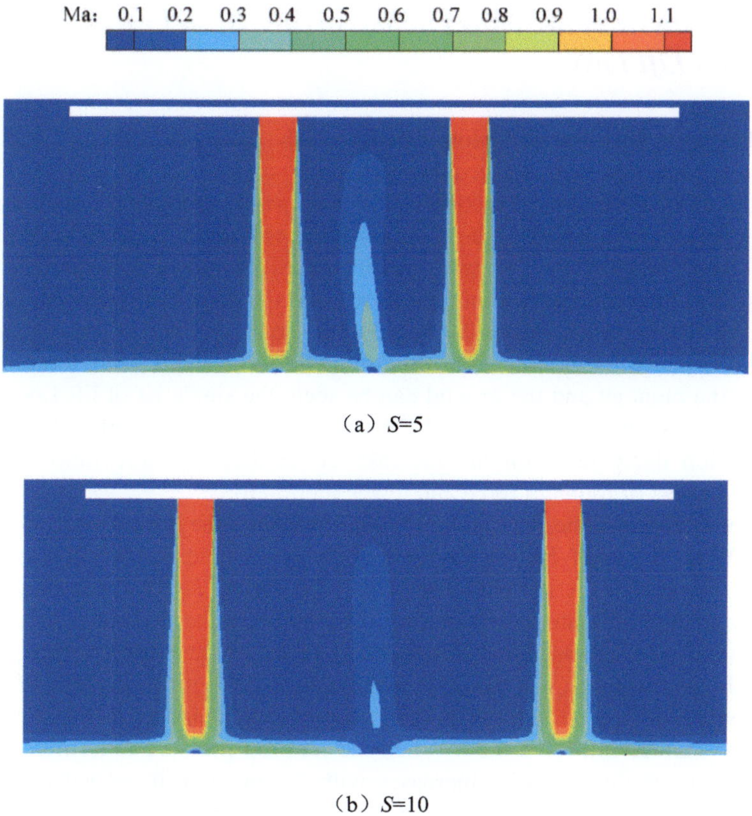

Fig. 4.55 Mach number distribution of the symmetry plane at $S = 5$ and $S = 10$ ($H = 7$, $NPR = 2.25$, R = 1, U = 0). (**a**) $S = 5$, (**b**) $S = 10$

reduces the pressure around the jet and makes the value of the downward suction force increase.

Figure 4.55 shows the Mach number distribution on the symmetry surface, it can be seen from the figure, when the jet spacing is small, the fountain area airflow velocity is larger, the fountain area airflow momentum increases, the fountain force on the flat plate lower surface is increased, on the contrary, the increase of the jet spacing attenuates the velocity and momentum of the airflow in the fountain area, so that the fountain force is reduced. From the pressure distribution and force on the lower surface of the flat plate, it can be seen that the influence of S on the dual-jet ground interference effect is mainly reflected in the change of the fountain force, S increases, the fountain force decreases, the lower surface of the flat plate is subjected to the downward force increases, and the lift loss increases.

4.6.4 Analysis of the Effect of Jet Stream Drop Pressure Ratio on Lift Loss

Table 4.8 shows the force on the lower surface of the plate when *NPR* = 1.5 and *NPR* = 3. From the table, it can be seen that the combined force on the lower surface of the flat plate is in the same direction as the downward suction force, and when the *NPR* increases from 1.5 to 3, the fountain force and the downward suction force both increase, in which the fountain force increases about 3.13 times and the downward suction force increases about 2.4 times, and the combined force on the lower surface of the flat plate increases, because the lower surface of the flat plate is mainly dominated by the increase of the fountain force and the interference effect between the monojet and the ground can be seen, the single jet of lift Loss coefficient decreases with the increase of *NPR*, so the lift loss coefficient under the dual-jet condition decreases with the increase of jet drop pressure ratio under the combined effect of fountain force and lower suction force.

Figure 4.56 shows the static pressure distribution on the lower surface of the flat plate at *NPR* = 1.5 and *NPR* = 3. Comparing the hydrostatic pressure distribution on the lower surface of the flat plate for *NPR* = 1.5 and *NPR* = 3, it can be seen that when the jet drop pressure ratio increases, the pressure in the fountain area increases, the fountain force increases, and at the same time, the range of the low-pressure area around the jet increases, and the downward suction force increases. The reason for the increase in fountain force can be seen from the Mach number distribution on the symmetry surface in Figure 4.57. When the jet drop pressure ratio increases, the fountain zone airflow velocity increases with the increase in jet velocity, which leads to an increase in fountain force, and the reason for the increase in down

Table 4.8 Forces on the lower surface of the flat plate at *NPR* = 1.5 and *NPR* = 3 (H = 7, S = 7.5, R = 1, U = 0)

	NPR = 1.5	NPR = 3
F_f/N	41.04	169.64
F_i/N	−79.92	−271.88
$F_f + F_i$/N	−38.88	−102.24
C_f	0.047	0.041

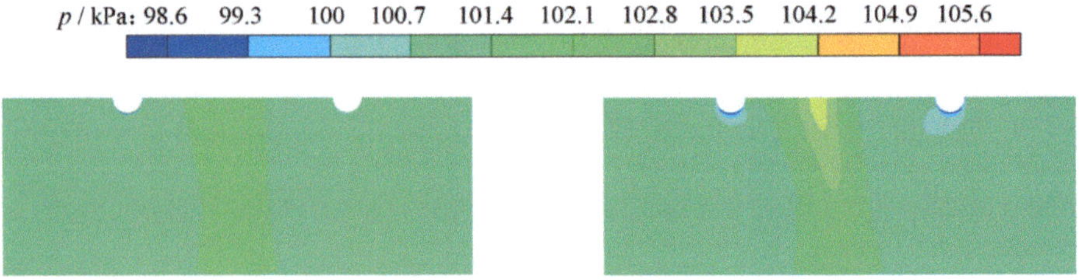

(a) *NPR*=1.5 (b) *NPR*=3

Fig. 4.56 Hydrostatic pressure distribution on the lower surface of a flat plate at *NPR* = 1.5 and *NPR* = 3 (H = 7, S = 7.5, R = 1, U = 0). (**a**) *NPR* = 1.5. (**b**) *NPR* = 3

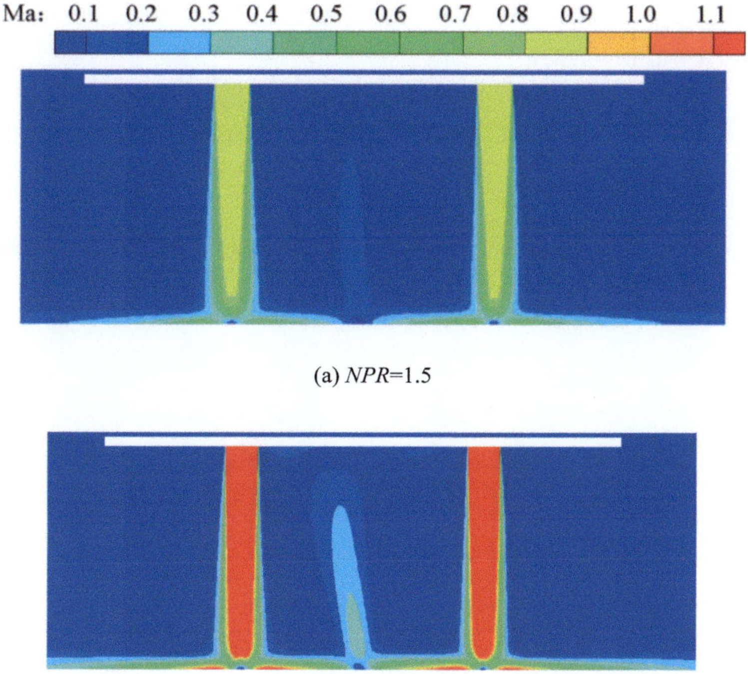

(a) *NPR*=1.5

(b) *NPR*=3

Fig. 4.57 Mach number distribution of the symmetry plane at *NPR* = 1.5 and *NPR* = 3 (H = 7, S = 7.5, R = 1, U = 0). (**a**) *NPR* = 1.5. (**b**) *NPR* = 3

suction is caused by the increase in jet velocity on the one hand, and by the interference between the high-speed fountain zone airflow and the jet on the other hand.

In summary, the fountain force is greatly influenced by the jet drop pressure ratio, when the jet drop pressure ratio increases, the force on the lower surface of the flat plate is dominated by the increase of the fountain force, therefore, when the jet drop pressure ratio increases, the lift loss coefficient is mainly influenced by the fountain force, which decreases with the increase of the jet drop pressure ratio.

4.6.5 Analysis of the Effect of Jet Flow Ratio on Lift Loss

Table 4.9 shows the force on the lower surface of the flat plate when *R* = 0.7 and *R* = 1.3. From the table, it can be seen that the fountain force on the lower surface of the flat plate is smaller than the lower suction force, when *R* increases from 0.7 to 1.3, the fountain force and the lower suction force are both reduced, in which the fountain force is reduced by 17.42% and the lower suction force by 26.46%, and the combined force on the lower surface of the flat plate decreases, and the coefficient of loss of lift decreases.

Table 4.9 Forces on the lower surface of the plate at R = 0.7 and R = 1.3 (H = 7, S = 7.5, *NPR* = 2.25, U = 0)

	R = 0.7	R = 1.3
F_f/N	106.63	88.05
F_i/N	−200.82	−147.69
$F_f + F_i$/N	−94.19	−59.64
C_f	0.044	0.039

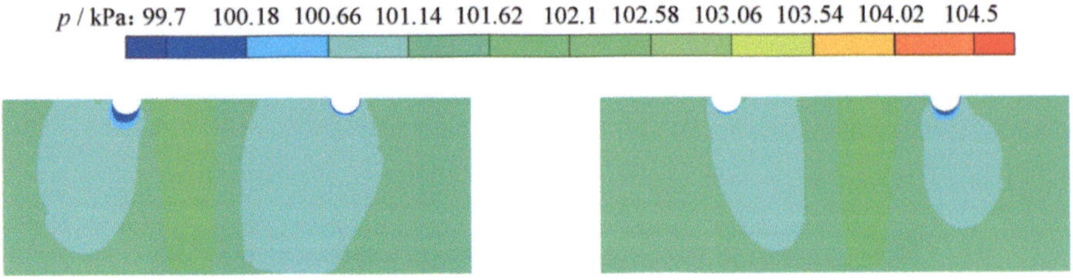

(a) R=0.7 (b) R=1.3

Fig. 4.58 Hydrostatic pressure distribution on the lower surface of a flat plate at R = 0.7 and R = 1.3 (H = 7, S = 7.5, *NPR* = 2.25, U = 0). (**a**) R = 0.7. (**b**) R = 1.3

Figure 4.58 shows the static pressure distribution on the lower surface of the flat plate at $R = 0.7$ and $R = 1.3$. As can be seen from the figure, when the back jet flow volume is larger, the high pressure area formed by the fountain flow is close to the front jet position, the airflow in the area around the front jet is formed under the effect of the jet's coiling and suction of the low-pressure area, which is affected by the position of the fountain area to move to the outside area of the front jet, due to the position of the fountain area is close to the front jet, the front jet interferes with the fountain area, and the degree of the front jet's coiling and suction is strengthened, the pressure around the jet area is reduced, and due to the increase in the drop pressure ratio of the back jet, the low-pressure area around the rear jet is increased. The pressure decreases, and due to the increase in the pressure ratio of the back jet, the low-pressure area around the back jet increases. When the rear jet flow is small, the high pressure area formed by the fountain flow is close to the rear jet position, and the region around the rear jet is formed into a low pressure area under the effect of jet coiling and suction, and the low pressure area is affected by the position of the fountain area to move to the outer region of the rear jet, and due to the position of the fountain area being close to the rear jet, the rear jet interferes with the fountain area, and the extent of the rear jet coiling and suction is enhanced, and the pressure in the area around the jet decreases, and due to the reduction of the pressure ratio of the rear jet drop, the range of the low pressure area around the rear jet increases. The pressure around the jet decreases, and the low-pressure area around the jet decreases due to the decrease in the pressure ratio of the jet. Figure 4.59 is the symmetry of the Mach number distribution on the surface, from the figure can be seen, when the back jet flow volume is larger, the intersection of the front and rear jets on the

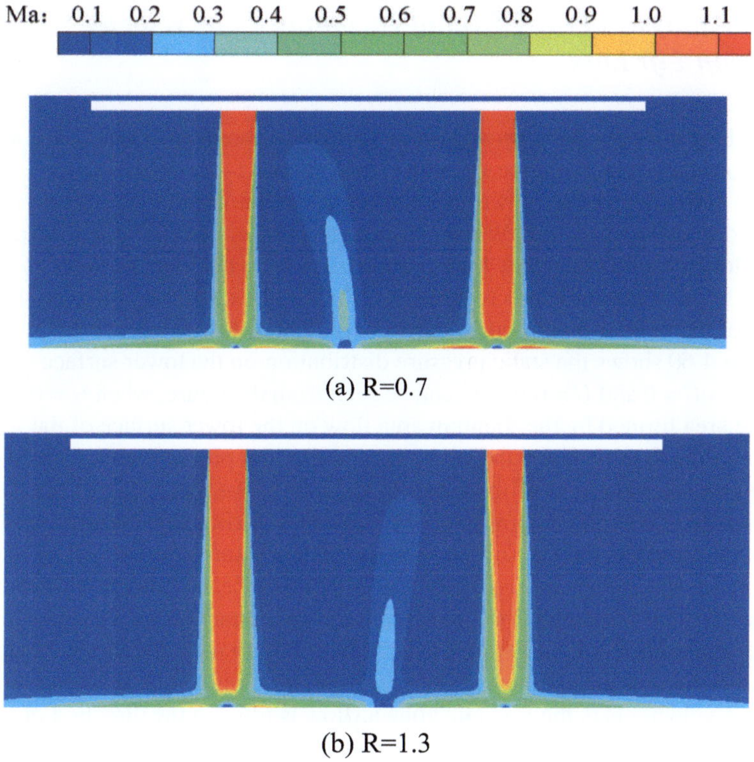

Fig. 4.59 Mach number distribution of the symmetry plane at R = 0.7 and R = 1.3 (H = 7, S = 7.5, *NPR* = 2.25, U = 0). (**a**) R = 0.7. (**b**) R = 1.3

ground in the direction of the front jet to move, the symmetry of the fountain area of the direction of the front jet tilt, when the amount of the back jet is smaller, the intersection of the front and rear jets on the ground in the direction of the back jet to move, symmetry of the fountain area of the direction of the jet to tilt the direction of the back jet.

From the force on the lower surface of the flat plate and the Mach number distribution of the symmetry plane, it can be seen that when the momentum ratio of the front and rear jets decreases, the pressure ratio of the rear jet drop increases, and the angle of inclination of the fountain area airflow to the direction of the cold flow increases, which leads to a relative decrease in the fountain force on the lower surface of the flat plate and a relative increase in the loss of lift. Therefore, the effect of the momentum ratio between the two jets on the loss of lift of the twin jets is mainly reflected in the fountain force on the lower surface of the flat plate. The change of the fountain force.

4.6.6 Analysis of the Effect of Incoming Flow Velocity on Lift Loss

Table 4.10 shows the force on the lower surface of the plate when $U = 0$ and $U = 0.03$. From the table, it can be seen that the fountain force on the lower surface of the plate is smaller than the lower suction force, and the combined force on the plate is in the downward direction. When U is increased from 0 to 0.03, the fountain force and the lower suction force are both reduced, in which the fountain force is reduced to 0, the lower suction force is reduced by 27.37%, the combined force on the lower surface of the plate is increased, and the coefficient of loss of lift force is increased.

Figure 4.60 shows the static pressure distribution on the lower surface of the flat plate when $U = 0$ and $U = 0.03$. As can be seen from the figure, when $U = 0$, the high pressure area formed by the fountain area flow on the lower surface of the flat plate appears in the area between the two jets, and the low pressure area formed by the coiling and suction effect of the front jet appears in the outer side of the front jet, and when $U = 0.03$, the incoming flow separates to form a low pressure area in the front of the flat plate's lower surface, and the high pressure area formed by the fountain flow on the lower surface of the flat plate disappears, and the fountain force becomes 0. The pressure of the low-pressure area near the jet increases and the value of the downward suction force decreases. Figure 4.61 shows the Mach number distribution on the symmetric surface, from which it can be seen that when the incoming velocity is 0, the fountain zone airflow is tilted in the direction of the cold flow, and when $U = 0.03$, the fountain zone airflow formed by the two jets is tilted

Table 4.10 Forces on the lower surface of the plate at U = 0 and U = 0.03 (H = 3, S = 10, R = 1, *NPR* = 2.25)

	U = 0	U = 0.03
F_f/N	108.49	0
F_i/N	−167	−121.29
$F_f + F_i$/N	−58.51	−121.29
C_f	0.037	0.0665

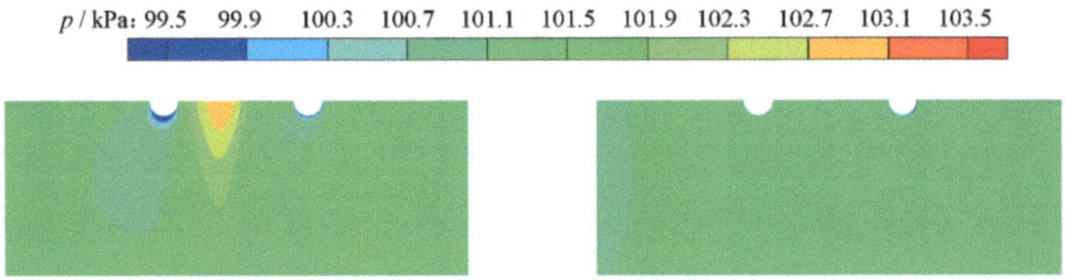

(a) U=0 (b) U=0.03

Fig. 4.60 Hydrostatic pressure distribution on the lower surface of a flat plate at U = 0 and U = 0.03 (H = 3, S = 10, R = 1, NPR = 2.25). (**a**) U = 0. (**b**) U = 0.03

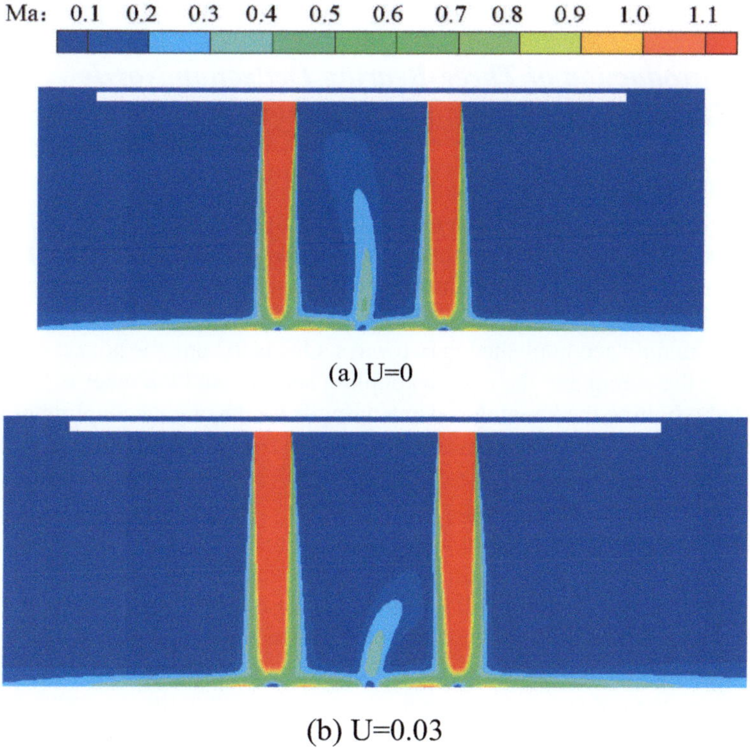

Fig. 4.61 Mach number distribution of the symmetry plane at U = 0 and U = 0.03 (H = 3, S = 10, R = 1, NPR = 2.25). (**a**) U = 0. (**b**) U = 0.03

backward along with the incoming flow, and the high velocity of the fountain zone does not flow to the lower surface of the flat plate under the influence of the incoming flow, due to the larger velocity of the incoming flow, the incoming flow interferes with the airflow of the fountain zone, the high velocity of the fountain zone does not flow to the lower surface of the flat plate under the influence of the incoming flow surface, the interference effect between the fountain flow and the jet flow on the lower surface of the flat plate disappears, so the downward suction caused by the high-speed airflow in the fountain area disappears, and the value of the downward suction decreases.

From the influence of the incoming flow on the pressure distribution on the lower surface of the plate, it can be seen that, on the one hand, the incoming flow can weaken the interference between the fountain flow and the jet flow so that the value of the downward suction force decreases, on the other hand, when the incoming flow is too large, the fountain force disappears, and the plate is mainly subjected to the action of downward suction force, and the loss of lift force increases.

4.7 Three-Bearing Deflection Nozzle Thermal Model Test

4.7.1 Introduction of Three-Bearing Deflection Nozzle Thermal State Experimental Equipment

1. Thermal test rigs

   The three-bearing deflection nozzle thermal test stand is shown in Fig. 4.62. The test stand consists of an upper clamp, a lower clamp and a main stand. The main frame consists of the engine support frame and the nozzle support frame, and the nozzle is fixed on the nozzle support frame in the form of a hanging, which is mainly accomplished in two ways. One is to hang the nozzle by welding between the clamp and the engine coupling section, and the other is to hang the nozzle by bolting the hanging rod extending from the engine coupling section to the main frame. The three-dimensional installation model of the engine and the three-bearing deflection nozzle on the test stand is shown in Fig. 4.63.
2. Six-component balance test rig

   The six-component balance test platform is mainly used for the measurement of six components, such as thrust, lift and lateral forces, as well as pitching moment, yaw moment and roll moment, with a measurement accuracy of 0.5% FS. The test platform mainly includes the balance body (as shown in Figure 4.64), balance support base, adapter plate, DC power supply, signal acquisition, and numerical processing system. Among them, the length of the six-component balance body is 300 mm, the width is 130 mm, the height is 80mm, and the main performance indexes are as follows:

$$\text{Thrust } F_x : 1000\text{N}; \text{Lift } F_y : 1000\text{N}; \text{Lateral force } F_z : 1000\text{N}$$

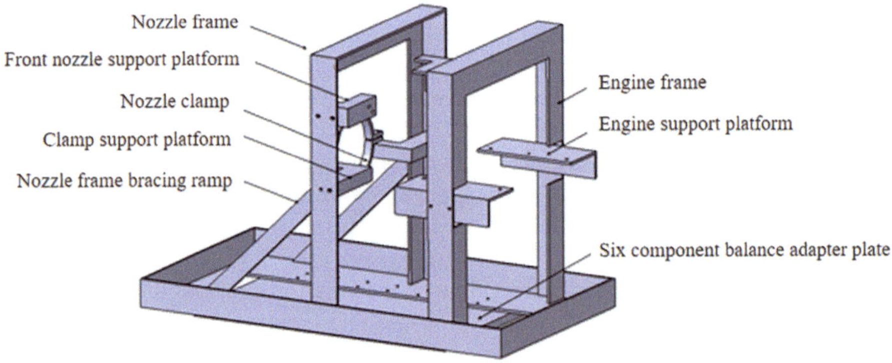

Fig. 4.62 3D model of the test stand

Fig. 4.63 Three-bearing deflection nozzle thermal test rig installation model

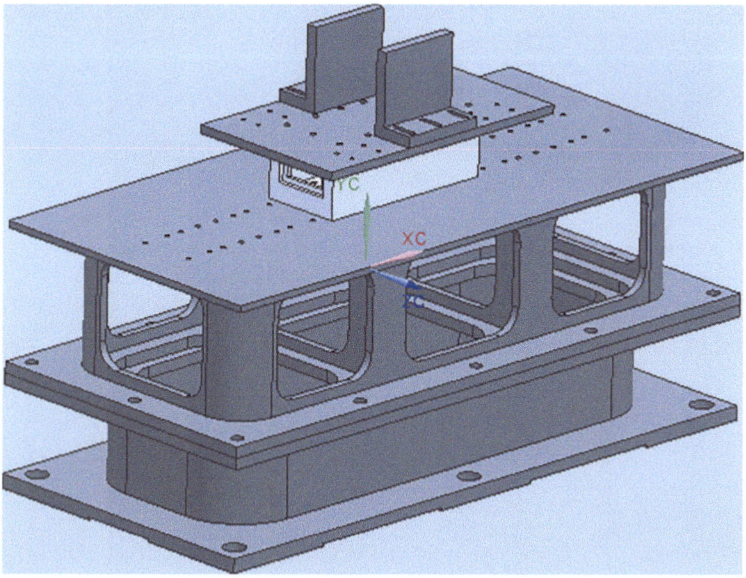

Fig. 4.64 Six-component balance body, support base and adapter plate

$$\text{Pitching moment } M_z: 30\,\text{kgm}; \text{Rolling moment } M_x: 10\,\text{kgm};$$
$$\text{Yawing moment } M_y: 20\,\text{kgm}$$

Data Acquisition System Model: Keithley 2700 with 7700 Control Measurement Module as shown in Fig. 4.65. This measurement identification integrates precision measurement, programmed switching and control functions in a compact chassis that can be rack or bench mounted.

3. AWT turbojet engines

The small turbojet engine used for the thermal experiments is the amt nike turbojet engine, whose profile dimensions are shown in Fig. 4.66.

According to the design point parameters and thrust performance of key components provided by the engine workbook (as shown in Fig. 4.67), the counting point parameters of each engine component were optimized and obtained, as shown in Table 4.11. Substituting this design point parameter into the overall engine performance calculation model, the engine performance under this design parameter was obtained, and this performance was compared with the test data,

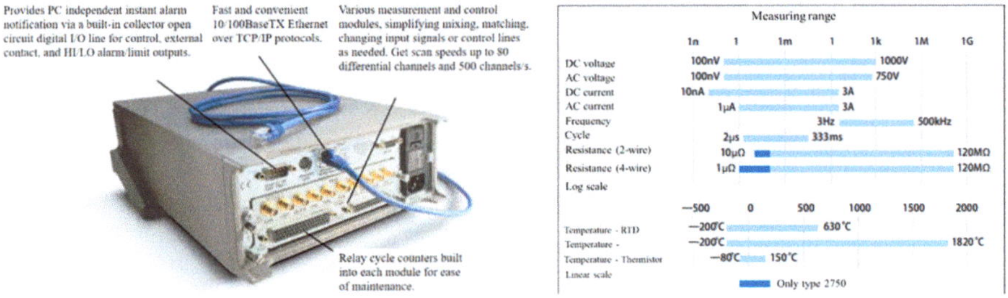

Fig. 4.65 Keithley measurement and control system and measurement range

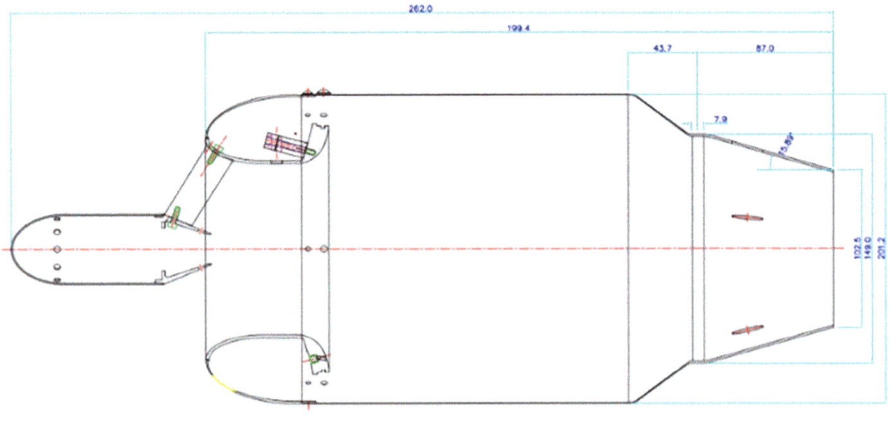

Fig. 4.66 Nike small turbojet engine

Specifications	Outlet area 0.007854 square meters		All data at STP +/- 2%
Engine diameter	201 mm	/	7,9 Inch
Engine length	524 mm	/	20,6 Inch
Engine weight	9150 gram	/	20,2 Lb
System airborne weight *	11300 gram	/	24,9 Lb
Thrust at max RPM at S.T.P.	784 N	/	176,2 Lbf
Maximum RPM	61.500	/	61.500
Thrust at Idle RPM	40 N	/	9 Lbf
Pressure ratio at max RPM	4:1	/	4:1
Mass flow	1250 Gr/sec	/	2.76 Lb/Sec
Normal EGT (internal EGT probe)	800 Deg C	/	1472 Deg F
Max EGT	875 Deg C	/	1607 Deg F
Fuel consumption 30.67g/s	1900 Gr/min	/	67 oz/min
Specific fuel consumption	40.36 gr/(Kn*sec)	/	1,42 lb/(lbf*hr)
Starting method	Direct kerosene starting system. (20-25 seconds starting time)		

Fig. 4.67 Engine performance parameters

Table 4.11 Engine component design point parameters

Component design parameters	(be) worth
Engine inlet air flow/kg/s	1.394
pressurization ratio	4
Compressor efficiency	0.753
turbine efficiency	0.725
Combustion chamber outlet temperature/K	1278
Combustion chamber efficiency	0.99
Combustion chamber total pressure recovery factor	0.95
Mechanical efficiency of the spindle	0.99
Total pressure recovery factor for turbine outlet culvert	0.95
Nozzle thrust factor	0.98
Nozzle flow coefficient	0.905
Nozzle outlet area/m^2	0.00786

Table 4.12 Engine design point performance

Engine design point performance parameters	Calculated value	Experimental value	Inaccuracies
Thrust/N	787	797	1.25%
Unit fuel consumption rate/g/(kN*s)	39.14	38.35	2.06%
Fuel flow rate/kg/s	0.0308	0.03067	0.4%
Nozzle outlet air velocity/m/s	563.74	558	1%

as shown in Table 4.12, and from the comparison between the calculation results and the test data, it can be seen that the overall engine performance calculation model is reliable.

4.7.2 *Three-Bearing Deflection Nozzle Thermal Experiment Results*

In order to compare the effect of the three-bearing deflector nozzle on the thrust performance of the engine, the thermal experimental results of the engine with the original nozzle are first given before the thermal experimental results of the three-bearing deflector nozzle are given.

1. Results of thermal experiments with the original nozzle

   As shown in Fig. 4.68, the engine with the original nozzle was mounted on an adapter frame, the thrust parameters were measured at several RPM state points, and the engine RPM and thrust were corrected to the corrected relationship under standard atmospheric pressure conditions at sea level (SLS) based on measured atmospheric temperatures and atmospheric pressures as follows:

$$\text{Corrected thrust} = \text{Measured thrust} \times \text{Standard atmospheric pressure at sea level} / \text{ambient pressure};$$

$$\text{Corrected RPM} = \text{Measured RPM} \times (\text{Sea Level Standard Atmospheric Temperature} / \text{Ambient Temperature})^{0.5}.$$

   As shown in Fig. 4.69, the experimental results are compared with the factory data of the AMT NIKE engine, it can be seen from the figure that all the thrust points obtained during this test fall on the thrust curve provided by the manufacturer, which can indicate the accuracy and reliability of the results of this test. In addition, it can be seen that in the physical maximum rotational speed, the thrust measured in this test data is greater than the given thrust by the manufacturer, mainly because the maximum converted rotational speed obtained in this test

(a) Side view (b) Side rear view

Fig. 4.68 Original nozzle test drive. (**a**) Side view. (**b**) Side rear view

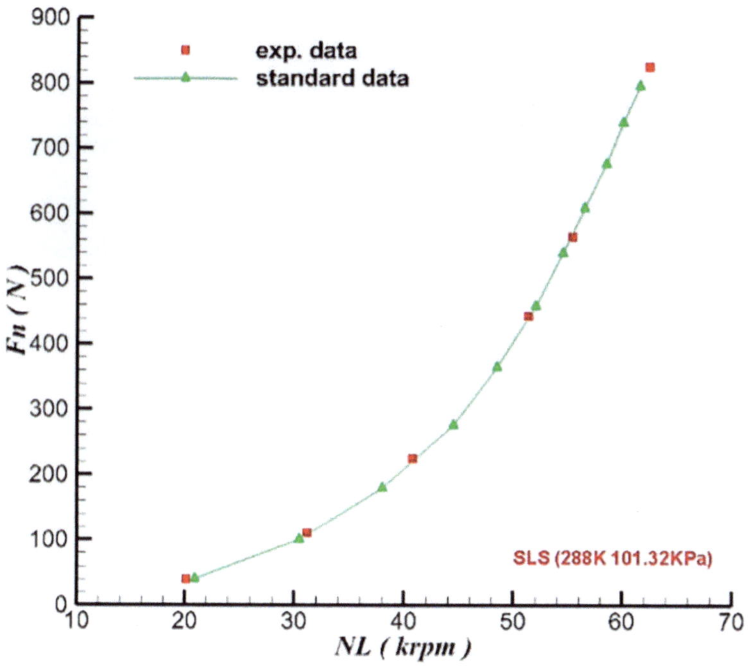

Fig. 4.69 Original nozzle test drive thrust data

(62.43 kPa) is greater than the maximum rotational speed of the manufacturer (61.5 kPa), according to the law that the thrust and rotational speed squared is approximately proportional to the relationship, it can be determined that the data obtained in this test is credible.

2. Experimental results of the thermal state of the three-bearing deflection nozzle at 30°vector angle

   The throttling characteristics of the engine were tested on a six-component balance measuring stand when the three-bearing deflected nozzle was at 30° vector angle, as shown in Fig. 4.70 are given for different viewpoints, and at this time, in order to ensure the stabilization of the thrust vector angle, the power supply of the drive motor needs to be switched on in order to keep the segments of the nozzle's cylinder stationary and not rotating.

   Table 4.13 and Figure 4.71 show, respectively, the variation of engine thrust with rotational speed for a three-bearing deflection nozzle at 30° vector angle. Table 4.13 also gives the corrected engine speed and corrected thrust.

3. Experimental results of thermal state at 60° vector angle for three-bearing deflection nozzle

   The throttling characteristics of the engine were tested on a six-component balance measuring stand at a 60° vector angle of the three-bearing deflector nozzle. As shown in Fig. 4.72, it gives a side view of the experiment, at which time, in order to ensure that the thrust vector angle is stabilized, it is necessary to turn

(a) Side view (b) Side and rear view

Fig. 4.70 Throttling characteristics test at 0° vector angle for three-bearing deflection nozzle. (**a**) Side view. (**b**) Side and rear view

Table 4.13 AMT NIKE engine with three-bearing deflection nozzle (30°) test run data

Physical speed (krpm)	Corrected RPM (krpm)	Thrust (N)	Corrective thrust (N)	Test drive environment Atmospheric temperature (K)	Test drive environment Atmospheric pressure (KPa)
19.5	19.67	32.60	34.12	283.1	96.8
29.5	29.75	84.57	88.52		
41.0	41.35	198.25	207.51		
50.0	50.43	348.94	365.23		
55.5	55.98	480.73	503.18		
62.0	62.53	689.13	721.31		

on the power supply of the drive motor in order to keep the segments of the nozzle's barrel at rest and not rotating.

Table 4.14 and Fig. 4.73 give, respectively, the variation of engine thrust with rotational speed for a three-bearing deflection nozzle at 90° vector angle. Table 4.14 also gives the corrected engine speed and corrected thrust.

4. Experimental results of thermal state at 90° vector angle for three-bearing deflection nozzle

Three-bearing deflection nozzle 90° vector angle, the throttle characteristics of the engine were measured on the six-component balance measuring stand, as shown in Fig. 4.74. The pictures are given in different viewpoints, respectively. At this time, in order to ensure the stability of the thrust vector angle, the power supply of the drive motor needs to be turned on in order to keep the segments of the nozzle's barrel at rest and not rotating.

Table 4.15 and Fig. 4.75 give, respectively, the variation of engine thrust with rotational speed for a three-bearing deflection nozzle at 90° vector angle. Table 4.15 also gives the corrected engine speed and corrected thrust. By comparing the engine thrust with a three-bearing deflector nozzle (90° vector angle) with the original nozzle at maximum operating conditions, it can be found that the thrust loss is about 14%.

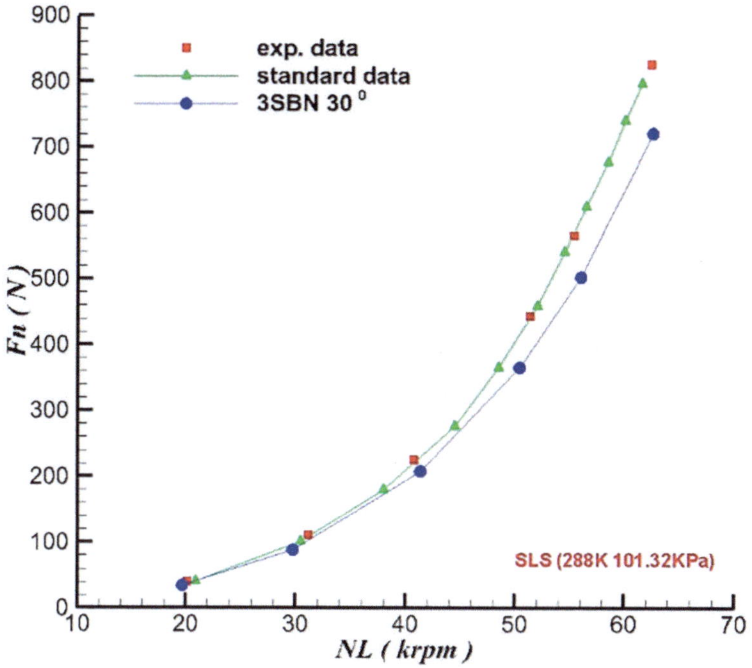

Fig. 4.71 Thrust performance comparison of engine with three-bearing deflector nozzle (30°) and original nozzle

Fig. 4.72 Throttling characteristics test for three-bearing deflector nozzle at 60° vector angle

Table 4.14 AMT NIKE engine with three-bearing deflector nozzle (60°) test run data

Physical speed (krpm)	Corrected RPM (krpm)	thrust (N)	Corrective thrust (N)	Test drive environment Atmospheric temperature (K)	Test drive environment Atmospheric pressure (KPa)
19.5	19.67	30.29	31.71	283.1	96.8
29.5	29.75	80.65	84.42		
40.0	40.34	178.49	186.82		
50.0	50.43	352.40	368.86		
55.0	55.47	471.74	493.77		
62.0	62.53	700.97	733.70		

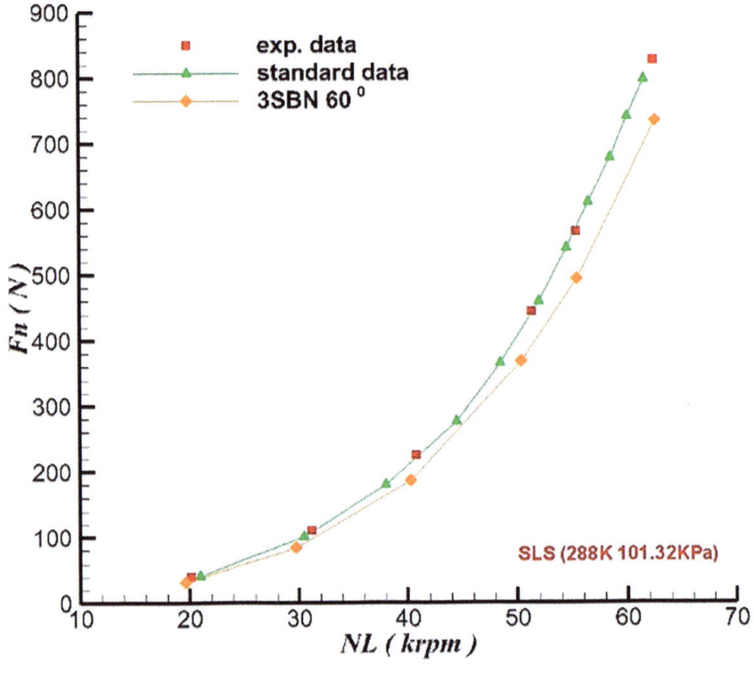

Fig. 4.73 Thrust performance comparison of engines with three-bearing deflector nozzles (60°) and original nozzles

5. Experimental results of the dynamic test run of the three-bearing deflection nozzle with multiple round trips of 90° deflection under large working conditions

   In order to further verify the reliability of the three-bearing deflector nozzle structure, repeated deflection verification tests were carried out over a long period of time and under maximum operating conditions. The process is as follows:

   (1) With the three-bearing deflection nozzle at 0°, push the engine's throttle lever to maximum;

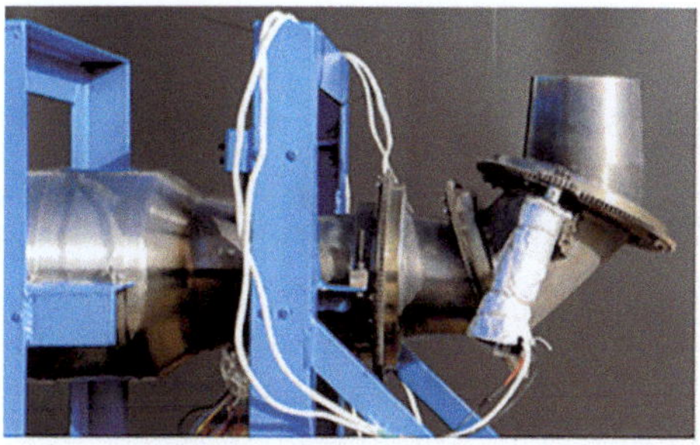

(a) Lateral attempt

(b) Side and rear attempts

Fig. 4.74 Throttling characteristics test for three-bearing deflector nozzle at 90° vector angle. (**a**) Lateral attempt. (**b**) Side and rear attempts

Table 4.15 AMT NIKE engine with three-bearing deflection nozzle (90°) test data

Physical speed (krpm)	Corrected RPM (krpm)	Thrust (N)	Corrective thrust (N)	Test drive environment Atmospheric temperature (K)	Test drive environment Atmospheric pressure (KPa)
19.5	19.67	31.64	33.12	283.1	96.8
31.5	31.77	94.41	98.82		
40.5	40.85	182.72	191.25		
50.0	50.43	335.03	350.67		
54.5	54.97	450.88	471.93		
62.0	62.53	686.74	718.81		

(2) A three-bearing deflection nozzle 90° deflection command is issued, and the nozzle angle is stabilized and held for 1 min;
(3) Issue a three-bearing deflection nozzle 0° deflection command for a short period of time;
(4) Repeat steps 2 and 3 three more times;
(5) A three-bearing deflection nozzle 90° deflection command is issued, and the nozzle angle is stabilized and held for 4 min;
(6) A three-bearing deflection nozzle 0° deflection command is issued, the engine decelerates and stops.

Figure 4.76 gives the variation of thrust in the Y direction during the round-trip deflection process, and it can be found that the maximum loss of thrust is about 15% after the standard atmosphere conversion. In addition, it can be seen that in addition to the thrust fluctuation in this working condition, the Y value at the fourth platform slightly decreased, which is due to the longtime of high temperature and high pressure work, resulting in a certain degree of deformation of the inner retaining ring of the first section of the bearing of the three-bearing deflector nozzle and thickening of the inner

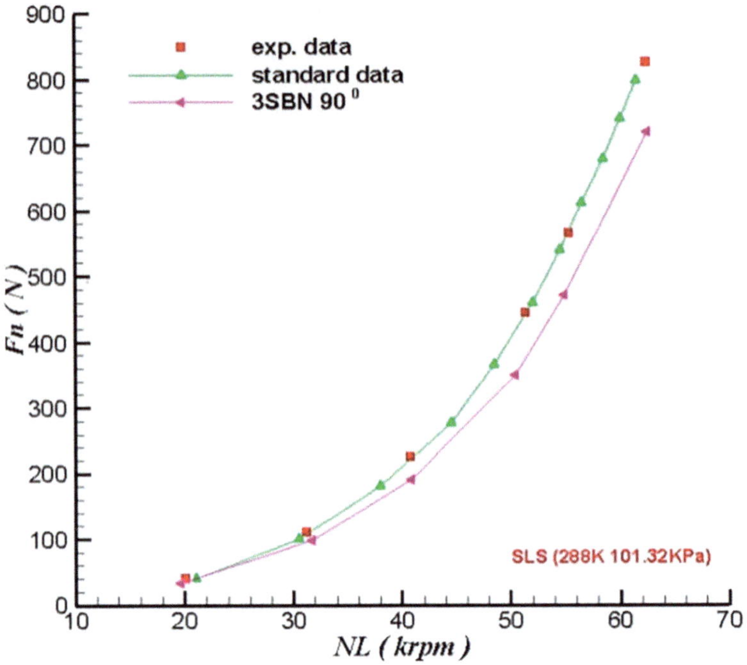

Fig. 4.75 Thrust performance comparison of engine with three-bearing deflector nozzle (90°) and original nozzle

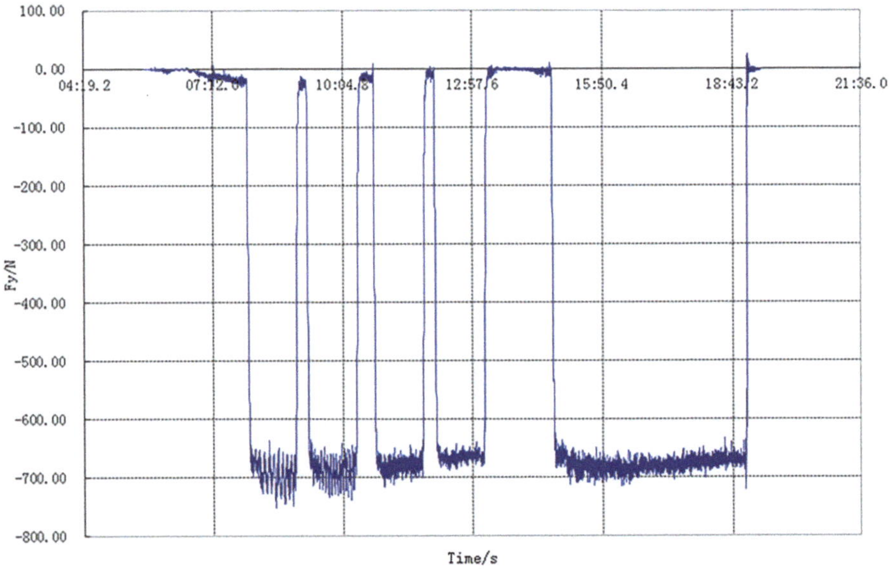

Fig. 4.76 Dynamic deflection of an engine with a three-bearing deflector nozzle to and from the engine at high operating conditions

retaining ring; on the other hand, it shows that in the pursuit of the material's (titanium alloy) lightness, it must be taken into account in terms of its resistance to high temperatures, and therefore it is recommended that in the use of this nozzle Therefore, it is recommended that the time of working at large angles and high temperatures should be controlled as much as possible when carrying out subsequent tests.

Chapter 5
Design Methods and Applications of Single Expansion Ramp Nozzle

5.1 Introduction

A hypersonic vehicle refers to an aircraft capable of sustained flight at Mach numbers greater than 5, powered by air-breathing engines or combined-cycle propulsion systems, operating within and across atmospheric layers (Liu 2004). Recognized as the third revolutionary advancement in aviation history following propeller and jet aircraft, it represents a critical development trend in aerospace for the 21st century. The exhaust nozzle, as the primary thrust-generating component of the engine, has been shown to contribute approximately 70% of total engine thrust at Mach 6 (Grarnland and Berens 1995). Thus, nozzle design quality directly impacts overall engine performance. Among nozzle configurations, the Single Expansion Ramp Nozzle (SERN) facilitates integrated propulsion-airframe design and is currently considered the optimal exhaust system for hypersonic vehicles. The SERN employs a highly integrated configuration between the vehicle aft-body and nozzle. While demonstrating excellent thrust coefficients at hypersonic speeds, its performance degrades significantly during transonic phases due to reduced nozzle pressure ratios. Under severe over-expansion conditions, the upper ramp surface not only fails to generate thrust but may even produce drag, leading to substantial deterioration of thrust coefficients (Christopher and Jaime 1991). Improving SERN performance at low Mach numbers has thus become a critical scientific challenge in hypersonic propulsion system integration. To address this, this chapter investigates SERN contour design and flow mechanisms for series- and parallel-configured TBCC engines. It explores active flow control via high-pressure secondary flow injection and passive flow control using cavity-based structures to enhance performance during over-expansion. The study further examines how key parameters of these active and passive control methods influence SERN performance and flow characteristics.

J. Shi et al., *Special Purpose Nozzles*,
https://doi.org/10.1007/978-981-95-2939-1_5

5.2 Single Expansion Ramp Nozzle Design Methodology

The aft-body design of hypersonic vehicles must meet the following requirements: generating sufficient thrust, as different aft-body configurations affect gas expansion and consequently influence the pressure distribution on the aft-body surfaces, necessitating an optimized configuration to produce the required thrust; avoiding plume interference with horizontal stabilizers; and minimizing structural mass increases caused by the lower and sidewalls of the aft-body (Deere and Asbury 1996). This section utilizes the method of characteristics to design the upper expansion ramp of the Single Expansion Ramp Nozzle (SERN), completes the SERN contour design, and numerically simulates its flow fields and performance under varying flight conditions.

5.2.1 Application of the Method of Characteristics in SERN Contour Design

The primary methodology for designing asymmetric high-expansion-ratio SERN contours for hypersonic vehicles is based on the method of characteristics. From a physical perspective, characteristic curves (simplified as "characteristics") are defined as trajectories along which physical disturbances propagate. Across these characteristics, dependent variables cannot be arbitrarily prescribed but must satisfy compatibility equations.
The general theory of the method of characteristics provides a numerical framework for analyzing steady two-dimensional (planar or axisymmetric) supersonic irrotational flow fields. By applying this theory, the governing flow equations (partial differential equations) can be transformed into characteristic and compatibility equations. For SERN design, the method of characteristics is used to solve two-dimensional irrotational supersonic flow problems, enabling the derivation of the nozzle expansion ramp contour.

The gas dynamic equations governing steady, two-dimensional, inviscid, irrotational flow of an ideal gas include:

$$\left(u^2 - a^2\right)u_x + \left(v^2 - a^2\right)v_y + 2uvu_y - \delta a^2 v / y = 0 \tag{5.1}$$

$$u_y - v_x = 0 \tag{5.2}$$

$$a^2 = a_0^2 - (\gamma - 1)V^2 / 2 \tag{5.3}$$

Here, u denotes the velocity component in the x-direction, v the velocity component in the y-direction, a the local speed of sound, a_0 the stagnation speed of sound,

γ the specific heat ratio, and V the total velocity magnitude. The parameter $\delta = 0$ represents planar flow, while $\delta = 1$ corresponds to axisymmetric flow.

By applying the method of characteristics, the governing partial differential equations of the flow can be transformed into characteristic equations and compatibility equations. For steady, two-dimensional, irrotational supersonic flow, these equations are expressed as follows:

$$(dy/dx)_{\pm} = \lambda_{\pm} = \tan(\theta \pm \alpha)(\text{Mach line}) \tag{5.4}$$

$$\begin{aligned}&(u^2 - a^2)du_{\pm} + \left(2uv - (u^2 - a^2)\lambda_{\pm}\right)dv_{\pm}\\&-(\delta a^2 v / y)dx_{\pm} = 0(\text{Along the Mach line})\end{aligned} \tag{5.5}$$

Here, θ denotes the flow angle, α the Mach angle, λ the characteristic slope, and x, y the two-dimensional coordinates. The C_+ and C_- characteristics (corresponding to the "+" and "−" signs in Eq. 5.4) represent the Mach lines of the flow, as illustrated in Fig. 5.1. In Eq. (5.5), the subscript ± indicates that the differentials du, dv, and dx are determined along the C_+ and C_- characteristics, respectively.

The modified Euler predictor-corrector method is applied to the finite difference grid illustrated in Fig. 5.2, where solid lines represent characteristic lines and dashed

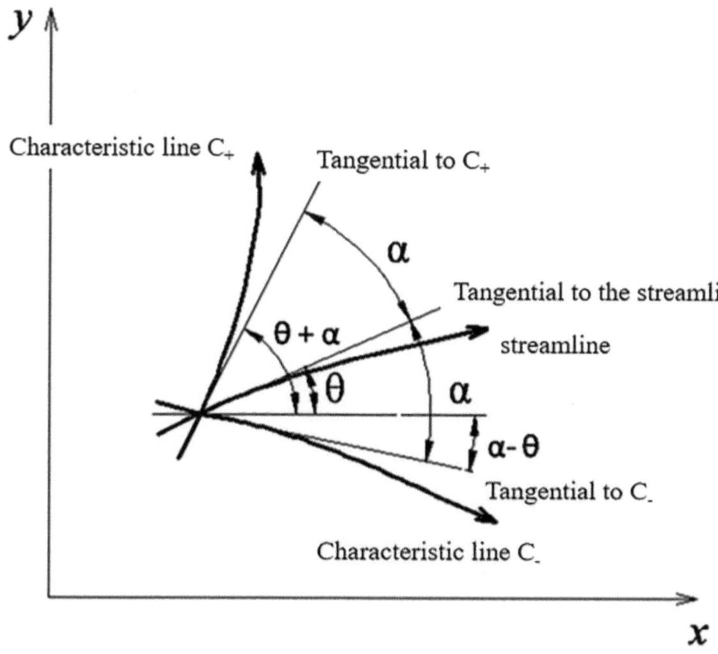

Fig. 5.1 Schematic of characteristic lines (Mach Lines) in steady two-dimensional irrotational supersonic flow

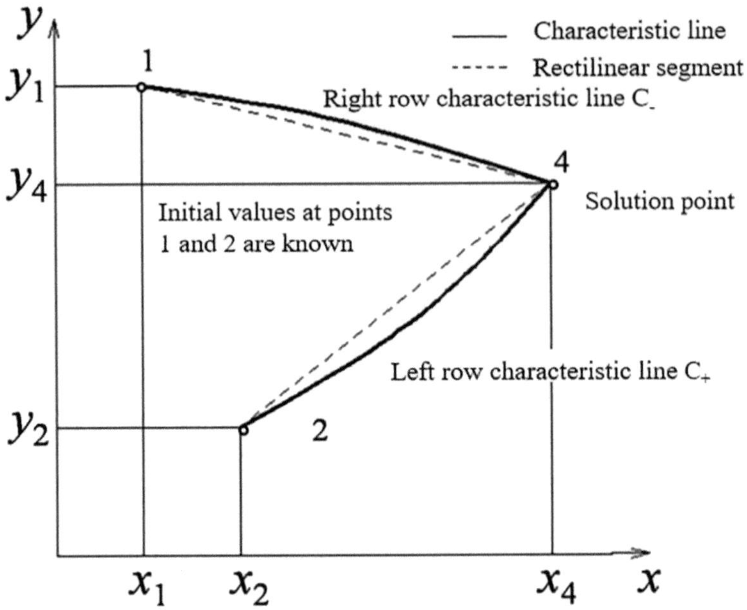

Fig. 5.2 Application of the method of characteristics to the finite difference grid for steady two-dimensional irrotational supersonic flow

lines denote straight-line segments connecting nodal points. In the computational implementation, these straight-line segments are substituted for the actual characteristic lines. This substitution enables the transformation of the characteristic equations (Eq. 5.4) and compatibility equations (Eq. 5.5) into finite difference formulations, as expressed in the following equations.

$$\Delta y_{\pm} = \lambda_{\pm} \Delta x_{\pm} \tag{5.6}$$

$$Q_{\pm} \Delta u_{\pm} + R_{\pm} \Delta v_{\pm} - S_{\pm} \Delta x_{\pm} = 0 \tag{5.7}$$

$$\lambda_{\pm} = \tan\left(\theta \pm \alpha\right) \tag{5.8}$$

$$Q_{\pm} = u_{\pm}^2 - a_{\pm}^2 \tag{5.9}$$

$$R_{\pm} = 2u_{\pm} v_{\pm} - \left(u_{\pm}^2 - a_{\pm}^2\right) \lambda_{\pm} \tag{5.10}$$

$$S_{\pm} = \delta a_{\pm}^2 v_{\pm} / y_{\pm} \tag{5.11}$$

Here, the “+” and “−” symbols denote the C_+ and C_- characteristic lines.

The computational methodology for interior points within the finite difference grid of the method of characteristics (as illustrated in Fig. 5.2) is presented here.

The governing equations for determining finite difference coefficients, based on the averaged parameter method, are provided below. Once the computational procedure for interior points is established, an analogous approach can be applied to handle wall boundary points.

$$y_4 - \lambda_+ x_4 = y_2 - \lambda_+ x_2 \tag{5.12}$$

$$y_4 - \lambda_- x_4 = y_1 - \lambda_- x_1 \tag{5.13}$$

$$Q_+ u_4 + R_+ v_4 = T_+ \tag{5.14}$$

$$Q_- u_4 + R_- v_4 = T_- \tag{5.15}$$

$$T_+ = S_+ (x_4 - x_2) + Q_+ u_2 + R_+ v_2 \tag{5.16}$$

$$T_- = S_- (x_4 - x_1) + Q_- u_2 + R_- v_1 \tag{5.17}$$

$$V_\pm = \sqrt{u_\pm^2 + v_\pm^2} \tag{5.18}$$

$$\theta_\pm = \tan^{-1}(v_\pm / u_\pm) \tag{5.19}$$

$$a_\pm = a(V_\pm) \tag{5.20}$$

$$M_\pm = V_\pm / a_\pm \tag{5.21}$$

$$\alpha_\pm = \sin(1/M_\pm) \tag{5.22}$$

$$u_+ = u_2, u_- = u_1, v_+ = v_2, v_- = v_1, y_+ = y_2, y_- = y_1 \left(\text{Prediction method}\right) \tag{5.23}$$

$$u_+ = \frac{u_2 + u_4}{2}, u_- = \frac{u_1 + u_4}{2}, v_+ = \frac{v_2 + v_4}{2}, v_- = \frac{v_1 + v_4}{2}, y_+ = \frac{y_2 + y_4}{2}, y_- = \frac{y_1 + y_4}{2} \left(\text{Correction method}\right) \tag{5.24}$$

Figures 5.3, 5.4, and 5.5 illustrate three distinct types of unit processes encountered when applying the method of characteristics. The computational approaches for these unit processes are analogous to the methodology for interior points.

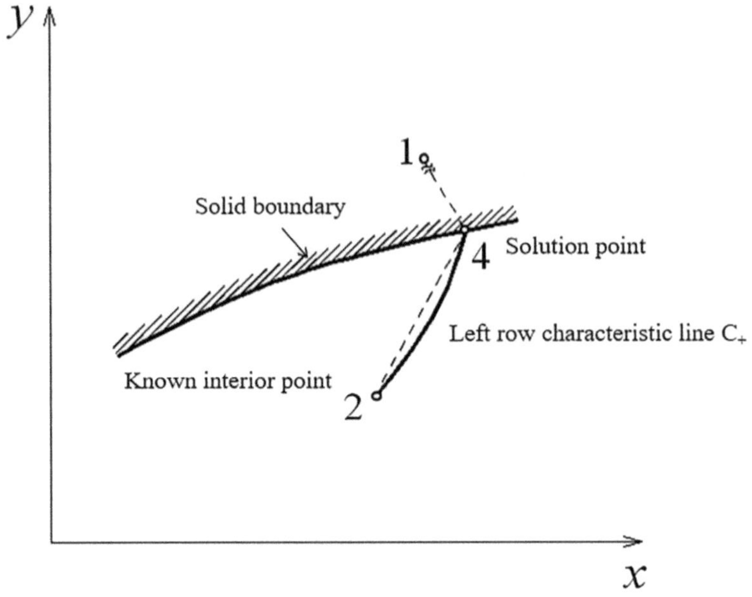

Fig. 5.3 Unit process for direct wall points

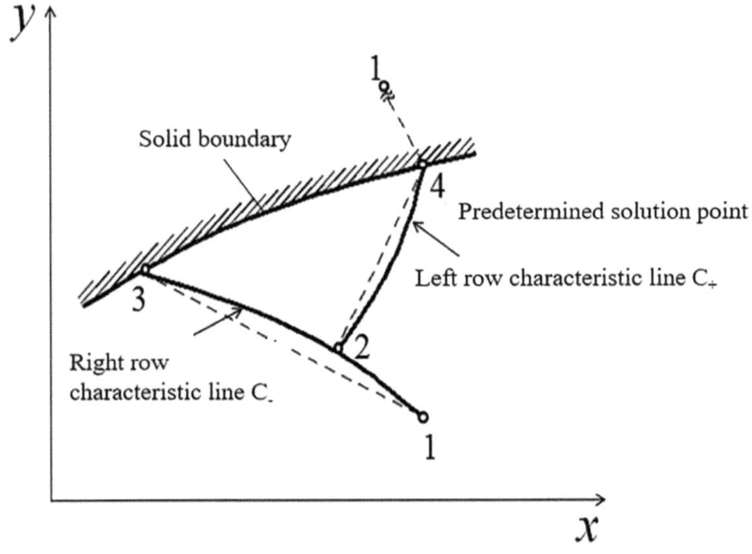

Fig. 5.4 Unit process for inverse wall points

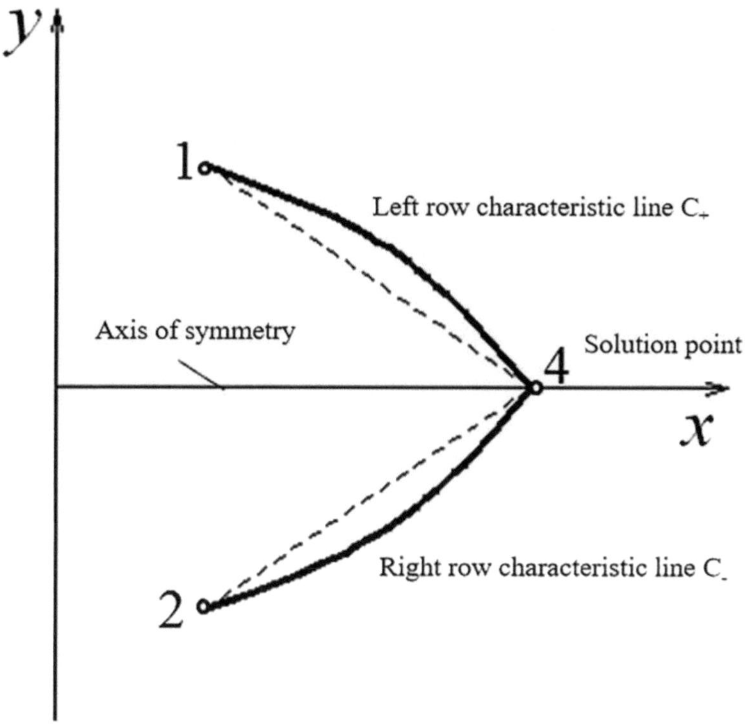

Fig. 5.5 Unit process for symmetry axis points

Building on the established procedure for interior points, similar techniques can be extended to resolve wall boundary points, which include direct wall points and inverse wall points.

Figure 5.3 depicts a typical direct wall point flow scenario. Here, the left-running C^+ characteristic originates from a known interior point (Point 2) and intersects the wall at Point 4. Since Point 1 lies outside the physical flow domain, the C^- characteristic connecting Points 1 and 4 is represented by a dashed line. To determine the position and flow parameters of the direct wall point (Point 4), only one characteristic line (C^+) and its corresponding compatibility equation are available. However, two additional constraints arise from the wall boundary conditions:$y = y(x)$,$dy/dx = \tan(\theta) = v/u$. By combining these boundary conditions with Eqs. (5.12) and (5.14), the position and flow parameters at Point 4 can be uniquely resolved.

However, in regions with steep gradients of flow parameters—such as the throat region of the nozzle—the direct wall point method may result in unacceptably coarse spacing of solution points along the wall, compromising required accuracy. To address this, the inverse wall point method is employed. This approach predefines solution point locations along the wall and uses the method of characteristics to determine their flow parameters. As illustrated in Fig. 5.4, initial Points 1 and 3 are obtained from prior computational steps, while Point 4 is a predefined solution location. Point 2 is the intersection of the backward-extending left-running C_+ characteristic from Point 4 and the right-running characteristic (Line 1–3). Since the coordinates (x_4, y_4) of Point 4 are predetermined, only its flow parameters (u_4, v_4) need resolution via the method of characteristics. Assuming Point 2's position and flow parameters are known, the compatibility equation (5.14) combined with the equation: $dy/dx = \tan(\theta) = v/u$ can be solved to compute u_4 and v_4.

For axisymmetric flow, where the x-axis serves as the symmetry axis, Fig. 5.5 illustrates a typical symmetry axis point (Point 4). Assuming Point 1 lies on the C^- characteristic passing through Point 4, a mirror point (Point 2) can be defined symmetrically below the x-axis, as shown. This configuration renders Point 4 analogous to an interior point in Fig. 5.2. In this scenario, $y_4 = v_4 = \theta_4 = 0$ the symmetry conditions simplify the unit process for symmetry axis points. By combining the right-running characteristic (Line 1-4) with $y_4 = v_4 = \theta_4 = 0$ constraints, Eqs. (5.13) and (5.15) can be solved simultaneously to determine x_4 and u_4.

5.2.2 *SERN Contour Design Methodology for Series-Configured TBCC Engines*

When determining the flow parameters in the divergent section of the nozzle, the primary source of error arises from approximations used to resolve the flow field near the nozzle throat (i.e., the throat region). A schematic of the throat region geometry is shown in Fig. 5.6. Among methods for analyzing two-dimensional nozzle throat flow fields, Sauer's approach provides the simplest framework.

To initiate the method of characteristics for solving the two-dimensional supersonic flow field, a starting line (e.g., Line OT in Fig. 5.6) must be defined where the Mach number exceeds unity ($Ma > 1$) across the throat. Using this method, all aerodynamic parameters (velocity, pressure, Mach number) in the divergent section can be systematically determined along a line suitable for initiating numerical computations (Zucrow and Hoffman 1984).

Once the initial value line (OT) is established based on the throat region flow field, the various unit processes of the method of characteristics can be applied to resolve the downstream flow. In Fig. 5.7, the curve TC represents the upper wall contour downstream of the throat, selected as a circular arc segment termed the initial expansion segment of the upper wall. Due to the steep flow parameter gradients in the throat region, the flow parameters along the upper expansion ramp's initial segment (TC) are resolved using the inverse wall point method. Flow parameters along the x-axis are determined by the intersection of right-running characteristics emitted from points on TC with the horizontal lower wall. The design Mach number (Ma_D) is achieved along the x-axis, with the corresponding right-running characteristic (SK) defining the onset of Ma_D. At this stage, the contour AK serves as the lower wall.

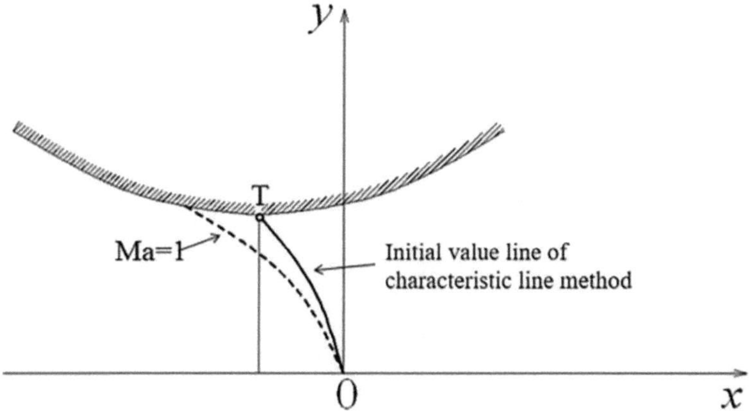

Fig. 5.6 Schematic of the throat region geometry and starting line (OT) for the method of characteristics

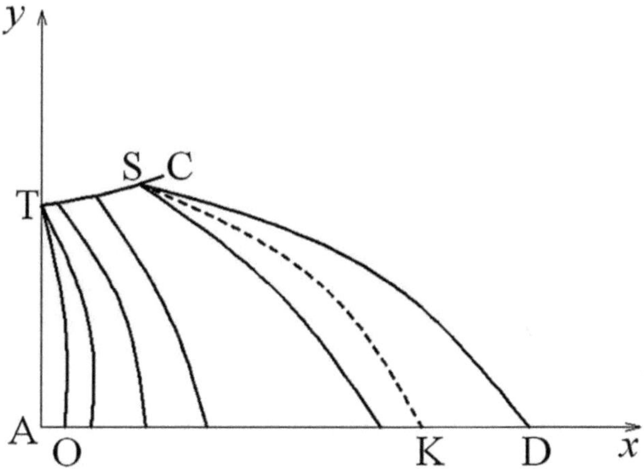

Fig. 5.7 Determination of the expansion ramp contour

To achieve a parallel and uniform exhaust flow at the nozzle exit with Ma = Ma_D and $\theta = 0$, all characteristic lines in this region must be straight lines inclined at an angle $\alpha_D = \sin^{-1}(1/Ma_D)$, as shown in Fig. 5.8. A straight characteristic line (KF) angled at α_D relative to the x-axis extends from Point K downstream to Point F, ensuring that the mass flow rate through the KF line equals the choked mass flow rate through the initial line TJ.

The remaining challenge lies in determining the flow field within the turning region R and the shape of the curved wall contour SF. This constitutes a Goursat problem—an initial-value problem with data prescribed on two characteristics of different families. Figure 5.8 illustrates how the unit process for interior points is applied starting from the KF line to resolve the supersonic flow field in region

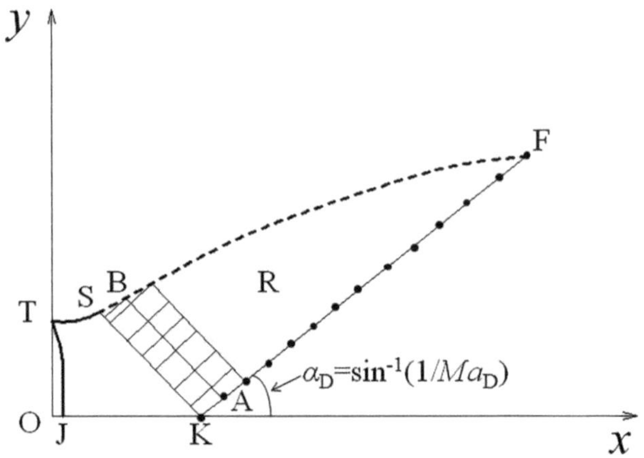

Fig. 5.8 Schematic of the initial expansion zone in nozzle design

R. Each new wall point on the curved contour SF is determined by mass balance along the corresponding characteristic lines originating from initial points on KF. This process is repeated for subsequent points along KF until the entire curved contour SF is defined.

In practice, nozzles designed to produce parallel uniform exhaust flows are typically elongated, with a significant portion of the downstream contour resembling a near-cylindrical cross-section. For such long nozzles, boundary layer growth cannot be neglected. Consequently, the boundary layer displacement thickness must be calculated, and the inviscid contour SF is offset outward by this thickness to obtain the actual nozzle wall geometry. This correction ensures that the final contour SF satisfies the requirement of being an inviscid streamline while accounting for viscous effects.

In this study, empirical formulas were applied to perform boundary layer correction on the nozzle contour (Alaa et al. 2005). The boundary layer thickness at the nozzle throat (ϕ_{throat}) was assumed to be zero ($\phi_{\text{throat}} = 0$), and the boundary layer thickness along the divergent section was modeled to increase linearly with axial position xx, governed by $d\varphi/dx = \tan \varepsilon$, where ε denotes the correction angle. The values of ε, selected based on the nozzle exit Mach number (Ma), are provided in Table 5.1. This approach ensures that the viscous effects are systematically integrated into the inviscid nozzle design, balancing aerodynamic performance with practical geometric constraints.

The contraction section is a critical component of the nozzle, designed to ensure uniform flow reaching the throat—a requirement for supersonic nozzle performance. To achieve a homogeneous throat flow field, the contraction profile must

Table 5.1 Relationship between correction angle ε and nozzle exit mach number Ma

Ma	2	4	6	8	10
ε(°)	0.5	0.5	0.7	1.5	2.0

progressively reduce transverse pressure gradients and radial velocity components generated at the inlet, ensuring these diminish to negligible levels at the throat. The structural configuration and manufacturing precision of the contraction section directly influence divergent section flow uniformity, turbulence control, and boundary layer management. Based on global nozzle design practices, contraction profiles such as double-arc curves, Vitoshinsky curves, bicubic curves, or quintic curves are commonly employed.

The contraction length must balance between minimizing weight/cost (avoiding excessive length) and preventing flow non-uniformity or separation (avoiding insufficient length). A typical contraction length ranges from 0.5~1.0D_0, where D_0 is the inlet diameter. Performance is governed by the contraction ratio $n = D_0/D_t$, (with D_t as the throat diameter) and the curvature profile.

For this design, the Vitoshinsky curve—derived under the assumption of ideal incompressible axisymmetric flow—is adopted. Its governing equation is:

$$r = \frac{r_e}{\sqrt{1-\left[1-\left(\frac{r_e}{r_0}\right)^2\right]\frac{\left(1-\frac{x^2}{l^2}\right)^2}{\left(1+\frac{x^2}{3l^2}\right)^3}}} \tag{5.25}$$

Given the inlet radius r_0, exit radius r_e (throat radius), and contraction length l, Contraction section profile with Vitosinsky curve is basically determined.

Assume a design point flight altitude of ALT = 27.5 km and a flight Mach number of Ma = 4.5. SERN consists of a stable section, a rectangular contraction section, a rectangular expansion section, and a single expansion ramp extension section. The nozzle inlet parameters are provided by the overall engine calculation, and the nozzle width is determined from the engine chamber exit diameter.

SERN geometric and aerodynamic parameters: the SERN throat area at = 1.2084 m^2 was determined from the engine exit flow rate, total temperature, and total pressure at the design point (m_{g8} = 331.23 kg/s, T_8^* = 2000 K, P_8^*=303435 pa), and from the SERN width W_n = 2.366 m, the nozzle throat height, H_t = 0.511 m, and the throat width to height ratio, 4.63, were obtained. According to the environmental parameters, the free flow pressure at 27.5 km altitude is 1712.4 pa, and the nozzle design point drop pressure ratio *NPR* = 177.

The parameters of the SERN critical section were obtained from the above calculations, and the nozzle design was performed to obtain the two-dimensional profile of the SERV, as shown in Fig. 5.9. The upper and lower outer sections were added to the nozzle internal profile, taking into account the effects of the external flow. The structural parameters of the SERV are as follows: the internal expansion ratio A_l/A_t = 1.79, and the external expansion ratio A_e/At = 7.32 (where: A_l is the area at the exit of the lower nozzle ramp, A_e is the area at the exit of the upper nozzle expansion ramp, and A_r is the area of the upper nozzle expansion ramp extension);

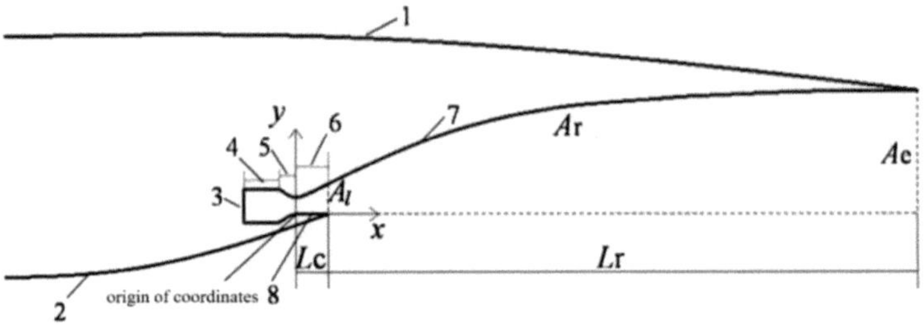

1 upper outer wall of nozzle; 2 Lower outer wall of nozzle; 3 nozzle inlet; 4 stable section; 5 rectangular shrink section; 6 Initial expansion section of upper expansion ramp; 7 upper expansion ramp extension; 8 Lower Expansion Surface Segment

Fig. 5.9 SERN 2D geometry. (1) upper outer wall of nozzle; (2) Lower outer wall of nozzle; (3) nozzle inlet; (4) stable section; (5) rectangular shrink section; (6) Initial expansion section of upper expansion ramp; (7) upper expansion ramp extension; (8) lower expansion surface segment

The ratio of the upper expansion ramp extension length L_r to throat height H_t is 36.32, and the lower expansion ramp length L_C to throat height H_t is 1.98.

5.2.3 Design Method of Tail Nozzle for Parallel TBCC Engine

The SERN external profile adopted by the tail nozzle of the parallel TBCC engine can be well blended with the undersurface of the hypersonic vehicle, allowing the SERN to achieve not only a very high drop pressure ratio, but also good thrust performance within a wide flight envelope. At flight Ma = 3, the turbine engine was shut down and the ramjet began to work alone, and as the flight Mach number continued to rise, the vehicle's requirement for a ramjet drop pressure ratio gradually increased, so the size and profile of the ramjet tail nozzle dictated the tail nozzle for the entire parallel TBCC engine, so it was first necessary to determine the SERN profile of the ramjet with the design point parameters of the ramjet tail nozzle, as shown in Fig. 5.10.

The turbojet nozzle drop pressure ratio is less than the ramjet nozzle, so once the ramjet nozzle profile is determined, simply place the turbojet tail nozzle on the ramjet SERN's upper expansion ramp, as shown in Fig. 5.11, the key is how to determine the relative positions of the turbojet nozzle and the ramjet nozzle. In order to make the mode conversion of parallel TBCC engine go smoothly, the tail nozzle must have good thrust performance at the stage change point. Based on this, an optimal relative position of the turbonozzle and ramjet nozzle is found by numerical simulation, which enables the thrust performance of parallel TBCC engine with tail nozzle to be optimal during mode conversion.

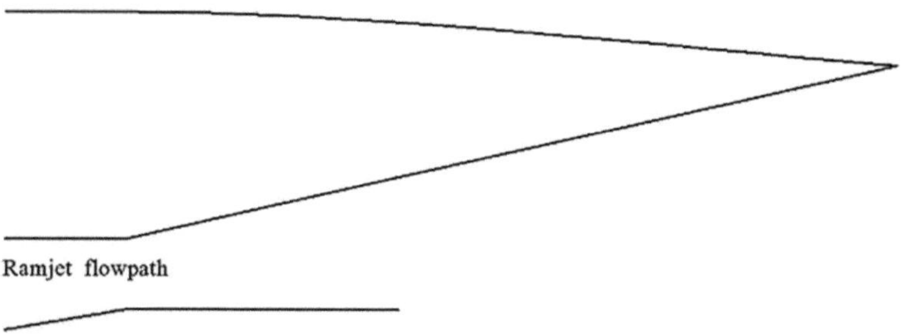

Fig. 5.10 Ramjet nozzle profile diagram

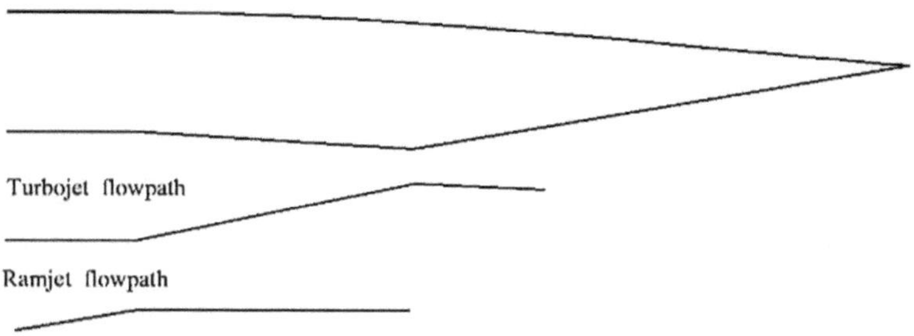

Fig. 5.11 Tail nozzle diagram for a parallel TBCC engine

Within a wide flight envelope, the tail nozzle for a parallel TBCC engine must have an adjustable throat and exit area for good thrust performance. The tail nozzle for a parallel TBCC engine consists of a turbine engine nozzle and a ramjet nozzle, so both the turbine engine nozzle throat and exit area, and the ramjet nozzle throat and exit area must be adjustable. For ramjets, the adjustment of the nozzle throat area is important, which severely affects the ramjet's performance, and the exit area can be automatically adjusted by the SERN's adaptive compensation, so a separate adjustment mechanism is not required; For a turbine engine, the nozzle upper expansion surface is fused with the ramjet nozzle upper expansion surface, so its nozzle configuration is also asymmetric. In order to guarantee the performance of the turbine engine nozzle at high flight Mach numbers, a length L was adapted to the lower wall of the turbine engine tail nozzle, and two adjustable mechanisms were set so that the turbine engine tail nozzle throat was adjustable from the exit area (Zhang et al. 2012a, b), so the adjustment scheme of the tail nozzle for a parallel TBCC engine is shown in Fig. 5.12.

The turbojet nozzle drop pressure ratio is less than the ramjet nozzle, so the design of a tailpipe for a parallel TBCC engine can be accomplished by placing the turbojet tailpipe on the ramjet SERN's upper expansion ramp. As shown in Fig. 5.13, the length of the expansion surface on the ramjet tail nozzle L = 1189.5 mm, the nozzle throat height H_t = 60 mm, and the width W_n = 296.4 mm; A length L_r = 197 mm was adapted for the lower wall of the turbonozzle, and the width of the turbojet was consistent with the ramjet width W_n = 296.4 mm, so the turbonozzle

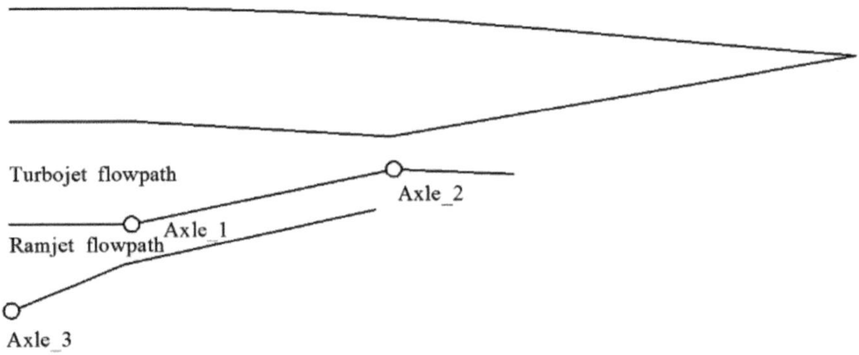

(a) Turbine engines working separately

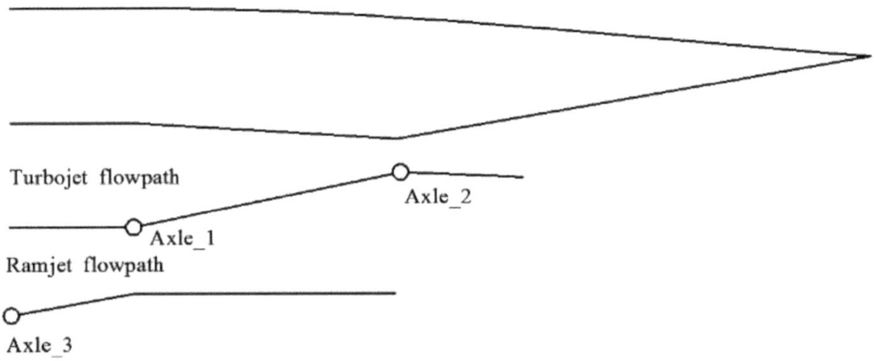

(b) Turbine and ramjet engines work together

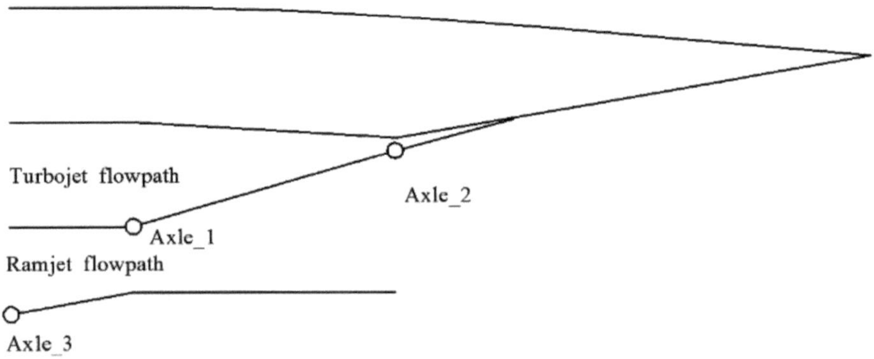

(c) Ramjet operated separately

Fig. 5.12 Tailpipe adjustment scheme for parallel TBCC engines. (**a**) Turbine engines working separately. (**b**) Turbine and ramjet engines work together. (**c**) Ramjet operated separately

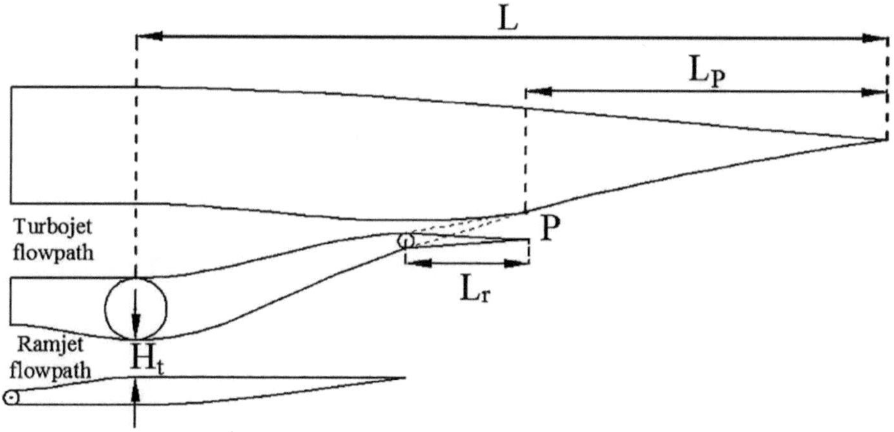

Fig. 5.13 Tail nozzle structure diagram for parallel TBCC engines

throat height was 20.26 mm. Next, reasonably arrange the relative positions of the turbojet nozzle and ramjet nozzle, assuming that the exit of the turbojet nozzle is at the point P of the expansion surface on the ramjet nozzle, the distance from the point P to the end of the expansion surface on the ramjet is L_p, with the value of $(L - L_p)/L$ as a measure of the relative positions of the turbojet nozzle and ramjet nozzle.

Through numerical simulation, an optimal P point position is found, which enables the thrust performance of the tail nozzle of the parallel TBCC engine to be optimal during mode conversion. As shown in Figs. 5.14 and 5.15, the turbonozzle performs optimally at $(L - L_p)/L = 0.62$ and the ramjet tail nozzle performs optimally at $(L - L_p)/L = 0.42$ for thrust. Figure 5.16 gives the axial thrust coefficient of a tailpipe for a parallel TBCC engine at different relative positions, with two peaks in the axial thrust coefficient of a tailpipe for a parallel TBCC engine due to the different thrust performance optimum positions of the turbine nozzle and ramjet nozzle (as shown in Figs. 5.14 and 5.15). From Fig. 5.16, the tail nozzle thrust performance is optimal when $(L - L_p)/L = 0.52$, so the parallel TBCC engine takes $(L - L_p)/L = 0.52$ with the tail nozzle.

5.3 Flow Mechanism of Single Expansion Ramp Nozzle

After the SERN nozzle profile is obtained, this section simulates the flow field of tandem and parallel single expansion ramp nozzles in both design and off design regimes based on CFD techniques, and analyzes the performance and flow characteristics of the designed SERN in both design and off design regimes in detail.

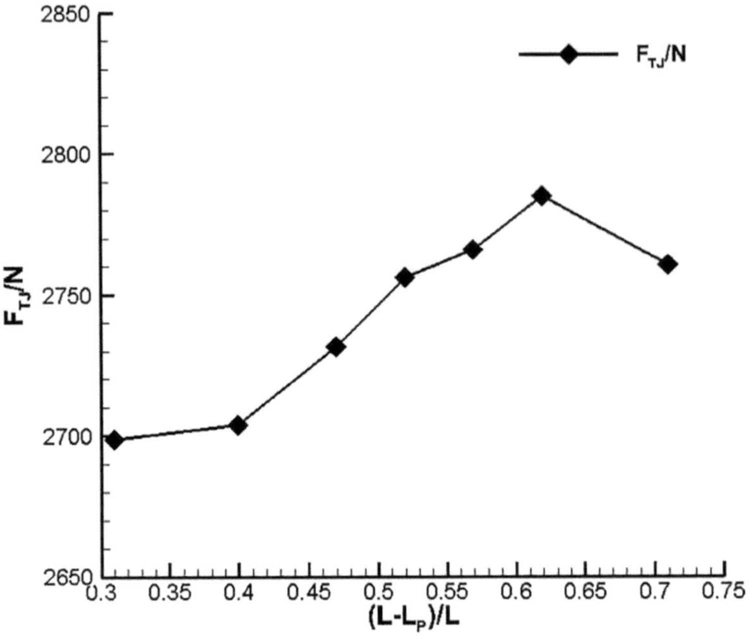

Fig. 5.14 Turbine nozzle axial thrust at different relative positions

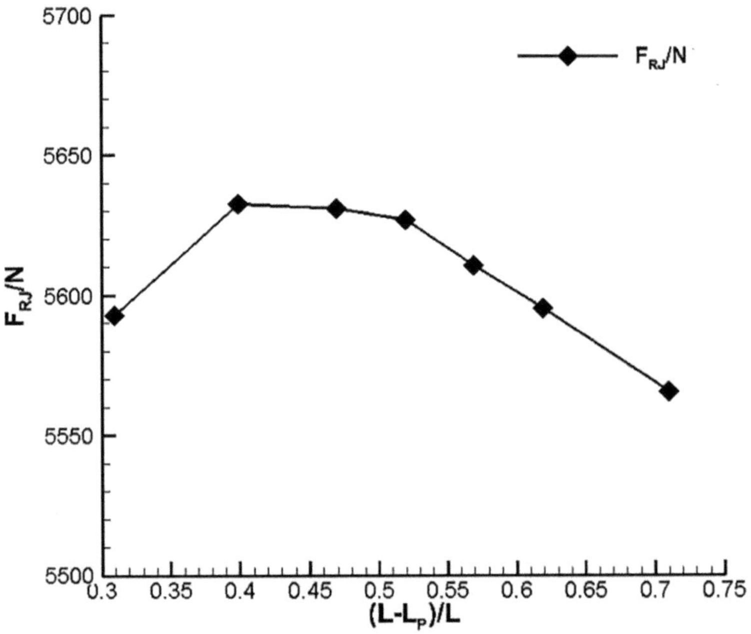

Fig. 5.15 Ramjet axial thrust at different relative positions

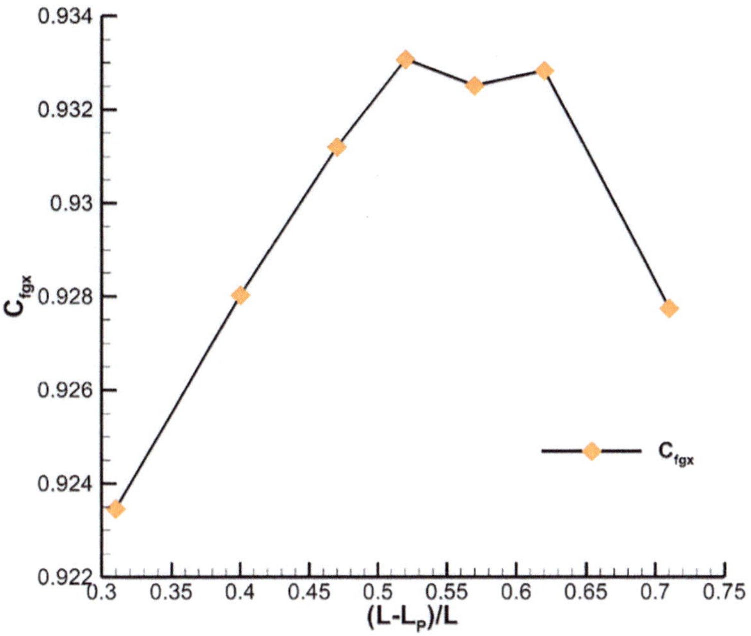

Fig. 5.16 Tail nozzle axial thrust coefficient at different relative positions

5.3.1 SERN Flow Field Characteristics for Tandem TBCC Engines

Numerical simulation of the designed SERN at the design point. Design points for the prototype SERN: NPR_D = 177, Ma = 4.5, Total nozzle inlet temperature T* = 2000 K, flight altitude 27.5 km, ambient static temperature at this altitude T = 222.4 K, total temperature T* = 1131.20 K, static pressure P infinity = 1712.4 Pa. The Mach number distribution and streamline distribution of the SERN at the design point are shown in Fig. 5.17. The nozzle exit velocity direction is parallel to the axial direction, and the ratio of the actual nozzle exit pressure, P, to the ambient pressure, P infinity, varies in the range 0.965–0.998, and the nozzle can be approximated to reach a fully expanded state. The results show that the designed SERN expansion ramp profile is reasonable. Visible from the local flow line at the trailing edge of the nozzle's lower ramp, as shown in Fig. 5.18, the interaction of the nozzle's internal and external flow is visible when the high-speed outflow meets the nozzle's internal flow at the trailing edge of the lower ramp, due to the greater slope of the nozzle's lower outer wall. In the process of hypersonic vehicle exhaust system design, the integrated design of exhaust system and vehicle is a problem to be considered.

The nozzle upper expansion ramp pressure distribution is shown in Fig. 5.19, where the longitudinal coordinate is P/P* and the transverse coordinate is $(X\text{-}X_C)$/

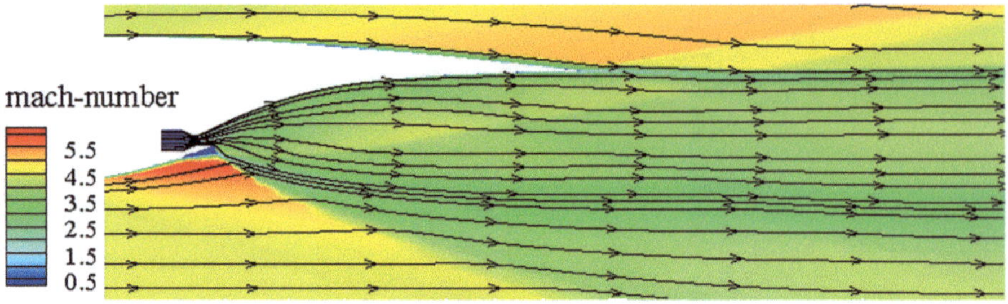

Fig. 5.17 Mach number and streamline chart of SERN at design point

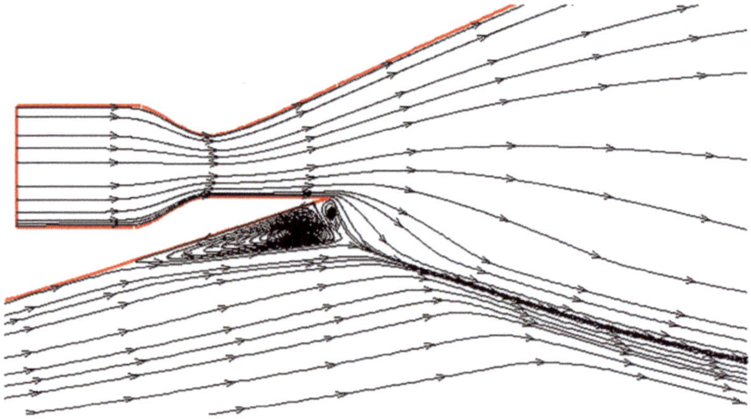

Fig. 5.18 Local streamlines near the trailing edge of the SERN lower ramp

Fig. 5.19 Expansion ramp pressure coefficient distribution on SERN

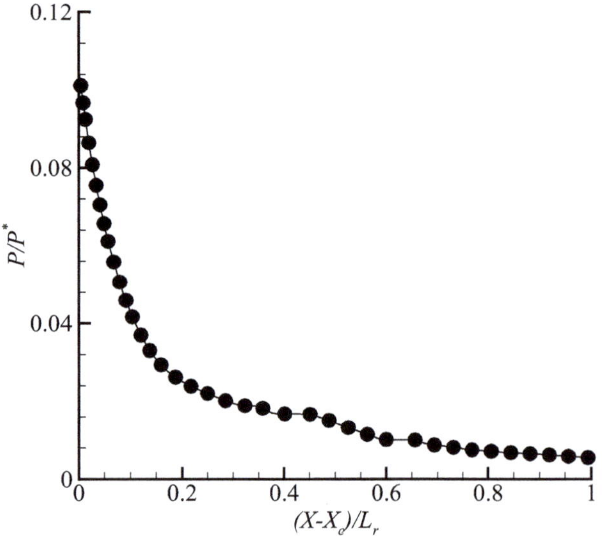

L_r, where X is the abscissa of each point on the upper expansion ramp extension, X_C is the starting coordinate of the upper expansion ramp extension, L_r is the axial length of the upper expansion ramp extension, and the trailing edge point of the lower ramp section is the starting point of the abscissa, dimensionless with the axial length L_r of the upper expansion ramp extension. At the nozzle design point, the airflow expands freely within the nozzle, and the upper expansion ramp pressure gradually decreases to full expansion. At the design point, the expansion ramp pressure distribution on the nozzle is good, and the nozzle performance is good.

Numerical simulations of the internal and external flow fields of the SERN were performed at non design points to further understand the designed SERN. Assuming that the total gas temperature at the nozzle inlet of the vehicle at each flight regime is constant at 2000K, the nozzle inlet total pressure P* is calculated from the engine population, and the nozzle drop pressure ratio P*/P infinity at each flight regime is derived from the ambient pressure P infinity. In other flight regimes, parameters for each flight regime: flight altitude H, flight Mach number Ma, drop pressure ratio *NPR*, inlet total pressure P*, and ambient pressure P infinity are shown in Table 5.2.

The Mach number distribution of the internal and external flow of the nozzle for SERN at flight Mach 2.5, 3.0, 3.5, and 4.0, respectively, is shown in Fig. 5.20. As the flight Mach number gradually decreases, the nozzle drop pressure ratio gradually decreases and moves away from the design drop pressure ratio, and the nozzle enters an over expanded state. As shown in Fig. 5.20a, at this point the nozzle drop pressure ratio differs most from the design drop pressure ratio, and a oblique shock wave is emitted at the trailing edge of the SERN lower ramp, which acts on the upper expansion ramp through the flow field within the nozzle, at which point the shock wave is weak enough to separate the flow at the upper expansion ramp. It can be seen that as the Mach number of flight decreases, the interaction of the nozzle external flow with the flow at the trailing edge of the lower ramp decreases, and the shear layer of the nozzle internal flow with the external flow deflects upwards if the separation region decreases.

The pressure distribution of the expansion ramp on the SERN at different flight Mach numbers is shown in Fig. 5.21, and the dimensionless definition of the abscissa is shown in Fig. 5.19. As the Mach number of flight approaches a design point such as Ma = 4.0, the nozzle internal flow expands gradually along the upper expansion ramp and the pressure decreases gradually. As the flight Mach number gradually moves away from the design point, due to the oblique shock wave action from the trailing edge of the lower ramp of the nozzle acting on the upper expansion ramp,

Table 5.2 Aerodynamic parameters of nozzle in various flight regimes

Flight status	1	2	3	4
H(m)	18300	20700	24700	25900
Ma	2.5	3.0	3.5	4.0
P*	99258	139306	144873	221635
P_0	7140.9	4905.1	2629.3	2185.7
NPR	13.9	28.4	55.1	101.4

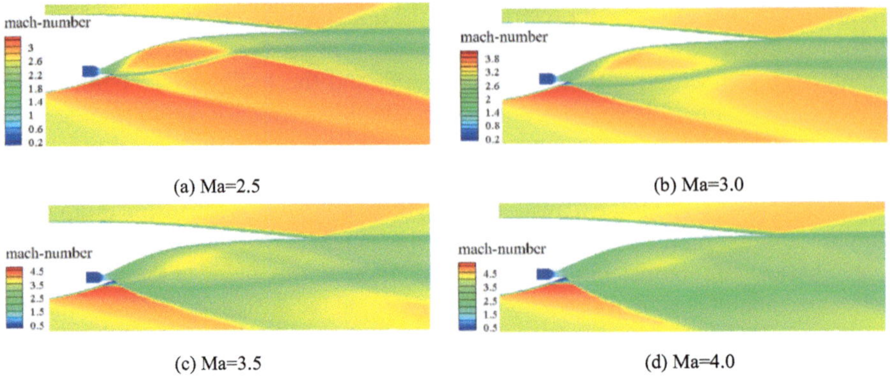

Fig. 5.20 SERN Mach number distribution at different flight Mach numbers. (**a**) Ma = 2.5. (b) Ma = 3.0. (**c**) Ma = 3.5. (**d**) Ma = 4.0

Fig. 5.21 Expansion ramp pressure distribution on nozzle at different flight Mach numbers

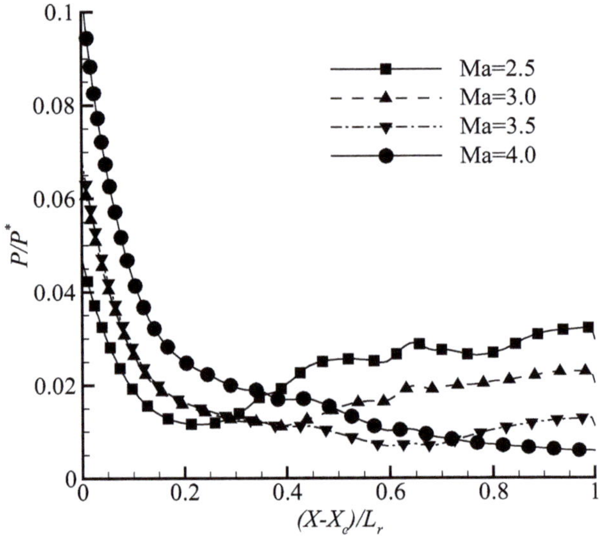

the upper expansion ramp pressure changes from decreasing first to increasing later, and as the flight Mach number gradually decreases, the position of the pressure rise point of the upper expansion ramp moves upstream of the nozzle.

Figures 5.22 and 5.23 show changes in nozzle flow coefficient and axial thrust coefficient at different flight Mach numbers, respectively. It can be seen that the flow coefficient of the nozzle remains essentially unchanged at different flight Mach numbers. The axial thrust coefficient of the nozzle nozzle drops sharply with the decrease of flight Mach number, which is 0.7233 at Mach 2.5. Because as the flight

Fig. 5.22 Changes in C_d at different flight Mach numbers

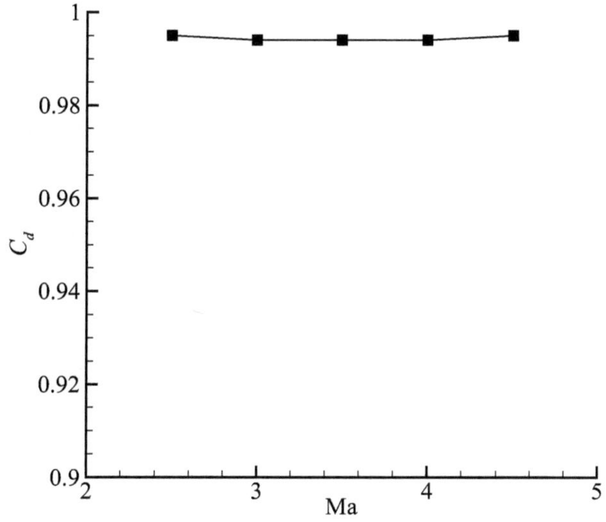

Fig. 5.23 Changes in C_{fgx} at different flight Mach numbers

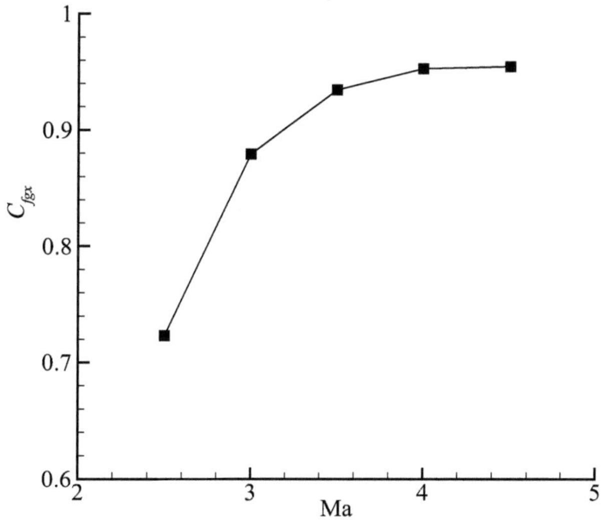

regime moves away from the design regime, the nozzle drop pressure ratio decreases, and the internal flow is in an over expanded state, reducing the pressure integral on the upper expansion ramp of the nozzle, which is an important factor in affecting the nozzle axial thrust coefficient, thus causing the nozzle axial thrust coefficient to decrease.

5.3.2 Flow Field Characteristics of the Tail Nozzle for the Parallel TBCC Engine

In this section, the tail nozzles of a parallel TBCC engine operating at typical flight regimes of Ma = 0.5 (ramjet gradually opens), Ma = 0.9 (ramjet starts firing), Ma = 1.3, Ma = 2, and Ma = 3 (TBCC tail nozzle stage change point) are simulated numerically, and the tail nozzle flow characteristics at different flight regimes are analyzed.

1. Mach number flown Ma = 0.5
   Figure 5.24a shows a Mach number flow field diagram of a tailpipe for a parallel TBCC engine at flight Ma = 0.5, in which the turbonozzle has a drop pressure ratio of $NPR_T = 6.08$, the turbonozzle is in an over expansion operating condition, and a shock wave is generated at the trailing edge of its lower wall, which passes through the flow field and intersects the upper expansion surface of the turbonozzle, and interacts with the upper expansion surface boundary layer causing the boundary layer to separate, with the front end of the separation zone inducing a separation shock wave, the presence of which deflects the flow, as shown in Fig. 5.24b. As can be seen from Fig. 5.24a, the structure of a shock train occurs within the turbine nozzle, due to the flow being in a state of severe overexpansion, which continues to expand after shock compression, followed by recompression and reexpansion, such that the reciprocating process forms a shock train. At the same time, the ramjet did not fire, and to reduce drag the ramjet was left open, at which point the ramjet's drop pressure ratio $NPR_R = 1.09 < 1.893$, but supersonic flow phenomena occurred within the ramjet due to the ejecting action of the turbojet flow. As can be seen from Fig. 5.24b, the excessive expansion angle of the ramjet exit gives a large area of separation zone to the expansion surface on the nozzle, but the ejector action of the turbine nozzle causes the flow within the ramjet to quickly rendezvous with the flow within the turbine nozzle and flow to the rear (Jiang 2006; Doyle Knight et al. 2003).
2. Mach number flown Ma = 0.9
   Figure 5.25a shows the Mach number flow field diagram of the tail nozzle for a parallel TBCC engine at flight Ma = 0.9, in which the drop pressure ratio NPR_T

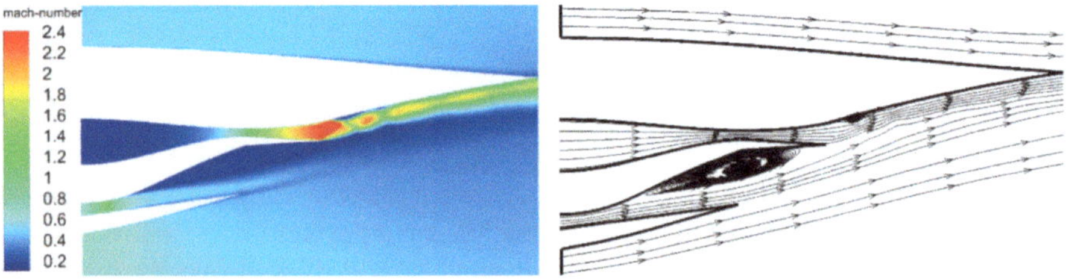

(a) Mach number cloud chart (b) Streamlined graph

Fig. 5.24 Flow field diagram of the tail nozzle at Mach = 0.5 in flight. (**a**) Mach number cloud chart. (**b**) Streamlined graph

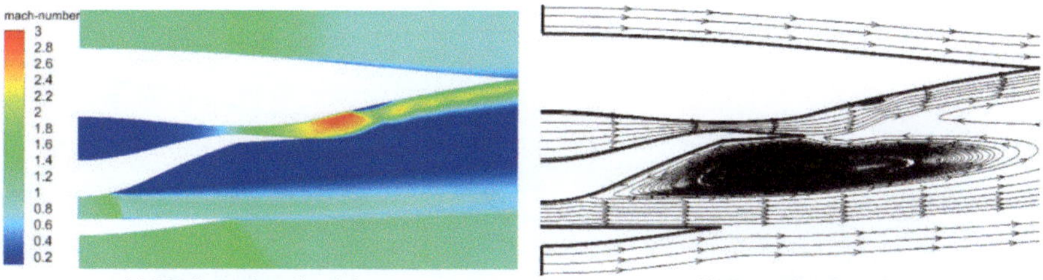

(a) Mach number cloud chart (b) Streamlined graph

Fig. 5.25 Flow field diagram of the tail nozzle at Mach = 0.9 in flight. (**a**) Mach number cloud chart. (**b**) Streamlined graph

of the turbine nozzle is = 8.36, and the degree of over expansion of the flow within the turbine nozzle is reduced, the shock angle is reduced, and the separation zone and separation shock shock waves are gradually away from the throat due to the increase in the drop pressure ratio relative to the Ma = 0.5 regime. At the same time, the ramjet was fully open in preparation for the ramjet ignition, at which point the drop pressure ratio changed little NPR_R = 1.43 relative to the Ma = 0.5 state, but because the ramjet was fully open, the nozzle exit expansion angle was further increased, and the separation zone of the upper expansion surface was enlarged (as in Fig. 5.25b), resulting in a difficult confluence of the turbojet flow and ramjet flow.

3. Mach number flown Ma = 1.3
   Figure 5.26a shows a Mach number flow field diagram of a tailpipe for a parallel TBCC engine at flight Ma = 1.3. As the flight Mach number continues to increase, the drop pressure ratio (NPR_T = 13.74) of the turbine nozzle increases, the degree of over expansion of the flow within the nozzle further decreases, the angle of the shock wave generated by the over expansion decreases, the shock wave intersects the upper expansion surface of the nozzle and is reflected downstream, the separation zone largely disappears, and the shock train structure within the nozzle due to constant expansion compression also disappears. At the same time, the ramjet had ignited, and the ramjet drop pressure ratio NPR_R = 2.28, relative to the Ma = 0.9 regime, the degree of over expansion of the flow within the ramjet was reduced by the increase in the drop pressure ratio, and the separation zone of the expansion ramp on the ramjet was significantly reduced. Inside the ramjet, the shock wave generated by over expansion intersects the separation shock, allowing the flow within the nozzle to converge, as shown in Fig. 5.26b, and the ramjet flow rapidly rendezvous with the turbine nozzle flow after exiting due to the ejector action of the turbine nozzle.
4. Mach number flown Ma = 2
   Figure 5.27a shows a Mach number flow field diagram of the tail nozzle for a parallel TBCC engine at flight Ma = 2. As the flight Mach number continues to increase, the drop pressure ratio (NPR_T = 23.71) of the turbine nozzle increases, further reducing the degree of over expansion of the flow within the nozzle, essentially similar to Fig. 5.26a, except that the shock position is further aft. At

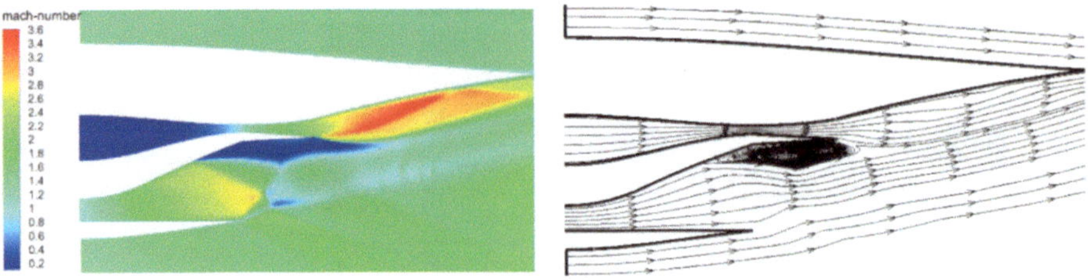

(a) Mach number cloud chart (b) Streamlined graph

Fig. 5.26 Flow field diagram of the tail nozzle at Mach = 1.3 in flight. (**a**) Mach number cloud chart. (**b**) Streamlined graph

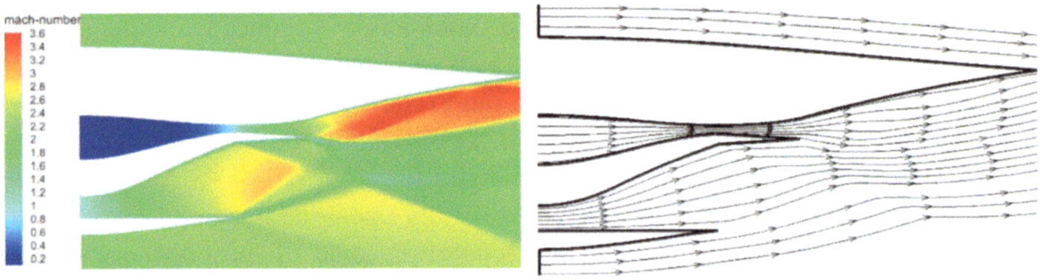

(a) Mach number cloud chart (b) Streamlined graph

Fig. 5.27 Flow field diagram of the tail nozzle at Mach = 2 in flight. (**a**) Mach number cloud chart. (**b**) Streamlined graph

the same time, the ramjet was in working condition, with the ramjet's drop pressure ratio NPR_R = 5.81, relative to the Ma = 1.3 condition, the degree of over expansion of the flow in the ramjet decreased due to the increase in the drop pressure ratio, and the separation zone of the expansion ramp on the ramjet basically disappeared. A shock wave due to over expansion on the lower wall of the ramjet intersects a shock wave due to a sudden change in profile on the upper wall, allowing the flow within the ramjet to pool, as shown in Fig. 5.27b.

5. Mach number flown Ma = 3
   Figure 5.28a shows a Mach number flow field diagram of the tail nozzle for a parallel TBCC engine at flight Ma = 3, with the drop pressure ratio (NPR_T = 30.30) of the turbine nozzle increasing as the flight Mach number continues to increase, and the flow within the nozzle expands almost completely. At the same time, the ramjet was in operation, with the ramjet's drop pressure ratio NPR_R = 20.82, and only shock waves due to sudden changes in the profile were present within the ramjet, as shown in Fig. 5.28, with no significant shock waves and separation zones throughout the flow field.

Figures 5.29 and 5.30 show the pressure distribution on the expansion surfaces on the turbojet nozzle and on the ramjet nozzle under different flight regimes, respectively, where the abscissa is the axial distance from each point on the upper wall of

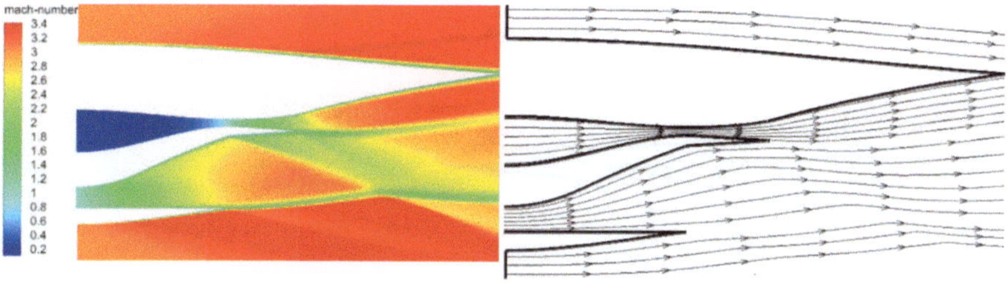

(a) Mach number cloud chart (b) Streamlined graph

Fig. 5.28 Flow field diagram of the tail nozzle at Mach = 3 in flight. (**a**) Mach number cloud chart. (**b**) Streamlined graph

the turbojet and ramjet to the ramjet throat, and the ordinate is the dimensionless pressure on the upper wall of the turbojet (ratio of the static pressure on the upper wall of the turbojet to the total pressure at the inlet of the turbojet) and the dimensionless pressure on the upper wall of the ramjet (ratio of the static pressure on the upper wall of the ramjet to the total pressure at the inlet of the ramjet), respectively. For a turbine nozzle, as the Mach number of flight increases, the nozzle's drop pressure ratio increases, the degree of over expansion of the flow within the nozzle decreases, and the shock wave generated by the over expansion is gradually pushed out of the nozzle, with the location of the pressure spike moving backwards until at Ma = 3, the flow within the nozzle expands almost completely, with no pressure spike occurring. As can be seen from Fig. 5.29, multiple surges of pressure at the expansion surfaces on the turbine nozzles occurred at flight regimes of Ma = 0.5 and Ma = 0.9, which were caused by shock trains, which disappeared after Ma = 1.3 as the degree of over expansion diminished. For ramjets, as the Mach number of flight increases, the drop pressure ratio of the nozzle increases, the degree of over expansion of the flow within the nozzle decreases, the shock wave generated by the over expansion gradually moves back, and the location of the pressure spike moves back. From Fig. 5.30, it can be seen that the pressure distributions of Ma = 2 and Ma = 3 almost coincide, which is mainly due to a reduction in the degree of over expansion, the shock wave being pushed outside the ramjet, and the pressure surges of both Ma = 2 and Ma = 3 are caused by the shock wave generated by a sudden change in the profile.

Changes in the flow coefficient C_d, axial thrust coefficient C_{fgx}, and thrust vectoring angle delta-p of the tailpipe for a parallel TBCC engine in different flight regimes are shown in Fig. 5.31. As can be seen from the figure, as the Mach number of flight increases, the flow coefficient of the tail nozzle basically remains above 0.98, and both the axial thrust coefficient and thrust vectoring angle decrease first and then increase. Combined with Mach number cloud charts and streamlines, the ramjet nozzle opens gradually as flight Ma increases from 0.5 to 0.9, and the separation zone of the expansion surface on the ramjet nozzle increases gradually, thus reducing the nozzle axial thrust coefficient from 0.808 to 0.668, and the thrust vector angle from -18.57° to −46.85°; As the flight Ma increases to 1.3, although the

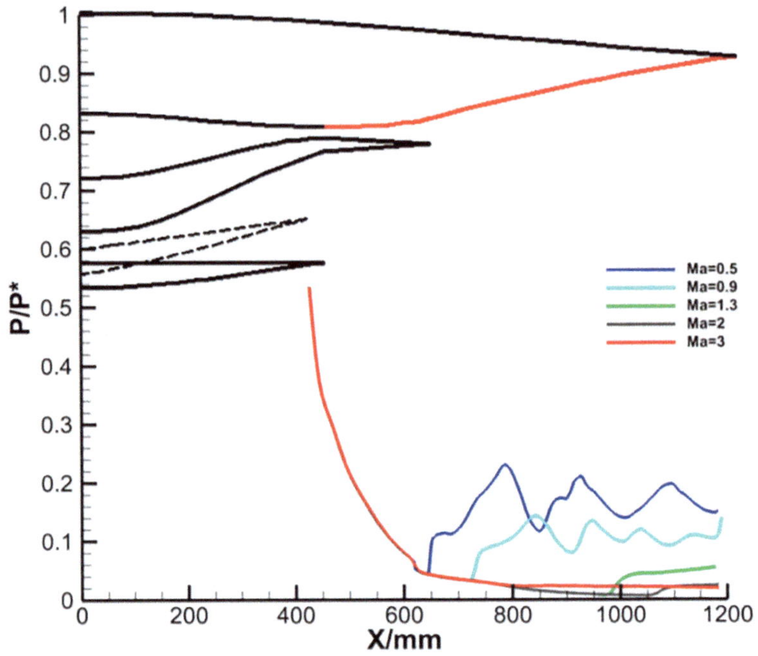

Fig. 5.29 Pressure distribution on the expansion surface of the turbine nozzle

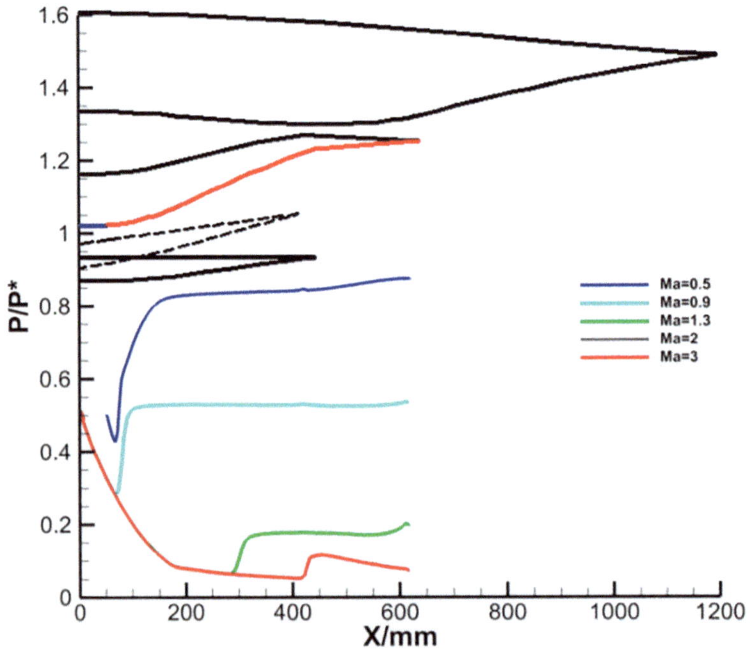

Fig. 5.30 Expansion Surface Pressure Distribution on Ramjet

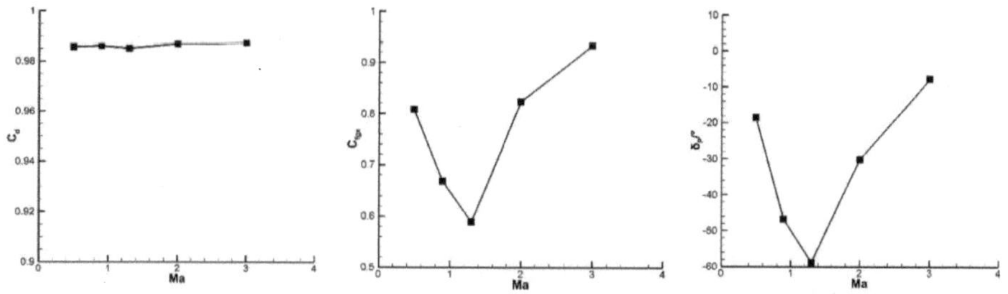

(a)Changes in C_d (b)Changes in C_{fgx} (c)Changes in δ_p

Fig. 5.31 Changes in C_d, C_{fgx}, and δ_p under different flight regimes. (**a**) Changes in C_d. (**b**) Changes in C_{fgx}. (**c**) Changes in δ_p

separation zone inside the nozzle is significantly reduced, the increase in the Mach number of the external flow causes a large increase in the nozzle drag, which in turn affects the overall nozzle performance; As flight Ma increased from 1.3 to 3, the nozzle drop pressure ratio increased, the nozzle over expansion decreased, the separation zone disappeared, and the nozzle performance increased gradually, the nozzle axial thrust coefficient increased from 0.589 to 0.934, and the thrust vector angle from −58.82° to −7.79°.

From the above analysis, it can be seen that the performance of the tail nozzle for the parallel TBCC engine deteriorated dramatically in subsonic and transonic flight regimes, so further performance improvements are urgently needed.

5.4 SERN Performance Improvement for Tandem TBCC Engines Based on an Active Flow Control Approach

In recent years, scholars at home and abroad have conducted a number of exploratory studies in response to the problem of transonic performance degradation of the SERN for tandem TBCC engines, and have proposed a corresponding mechanical regulation method, that is, using the mechanical structure to adjust the throat position and area ratio of the nozzle to improve the performance of the nozzle under different operating conditions. Although the mechanical method can achieve an improvement in nozzle performance, the size of the SERN is large due to the application of air breathing hypersonic flight techniques to new generation vehicles and hypersonic transport aircraft, and there are many problems with its cooling and reliability. Therefore, researchers in various countries make full use of the flow control theory of fluid mechanics, hoping to improve the transonic performance of SERN with flow control techniques. There are currently two main categories of aerodynamic methods, active control and passive control, in which passive control includes the ventilated ejector method (Zhang et al. 2012a, b) and the passive cavity control method (Jürgen and Alfred 1998); Active controls are the external combustion method (Yungster and Trefny 1994) and the secondary flow injection method (Eric and Dan 2004). This section investigates the flow mechanism of active flow control based on secondary flow injection for performance improvement in the SERN over expansion operating regime, and discusses the effects of secondary flow injection geometric and aerodynamic parameters on nozzle performance.

5.4.1 SERN Flow Mechanism for Tandem TBCC Engines with Secondary Flow Injection

1. Geometric model and dimensions

The nozzle studied in this section was selected from the Langley Research Center's wind tunnel test nozzle (Capone et al. 1992), which had an external area ratio of 2.32, a design point external drop pressure ratio of 13.77, an internal area ratio of 1.47, a design point internal drop pressure ratio of 6.00, a nozzle length to throat height ratio of 15.03 L_n/H_t, an upper expansion ramp (Ramp) length to throat height ratio of 6.06 L_r/H_t, and a lower ramp (Cowl) length to throat height ratio of 1.21 L_c/H_t. In order to investigate the effect of secondary flow injection on the SERN performance, a converging secondary flow nozzle was opened in the lower oblique plate. The SERN shape is shown in Fig. 5.32, in which the trailing edge of the lower oblique plate was locally enlarged.

The main control parameters for secondary flow injection are: the relative position of the secondary flow nozzle (expressed in X_L, defined as the ratio of the secondary flow injection position to the length of the lower wall of the turbine nozzle), the angle of secondary flow injection (expressed in α, defined as the angle

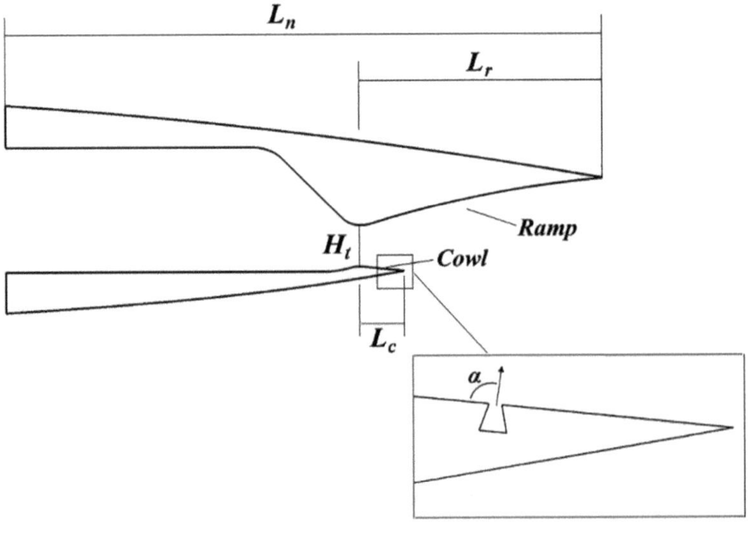

Fig. 5.32 Structure diagram of nozzle studied

between the direction of secondary flow injection and the upstream wall of the secondary flow nozzle), the ratio of secondary flow injection flow (expressed in W, defined as the ratio of secondary flow rate to the main flow rate of the turbine nozzle), and the ratio of secondary flow injection pressure (expressed in *SPR*, defined as the ratio of the total pressure of the secondary flow inlet to the total pressure of the turbine nozzle inlet), etc.

2. Numerical results and analysis of the SERN flow field with secondary flow injection

The performance of the nozzle is strongly related to the separation zone as well as the location of the separation zone when the SERN enters the over expansion regime, and Fig. 5.33 gives a map of Mach number and pressure distribution in the over expansion regime of the prototype nozzle under study. It can be seen that the flow inside the nozzle is obviously in an over expansion state at low drop pressure ratio, the trailing edge of the lower oblique plate will produce a oblique shock wave, and a large area of separation zone occurs at the upper expansion ramp, resulting in a sharp decrease in nozzle performance.

The nozzle performance is improved by ejecting secondary flow, with Mach number and pressure profiles of the nozzle at secondary flow injection shown in Fig. 5.34, and a scaled up view of the flow field near the jet orifice shown in Fig. 5.35. It can be seen that when the secondary flow is injected into the main stream, the pressure of the jet is greater than that of the surrounding airflow, and a series of expansion waves are formed near the jet orifice, the jet expands rapidly after passing through the expansion wave system, the over expanded jet is recompressed by the surrounding cylindrical shock wave and Mach disk, the airflow continues to expand after passing through the Mach disk, and is blown downstream by the main stream. Due to the injection of the secondary flow, the

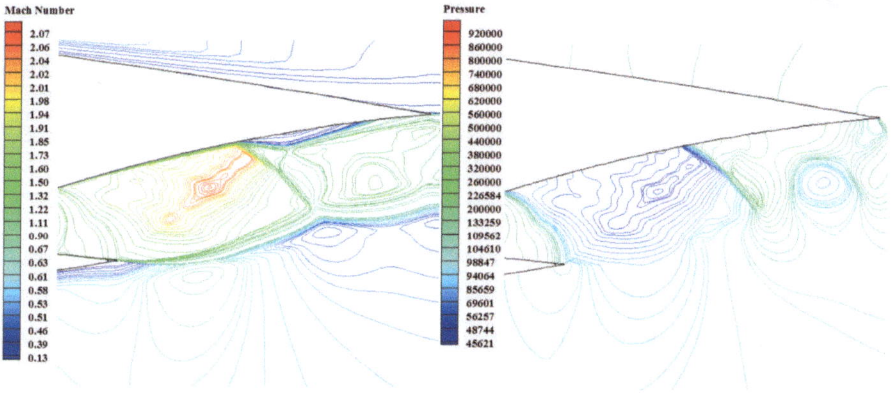

(a) Mach number distribution (b) pressure distribution

Fig. 5.33 Mach number and pressure distribution at over expansion of prototype nozzle. (**a**) Mach number distribution. (**b**) Pressure distribution

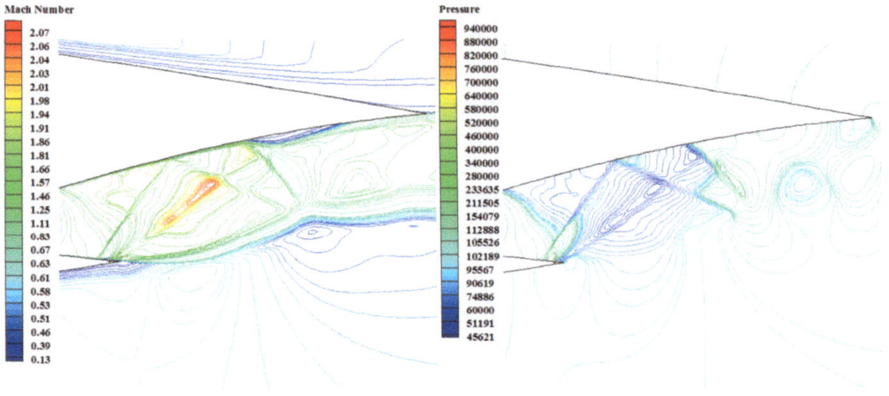

(a) Mach number distribution (b) pressure distribution

Fig. 5.34 Mach number and pressure distribution of nozzle in secondary flow injection regime. (**a**) Mach number distribution. (**b**) Pressure distribution

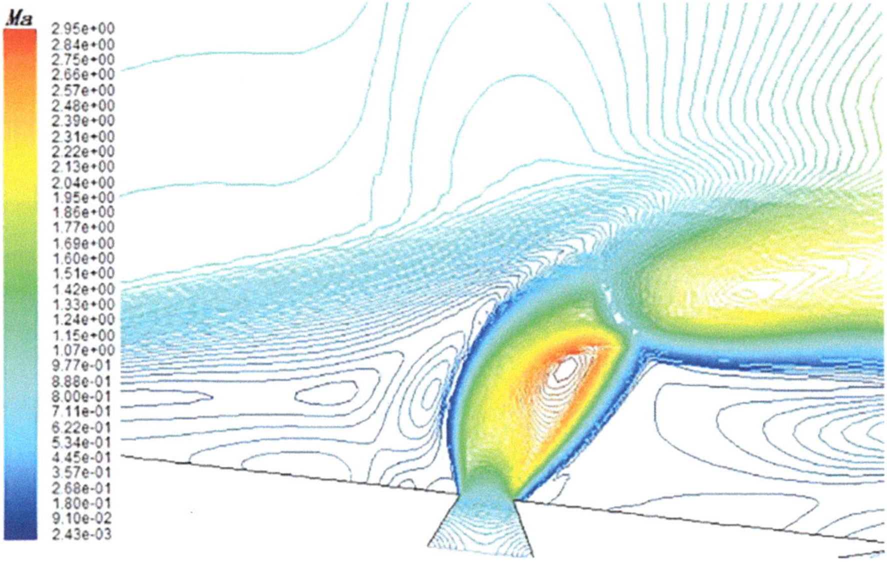

Fig. 5.35 Flow field structure near the jet jet orifice

main flow is blocked upstream of the secondary flow nozzle, forming a bow shock upstream of the secondary flow nozzle, which increases the pressure ahead of the secondary flow nozzle, and the disturbance propagates upstream through the wall boundary layer, creating a wedge like separation region near the wall upstream of the secondary flow nozzle, which in turn blocks the incoming flow, forming a separation shock, which intersects to form a "λ wave".

The mechanism of boundary layer separation caused by shock wave interaction with the boundary layer is analyzed by combining the literature and the calculation results in this section. When the shock wave intersects the wall boundary layer, the normal shock strength along the wall is different, and the pressure change before and after the shock wave is different, and the normal pressure gradient within the boundary layer is no longer zero. This pressure imbalance increases the thickness of the boundary layer before and after the shock wave, which provides conditions for boundary layer separation. In the subsonic region of the boundary layer, the pressure rise behind the shock wave is transmitted upstream, which makes the flow in the subsonic region backflow upstream; But in the supersonic flow outside the boundary layer, changes in downstream parameters are not transmitted to the upstream, and the flow is still flowing downstream, at which point the flow velocity in the boundary layer is distributed from negative to positive along the normal direction of the wall, which causes the flow vortices to separate from. The separation point is at some point upstream of the shock, and the closer to the wave surface, the thicker the boundary layer.

Figure 5.36 shows the change of pressure on the upper expansion ramp before and after secondary flow injection, corresponding to Fig. 5.34. It can be seen from Fig. 5.34 that ejecting the secondary flow on the lower oblique plate introduces

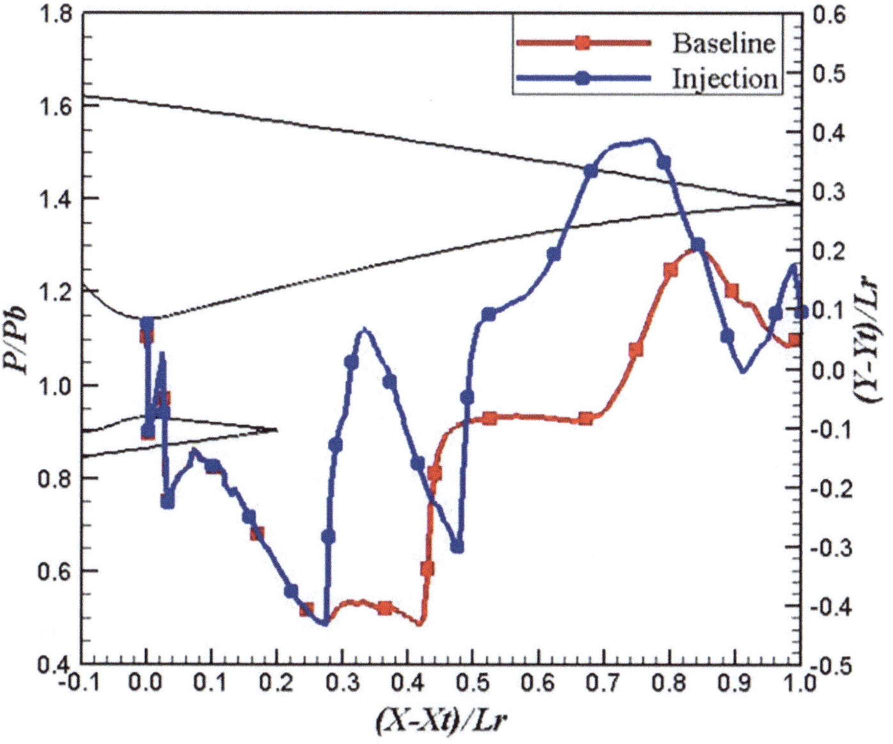

Fig. 5.36 Effect of secondary flow injection on pressure distribution on the upper expansion ramp

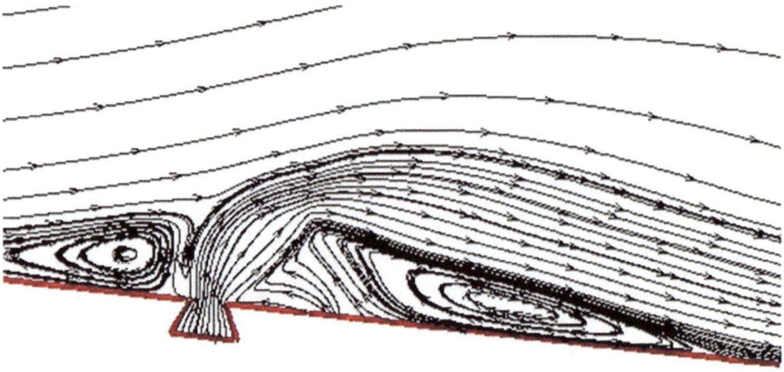

Fig. 5.37 Streamline diagram near the secondary flow jet orifice

a shock wave upstream of the injection position, which interacts with the boundary layer on the upper expansion wall, raising the pressure in this region; In addition, secondary flow injection causes the pressure to increase in the region downstream of the upper expansion ramp. The increase in the upper expansion wall pressure is responsible for the improved nozzle performance, and is a focus of this section's study. The change in the upper expansion wall pressure is discussed in more detail in the latter two sections.

Figure 5.37 gives a flow line diagram near the secondary flow jet orifice, as seen when the secondary flow jet orifice is further from the trailing edge of the lower ramp. At this time, the secondary flow cannot attach to the wall, but at some point downstream of the jet hole, the secondary flow is compressed by the main flow and then attaches to the wall, which results in the formation of a separation zone downstream of the secondary flow. It can be seen that within the separation zone downstream, there is only one vortex rotating clockwise. The size and relative position of the vortices vary with the change of secondary flow injection parameters. Separation occurs in the boundary layer upstream of the secondary flow, and the separation region contains two counter rotating separation vortices, the first immediately adjacent to the secondary flow nozzle, rotating counterclockwise, and the influence region is very small; The second, located upstream of the first, rotates clockwise and has a very large area of influence, which is a major feature of the flow field in the upstream separation zone.

Figure 5.38 gives the wall pressure distribution of the lower oblique plate, where Fig. 5.38 corresponds to Fig. 5.37. It can be seen that a high-pressure region forms upstream of the nozzle due to the interaction of the boundary layer with the shock wave; Due to the intense expansion of the gas and jet ejection, there is a low-pressure region downstream of the nozzle where the pressure is almost constant, but as the flow continues downstream, the pressure begins to rise and the low-pressure region ends. The combined action of high and low pressure regions is the characteristic of the pressure distribution near the secondary flow.

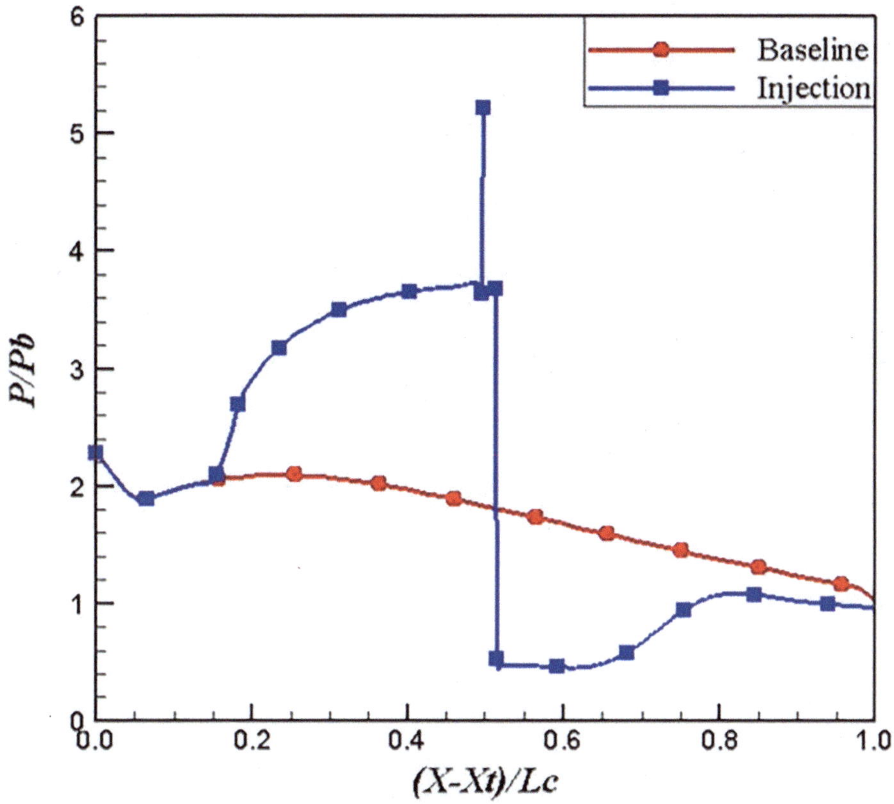

Fig. 5.38 Pressure Distribution on the Lower Slanted Plate Wall

5.4.2 *Effects of Geometric Parameters on SERN Performance for Tandem TBCC Engines*

1. Effect of secondary flow injection angle on SERN flow field and performance

At the lower ramp of the nozzle, the secondary flow is injected, and the injection flow ratio W = 1.93%, the injection total pressure ratio *SPR* = 1, the injection position $X_L = 0.868$, and the outflow Mach number Ma = 0.6 are kept constant during injection; At *NPR* equal to 3.97 and 7.00 respectively, adjust the secondary flow nozzle to make the injection angle α equal to 30°, 45°, 75°, 90°, 105°, 135°, and study the effect of secondary flow injection angle on SERN performance.

Figures 5.39 and 5.40 show the nozzle Mach number distribution at different injection angles, and Fig. 5.41 shows the pressure distribution at the upper expansion ramp at different injection angles, and it can be seen that at *NPR* = 3.97, with the injection angle port reduced from 135 ° to 45 °, the shock becomes stronger and the shock angle increases, expanding the separation zone produced by the shock

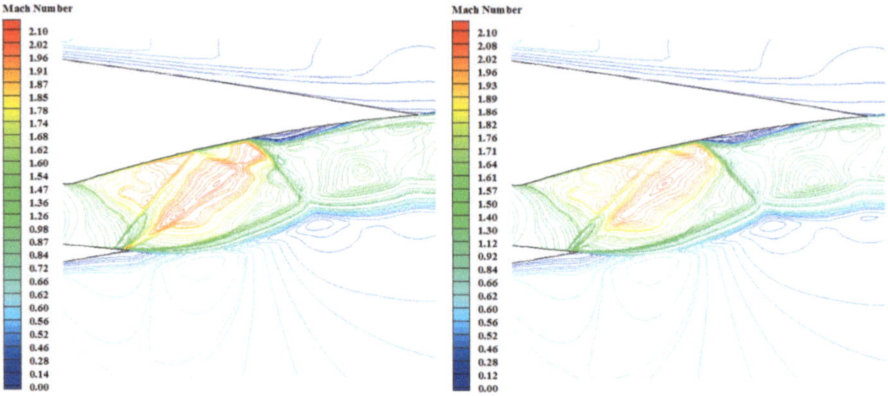

(a)α=45°,*NPR*=3.97 (b)α=135°,*NPR*=3.97

Fig. 5.39 Mach number distribution at different injection angles at *NPR* = 3.97. (**a**) α = 45°, *NPR* = 3.97. (**b**) α = 135°, *NPR* = 3.97

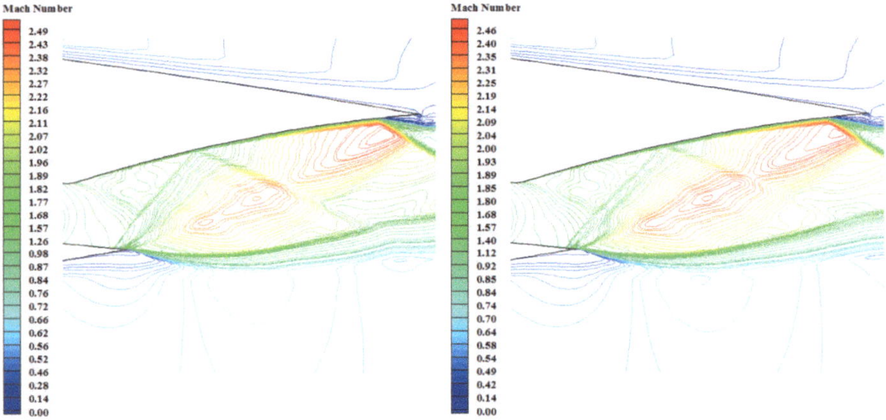

(a)α=45°,*NPR*=7.00 (b)α=135°,*NPR*=7.00

Fig. 5.40 Mach number distribution at different injection angles at *NPR* = 7.00. (**a**) *α* = 45°, *NPR* = 7.00. (**b**) α = 135°, *NPR* = 7.00

interacting with the boundary layer at the lower ramp as the secondary flow causes the shock to move forward. The interaction region between the shock and the boundary layer at the upwelling expansion ramp expands and moves upstream. the change of the upstream interference region affects the change of the upwelling expansion ramp pressure, making the peak of the upwelling high-pressure region larger and moving upstream; The downstream separation zone of the upper expansion ramp decreases slightly with the enlargement of the upstream separation zone, and the downstream pressure distribution increases slightly.

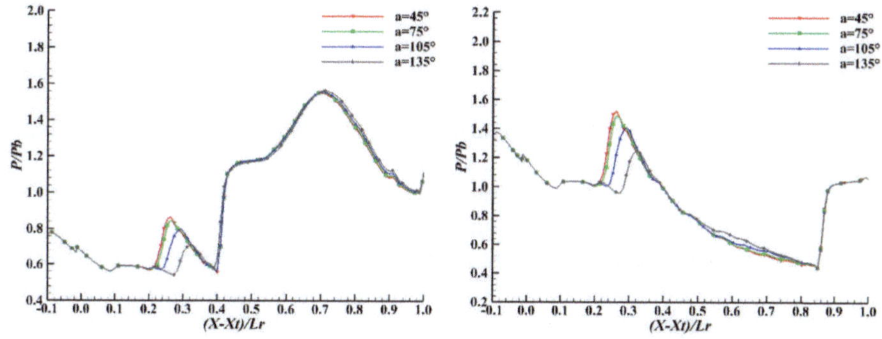

(a) *NPR*=3.97 (b) *NPR*=7.00

Fig. 5.41 Upper expansion wall pressure distribution at different ejection angles. (**a**) *NPR* = 3.97. (**b**) *NPR* = 7.00

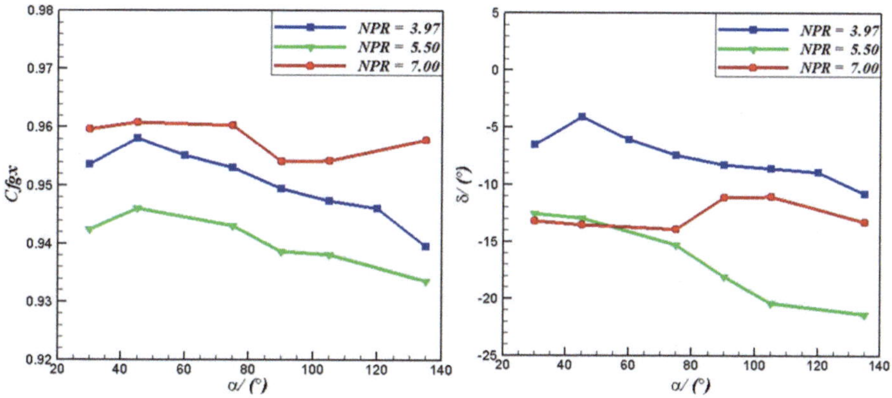

Fig. 5.42 Effect of secondary flow injection angle change on nozzle performance

At *NPR* = 7.00, as the injection angle port decreases from 135° to 45°, the shock generated by the secondary flow, as well as the change in the interference region and pressure distribution upstream of the upper expansion ramp, are similar to those of the above two *NPR*s; In terms of separation zone and pressure distribution downstream of the upper expansion ramp, when α decreases from 135° to 45°, the separation zone at the trailing edge is less affected, and the pressure within this zone is almost constant.

Figure 5.42 shows the effect of the injection angle on nozzle performance, with the axial thrust coefficient of the nozzle increasing from 0.9396 to 0.9581 as α decreases from 135° to 45° at *NPR* = 3.97, while the thrust vector angle increases from −10.79° to −4.1°, the non axial force in the direction down decreases, the deflection moment decreases, and the nozzle performance improves, and the nozzle performance reaches its optimum when $\alpha = 45°$, and begins to decline again when α continues to decrease to 30°.

At *NPR* = 7.00, the change rule of nozzle performance with α is not obvious, and the change of performance is closely related to the change of separation zone at the trailing edge of the upper expansion ramp. The maximum value 0.9608 and minimum value 0.9541 of the nozzle's axial thrust coefficient occur at muzzle $\alpha = 45°$ and $\alpha = 90°$, respectively, and the maximum value -11.05° and minimum value -13.9° of the thrust vector angle occur at $\alpha = 105°$ and $\alpha = 75°$, respectively.

Considering the change of nozzle performance with injection angle at different *NPR*s, the following conclusions are drawn: the maximum values of the axial thrust coefficient of the nozzle appear at $\alpha = 45°$ at all three *NPR*s; At *NPR* = 3.97, the maximum thrust vectoring angle of the nozzle occurs at $\alpha = 45°$, and at *NPR* = 5.50, the thrust vectoring angle of the nozzle at $\alpha = 45°$ is very close to the maximum. It can be considered that the nozzle performs optimally at $\alpha = 45°$.

2. Effect of secondary flow injection location on SERN flow field and performance

The nozzle lower ramp ejected the secondary flow, keeping the injection flow ratio W = 3.93%, the injection total pressure ratio *SPR* = 2, the injection angle $\alpha = 90°$, and the outflow Mach number Ma = 0.6 constant while ejecting: adjusting the secondary flow nozzle so that the injection position X_L was equal to 0.934, 0.868, 0.769, 0.636, 0.504, and 0.339, respectively, at *NPR* = 3.97 and 7.00, studied the effect of the secondary flow injection position on the SERN performance.

Figures 5.43 and 5.44 show the nozzle Mach number distribution at different injection positions, Fig. 5.45 shows the pressure distribution at the upper expansion ramp at different injection positions, and it can be seen that at *NPR* = 3.97, as the secondary flow injection position moves upstream, X_L gradually changes from 0.934 to 0.504, the shock wave generated by the secondary flow injection moves upstream, and the interference region induced by the shock wave at the upper expansion ramp moves upstream. The change in the interference region affects the change in the upper expansion ramp pressure, causing the peak in the upstream high-pressure region to become larger and move upstream: in the range of X_L from 0.934 to 0.769, the separation region downstream of the upper expansion ramp The distribution of and pressure changes little, but when X_L changes to 0.636, the downstream separation region of the upwelling expansion ramp suddenly expands, and the pressure in the downstream region drops significantly; When X_L was reduced to 0.339, the secondary flow injection began to affect the flow at the nozzle throat, and the shock wave strength upstream of the jet orifice was relatively weak, which resulted in essentially the same flow field structure at the three *NPR*s, which differed greatly from the flow field structure studied in this section and was not discussed in detail.

At *NPR* = 7.00, as the secondary flow injection position moves upstream, the shock wave generated by the jet, as well as the change of the interference region and pressure distribution upstream of the upper expansion ramp, are similar to those of the above two *NPR*s; In terms of separation zone and pressure distribution downstream of the upper expansion ramp, when X_L equals 0.868, a larger sepa-

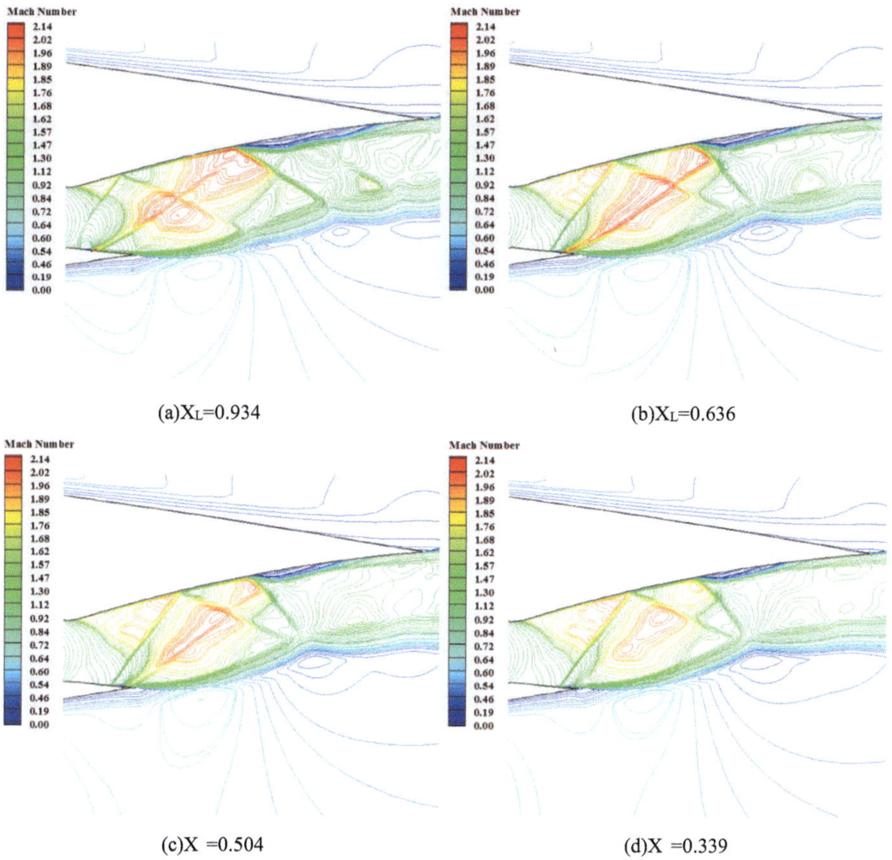

Fig. 5.43 Mach number distribution at different injection positions at *NPR* = 3.97. (**a**) $X_L = 0.934$. (**b**) $X_L = 0.636$. (**c**) $X_L = 0.504$. (**d**) $X_L = 0.339$

ration zone occurs at the trailing edge, and the range of the high-pressure zone within this zone increases.

Figure 5.46 shows the effect of the secondary flow injection position on nozzle performance, with minor changes in the axial thrust coefficient and thrust vector angle of the nozzle as the secondary flow injection position X_L decreases from 0.934 to 0.769 at *NPR* = 3.97, with a large area of separation downstream of the upper expansion ramp and reduced pressure in the downstream region responsible for the nozzle performance degradation when X_L changes to 0.636.

At *NPR* equal to 5.50 and 7.00, nozzle performance can be considered optimal at X_L = 0.934, with nozzle performance decreasing as the jet orifice moves upstream. At *NPR* equal to 5.50, the axial thrust coefficient C_{fgx} decreases by a maximum of 0.0158 and the thrust vector angle delta-v decreases by a maximum of 8.53°; At *NPR* equal to 7.00, the axial thrust coefficient C_{fgx} decreases by a maximum of 0.0104 and the thrust vector angle delta decreases by a maximum of 2.05°.

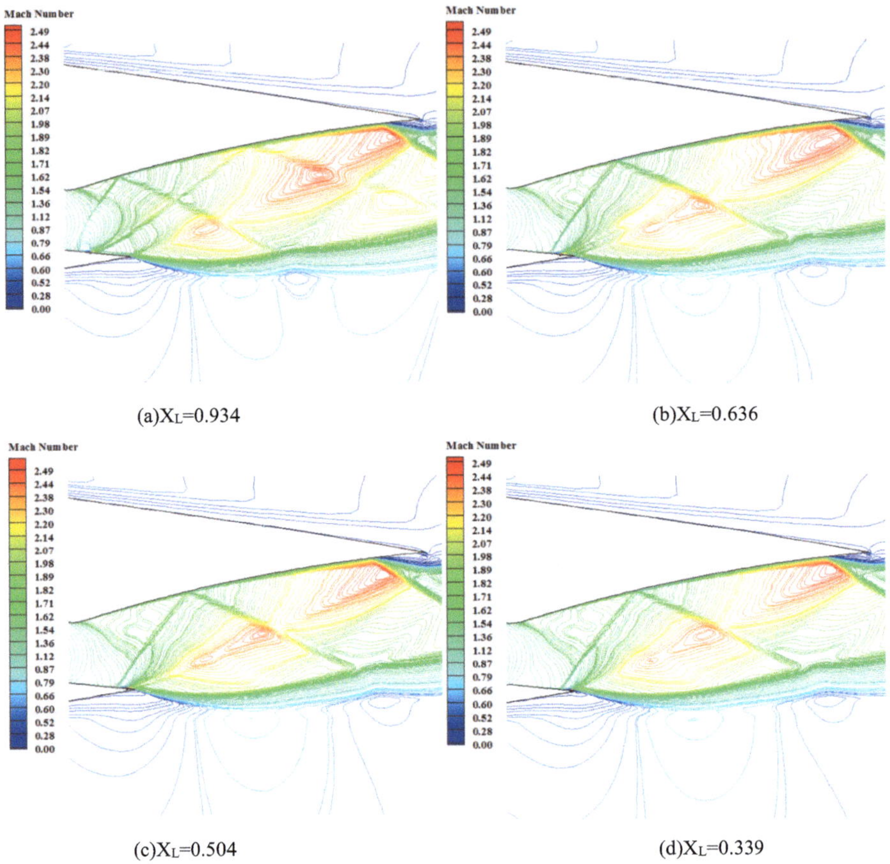

Fig. 5.44 Mach number distribution at different injection positions at *NPR* = 7.00. (**a**) $X_L = 0.934$. (**b**) $X_L = 0.636$ (**c**) $X_L = 0.504$. (**d**) $X_L = 0.339$

Considering the nozzle performance changes with injection position at different *NPR*s, the following conclusions can be drawn: when the injection position is close to the trailing edge of the lower ramp, the nozzle performance is higher; When it is not possible to open a jet hole at the trailing edge of the lower ramp plate, the jet hole may be moved properly in the upstream direction, which does not cause a large decrease in performance to a certain extent.

5.4.3 Effects of Aerodynamic Parameters on SERN Performance for Tandem TBCC Engines

1. Effect of secondary flow jet flow rate on SERN flow field and performance

A secondary flow is injected at the lower ramp of the nozzle, maintaining the overall injection pressure ratio when injected *SPR* = 1, Injection angle $\alpha = 90°$, injec-

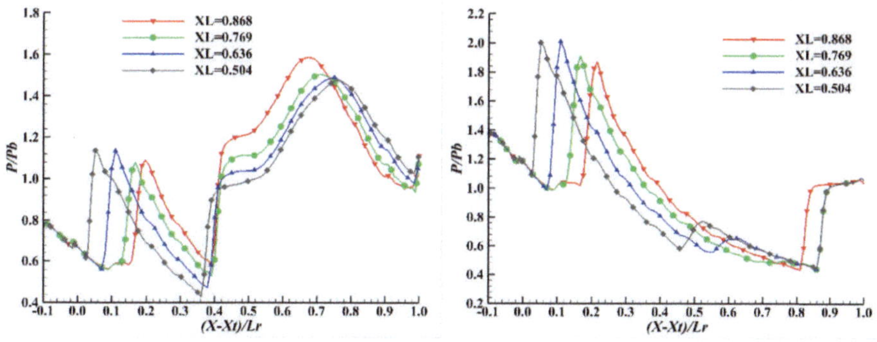

(a) *NPR*=3.97 (b) *NPR*=7.00

Fig. 5.45 Upper expansion wall pressure distribution at different ejection locations. (**a**) *NPR* = 3.97. (**b**) *NPR* = 7.00

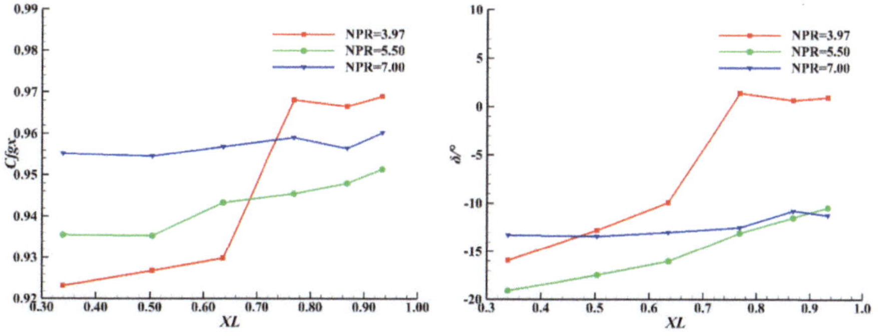

(a) Thrust coefficient (b) vector angle

Fig. 5.46 Effect of change in position of secondary flow injection on nozzle performance. (**a**) Thrust coefficient. (**b**) vector angle

tion position $X_L = 0.868$, outflow Mach number Ma = 0.6 unchanged; With *NPR* equal to 3.97 and 9.00 respectively, the secondary flow nozzles were adjusted so that the flow ratios W were equal to 1.93%, 3.86%, 4.70%, 6.21% and 7.53%. The effect of secondary flow injection flow on the SERN performance was investigated.

Figures 5.47 and 5.48 show the nozzle Mach number distribution at different flows, and Fig. 5.49 shows the pressure distribution at the upper expansion ramp at different flows. It can be seen that at *NPR* = 3.97, as the secondary flow rate ratio increases from 1.93% to 7.53%, the shock wave generated by the secondary flow injection at the lower oblique plate moves upstream, becoming stronger and the shock angle increases. The interference region induced by shock waves at the upper expansion ramp begins to appear flow separation, and moves upstream,

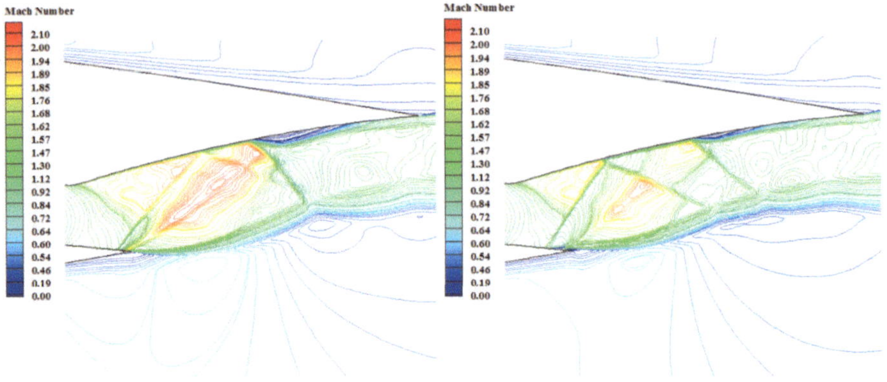

(a)*W*=1.93% (b)*W*=7.53%

Fig. 5.47 Mach number distribution at different jet flows at *NPR* = 3.97. (**a**) *W* = 1.93%. (**b**) *W* = 7.53%

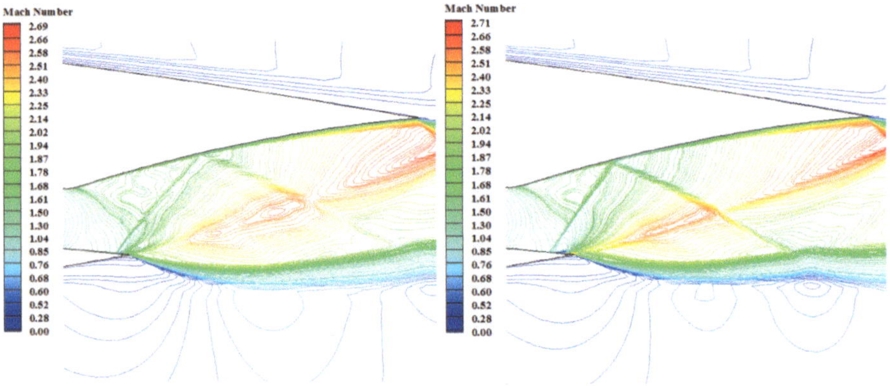

(a)*W*=1.93% (b)*W*=7.53%

Fig. 5.48 Mach number distribution at different jet flows at *NPR* = 9.00. (**a**) *W* = 1.93%. (**b**) *W* = 7.53%

increasing the separation region area. The change of the upstream separation region affects the change of the upper expansion ramp pressure, making the area and peak of the upstream high-pressure region rapidly increase at the same time. With the increase of secondary flow rate, the area of separation zone downstream of the upper expansion ramp decreases. The downstream pressure of the upper expansion ramp begins to increase with the increase of secondary flow rate, reaching a maximum at W = 4.70%, and decreases as secondary flow rate continues to increase.

At other *NPR*s, the shock generated by the secondary flow injection and the separation zone and high pressure zone upstream of the upper expansion ramp varied

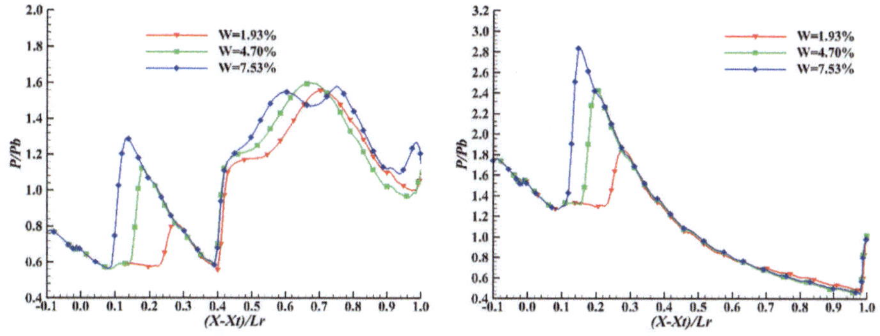

(a) *NPR*=3.97 (b) *NPR*=9.00

Fig. 5.49 Upper expansion wall pressure distribution at different jet flows. (**a**) *NPR* = 3.97. (**b**) *NPR* = 9.00

with the secondary flow rate in essentially the same way as at *NPR* = 3.97, and the peak value of the high pressure zone upstream increased with *NPR*. However, at different *NPR*, the effect of secondary flow rate on the downstream pressure distribution of the upper expansion ramp varies greatly. With the increase of *NPR*, the increase of downstream pressure caused by secondary flow injection decreases gradually, and at *NPR* = 9.00, the downstream pressure distribution has little change, which is related to the flow separation of the nozzle itself at different *NPR*. With the increase of *NPR*, the flow separation degree of the nozzle decreases gradually and disappears, and the effect of secondary flow injection on the flow separation downstream of the upper expansion ramp decreases.

Figure 5.50 shows the effect of secondary flow jet flow on nozzle performance, with changes in nozzle performance directly related to upper expansion ramp flow separation and changes in pressure. At *NPR* = 3.97, as the secondary flow rate ratio increases from 1.93% to 4.70%, the axial thrust coefficient increases from 0.9494 to 0.9701, while the thrust vector angle increases from −8.3°. Increasing to 1.38°, the non axial force downward decreases, the deflection moment decreases, and the nozzle performance is improved. The nozzle performance is optimal when W = 4.70%, and the axial thrust coefficient of the nozzle increases by 0.0403 and the thrust vector angle increases by 16.09° relative to the non injected secondary flow. As W increases to 6.21% and 7.53%, nozzle performance begins to decline again, and the decrease in pressure downstream of the up expansion ramp relative to W = 4.70% may be responsible for the performance degradation.

Table 5.3 gives the maximum values for nozzle performance improvement at different *NPR*s, and it can be seen that the improvement in nozzle performance by secondary flow injection fades as *NPR* increases, with even a slight decrease in the C_{fgx} value of the nozzle at *NPR* = 9.00, and the case at *NPR* = 9.00 is no longer addressed in later discussions of other injection parameters. In addition, with the increase of *NPR*, the W corresponding to the optimal performance of the

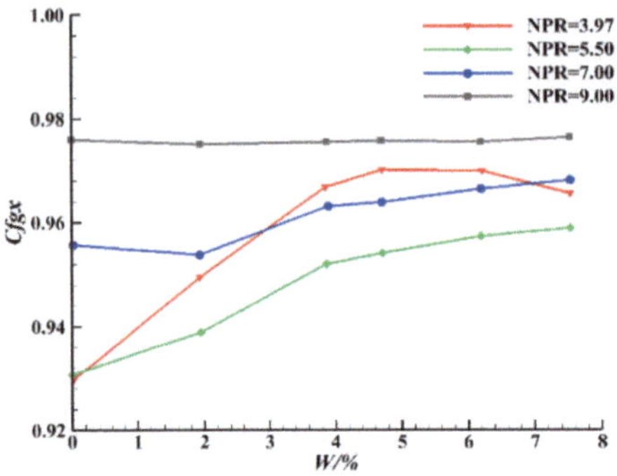

(a) Thrust coefficient

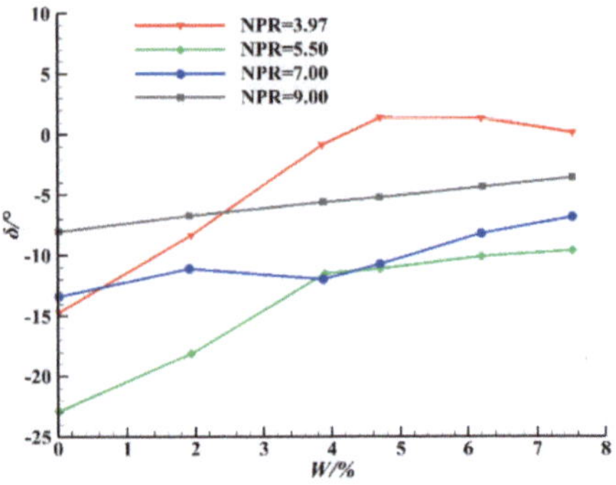

(b) Vector angle

Fig. 5.50 Effect of secondary flow rate change on nozzle performance. (**a**) Thrust coefficient. (**b**) Vector angle

Table 5.3 Improvements in nozzle performance at different *NPR*s

NPR	Performance		
	C_{fgx} improvement at optimum performance	δ improvement at optimum performance	W improvement at optimum performance
NPR = 3.97	0.0403	16.09°	4.70%
NPR = 5.50	0.0283	13.31°	7.53%
NPR = 7.00	0.0125	6.53°	7.53%
NPR = 9.00	-0.0002	4.34°	7.53%

nozzle is increased, and the secondary flow injection flow required to achieve the optimal performance is increased.

2. Effect of secondary jet total pressure on the SERN flow field and performance

Jet the secondary flow at the lower ramp of the nozzle, keeping the injection angle $\alpha = 90°$, injection position $X_L = 0.868$, and outflow Mach number Ma = 0.6 constant while ejecting; Keeping the flow ratio W = 1.93% constant at *NPR* = 3.97 and 7.00, adjust the secondary flow so that the injection total pressure ratio *SPR* is equal to 0.8, 1.0, 1.3, 1.7 and 2.0. In order to keep the secondary flow rate constant, the area of the secondary flow injection hole was reduced while increasing the total pressure ratio of the secondary flow. The effects of the total pressure of secondary flow injection on the SERN performance at different *NPR*s were studied by the above methods.

Figures 5.51 and 5.52 show the nozzle Mach number distribution at different total pressures, and Fig. 5.53 shows the pressure distribution at the upper expansion ramp at different total pressures. The Mach number distribution and pressure distribution near the injection hole are very different under different conditions of total injection pressure, but the change of the whole nozzle flow field is very small, the strength and location of the shock wave and the size and location of the upper expansion ramp separation zone are almost unchanged; The change of the total pressure caused by the change of the upper expansion ramp pressure is

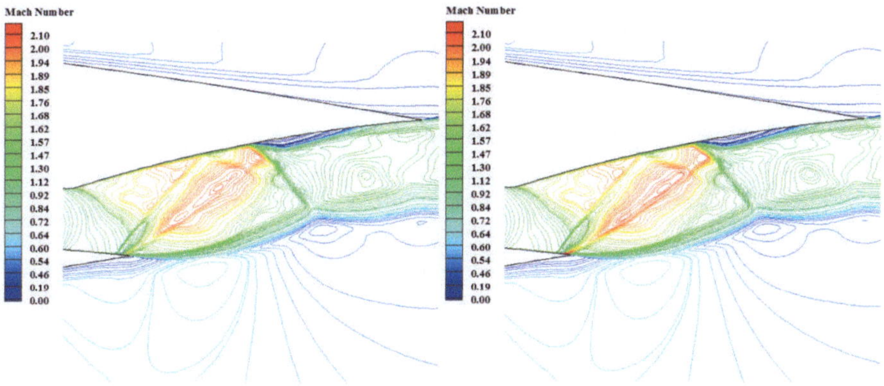

(a)*SPR*=0.8 (b)*SPR*=2.0

Fig. 5.51 Mach number distribution at different total injection pressures at *NPR* = 3.97. (**a**) *SPR* = 0.8. (**b**) *SPR* = 2.0

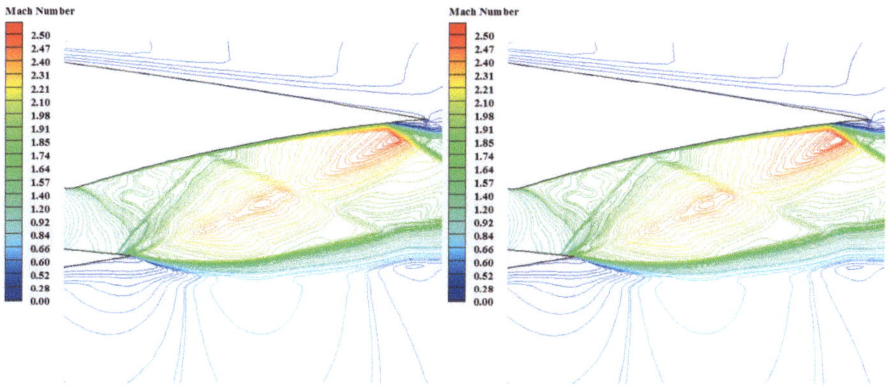

(a)*SPR*=0.8 (b)*SPR*=2.0

Fig. 5.52 Mach number distribution at different total injection pressures at *NPR* = 7.00. (**a**) *SPR* = 0.8. (**b**) *SPR* = 2.0

similarly small, the size, location and peak value of the high-pressure zone are almost unchanged, only small changes occur under special operating conditions.

Figure 5.54 shows the effect of the secondary flow injection total pressure on nozzle performance, and the corresponding nozzle performance change is small because the change in the injection total pressure has little effect on the nozzle flow field. As can be seen from Table 5.4, the axial thrust coefficient C_{fgx} and thrust vectoring angle delta of the nozzle vary very little at each *NPR*, and the variations in both performance parameters are relatively large at *NPR* = 7.00. In the previous literature on jet flow disturbances, the study of the effect of total jet pressure is often

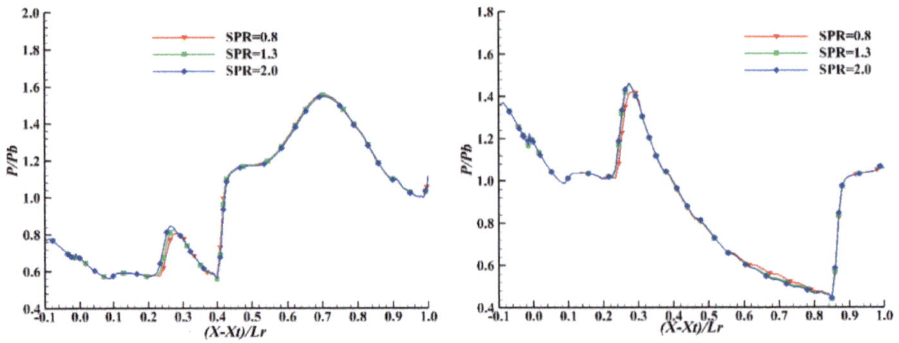

(a) *NPR*=3.97 (b) *NPR*=7.00

Fig. 5.53 Upper expansion wall pressure distribution at different total ejection pressures. (**a**) *NPR* = 3.97. (**b**) *NPR* = 7.00

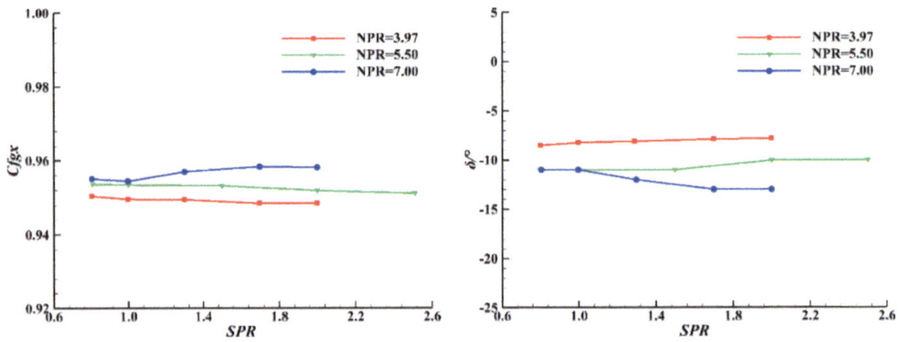

(a) Thrust coefficient (b) Vector angle

Fig. 5.54 Effect of change in total pressure of secondary flow injection on nozzle performance. (**a**) Thrust coefficient. (**b**) Vector angle

Table 5.4 Effect of total injection pressure change on nozzle performance

NPR	Performance	
	Max change in C_{fgx} due to change in total pressure	Max change in δ due to change in total pressure
NPR = 3.97	0.0021	0.77°
NPR = 5.50	0.0031	0.96°
NPR = 7.00	0.0039	1.67°

associated with the jet flow rate, which basically considers the total jet pressure to be an important parameter affecting the flow field and performance. This section separates the study of total jet pressure from the jet flow rate and other parameters, and discusses the effect of total jet pressure separately. It is found that the total secondary jet pressure has less effect on the nozzle flow field and performance in the application of improving the SERN performance, and the choice of total jet pressure only requires that the secondary flow be injected into the main flow at sonic speed.

5.5 Tailpipe Performance Improvement for a Parallel TBCC Engine Based on an Active Flow Control Approach

The performance of a tailpipe for a parallel TBCC engine worsens significantly at subsonic and transonic speeds. How to improve the performance of a TBCC tailpipe at operating conditions is a concern. An active flow control method based on secondary flow injection, which ejects secondary flow at the middle bulkhead of the tail nozzle (the lower wall of the turbine nozzle), is used to improve the flow field quality and improve the nozzle performance. The interaction mechanism between secondary flow injection and the supersonic main flow of the nozzle is studied, the effect of secondary flow injection on the flow field performance in the nozzle and the flight regime adapted to it are analyzed, and the effect of the changes in secondary flow injection geometry and aerodynamic parameters on the thrust and thrust vector performance of the tail nozzle for the parallel TBCC engine is identified.

5.5.1 Tail Nozzle Flow Mechanism for Parallel TBCC Engines with Secondary Flow Injection

This section focuses on the calculation and analysis of the flow field and performance of the tailpipe for the parallel TBCC engine with secondary flow injection, analyzes the effect of the introduction of secondary flow on the flow field inside and outside the tailpipe for the parallel TBCC engine, and explores the flow control mechanism of secondary flow; The suitable flight regime of secondary flow injection was studied, and the effect of the change of geometry and aerodynamic parameters of secondary flow injection on the thrust and thrust vectoring performance of the tailpipe for a parallel TBCC engine was discussed.

1. Tail nozzle geometry model for parallel TBCC engines with secondary flow injection

In the subsonic and transonic flight phases, thrust of the parallel TBCC engine is mainly provided by the turbine engine, which operates the turbonozzle in an over expansion condition, producing complex structures such as shock waves and separation zones within the nozzle, and the performance of the nozzle drops dramatically. The performance of the nozzle is improved by introducing secondary flow at the middle bulkhead of the tail nozzle (the lower wall of the turbine nozzle). Figure 5.55 is the geometric model of the tail nozzle for a parallel TBCC engine with a secondary flow jet orifice, which is a converging orifice, to ensure that the secondary flow is injected into the main flow at the same Mach number (Ma = 1). Detailed parameter definitions and values are given in Sect. 5.2.3.

2. Wake nozzle flow field characteristics for a parallel TBCC engine with secondary flow injection

At transonic speeds, as shown in Fig. 5.56a (Ma = 0.9), the turbine nozzle is in a state of over expansion with significant flow separation on the upper wall, and the ramjet nozzle has a large area of separation zone on its upper wall due to excessive expansion angle of the wall, resulting in a sharp drop in nozzle performance.

A secondary flow injection method is used to improve nozzle performance, and Fig. 5.56b gives a cloud map of the Mach number distribution of the nozzle in the secondary flow injection regime at flight Ma = 0.9. From the figure, it can be found that a bow shock wave forms in the supersonic flow field upstream of the secondary flow jet hole in the turbine nozzle due to secondary flow obstruction,

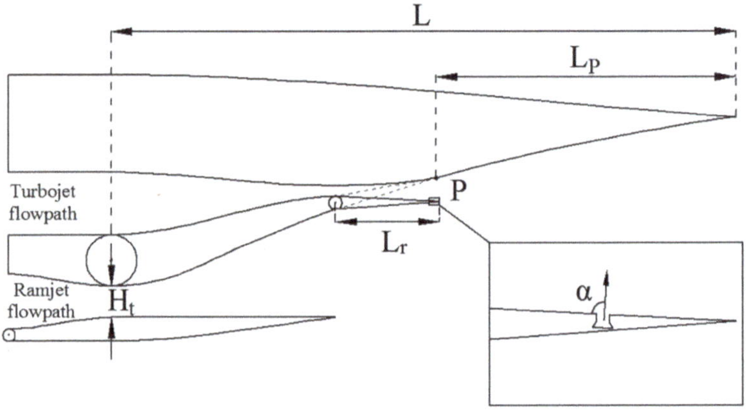

Fig. 5.55 TBCC tail nozzle with secondary flow injection

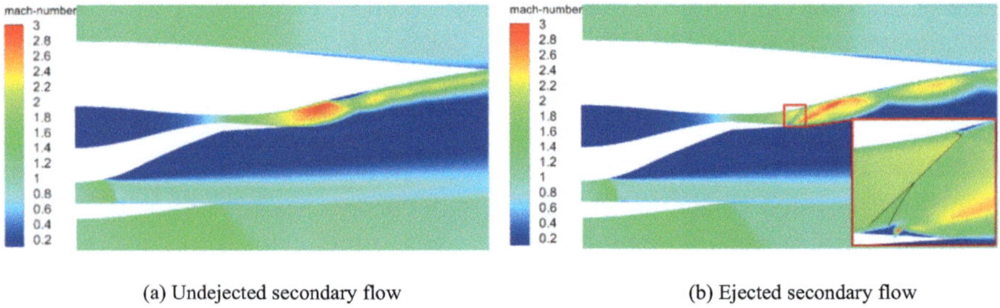

Fig. 5.56 Cloud diagram of Mach number distribution in the tail nozzle for the parallel TBCC engine. (**a**) Undejected secondary flow. (**b**) Ejected secondary flow

which interacts with the boundary layer on the lower wall of the turbine nozzle, causing the boundary layer to separate, inducing a separation shock wave ahead of the separation zone, and the intersection of the two shock waves forms the "λ" shock wave shown in Fig. 5.57b. The bow shock passes through the flow field and intersects the upper expansion surface of the turbine nozzle, causing local separation of the upper expansion surface boundary layer and the formation of a separation shock in the main flow. Comparing Fig. 5.56a, it can be found that the separation zone due to over expansion downstream of the upper expansion surface of the turbine nozzle almost disappears, while the flow field in the ramjet nozzle changes insignificantly.

Figure 5.57a, b are Mach number cloud charts and streamlines at the secondary flow jet orifice at flight Ma = 0.9. The under expanded secondary flow expands rapidly after ejecting into the main flow and creates compression waves around it, above which a Mach disk forms, and the over expanded jet flows downstream with the main flow after repressurization through the Mach disk. A bow shock is generated in the main stream due to the blocking effect of the secondary flow, and the static pressure of the main stream surges after passing the bow shock, creating a somewhat strong counter pressure gradient that creates a clockwise separation vortex near the wall (Fig. 5.57 (b) Red 1 position). Between the separation vortices and the secondary flow jets, a counterclockwise induced vortex is generated due to the ejector action of the secondary flow (Fig. 5.57b Red 2 position). The secondary flow ejects into the main stream, achieving a certain jet depth in the main stream, and continuing with the main stream towards the downstream phase, the flow is compressed to form a compression wave due to wall constraints, and creates a clockwise closed separation zone (Fig. 5.57b Red Position 3).

Figure 5.58 shows the changes in the pressures on the upper, lower, and upper expansion surfaces of the turbine nozzle before and after secondary flow injection, where the abscissa X is the axial distance from each point of the upper expansion surface to the ramjet throat, and the ordinate is the dimensionless pressure on the upper and lower walls of the turbine nozzle (ratio of static pressure on the upper and

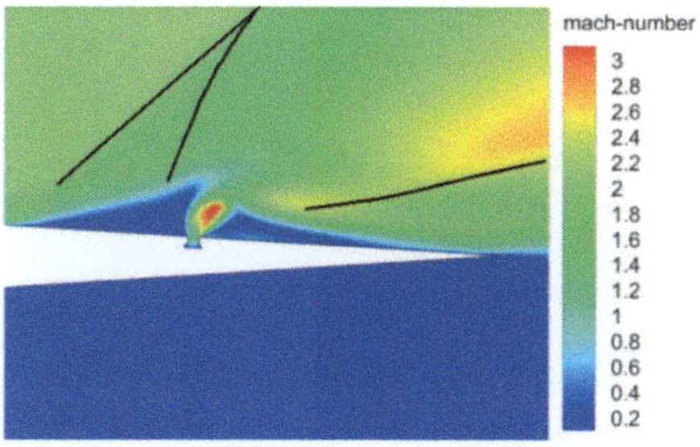

(a) Mach Number Cloud Chart at Ma=0.9

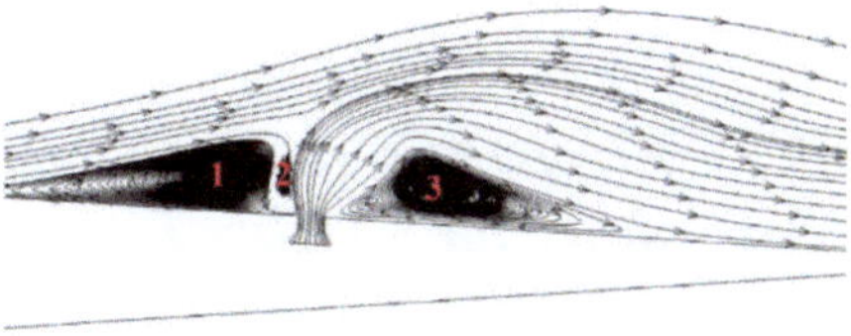

(b) Streamline Chart at Ma=0.9

Fig. 5.57 Flow field diagram at the secondary flow jet orifice. (**a**) Mach Number Cloud Chart at Ma = 0.9. (**b**) Streamline Chart at Ma = 0.9

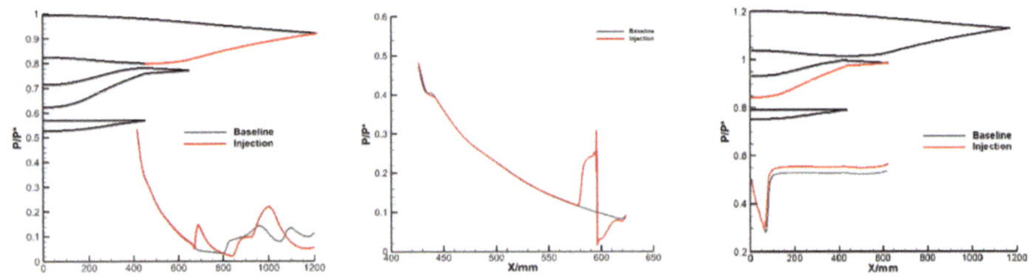

(a) Turbine nozzle upper wall (b) Turbine nozzle lower wall (c) Ramjet upper wall

Fig. 5.58 Wall pressure distribution before and after secondary flow injection. (**a**) Turbine nozzle upper wall. (**b**) Turbine nozzle lower wall. (**c**) Ramjet upper wall

lower walls of the turbine nozzle to the total pressure at the turbine nozzle inlet) and the dimensionless pressure on the upper wall of the ramjet nozzle (ratio of static pressure on the upper wall of the ramjet to total pressure at the ramjet nozzle inlet), respectively. The "λ" shock generated by the secondary flow injection creates a high-pressure region upstream of the secondary flow injection port, and the ejecting action of the secondary flow creates a low-pressure region downstream of the secondary flow injection port, as shown in Fig. 5.58b. The bow shock generated by the secondary flow injection interacts with the expanding wall boundary layer on the turbine nozzle to create a new separation zone and forms a separation shock in the main flow, which intersects the bow shock and in turn creates a reflected shock, which, under the influence of a complex wave system, locally surges the static pressure on the upper wall of the turbine nozzle, as shown in Fig. 5.58a. The rise in pressure weakens the degree of over expansion of the gas in the nozzle, improves the separation zone downstream, and increases the pressure downstream of the turbine nozzle. Increased pressure downstream of the turbine nozzle improves the low-pressure environment of the ramjet channel due to airflow ejection, raising the overall pressure distribution on the expanding surface of the ramjet nozzle, as shown in Fig. 5.58c. The increase of the expansion surface pressure on the turbine nozzle and ramjet nozzle after secondary flow injection is the main reason for the improvement of nozzle performance. Calculations show that the nozzle axial thrust coefficient with secondary flow injection is raised from 0.668 to 0.717, and the thrust vectoring angle is improved from −46.85° to −34.09°.

5.5.2 *Effects of Geometric Parameters on Tail Nozzle Performance for Parallel TBCC Engines*

1. Effect of the relative position of the secondary flow nozzle on the flow field and performance of the tail nozzle

This section focuses on the effect of the secondary flow injection position on the tail nozzle performance for a parallel TBCC engine at Mach = 0.5 in flight. The effect of secondary flow injection position X_L on the performance of a tailpipe for a parallel TBCC engine was investigated, with secondary flow injection flow $W = 4.8\%$, secondary flow injection pressure ratio $SPR = 1$, and secondary flow injection angle $\alpha = 90°$, maintained at various flight regimes by secondary flow injection at the tailpipe intermediate bulkhead, where X_L = 0.5912, 0.7866, 0.8591, 0.9184, 0.9521.

Figure 5.59 is a Mach number cloud diagram of a tailpipe for a parallel TBCC engine at different secondary flow injection positions at flight Ma = 0.5. It can be found that the turbonozzle is in an over expanded working condition when the secondary flow is not injected, and the shock wave emitted at the trailing edge of the lower wall interacts with the upper expansion surface of the nozzle to create a separation zone. Supersonic flow occurs in the ramjet nozzle due to the ejecting action of the turbonozzle, the expansion angle at the exit of the ramjet nozzle is too large, and a large area of separation zone occurs on the upper expansion surface. As shown in Fig. 5.59b, with secondary flow injection at the position X_L = 0.5912 on the lower wall of the turbine nozzle, a closed separation occurs on the lower wall of the turbine nozzle and the separation zone on the upper wall of the turbine nozzle increases. As the position of secondary flow injection shifted back, open separation occurred on the lower wall of the turbonozzle, the degree of over expansion of the turbonozzle decreased significantly, and the downstream separation zone disappeared gradually, with no significant change in the internal flow field of the ramjet.

Figure 5.60 shows the pressure distribution on the upper expansion surface and on the lower wall of the nozzle at different injection positions, respectively, where the abscissa is the axial distance from each point on the nozzle wall to the ramjet throat, and the ordinate is the dimensionless pressure on the upper and lower walls of the turbine nozzle (ratio of static pressure on the upper and lower walls of the turbine nozzle to the total pressure at the turbine nozzle inlet) and the dimensionless pressure on the upper wall of the ramjet nozzle (ratio of static pressure on the upper wall of the ramjet to total pressure at the ramjet nozzle inlet), respectively. From Figs. 5.60a and 5.61b, it can be found that the bow shock generated by the secondary flow injection, as well as the separation shock induced by the separation zone, moves back with the position of the secondary flow injection shifted back. It can be seen from Fig. 5.60a that the secondary flow injection has a large improvement in the expansion surface pressure on the

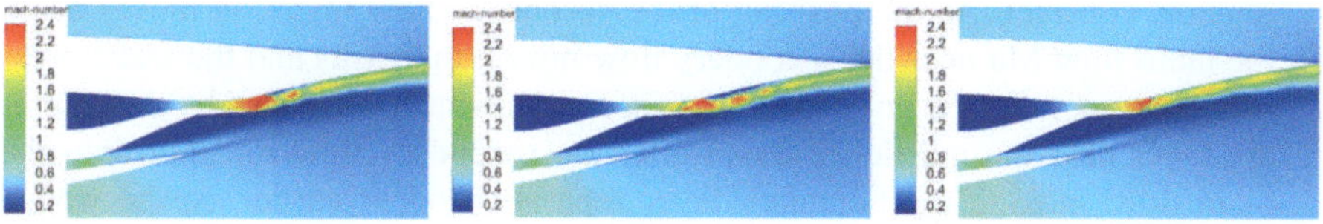

(a) Undejected secondary flow (b) X_L=0.5912 (c) X_L=0.9521

Fig. 5.59 Mach number distribution at different injection positions. (**a**) Undejected secondary flow. (**b**) X_L = 0.5912. (**c**) X_L = 0.9521

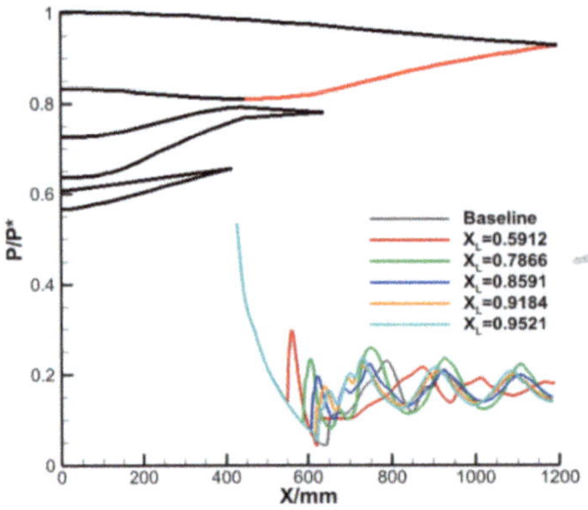

(a) Pressure distribution on the expansion surface of the turbine nozzle

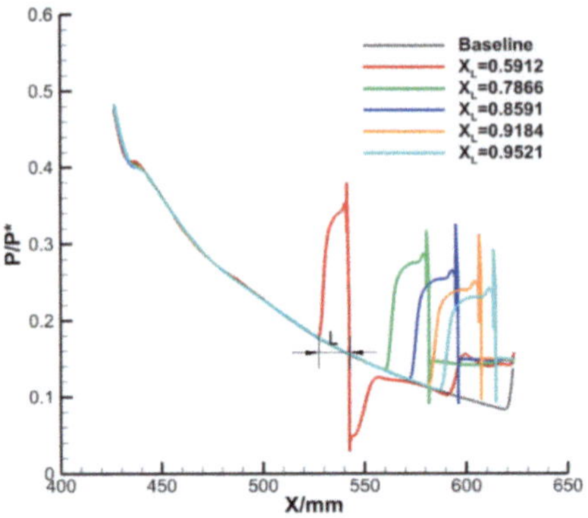

(b) Pressure distribution on the lower wall of the turbine nozzle

Fig. 5.60 Pressure distribution on various wall faces at different secondary flow injection locations. (**a**) Pressure distribution on the expansion surface of the turbine nozzle. (**b**) Pressure distribution on the lower wall of the turbine nozzle

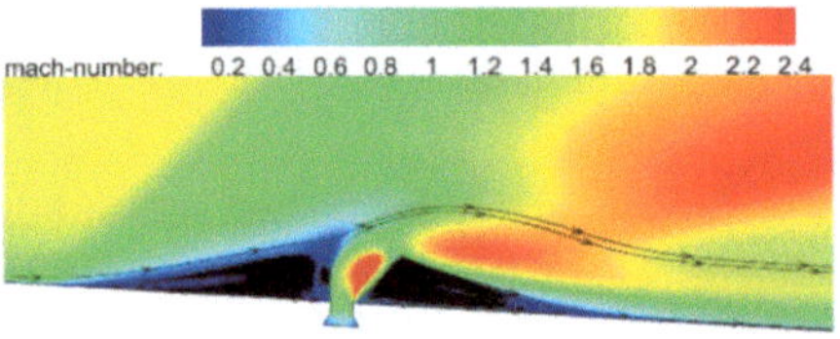

(a)X_L=0.5912

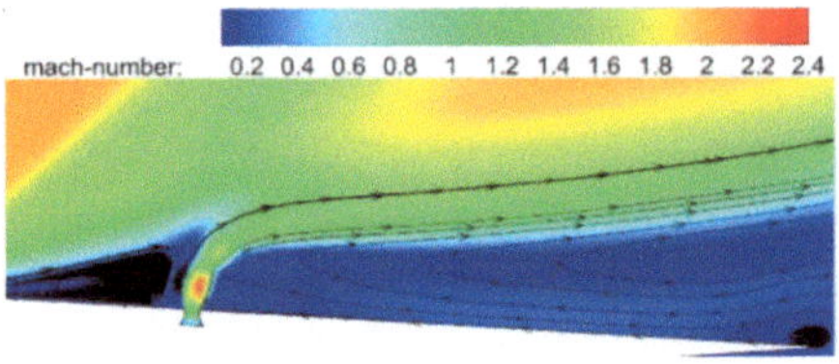

(b)X_L=0.7866

Fig. 5.61 Local enlarged view of the secondary flow injection port at different secondary flow injection positions. (**a**) $X_L = 0.5912$. (**b**) $X_L = 0.7866$

turbine engine. L in Fig. 5.60b represents the length of the wedge shaped separation region (as regions 1, 2 of Fig. 5.57b) created upstream of the secondary flow injection port at the secondary flow injection position $X_L = 0.5912$, which gradually increases as the secondary flow injection position moves aft. It can also be found from the figure that there were two pressure surges downstream of the secondary flow injection port when injection position $X_L = 0.5912$; When $X_L = 0.7866$–0.9521, the pressure downstream of the secondary flow injector remains constant. As the position of the secondary flow injection moves back, the main flow Ma near the secondary flow nozzle increases and the static pressure decreases. On the one hand, the pressure gradient experienced by the boundary layer near the nozzle increases after the secondary flow injection into the main flow, making the boundary layer separation occur earlier, so the wedge shaped separation zone in front of the secondary flow nozzle lengthens and increases, and the shock strength increases; On the other hand, in the over expanded state, the secondary flow is injected into the main stream, achieving a certain jet depth in the main stream, and continuing with the main stream towards the downstream of the phase, the flow is compressed to form a compression wave due to wall constraints, and creates a closed separation zone. The nozzle is in an over expanded working condition, and a shock wave is generated near the trailing edge on the lower wall of the nozzle, so there are two pressure surges downstream of the secondary flow nozzle when $X_L = 0.5912$. When $X_L > 0.7866$, the main flow has largely expanded to atmospheric pressure near the secondary flow nozzle, the acceleration of the flow is not obvious, and the impact of the main flow on the secondary flow is weakened, so the downstream of the secondary flow nozzle changes from a previous closed separation to an open separation, as shown in Fig. 5.62. The jet action of the turbonozzle causes a supersonic flow in the ramjet nozzle, which is in a state of severe over expansion, and a shock wave near the nozzle throat causes a sudden rise in the pressure on the upper wall of the nozzle.

Figure 5.62 shows the axial thrust coefficient and thrust vectoring angle of the tail nozzle for a parallel TBCC engine as a function of the position of the secondary flow injection, and it can be found that both the thrust coefficient and thrust vectoring angle of the nozzle have a tendency to rise first and then decrease. When $X_L = 0.5912$, a closed separation occurs downstream of the secondary flow nozzle and a large area separation occurs on the upper expansion surface of the turbine nozzle, seriously affecting the nozzle performance; As the position of the secondary flow nozzle moves back, the main flow Mach number at the jet position increases and the main flow static pressure decreases, equivalent to a reduction in the jet's back pressure, which in turn causes the Mach disk and bow shock to become larger and stronger, increasing the pressure downstream of the turbojet nozzle, reducing the separation zone due to over expansion, reducing the drag of the nozzle, and thus improving the nozzle's performance. As shown in Fig. 5.62, nozzle thrust performance is optimal at $X_L = 0.8591$, with a 3.69%

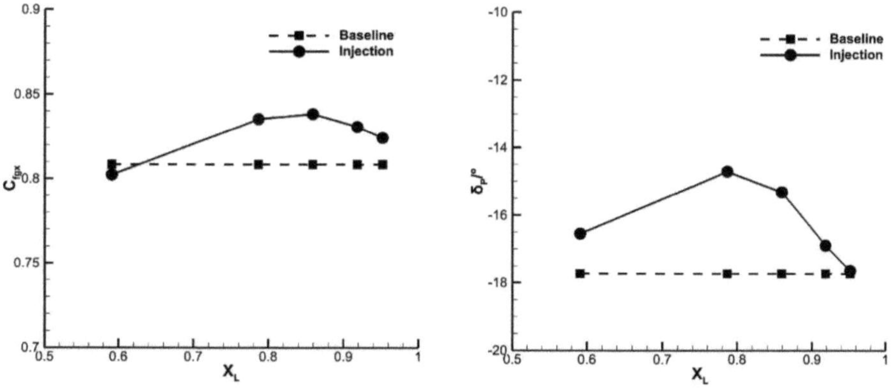

(a) Axial thrust coefficient (b) Thrust vector angle

Fig. 5.62 Effect of secondary flow injection position on C_{fgx} and δ_p. (**a**) Axial thrust coefficient. (**b**) Thrust vector angle

increase in nozzle axial thrust coefficient with secondary flow injection compared to the baseline nozzle, and a thrust vectoring angle improved from −17.73° to −15.32° (−14.70° at $X_L = 0.7866$). But as the position of the secondary flow nozzle continued to move backwards, the shock continued to strengthen, the separation zone produced by the shock interacting with the boundary layer was too large, and the loss of flow past the shock was more severe, which in turn caused a decrease in nozzle performance.

2. Effect of Secondary Flow Injection Angle on the Wake Nozzle Flow Field and Its Performance

This section focuses on the effect of the secondary flow injection angle on the tail nozzle performance for a parallel TBCC engine at Mach = 0.5 in flight. The effect of secondary flow injection angle α (the angle between the direction of secondary flow injection and the upstream wall of the secondary flow nozzle) on the performance of the tail nozzle for a parallel TBCC engine was studied, with secondary flow injection position $X_L = 0.8591$, secondary flow injection flow W = 4.8%, and secondary flow injection pressure drop ratio *SPR* = 1, at each flight regime, by a secondary flow injection at the middle bulkhead of the tail nozzle, where α = 30°, 45°, 70°, 90°, 110°, 135°, 150°.

Figure 5.63 is a Mach number cloud diagram of a tailpipe for a parallel TBCC engine at flight Mach = 0.5 at different secondary flow injection angles. The momentum of the secondary flow in both the vertical and main directions varies with the increase of the injection angle of the secondary flow, directly affecting the jet trajectory, jet depth, and the intensity of the secondary flow induced shock wave and the size of the separation zone. As shown in Fig. 5.63, as α increases, the separation zone upstream of the upper expansion surface due to shock wave interaction with the boundary layer and the wedge shaped separation zone upstream of the secondary flow nozzle gradually decrease, and downstream of the upper expansion surface of the turbine nozzle gradually decrease, and when α > 90°, the separation zone downstream essentially disappears, with no significant change in the flow field within the ramjet.

Figure 5.64 shows the pressure distribution on the upper expansion surface and on the lower wall of a turbine engine nozzle at different injection angles, respectively, where the abscissa is the axial distance from each point on the nozzle wall to the ramjet throat, and the ordinate is the dimensionless pressure at each wall point, respectively. From Fig. 5.64a, b, it can be seen that the momentum of the secondary

(a) Undejected secondary flow (b)α=30° (c)α=90°

Fig. 5.63 Mach number distribution at different injection angles. (**a**) Undejected secondary flow. (**b**) α = 30°. (**c**) α = 90°

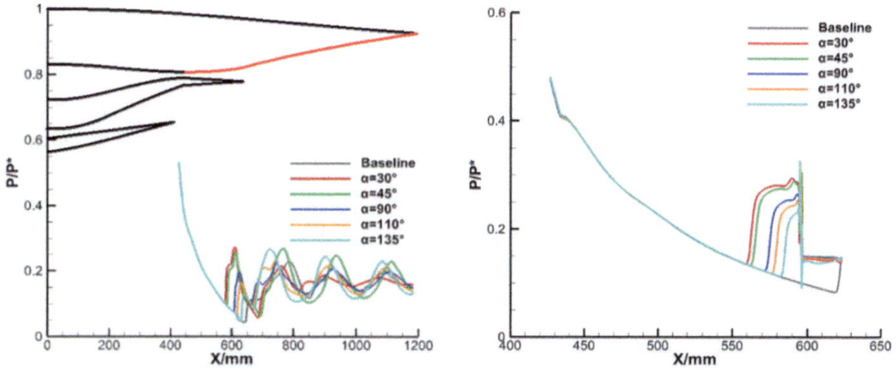

(a) Pressure distribution on the expansion surface of the (b) Pressure distribution on the lower wall of the

Fig. 5.64 Wall pressure distribution at different secondary flow injection angles. (**a**) Pressure distribution on the expansion surface of the turbine nozzle. (**b**) Pressure distribution on the lower wall of the turbine nozzle

flow in both the vertical and main flow directions changes as the secondary injection angle increases, the bow shock induced by the secondary flow weakens and moves downstream, the wedge shaped separation zone upstream of the secondary flow injector becomes smaller, and the downstream of the secondary flow injector always exhibits open separation.

Figure 5.65 shows the axial thrust coefficient and thrust vectoring angle of the tailpipe for a parallel TBCC engine as a function of the secondary flow injection angle. It can be found that the thrust coefficient and thrust vectoring angle of the nozzle both rise first and then decrease. At $\alpha = 30°$, the secondary flow induced shock wave is stronger and the separation zone is larger, as shown in Fig. 5.64a, b, where the effect of pressure gain from the secondary flow is mostly offset by the effect of total energy penalty from the separation zone, so the overall thrust performance improvement is less. As the secondary flow injection angle increases, the shock wave induced by the secondary flow gradually weakens, the separation zone decreases, and the difference between the pressure gain effect brought by the shock wave and the total energy loss effect brought by the separation zone gradually increases, so the thrust and thrust coefficient gradually increase, and when $\alpha = 90°$, the thrust performance is optimal, the axial thrust coefficient increases by 3.69%, and the thrust vector angle is improved from −17.73° to −15.32° (when $\alpha = 45°$, the thrust vector angle reaches a peak of −14.60°). But as the secondary flow injection angle continues to increase, the shock strength continues to weaken, the shock's raising of the main static pressure decreases, and the improvement of the secondary flow on the degree of over expansion of the turbine nozzle weakens. The difference between the effect of pressure gain brought by the shock and the effect of total energy loss brought by the separation zone gradually decreases, which in turn results in a decrease in nozzle performance.

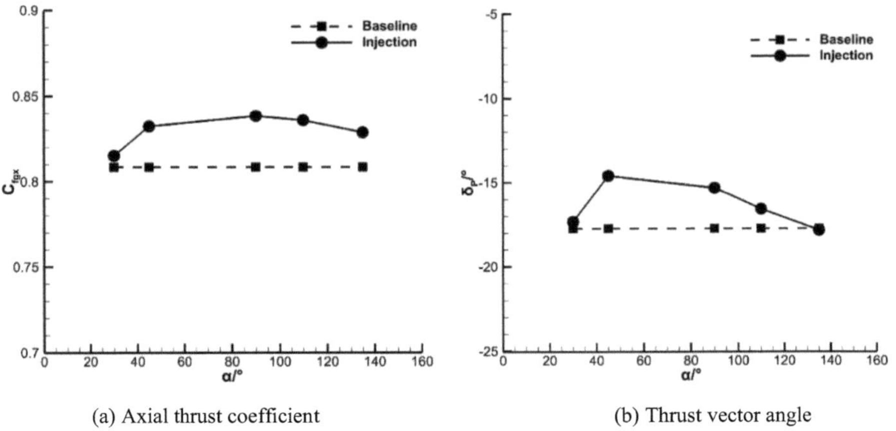

Fig. 5.65 Effect of secondary flow injection angle on C_{fgx} and δ_p.(**a**) Axial thrust coefficient. (**b**) Thrust vector angle

5.5.3 Effect of Aerodynamic Parameters on Tail Nozzle Performance for Parallel TBCC Engines

1. Effect of secondary flow injection flow ratio on the tail nozzle flow field and its performance

This section focuses on the effect of secondary flow injection flow on the performance of the tail nozzle for a parallel TBCC engine at Mach 0. 5 in flight. In all flight regimes, the secondary flow injection was performed by the tail nozzle intermediate bulkhead, keeping the secondary flow injection position $X_L = 0.8591$, the secondary flow injection pressure ratio $SPR = 1$, and the secondary flow injection angle $\alpha = 90°$ constant. The effect of the secondary flow injection flow W on the tail nozzle performance for a parallel TBCC engine was studied, where W = 0.94%, 2.90%, 4.80%, 6.79%, 8.74% (varying the injection flow by changing the size of the injection hole).

Figure 5.66 is a Mach number cloud diagram of a tailpipe for a parallel TBCC engine at flight Mach = 0.5 with different secondary flow injection flows. It can be found that the turbonozzle is in over expansion working condition when the secondary flow is not injected, and the shock wave emitted at the trailing edge of the lower wall interacts with the upper expansion surface of the nozzle to create a separation zone. Supersonic flow occurs in the ramjet nozzle due to the ejecting action of the turbonozzle, the expansion angle at the exit of the ramjet nozzle is too large, and a large area of separation zone occurs on the upper expansion surface. As shown in Fig. 5.66b, at W = 0.94%, the momentum ratio of the secondary flow to the main flow is small, and the disturbance of the secondary flow has less influence, so the bow shock induced by the secondary flow does not intersect the expansion surface on the turbine nozzle at this time. With the increase of secondary flow injection rate, the separation zone caused by bow shock interacting with the upper wall upstream of the turbine nozzle gradually increases, and the separation zone due to over expansion downstream of the turbine nozzle gradually disappears. When W>2.90%, open separation occurs downstream of the secondary flow injection port. This is mainly due to the increase of the momentum ratio of the secondary flow to the main flow with the increase of the secondary flow jet flow rate, the jet depth of the secondary flow and the bow shock enhancement, which on the one hand increases the upper wall separation region of the turbine nozzle upstream; On the other hand, raising the static pressure of the main flow reduces the over expansion of the nozzle and the downstream separation zone. There was no significant change in the flow field within the ramjet nozzle.

Figure 5.67 shows the pressure distribution on the upper expansion surface and on the lower wall of a turbine engine nozzle at different jet flows, where the abscissa is the axial distance from each point on the nozzle wall to the ramjet throat, and the ordinate is the dimensionless pressure at each wall point, respectively. From Fig. 5.67a, b, it can be seen that as the secondary flow rate increases, the bow shock becomes stronger and moves upstream, and the wedge separation region upstream of the secondary flow injection port gradually increases. When W = 0.94%, the pressure downstream of the secondary flow injection port rises twice, and when W = 2.90%–8.74%, the pressure downstream of the secondary flow injection port remains constant. From Fig. 5.67c, it can be found that at Mach 0. 5 in flight, the secondary flow injection at the intermediate bulkhead has little effect on the pressure distribution on the upper wall of the ramjet.

As the secondary jet flow rate increases, the momentum ratio of the secondary flow to the main flow increases, the penetration depth of the secondary flow strengthens, the secondary flow perturbs the main flow more violently, the bow shock intensifies, the separation zone generated by the bow shock interacting with the upper and lower nozzle walls increases, and the separation shock induced before the separation zone increases, which in turn increases the amplitude of the pres-

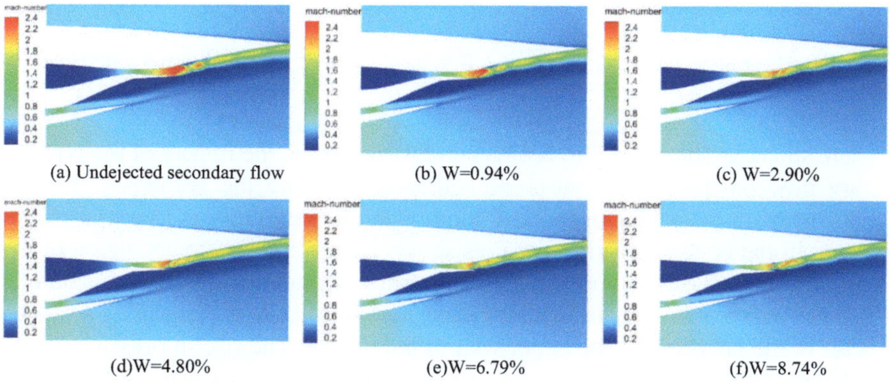

Fig. 5.66 Mach number distribution at different jet flows. (**a**) Undejected secondary flow. (**b**) W = 0.94%. (**c**) W = 2.90%. (**d**) W = 4.80%. (**e**) W = 6.79%. (**f**) W = 8.74%

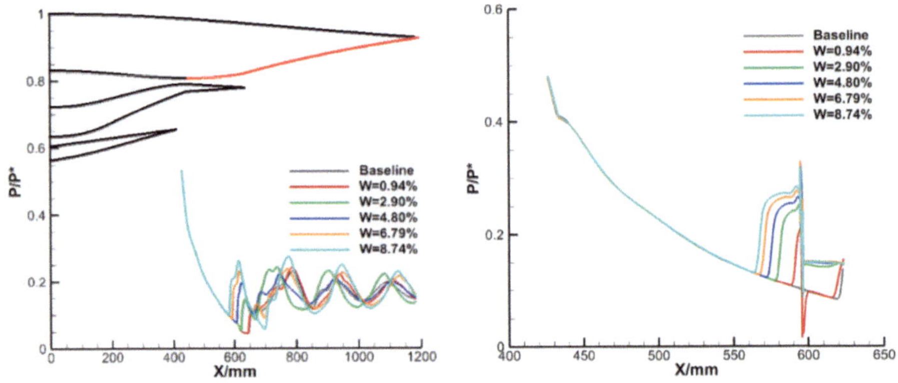

(a) Pressure distribution on the expansion surface of the turbine nozzle

(b) Pressure distribution on the lower wall of the turbine nozzle

Fig. 5.67 Wall pressure distribution at different secondary flow jet flows. (**a**) Pressure distribution on the expansion surface of the turbine nozzle. (**b**) Pressure distribution on the lower wall of the turbine nozzle

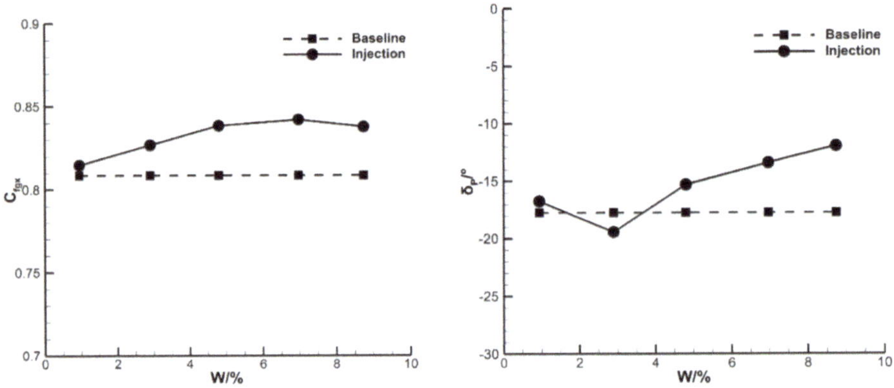

(a) Axial thrust coefficient

(b) Thrust vector angle

Fig. 5.68 Effect of secondary flow injection flow on C_{fgx} and δ_p. (**a**) Axial thrust coefficient. (**b**) Thrust vector angle

sure spike, as shown in Fig. 5.67a. When W > 2.90%, the impact of the secondary flow on the main flow is strengthened, allowing the secondary flow jet downstream to exhibit open separation, where the pressure is essentially constant in the open separation zone, as shown in Fig. 5.67b. The jet action of the turbonozzle causes a supersonic flow in the ramjet nozzle, which is in a state of severe over expansion, and a shock wave near the nozzle throat causes a sudden rise in the pressure on the upper wall of the nozzle. The rise in pressure downstream of the turbine nozzle gives a slight increase in the expansion surface pressure distribution on the ramjet.

Figure 5.68 shows the axial thrust coefficient and thrust vectoring angle of a tailpipe for a parallel TBCC engine as a function of secondary flow injection flow. It can be found that the thrust coefficient of the nozzle presents a trend of first rising and then descending, while the thrust vector angle presents a trend of first descending and then ascending. As the secondary flow injection rate increases, the shock wave induced by secondary flow intensifies, raising the pressure downstream of the turbojet nozzle, reducing the separation zone due to over expansion, reducing the drag of the nozzle, and thus improving the nozzle performance. As shown in Fig. 5.68a, nozzle thrust performance is optimal at W = 6.79%, with a 4.14% increase in nozzle axial thrust coefficient with secondary flow injection compared to the baseline nozzle. However, as the secondary flow jet flow continued to increase, the shock continued to enhance, the separation region produced by the shock interaction with the boundary layer was too large, and the loss of flow through the shock was more serious, which led to a decrease in nozzle performance. At W = 2.90%, it is known from the pressure distribution in Fig. 5.67b that the closed separation downstream of the secondary flow injection port transforms into an open separation, with the turbine exit flow deflecting substantially upwards, so at this flow the nozzle thrust vectoring angle drops slightly, as in Fig. 5.69b. With the increase of the secondary flow injection flow rate, the shock wave causes the enhancement of the static pressure rise gain, and the nozzle thrust vector increases gradually. When W = 8.74%, the thrust vector angle is improved from −17.73° to −11.93°.

2. Effect of secondary flow injection pressure ratio on the tail nozzle flow field and its performance

This section focuses on the effect of secondary flow injection pressure ratio on the performance of a tailpipe for a parallel TBCC engine at Mach 0.5 in flight. The effect of secondary flow injection pressure drop ratio *SPR* on the performance of a tailpipe for a parallel TBCC engine was investigated, with secondary flow injection flow W = 4.80% (by changing the size of the orifice), secondary flow injection position $X_L = 0.8591$, and secondary flow injection angle $\alpha = 90°$ constant, at each flight state, by performing secondary flow injection from the tailpipe intermediate bulkhead, where *SPR* = 0.5, 0.8, 1, 1.5, 2.

Figure 5.69 is a Mach number cloud diagram of a tailpipe for a parallel TBCC engine at flight Mach = 0.5 with different secondary flow injection pressures (keeping flow constant by changing the size of the orifice), from which it can be found that with secondary flow injection in this flight state, the separation zone due to over expansion disappears downstream of the nozzle, but the ramjet flow field is essentially unchanged. With a constant secondary flow rate, increasing the secondary flow drop pressure ratio increases the secondary flow's ability to do work, but the momentum ratio of the secondary flow to the main flow is constant, and the penetration depth of the secondary flow is constant, so the intensity of the shock wave induced by the secondary flow is constant, and the improvement of the over expansion range in the nozzle is basically consistent, so the flow fields in both different flight regimes remain constant.

Figure 5.70 shows the pressure distribution on the upper expansion surface, lower nozzle wall, and upper ramjet wall of a turbine engine nozzle at different secondary jet pressures at Mach 0.5 in flight, where the abscissa is the axial distance

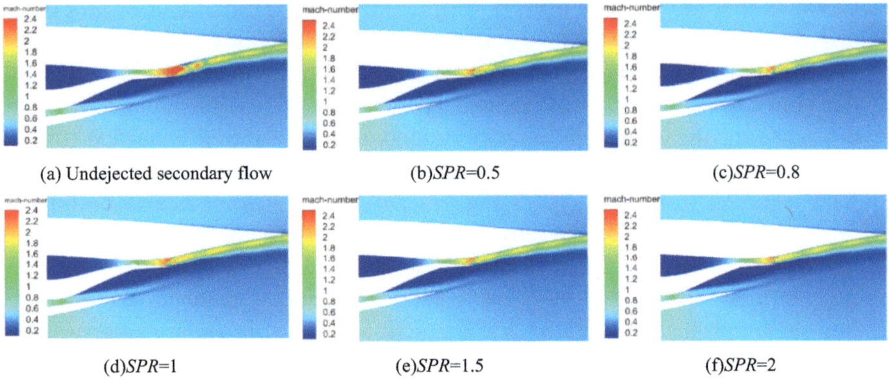

Fig. 5.69 Mach number distribution at different injection pressures (Ma = 0.5). (**a**) Undejected secondary flow. (**b**) *SPR* = 0.5. (**c**) *SPR* = 0.8. (**d**) *SPR* = 1. (**e**) *SPR* = 1.5. (**f**) *SPR* = 2

from each point on the nozzle wall to the ramjet throat, and the ordinate is the dimensionless pressure at each wall point. While maintaining a constant secondary jet flow rate, although the secondary flow pressure ratio increases, the secondary exit velocity remains constant at Ma = 1 (converging nozzle), so the momentum ratio of the secondary flow is constant from that of the main flow, the depth of the secondary flow jet is constant, the strength of the bow shock induced by the secondary flow is constant, and the size of the separation zone induced by the shock wave is essentially the same, as shown in Fig. 5.70a. At Mach 0.5 in flight, the turbine nozzle is in a severely over expanded operating condition, subject to shock waves at the trailing edge of the nozzle's lower wall, with open separation downstream of the secondary flow nozzle, as shown in Fig. 5.70b. It can also be found from Fig. 5.70b that there are higher peak points at the edge of the secondary flow nozzle, mainly due to the higher pressure at the exit of the secondary flow as the secondary flow pressure ratio increases, which in turn causes a higher pressure peak at the edge of the pressure exit.

Figure 5.71 shows the axial thrust coefficient and thrust vectoring angle of a tailpipe for a parallel TBCC engine as a function of secondary flow injection pressure at flight Ma = 0.5. From the diagram, it can be found that the axial thrust coefficient of the nozzle decreases slightly as the secondary flow injection pressure increases. Although the secondary flow injection pressure increases, due to the constant momentum ratio of the secondary flow to the main flow, the improvement effect of the secondary flow on the nozzle performance is basically consistent, so the actual thrust of the nozzle is essentially the same. As defined by the nozzle thrust coefficient (ratio of actual thrust to ideal thrust), an increase in secondary flow pressure necessarily increases the ideal thrust, so the thrust coefficient decreases slightly, as shown in Fig. 5.71a. Because the change in secondary flow injection pressure changes the turbine nozzle pressure as well as the ramjet pressure very little, the actual axial thrust changes little from the vertical lift, so the thrust vector angle is essentially unchanged, as shown in Fig. 5.71b.

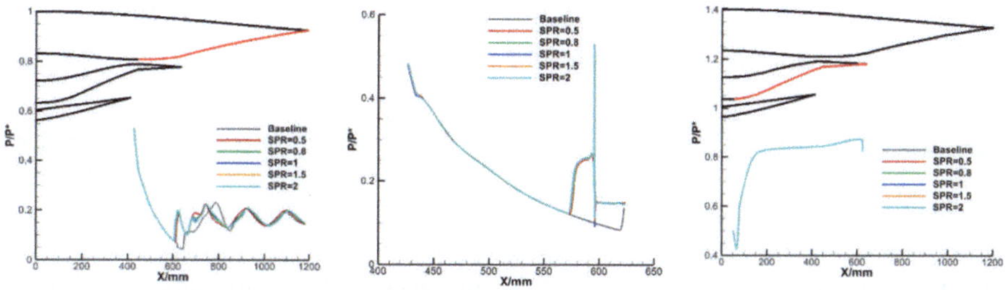

(a) Pressure distribution on the expansion surface of the turbine nozzle

(b) Pressure distribution on the lower wall of the turbine nozzle

(c) Pressure distribution on the expansion surface of the ramjet nozzle

Fig. 5.70 Pressure distribution across walls at different secondary flow injection pressures (Ma = 0.5). (**a**) Pressure distribution on the expansion surface of the turbine nozzle. (**b**) Pressure distribution on the lower wall of the turbine nozzle. (**c**) Pressure distribution on the expansion surface of the ramjet nozzle

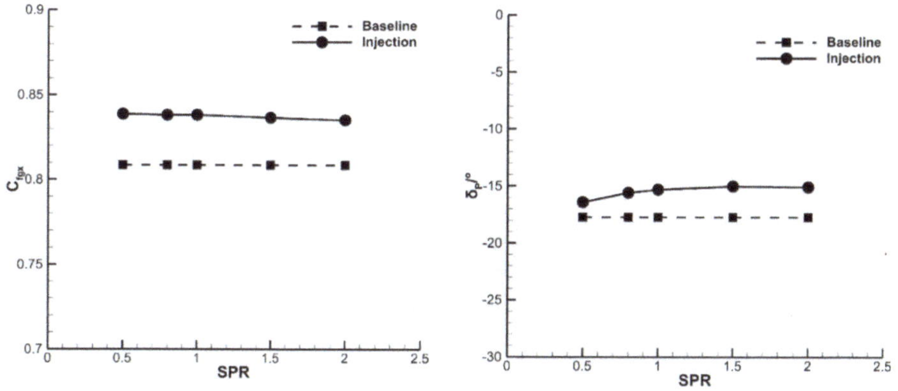

(a) Axial thrust coefficient

(b) Thrust vector angle

Fig. 5.71 Effect of secondary flow injection pressure on C_{fgx} and δ_p (Ma = 0.5). (**a**) Axial thrust coefficient. (**b**) Thrust vector angle

5.6 SERN Performance Improvement for Tandem TBCC Engines Based on Passive Flow Control Method

In order to improve the performance of SERN in the over expanded state, a passive flow control method based on passive cavity was proposed, in which a passive cavity structure was added to the expansion ramp on the nozzle to study its effect on the nozzle flow field and performance, as well as its applicable range.

5.6.1 Study on Flow Mechanism of Passive Cavity to Improve SERN Performance for Tandem TBCC Engines

This section focuses on the calculation and analysis of the flow field and performance of SERN with passive cavity structure, and the influence of passive cavity structure on the flow field inside and outside SERN and its flow control mechanism. At the same time, the adaptability of the passive cavity, that is, the range of its suitable drop pressure ratio, is studied, and the influence of the geometric parameters of the passive cavity on the flow control is discussed.

1. SERN geometric model with passive cavity

The high expansion ratio SERN with a passive cavity is shown in Fig. 5.72. Based on the high expansion ratio prototype SERN, the upper wall of the nozzle is modified by splitting the upper wall into three parts, the matrix, the passive cavity, and the orifice plate, where the passive cavity and orifice plate combine to form the passive cavity structure, as shown in Fig. 5.73, which combines with

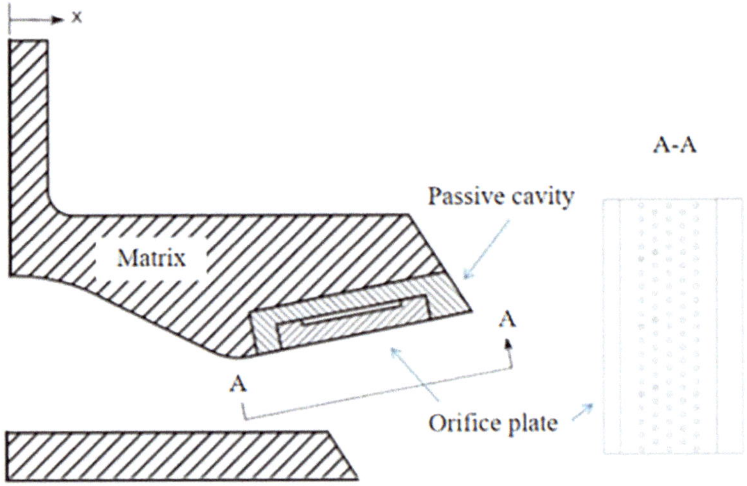

Fig. 5.72 SERN with passive cavity

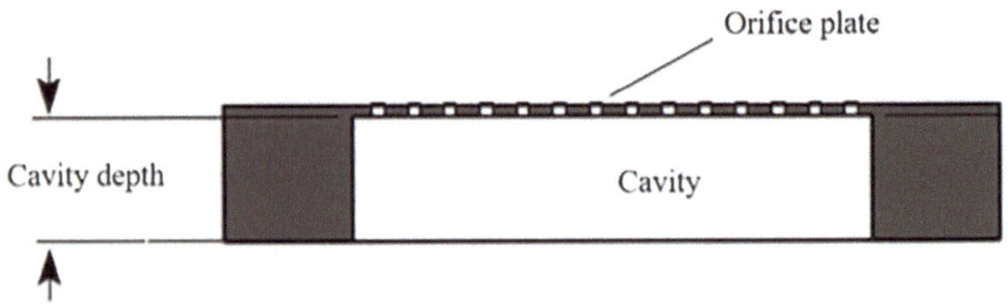

Fig. 5.73 Passive cavity structure

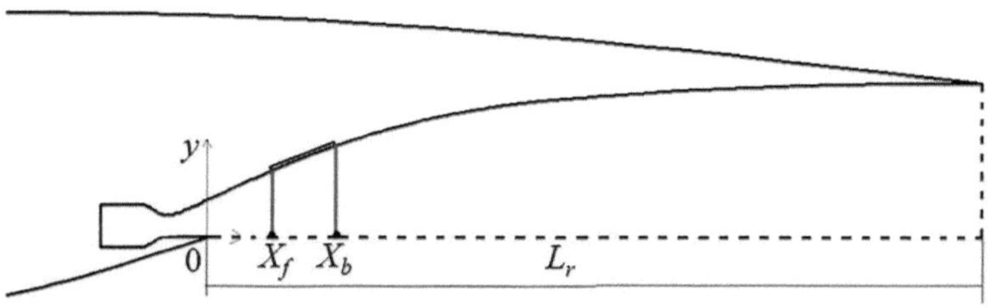

Fig. 5.74 SERN with passive cavity

the nozzle matrix to form the upper expansion ramp of the SERN. Several main parameters of the passive cavity include: aperture ratio, aperture, and cavity depth. Passive cavity mainly studies the change of cavity depth, while orifice plate studies the change of aperture and aperture.

Adding a passive cavity structure into the upper expansion ramp of the prototype SERN. The designed SERV with passive cavity is shown in Fig. 5.74, based on the internal flow of the SERN in the severe over expansion state, with the leading and trailing edge points of the passive cavity structure at the design positions on the upper expansion ramp respectively: $X_f/L_r = 0.0811$, $X_b/L_r = 0.1622$, Where X_f and X_b are the coordinates of the leading and trailing edge points of the passive cavity, respectively, and Lr is the axial length of the expansion ramp extension on the nozzle.

2. Comparison of SERN with passive cavity with prototype SERN

The flow characteristics and performance of prototype SERN and SERN with passive cavity were compared, in which the SERN with passive cavity has an orifice opening ratio of 4.93% (the orifice opening ratio is defined as the ratio of the area of the opening to the total area of the orifice plate), an aperture of 20 mm, and a cavity depth of 100 mm. The vehicle climb stage at low Mach number (Ma = 0.5) was selected for the calculation condition, where the SERN drop pressure ratio (*NPR* = 8) was far lower than the design drop pressure ratio, and the nozzle was in a severe over expansion state.

When the SERN enters an over expanded state, the performance of the nozzle is strongly related to the separation zone and the location of the separation zone. The prototype SERN Mach number distribution is shown in Fig. 5.75a, where the nozzle in the over expanded state creates a oblique shock wave at the trailing edge of the lower ramp, which passes through the jet stream and intersects the SERN upper expansion ramp, creating an incident shock wave that interferes with the boundary layer, causing the upper expansion ramp to flow separate and create an induced shock wave, with a large area of separation zone at the upper expansion ramp, resulting in a sharp decrease in nozzle performance. As shown in Fig. 5.75b, a passive cavity was added to the prototype SERN, which connected the upstream and downstream pressures of the induced shock, causing a change in the flow regime near the upwelling expansion ramp induced shock. The low-speed boundary layer flow behind the shock wave flows into the passive cavity under the pressure differential pressure before and after the shock wave,

(a) Prototype SERN Mach distribution

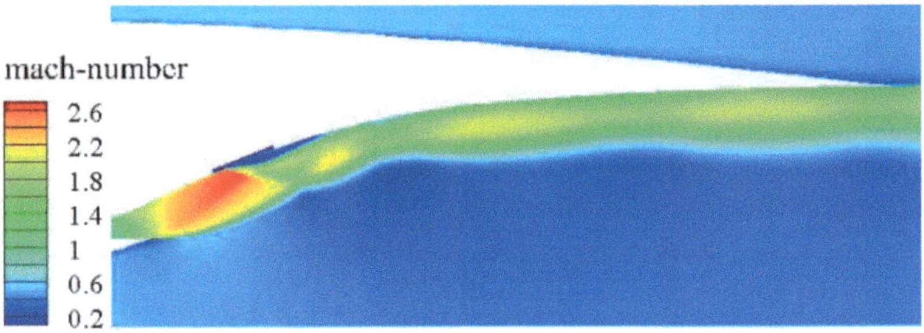

(b) SERN Mach distribution with passive cavity

Fig. 5.75 Mach number distribution diagram of prototype SERN and SERN with passive cavity. (**a**) Prototype SERN Mach distribution. (**b**) SERN Mach distribution with passive cavity

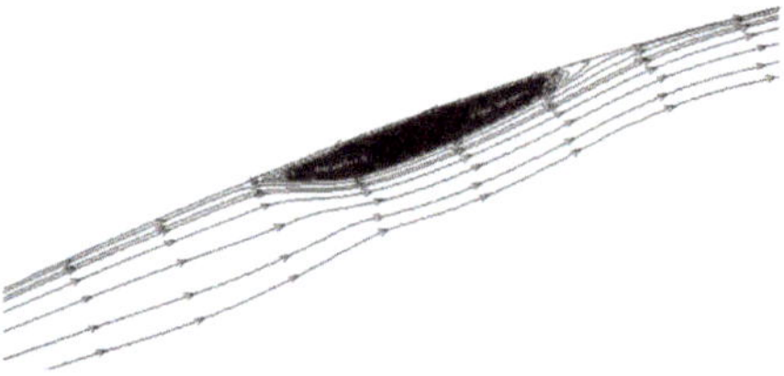

(a) Prototype SERN

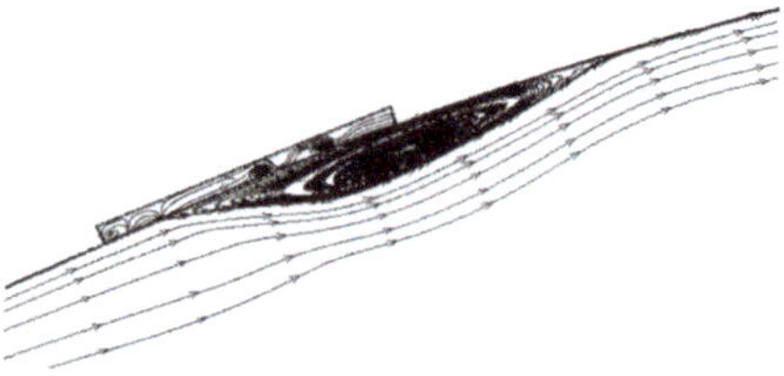

(b) SERN with passive cavity

Fig. 5.76 Local streamline distribution map. (**a**) Prototype SERN. (**b**) SERN with passive cavity

and the gas hole passing through the front of the shock wave is blown out of the cavity. The interaction with the boundary layer before the shock wave changes the flow pattern and pressure distribution at the root of the original shock wave, pushing the induced shock wave forward to the first hole of the passive cavity, at which point the separation region behind the shock wave increases accordingly. As shown in Fig. 5.76, the streamline distribution diagram near the passive cavity shows that the introduction of the passive cavity has the effect of enhancing flow separation and forming a stable reflow region.

Figure 5.77 shows the pressure coefficient distribution of the prototype SERN with an expansion ramp on the SERN with a passive cavity. The SERN lower wall

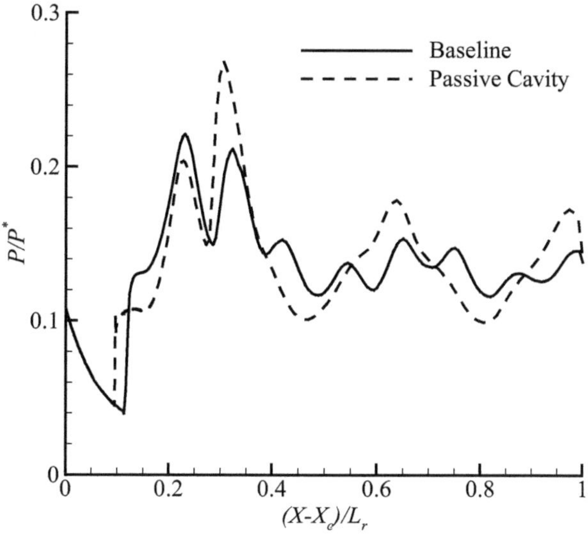

Fig. 5.77 Pressure distribution on expansion ramp on prototype nozzle and nozzle with passive cavity

Table 5.5 Prototype SERN and SERN with passive cavity performance parameters table

	Axial thrust coefficient C_{fgx}	Thrust vector angle δ(°)
Prototype SERN	0.7605	0.48
SERN with passive cavity	0.7738	0.54

exit oblique shock interacts with the upper expansion ramp boundary layer and induces the separation shock, causing the wall pressure to rise abruptly, leaving the pressure in the closed separation bubble formed downstream of the separation point essentially unchanged. After the separation bubble, the flow deflects towards the expansion surface to form a reattached shock that further increases the pressure, creating a confined shock wave separation due to the nozzle being in an over expanded state, a shock train structure in the nozzle, and multiple shock reflections from the upper wall, resulting in multiple surges of pressure. Compared with the prototype SERN, the position of the SERN shock with a passive cavity moves forward and the separation range expands, the second pressure peak after the separation zone increases significantly, after which the number of shock trains decreases, essentially reducing the total pressure loss of the flow, so the thrust performance of the SERN based on passive cavity flow control is improved.

As can be seen in Table 5.5, the adoption of passive cavity (4.93% open ratio, 20 mm aperture, and 100 mm cavity depth) passive flow control techniques (Dusa 1987) compared to the prototype SERN resulted in a 1.75% increase in the axial thrust coefficient without worsening the nose down moment. 1.75% is numerically small, but huge for the SERN axial thrust coefficient. Therefore, a passive cavity SERN with a certain configuration can improve the performance of the nozzle in the over expanded state.

Since the passive cavity structure is designed on the nozzle expansion section, it has little effect on the nozzle throat and the upstream flow field of the throat. The existence of the passive cavity structure has little effect on the nozzle flow coefficient. Due to the addition of the passive cavity structure, the effect on the flow over the upper and lower outer wall surfaces of the nozzle and where the flow inside and outside the nozzle is mixed is minor, and the difference between the tail drag X_b of the prototype SERN and the passive cavity SERN is less than 0.1%. The formula for the calculation of tail drag is:

$$X_b = -\int_M^9 (p - p_0) dA + X_f \tag{5.26}$$

Where: 9, M denotes the tail nozzle exit section and maximum section, respectively; X_f is the frictional drag.

Passive cavity structures are introduced at low flight Mach numbers and low drop pressure ratios. From the above, it can be seen that the existence of passive cavity structures can indeed improve the performance of SERN in the severe over expansion state at non design points, then how the passive cavity structure affects the performance of SERN when SERN is at design point, and the flow field and characteristic analysis of SERN with passive cavity at design point are calculated. The Mach number distribution of the SERN with a passive cavity is shown in Fig. 5.78. The flow field is essentially the same as the prototype SERN except for the low-speed return flow in the passive cavity. The expansion ramp pressure distribution on the nozzle is shown in Fig. 5.79. The existence of a passive cavity structure changes the pressure distribution near the passive cavity, resulting in some points with higher pressures, which affect the nozzle thrust coefficient. The performance of the prototype SERN with a passive cavity SERN is shown in Table 5.6. The presence of a passive cavity structure at the design point does not reduce the SERN's axial thrust coefficient and does not have the effect of a nose down moment.

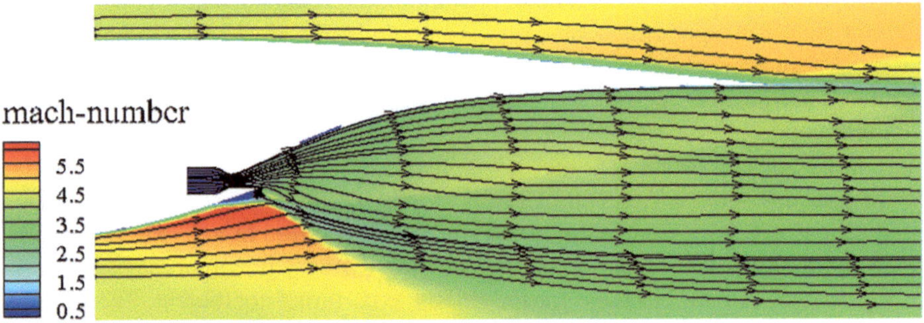

Fig. 5.78 SERN Mach number distribution with passive cavity

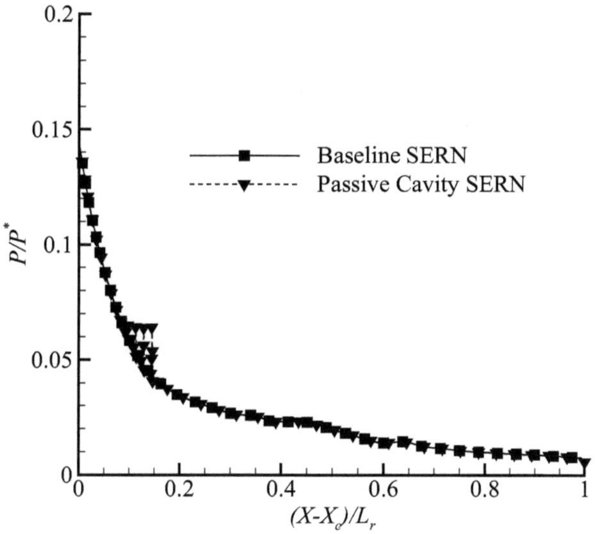

Fig. 5.79 Expansion ramp pressure distribution on SERN with passive cavity

Table 5.6 Prototype SERN and SERN with passive cavity performance parameters table

	Axial thrust coefficient C_{fgx}	Thrust vector angle δ(°)
Prototype SERN	0.9545	0.41
SERN with passive cavity	0.9603	0.40

5.6.2 *Passive Cavity to Improve the Adapted Operating Range of the SERV for Tandem TBCC Engines*

In order to obtain the applicable drop pressure ratio range of passive cavity flow control method, this section aimed at SERN with passive cavity structure designed in the previous section. The flow characteristics and performance parameters of SERN with passive cavity were calculated at Mach = 0.5, and the effects of passive cavity flow control methods on nozzle flow field and thrust performance were analyzed for high expansion ratio SERN with different drop pressure ratios in over expansion.

A passive cavity structure with a plate opening ratio of 4.93%, a 20 mm aperture, and a 100 mm cavity depth was selected, with an expansion ramp position of $X_f/L_R = 0.0811$ on the nozzle, $X_b/L_r = 0.1622$. The Mach number of the outer flow is 0.5, and the distribution of the SERN Mach number with a passive cavity at different drop pressure ratios is shown in Fig. 5.80, as seen from the figure. At the calculated drop pressure ratios, the trailing edge of the lower oblique plate emits free shock waves that pass through the nozzle flow field, hitting the upper expansion surface boundary layer, causing the boundary layer to separate and forming a separation shock wave ahead of the boundary layer separation zone, deflecting the flow

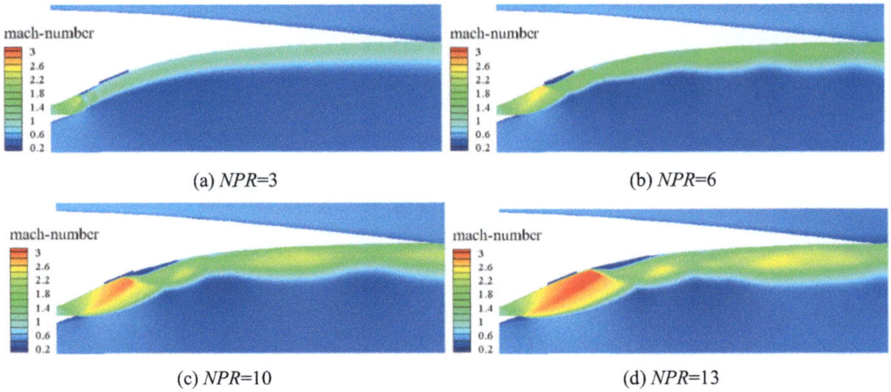

Fig. 5.80 Passive cavity SERN Mach number distribution at different drop pressure ratios. (**a**) $NPR = 3$. (**b**) $NPR = 6$. (**c**) $NPR = 10$. (**d**) $NPR = 13$

Fig. 5.81 Expansion ramp pressure distribution on passive cavity SERN at different drop pressure ratios

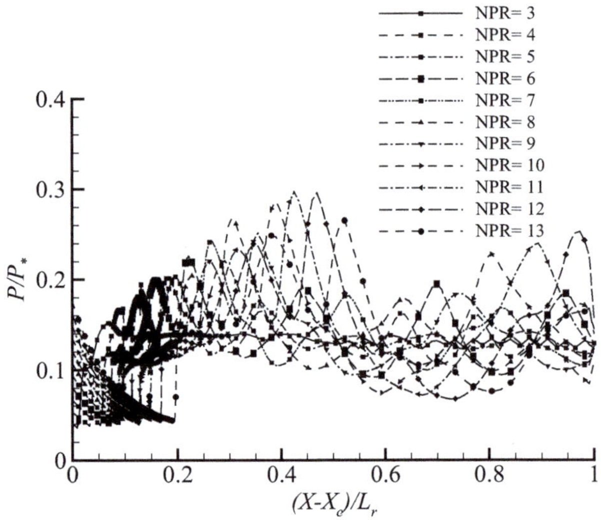

direction inward. As the drop pressure ratio increases, the free shock from the lower ramp is gradually pushed out of the nozzle, the shock angle becomes smaller, the separation shock is gradually away from the throat, and the separation region of the expansion ramp on the nozzle moves back gradually, and the separation region becomes larger. In the range of these drop pressure ratios, the existence of passive cavity structures affects the distribution of the expansion ramp pressure on the nozzle, and influences the nozzle thrust coefficient.

Expansion ramp pressure distribution on SERN with passive cavity at different drop pressure ratios is shown in Fig. 5.81. At lower drop pressure ratios, such as $NPR = 3$ and $NPR = 4$, the nozzle is in a state of over expansion due to the low nozzle drop pressure ratio, and the separation generated at the upper expansion ramp is ahead of the passive cavity structure, which can be seen from the pressure distribution, where the presence of the passive cavity structure has less effect on the upper expansion ramp pressure distribution of the nozzle. As the drop pressure ratio increases, such as in the drop pressure ratio range *NPR* = 5–12, at which point the separation region of the upper expansion ramp is within the region contained by the passive cavity, the presence of the passive cavity influences the distribution of the upper expansion ramp pressure, a second smaller separation region occurs after the first refluxing region, the pressure rises abruptly, and the rise is higher than that of the prototype SERN. As the drop pressure ratio increases, such as at *NPR* = 13, the upper expansion ramp separation region is essentially outside the passive cavity structure, which has little effect on the upper expansion ramp pressure distribution on the nozzle.

In the drop pressure ratio range *NPR* = 3–13, the passive cavity SERN increases the axial thrust coefficient by a percentage over the prototype SERN as shown in Table 5.7. The axial thrust coefficient of the SERN with passive cavity structure is higher than that of the prototype SERN except for the drop pressure ratio *NPR* = 6, and the maximum increase of the axial thrust coefficient of the SERN is 4.00% at *NPR* = 3. At *NPR* = 6, the return region of the expansion ramp on the nozzle coincides with the structure of the passive cavity. At this time, the existence of the passive cavity does not improve the nozzle performance, but reduces the axial thrust coefficient of the nozzle.

From the above, it can be seen that when *NPR* = 6, the passive cavity structure coincides with the position and size of the flow separation region at the expansion ramp on the nozzle. At this time, the existence of the passive cavity structure not only does not raise the SERN axial thrust coefficient, but reduces the SERN axial thrust coefficient. At this time, the existence of the passive cavity structure also makes the passive cavity SERN axial thrust coefficient lower than that of the prototype SERN. For this reason, the flow field and performance of SERN with passive cavity are calculated for the cases of 5.5, 5.8, 6.2 and 6.5 drop pressure ratios near *NPR* = 6 to further illustrate this problem. The local Mach number distribution of the nozzle in the fall pressure ratio range 5.5–6.5 is shown in Fig. 5.82. From the figure, it can be seen that the separation region at the expansion ramp on the nozzle coincides with the passive cavity structure in the range of 5.5–6.5 drop pressure ratio, and the existence of the passive cavity structure has little effect on the size and location of the expansion ramp separation region on the nozzle. The axial thrust coefficients of the prototype SERN and SERN with passive cavity in the drop pressure ratio range 5.5 to 6.5 are shown in Table 5.8. When the size of the passive cavity structure coincides with the position and size of flow separation within the nozzle, the existence of the passive cavity structure does not improve the flow field

Table 5.7 SERN thrust coefficient increase percentage with passive cavity

NPR	3	4	5	6	7	8	9	10	11	12	13
%	4.00	3.58	2.8	−1.41	0.70	1.75	2.26	1.59	0.53	1.22	1.37

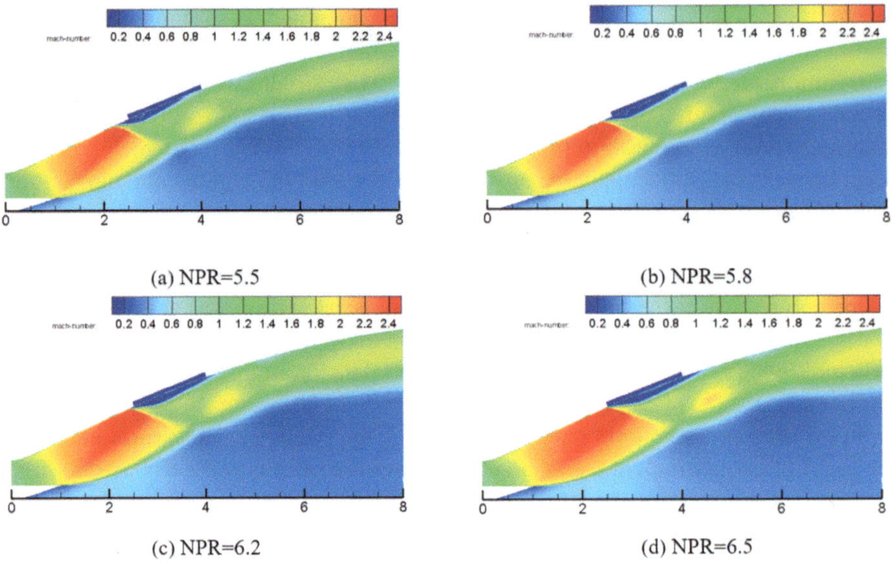

Fig. 5.82 Passive cavity SERN local Mach number distribution at different drop pressure ratios. (**a**) *NPR* = 5.5. (**b**) *NPR* = 5.8. (**c**) *NPR* = 6.2. (**d**) *NPR* = 6.5

Table 5.8 Comparison of thrust coefficient between prototype SERN and SERN with passive cavity

NPR	5.5	5.8	6.2	6.5
Prototype SERN	0.6516	0.6742	0.6941	0.7053
SERN with passive cavity	0.6475	0.6665	0.6828	0.6959

structure inside the nozzle, and the axial thrust coefficient of SERN with passive cavity is lower than that of the prototype SERN.

5.6.3 *Effects of Passive Cavity Geometry on SERV Performance for Tandem TBCC Engines*

The SERN with passive cavity structure and the passive cavity structure represented by Figs. 5.72 and 5.73 show that the plate opening ratio, aperture, and cavity depth of the passive cavity are important factors that affect the flow field and performance of the SERN with passive cavity. Therefore, the effects of the hole plate opening ratio, aperture and cavity depth on the passive cavity SERN are studied numerically.

1. Effect of orifice opening ratio on passive cavity SERN

First, the effect of orifice opening rate change on nozzle flow field of SERN with passive cavity under over expansion is analyzed, keeping the hole diameter at

Table 5.9 Number of holes at different orifice plate opening rates

Open hole ratio (%)	4.93	11.08	23.40	35.71	48.02	60.33	73.88	80.04
Number of holes	4	9	19	29	39	49	60	65

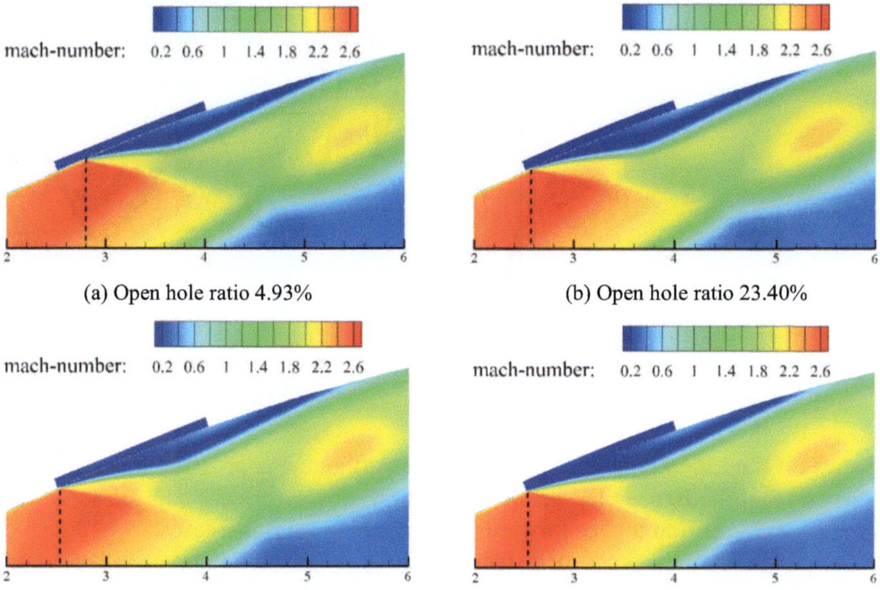

(a) Open hole ratio 4.93%

(b) Open hole ratio 23.40%

(c) Open hole ratio 4802%

(d) Open hole ratio 73.88%

Fig. 5.83 Local Mach number distribution near the SERN passive cavity in a passive cavity with different plate opening rates. (**a**) Open hole ratio 4.93%. (**b**) Open hole ratio 23.40%. (**c**) Open hole ratio 48.02%. (**d**) Open hole ratio 73.88%

20 mm and the cavity depth at 100 mm constant, with the opening rates at 4.93%, 11.08%, 23.40%, 35.71%, 48.02%, 60.33%, 73.88% and 80.04% respectively. The number of holes at different orifice opening rates is shown in Table 5.9. The SERN of passive cavity with different apertures was simulated numerically at Mach = 0.5 and *NPR* = 8. Figure 5.83 shows the local Mach number distribution near the passive cavity, and the performance of the nozzle at SERN in the over expansion regime is related to the size of the separation zone and the starting location of the separation zone. The structure of the flow field inside the passive cavity SERN with different orifice ratios is similar, and the existence of the passive cavity structure results in a stable region of return flow at the expansion ramp on the nozzle. The location of the first hole on the orifice plate varies for different orifice plate opening ratios, and as the orifice plate opening ratio increases, the location of the first hole is closer to the leading edge location of the orifice plate, so the flow with higher pressure behind the shock wave flows into the passive cavity, while the initial location of the flow exiting from the leading edge of the passive cavity varies, resulting in different positions and intensi-

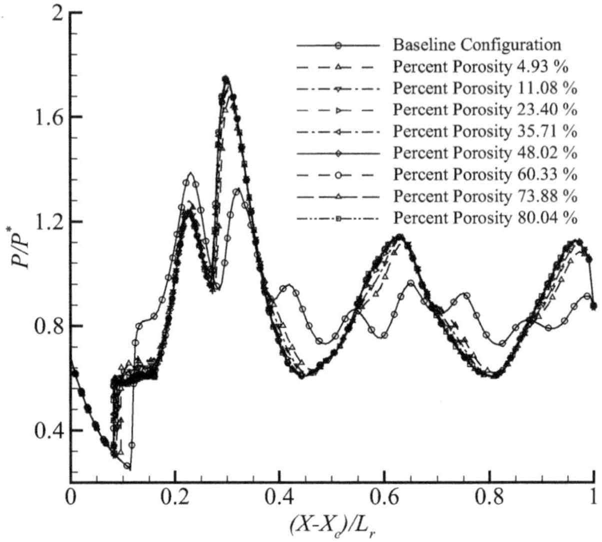

Fig. 5.84 Upper expansion ramp pressure distribution

ties for the formation of the leading section oblique compression wave. It can be seen from the figure that the shock position moves forward and the separation zone start position moves forward as the opening ratio increases.

Figure 5.84 shows the pressure distribution on the expansion ramp on the prototype SERN and passive cavity SERN with different orifice opening rates. Compared to the prototype SERN, the change of the upstream separation zone affects the change of the expansion ramp pressure on the nozzle due to the shift of the separation zone generated at the expansion ramp on the SERN with a passive cavity upstream and the increase of the separation zone area. With the increase of the orifice opening ratio, the separation starting position of the upper expansion ramp moves forward, the front section oblique compression wave weakens gradually, the pressure increase of the upper expansion ramp decreases gradually, and the upper expansion ramp pressure decreases gradually, resulting in the thrust coefficient decreasing gradually. From the pressure distribution on the passive cavity underside with different orifice opening ratios as shown in Fig. 5.85, starting at the passive cavity leading edge point X_f in the abscissa and dimensionless with the passive cavity axial length L in the abscissa, it can be more clearly seen that with the same aperture and cavity depth, as the opening ratio increases, the return flow at the passive cavity increases gradually, and the passive cavity underside pressure decreases gradually, which in turn continuously influences the upper expansion ramp pressure distribution, so that the passive cavity nozzle thrust coefficient varies with the orifice opening ratio as shown in Fig. 5.86, and as the orifice opening ratio increases, the passive cavity SERN axial thrust coefficient decreases.

2. Effect of aperture size of orifices on passive cavity SERV

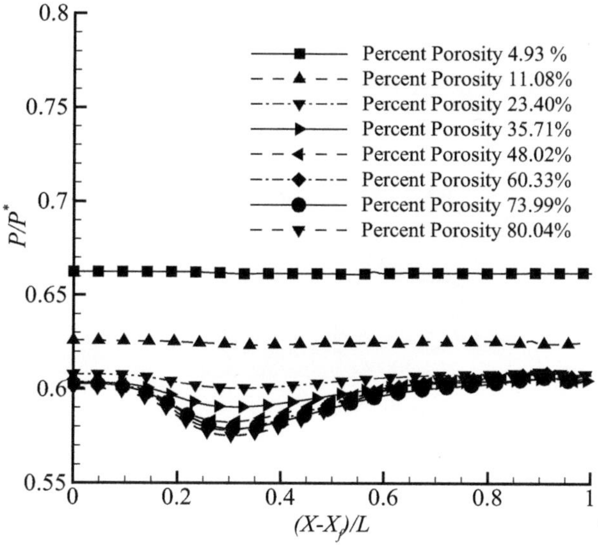

Fig. 5.85 Passive cavity bottom pressure distribution

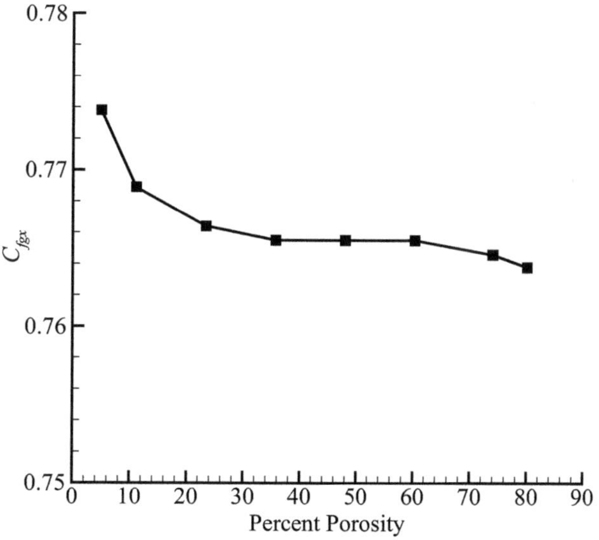

Fig. 5.86 Thrust coefficient changes with orifice opening ratio

The effect of the aperture change of the passive cavity orifice plate on the passive cavity SERN was analyzed, with the orifice plate opening rate of 35.71% and the cavity depth of 100 mm unchanged, and the hole diameters of 10.00 mm, 14.50 mm, 20.00 mm, 23.20 mm, 29.00 mm, and 36.25 mm, respectively. The passive cavity SERN with six apertures was numerically simulated at Mach = 0.5 for the outflow and *NPR* = 8 for the drop pressure ratio. The number of holes at different aperture sizes is shown in Table 5.10. The local Mach number distribution of passive cavity nozzles with different orifice plate apertures is shown in Fig. 5.87.

Table 5.10 Number of holes in different aperture size

Aperture(mm)	10.00	14.50	20.00	23.20	29.00	36.25
Number of holes	58	40	29	25	20	16

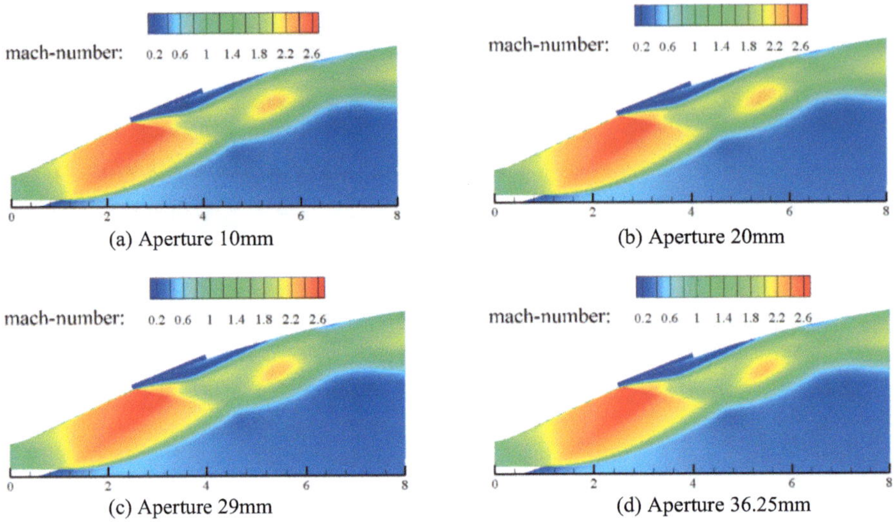

Fig. 5.87 Mach number distribution near SERN passive cavity with different apertures. (**a**) Aperture 10 mm. (**b**) Aperture 20 mm. (**c**) Aperture 29 mm. (**d**) Aperture 36.25 mm

At the same opening rate and cavity depth, the first hole of the orifice plate is approximately the same position from the edge of the orifice plate, so the position where the airflow exits the leading edge of the passive cavity is the same, the intensity of compression waves at the leading edge of the passive cavity is essentially the same, and the passive cavity SERN flow field is essentially the same for each orifice plate aperture.

The expansion surface pressure coefficient distribution on the SERN of passive cavities with different orifice plate apertures is shown in Fig. 5.88. At the same orifice opening rate and cavity depth, the flow field of the entire nozzle changes little, the size and location of the separation region of the upper expansion ramp changes little, and the size, location, and peak value of the high-pressure region are almost constant, which indicates that the flow field structure at the corresponding location of the passive cavity nozzle with different orifice apertures is consistent, and the upper expansion ramp pressure is almost constant. Therefore, the change of orifice size has little effect on the flow field structure and the pressure distribution of the upper expansion ramp in the passive cavity SERN. Figure 5.89 shows the passive cavity underside pressure distribution for different orifice plate aperture sizes. Under the same open hole ratio and cavity

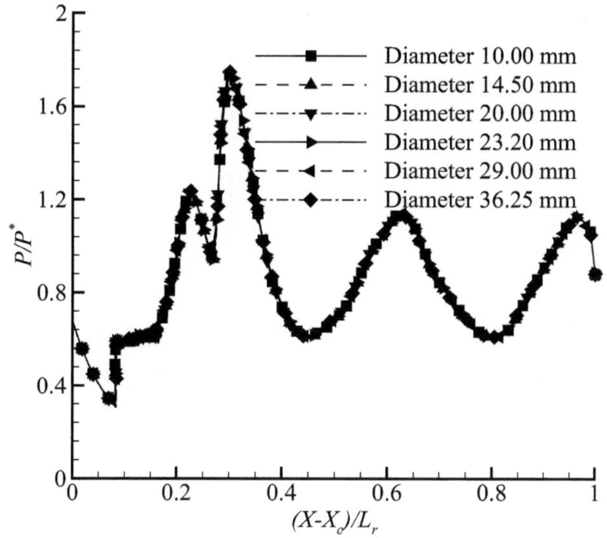

Fig. 5.88 Upper expansion ramp pressure distribution variation chart

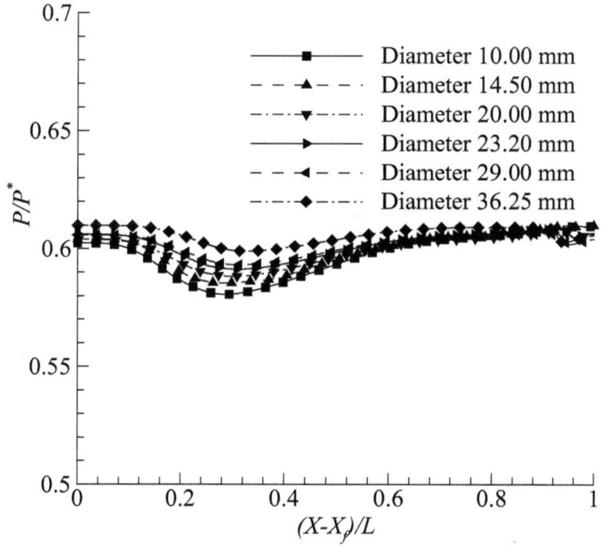

Fig. 5.89 Passive cavity bottom pressure distribution

depth, the change of back flow at the passive cavity with different aperture size is small, and the change of pressure distribution on the bottom surface is small. Therefore, at the same orifice plate opening rate of 35.71% and cavity depth of 100 mm, the size of the passive cavity orifice plate has less effect on the flow in the nozzle, that is, less effect on the nozzle thrust coefficient.

The relative position of the induced shock at the upper expansion ramp of a passive cavity SERN with different aperture sizes at the same opening rate and cavity

Table 5.11 Relative positions of induced shock waves

Aperture (mm)	10.00	14.50	20.00	23.20	29.00	36.25
Relative position (%)	8.20	8.24	8.30	8.33	8.39	8.46

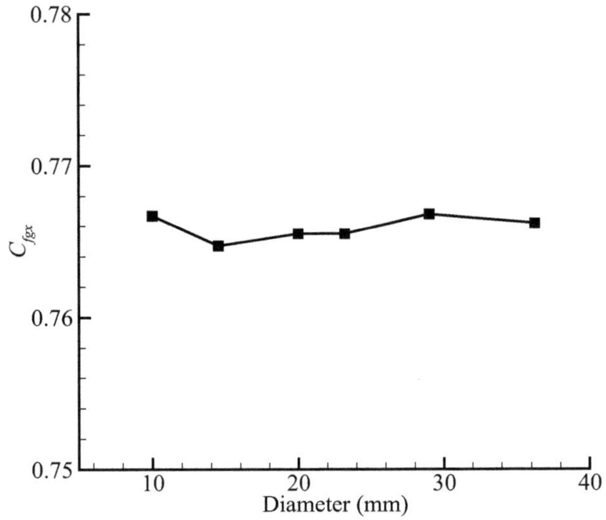

Fig. 5.90 Variation of nozzle axial thrust coefficient with aperture size

depth (i.e. the ratio of the shock position to the axial distance of the upper expansion ramp extension) is shown in Table 5.11. As can be seen from the table, as the aperture increases, the relative position of the channel oblique shock generated at the trailing edge of the lower oblique plate, which intersects the expansion ramp on the SERN, creates an incident shock that interacts with the boundary layer and creates an induced shock that moves towards the nozzle exit, but the relative position of the move is smaller, varying within 1%. The position, intensity and shock angle of the two shock waves near the passive cavity are essentially the same, and the structure of the shock train near it is also the same. For different aperture sizes, the flow is the first orifice at the leading edge of the orifice plate to separate the gas flow, and while the aperture size varies, since the opening rate is kept constant, meaning that the sum of the absolute values of the flow rate into and out of the passive cavity is constant, the different aperture sizes have the same effect on the flow near the unprovoked cavity. So the area of flow separation region is the same at the passive cavity, and the area of small separation region downstream of the passive cavity is also the same, and the overall internal flow of the nozzle changes less. Figure 5.90 shows how the SERN thrust coefficient with a passive cavity varies with the size of the orifice plate aperture, which has less effect on the nozzle thrust coefficient.

Table 5.12 Position change of the passive cavity on the nozzle expansion ramp

	1	2	3	4	5	6	7	8	9	10
X_f/L_r	3.90%	6%	8.11%	12%	16%	18%	20%	23%	25%	27%
X_b/L_r	12%	14%	16%	20%	24%	26%	28%	31%	34%	36%

5.6.4 Effect of Passive Cavity Axial Position on SERV Performance for Tandem TBCC Engines

The effects of the passive cavity structure on the nozzle flow field and performance at different positions of the expansion ramp on the SERN were analyzed. At some drop pressure such as *NPR* = 8, a passive cavity structure with an aperture of 20 mm and a depth of 100 mm is selected, leaving the passive cavity size unchanged. Changing the position of the passive cavity structure on the nozzle expansion ramp changes the passive cavity leading edge position X_f/L_R, and the passive cavity trailing edge position X_b/L_R changes accordingly so that remains unchanged. See Fig. 5.74 for the SERN structure with a passive cavity. The positions of the leading and trailing edge points of the passive cavity at different points of the expansion ramp on the SERN are shown in Table 5.12. From the position and size of the separation region of the prototype SERN at a drop pressure ratio of 8, it is known that the entire passive cavity structure at position 1 is essentially in front of the separation region, position 3 is the position chosen to study the structural parameter changes of the passive cavity, and position 9 and position 10 are essentially behind the separation region. For two kinds of passive cavity structures with apertures of 4.93% and 23.40%, the effect of different positions of the expansion ramp on the SERN is analyzed.

Considering the passive cavity structure with an aperture of 20 mm and a depth of 100 mm, the effect of the passive cavity structure on the SERN flow field and performance is calculated by different positions of the expansion ramp on the SERN according to the passive cavity structure at different positions of the expansion ramp on the SERN. Numerical simulations were performed at Mach 0.5 for the external flow and 8 for the nozzle drop pressure ratio. The local Mach numbers of the prototype SERN and SERN with passive cavity near the passive cavity are shown in Fig. 5.91. When the passive cavity is located closer to the throat, the separation region on the upper expansion ramp still exists. The existence of the passive cavity affects nozzle performance by changing the upper expansion ramp pressure. At 12% position, passive cavity structure basically coincides with the return current region, and SERN with the passive cavity has lower performance than prototype SERN. When the passive cavity structure gradually moves away from the nozzle throat position, after 18% position, the separation region of the upper expansion ramp decreases and vanishes, at which point the passive cavity acts to suppress flow separation.

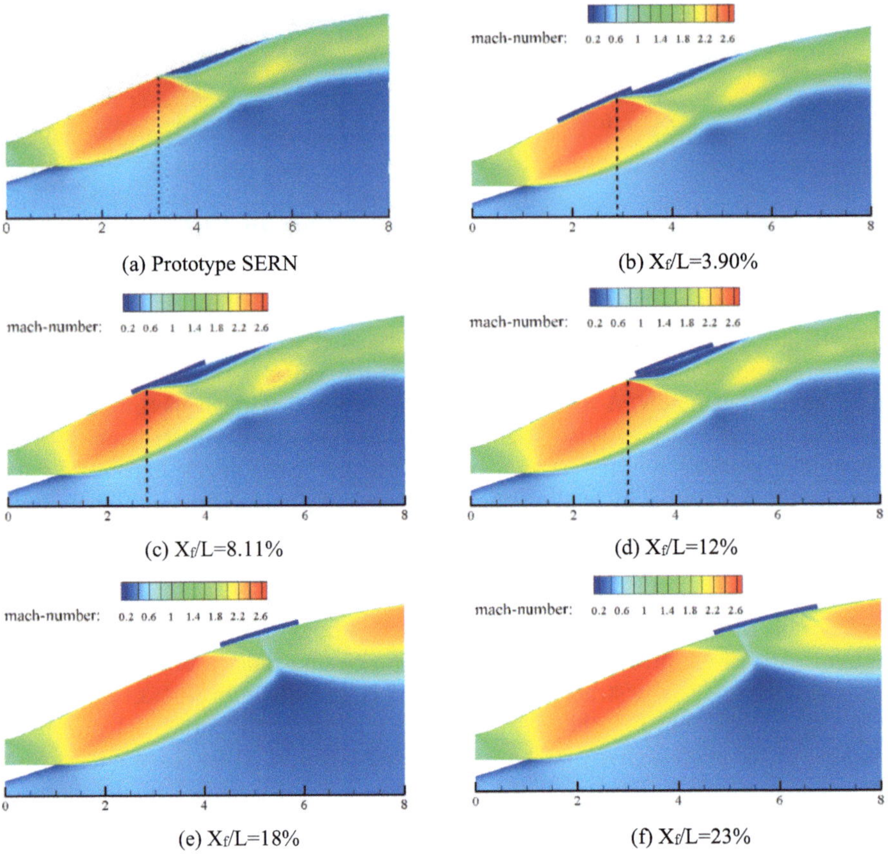

Fig. 5.91 Mach number distribution near the passive cavity at different locations on the expansion ramp on the SERN. (**a**) Prototype SERN. (**b**) $X_f/L = 3.90\%$. (**c**) $X_f/L = 8.11\%$. (**d**) $X_f/L = 12\%$. (**e**) $X_f/L = 18\%$. (**f**) $X_f/L = 23\%$

Figure 5.92 shows the expansion ramp pressure distribution on the nozzle as the passive cavity structure expands the ramp at different locations on the nozzle. When the separation region of the upper expansion ramp still exists, the pressure distribution of the upper expansion ramp and the flow field in the nozzle are basically the same. When the separation region of the upper expansion ramp disappears, the flow field inside the nozzle changes significantly, as does the pressure distribution of the upper expansion ramp, with fewer shock trains and higher shock peaks, significantly improving the nozzle performance. As shown in Fig. 5.93 and Table 5.13, the nozzle axial thrust coefficient varies and the axial thrust coefficient increases by a percentage as the passive cavity structure expands the ramp at different locations on the SERN. When the passive cavity structure acts to facilitate flow separation at the expansion ramp on the nozzle, there is an optimal position, position 2, where $X_f/L_R = 6\%$, the axial thrust coefficient of the nozzle increases by 3.10%. When the

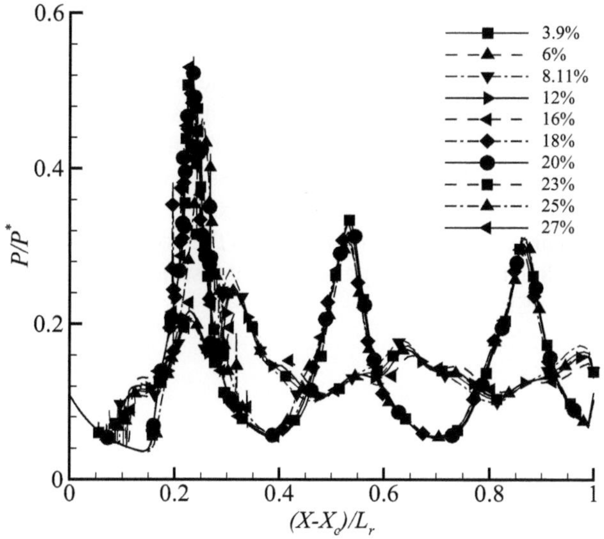

Fig. 5.92 Expansion ramp pressure distribution on SERN

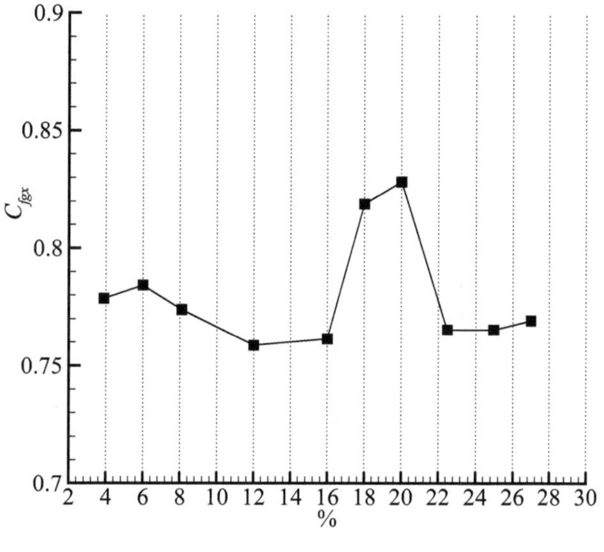

Fig. 5.93 SERN axial thrust coefficient

Table 5.13 Thrust coefficient improvement percentage of SERN with passive cavity structure at different locations of the expansion ramp on the nozzle

X_f/L_r	3.90%	6%	8.11%	12%	16%	18%	20%	23%	25%	27%
%	2.36	3.10	1.75	−0.25	0.11	7.64	8.89	0.61	0.61	1.13

passive cavity structure acts to suppress flow separation at the expansion ramp on the nozzle, there is also an optimal position, position 7, where $X_f/L_R = 20\%$, the axial thrust coefficient of the nozzle is increased by 8.89%. It is known that the thrust coefficient of the nozzle increases significantly when the passive cavity structure acts to suppress flow separation than when it acts to facilitate flow separation. At position 4, $X_f/L_R = 12\%$, the passive cavity structure did not play a positive role when the size and position of the passive cavity structure coincided with the separation area at the expansion ramp on the nozzle, at which point the SERN axial thrust coefficient with the passive cavity structure was slightly lower than the prototype SERN.

Chapter 6
Coupled Design and Calculation of Special-Purpose Nozzles and Aero-Engine Systems

6.1 Introduction

In contrast to conventional nozzles, special-purpose nozzles are often characterized by unconventional geometric configurations or are designed to operate under specific and complex inlet conditions. These unique features not only increase the dimensionality of the nozzle design space, but also significantly complicate the coupling mechanisms between nozzle geometry and the overall engine operating conditions. For conventional nozzles, the primary influence on overall engine performance is typically limited to variations in throat and exit areas. However, in the case of special-purpose nozzles, geometric parameters have a more intricate impact on the engine operating point. Additionally, the performance behavior of special-purpose nozzles under nonstandard inlet conditions diverges notably from that of conventional configurations, rendering it inappropriate to analyze engines equipped with such nozzles using conventional nozzle performance characteristics (Zhang 2005). Hence, the development of a coupled simulation model that integrates special-purpose nozzle behavior with complete engine system dynamics is essential for accurately evaluating their influence on overall engine performance.

The traditional performance simulation models of aircraft engines are typically zero-dimensional, where the operational characteristics of components are represented using performance maps. This chapter utilizes nozzle performance maps to couple the three-dimensional computational fluid dynamics (CFD) simulation results of special-purpose nozzles with the overall engine performance simulation model. Initially, a three-dimensional CFD simulation is performed for the special-purpose nozzle to obtain characteristic data points. These data points are subsequently used to construct a surrogate model for the nozzle's performance characteristics. Finally, the surrogate model is integrated with the overall engine simulation model, resulting in a coupled simulation framework that incorporates the influence of the special-purpose nozzle on the overall engine operating conditions.

J. Shi et al., *Special Purpose Nozzles*,
https://doi.org/10.1007/978-981-95-2939-1_6

The chapter first introduces the methodology for overall aero-engine performance calculation. It then presents a surrogate model-based coupled calculation method for advanced components and engine systems. Finally, the proposed approach is applied to the coupled simulation of three representative special-purpose nozzles—namely, the serpentine nozzle, the three-bearing swivel nozzle, and the single expansion ramp nozzle—demonstrating its applicability and effectiveness in analyzing nozzle–engine interactions.

6.2 Object-Oriented Methodology for Overall Aero-Engine Performance Calculation

The aero-engine performance simulation system is widely used in the development, operation, and maintenance of engines, serving as an indispensable tool in the advancement of aero-engine technology. By considering the engine as an integrated system composed of components such as the intake ducts, compression modules, combustion chamber, expansion modules, and nozzle, and analyzing the working principles and interrelationships among these components, the overall engine performance can be effectively characterized. This is achieved by formulating numerical methods for performance calculation from both the component level and the overall engine level, thereby providing a fundamental basis for comprehensive performance analysis of aero-engines.

6.2.1 Computational Methods for Aero-Engine Components

6.2.1.1 Intake Duct

The aerodynamic and thermodynamic calculations of the engine begin with the intake duct. First, the atmospheric temperature T_0 and pressure P_0 are calculated using the International Standard Atmosphere (ISA) model based on the altitude. Then, using the Mach number Ma_0 and the total pressure recovery coefficient σ_{intake} of the intake duct, the total temperature T_{t2} and total pressure P_{t2} at the outlet of the intake duct can be calculated by Eqs. (6.1) and (6.2).

$$P_{t2} = P_0 \left(1 + \frac{\gamma - 1}{2} Ma_0^{\ 2}\right)^{\gamma / \gamma - 1} \sigma_{inlet} \tag{6.1}$$

$$T_{t2} = T_0 \left(1 + \frac{\gamma - 1}{2} Ma_0^{\ 2}\right) \tag{6.2}$$

6.2.1.2 Compression Process

When the airflow passes through the fan, booster stage, and high-pressure compressor, it undergoes the same compression process. Similar calculation method is therefore available for each component. Given the component's pressure ratio π_c, isentropic efficiency η_c or polytropic efficiency e_c, inlet mass flow rate $\overset{\infty}{W}_{in}$, total temperature T_{tin}, total pressure P_{tin}, specific humidity ψ_{in}, and oil-to-air ratio f_{rin} (zero for compression components), the specific enthalpy H_{in} and specific entropy S_{in} at the compressor inlet can be calculated according to Eqs. (6.3) and (6.4):

$$H_{in} = f\left(T_{tin}, f_{rin}, \psi_{in}\right)_h \tag{6.3}$$

$$S_{in} = f\left(T_{tin}, f_{rin}, \psi_{in}\right)_s \tag{6.4}$$

Given the isentropic efficiency η_c, the specific entropy $S_{out,\,s}$ at the compressor outlet under isentropic compression is calculated based on the pressure ratio π_c and the inlet specific entropy S_{in} by the following formula:

$$S_{out,s} = S_{in} + R\ln\left(\pi_c\right) \tag{6.5}$$

From the isentropic outlet specific entropy S_{out}, the outlet total temperature $T_{tout,\,s}$ is obtained by:

$$T_{tout,s} = f\left(S_{out,s}, f_{rout}, \psi_{out}\right)_t \tag{6.6}$$

The outlet specific enthalpy $H_{out,\,s}$ is derived from $T_{tout,\,s}$ by:

$$H_{out,s} = f\left(T_{tout,s}, f_{rout}, \psi_{out}\right)_h \tag{6.7}$$

The actual outlet total enthalpy H_{out} is then calculated by applying the isentropic efficiencyη_c:

$$H_{out} = H_{in} + \left(H_{out,s} - H_{in}\right)/\,\eta_c \tag{6.8}$$

The actual total temperature T_{tout} and total pressure P_{tout} at the outlet are then determined using the outlet specific enthalpy H_{out} and the pressure ratio π_c according to the following formulas:

$$T_{tout} = f\left(H_{out}, fr_{out}, \psi_{out}\right)_h \tag{6.9}$$

$$P_{out} = \pi_c P_{in} \tag{6.10}$$

Given the polytropic efficiency e_c, the actual specific entropy S_{out} at the compressor outlet is calculated using the pressure ratio π_c, inlet specific entropy S_{in}, and polytropic efficiency$e_c$by the following formula:

$$S_{out} = S_{in} + R\ln(\pi_c)/e_c \tag{6.11}$$

The actual total temperature T_{tout} at the compressor outlet is calculated from the actual specific entropy S_{out} at the outlet:

$$T_{tout} = f(S_{out}, fr_{out}, \psi_{out})_t \tag{6.12}$$

The actual total pressure P_{tout} at the outlet is calculated using the pressure ratio π_c and Eq. (6.10), and the specific enthalpy H_{out} at the outlet is then calculated from the actual total temperature T_{tout} using Eq. (6.12):

$$H_{out} = f(T_{tout}, f_{rout}, \psi_{out})_h \tag{6.13}$$

Finally, the work required for the compression process is calculated using the mass flow rate $\overset{\infty}{W}_{in}$ and the outlet total temperature T_{tout}:

$$L_c = \dot{W}_{in}(H_{out} - H_{in}) \tag{6.14}$$

6.2.1.3 Combustion Process

In the combustion chamber, the mixture of air and fuel undergoes a chemical reaction upon ignition, releasing heat and resulting in a rise in gas temperature. During the calculation process, the combustion efficiency η_b, the total temperature T_{tout} at the combustion chamber outlet, and the lower heating value L_{cv} of the fuel are specified. Based on the energy balance of the combustion chamber, a relationship can be derived between the fuel–air ratio f_r and the specific enthalpy H_{in} at the inlet and the specific enthalpy H_{out} at the outlet of the combustion chamber, the combustion efficiency η_b, and the fuel heating value L_{cv}:

$$f_r = \frac{H_{out} - H_{in}}{\eta_b L_{cv} - H_{out}} \tag{6.15}$$

The fuel mass flow rate $\overset{\infty}{W}_f$ is then calculated using the fuel–air ratio:

$$\dot{W}_f = \dot{W}_{in} f_b \tag{6.16}$$

At this point, the total mass flow rate $\overset{\infty}{W}_{out}$ at the outlet of the combustion chamber is the sum of the fuel mass flow rate $\overset{\infty}{W}_f$ and the inlet air mass flow rate $\overset{\infty}{W}_{in}$:

$$\dot{W}_{out} = \dot{W}_{in} + \dot{W}_f \tag{6.17}$$

6.2.1.4 Expansion Work Process

The expansion work process primarily occurs as the gas flow passes through the turbine. The exhaust gases drive the turbine, performing work, while the total temperature and total pressure of the flow decrease. Given the component's pressure ratio π_c, isentropic efficiency η_c or polytropic efficiency e_c, inlet mass flow rate $\dot{W}_{in}$, total temperature T_{tin}, total pressure P_{tin}, specific humidity ψ_{in}, and fuel–air ratio f_{rin}, the calculation process is essentially the inverse of the compression process, which is not detailed here. At the turbine inlet, given the total temperature T_{tin}, total pressure P_{tin}, specific humidity ψ_{in}, fuel–air ratio f_{rin}, isentropic efficiency η_t or polytropic efficiency e_t, and specific enthalpy drop ΔH across the turbine, the specific enthalpy H_{in} at the turbine inlet can be calculated from the total temperature T_{tin} at the inlet:

$$H_{in} = f\left(T_{tin}, f_{rin}, \psi_{in}\right)_h \tag{6.18}$$

Given the isentropic efficiency η_t, the isentropic outlet specific enthalpy $H_{out,\,s}$, total temperature $T_{tout,\,s}$, and specific entropy $S_{out,\,s}$ at the outlet can be calculated according to the specific enthalpy H_{in} and the enthalpy drop ΔH at the inlet as follows:

$$H_{out,s} = H_{in} - \Delta H / \eta_t \tag{6.19}$$

$$T_{tout,s} = f\left(H_{out}, f_{rout}, \psi_{out}\right)_t \tag{6.20}$$

$$S_{out,s} = f\left(T_{tout,s}, f_{rout}, \psi_{out}\right)_s \tag{6.21}$$

The turbine expansion ratio can be calculated from the difference in specific entropy between the inlet and outlet:

$$\pi_t = \exp\left(\frac{S_{out,s} - S_{in}}{R}\right) \tag{6.22}$$

The total pressure P_{tout} at the outlet is determined based on the expansion ratio π_t:

$$P_{tout} = P_{tin}\pi_t \tag{6.23}$$

Given the polytropic efficiency e_t, the actual specific enthalpy H_{out}, total temperature T_{tout}, and specific entropy S_{out} at the turbine outlet can be calculated from the specific enthalpy H_{in} and enthalpy drop ΔH at the inlet:

$$H_{out} = H_{in} - \Delta H \tag{6.24}$$

$$T_{tout} = f\left(H_{out}, f_{rout}, \psi_{out}\right)_t \tag{6.25}$$

$$S_{out} = f\left(T_{tout}, f_{rout}, \psi_{out}\right)_s \tag{6.26}$$

The turbine expansion ratio is then calculated based on the specific entropy difference between the inlet and outlet, and the polytropic efficiency e_t:

$$\pi_t = \exp\left(\frac{S_{out} - S_{in}}{e_t R}\right) \tag{6.27}$$

Finally, the total pressure P_{tout} at the outlet is calculated using Eq. (6.23).

6.2.1.5 Mixing Process in the Mixing Chamber

In the mixing chamber, the outlet parameters are calculated using the pipeline mixing theories. One of the most important considerations is ensuring the static pressure balance between the inner and outer inlets of the mixing chamber. During the calculation, the following parameters are given: the mass flow rate $\overset{\infty}{W}_{in\_sub}$, total temperature T_{tin_sub}, total pressure P_{tin_sub}, specific humidity ψ_{in_sub}, and fuel–air ratio f_{rin_sub} at the outer inlet; and the mass flow rate W_{in}, total temperature T_{tin}, total pressure P_{tin}, specific humidity ψ_{in}, and fuel–air ratio f_{rin} at the inner inlet. The specific enthalpy at the inlet is calculated from the total temperature at the inlet:

$$H_{in\_sub} = f\left(T_{tin\_sub}, f_{rin\_sub}, \psi_{in\_sub}\right)_h \tag{6.28}$$

$$H_{in} = f\left(T_{tin}, f_{rin}, \psi_{in}\right)_h \tag{6.29}$$

Based on the conservation of mass, energy, and momentum, the following control equations are obtained:

$$\dot{W}_{n\_sub} + \dot{W}_{in} = \dot{W}_{\mathrm{out}} \tag{6.30}$$

$$\dot{W}_{in\_sub} H_{in\_sub} + \dot{W}_{in} H_{in} = \dot{W}_{\mathrm{out}} H_{\mathrm{out}} \tag{6.31}$$

$$\dot{W}_{in\_sub} V_{in\_sub} + \mathrm{P}_{\mathrm{sin\_}sub} A_{in\_sub} + \dot{W}_{in} V_{in} + \mathrm{P}_{\mathrm{sin}} A_{in} = I_{out} \tag{6.32}$$

$$H_{out} = C_p T_{out} + V_{out}^2 / 2 \tag{6.33}$$

$$I_{out} = W_{out}V_{out} + P_{sout}A_{out} \tag{6.34}$$

$$\dot{W}_{out} = V_{out}A_{out}p_{sout} / RT_{sout} \tag{6.35}$$

Equations (6.30), (6.31) and (6.32) are used to determine the outlet values of W''_{out}, $H_{out}$and I_{out}. By solving the system of nonlinear equations represented by Eqs. (6.33), (6.34) and (6.35), where the left-hand side is known and the right-hand side contains three unknowns V_{out}, T_{sout} and P_{sout}, the closed-form solutions of these unknowns can be thus obtained through iteration.

The outlet total temperature T_{tout} is calculated from the outlet specific enthalpy H_{out} (where f_{rout} and ψ_{out} are obtained based on mass averaging):

$$T_{tout} = f\left(H_{out}, f_{rout}, \psi_{out}\right)_t \tag{6.36}$$

The outlet total pressure P_{tout} is then determined based on the outlet total temperature T_{tout}, static temperature T_{sout}, and static pressure P_{sout}:

$$P_{tout} = P_{sout}\left(T_{tout} / T_{sout}\right)^{\frac{\gamma}{\gamma-1}} \tag{6.37}$$

6.2.1.6 Nozzle Expansion and Acceleration Process

When the airflow passes through the nozzle, it undergoes expansion and acceleration before entering the atmospheric environment. Taking a convergent nozzle as an example, the calculation involves given parameters: the inlet mass flow rate W_{in}, total temperature T_{tin}, total pressure P_{tin}, specific humidity ψ_{in}, fuel–air ratio f_{rin}, environmental pressure P_0, and the nozzle contraction angle κ (as shown in Fig. 6.1).

The specific entropy S_{in} at the inlet is calculated from the total temperature T_{tin} at the inlet:

$$S_{in} = f\left(T_{tin}, f_r, \psi\right)_s \tag{6.38}$$

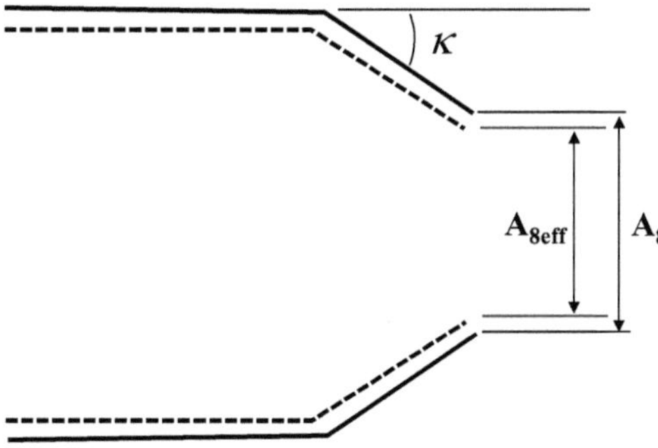

Fig. 6.1 Relationship between the geometric area and effective area at the nozzle exit

Since the expansion process of the airflow through the nozzle can be considered an isentropic expansion process, we have: $T_{tout} = T_{tin}$, $H_{out} = H_{in}$, and $S_{out} = S_{in}$.

Using the outlet total temperature T_{tout}, total pressure P_{tout}, inlet specific entropy S_{out}, and environmental pressure P_0, the outlet static temperature T_{sout} can be calculated by:

$$T_{sout} = f\left(\left(f\left(T_{tout}, f_{rout}, \psi_{out}\right)_s - R\log\left(P_{tout} / P_0\right)\right), f_{rout}, \psi_{out}\right)_s \tag{6.39}$$

The relationship between the specific enthalpy H_{out} at the nozzle exit and the kinetic energy per unit mass of the airflow allows for the calculation of the airflow velocity V_{out} at the nozzle exit:

$$V_{out} = \sqrt{2\left(H_{out} - f\left(T_{sout}, f_{rout}, \psi_{out}\right)\right)} \tag{6.40}$$

The Mach number Ma at the nozzle exit is calculated from the nozzle exit airflow velocity V_{out} and static temperature T_{sout}:

$$Ma = {V_{out}}\Big/{\sqrt{\gamma R T_{sout}}} \tag{6.41}$$

When the Mach number calculated from Eq. (6.41) is less than 1, i.e. $Ma < 1.0$, the nozzle is in a subcritical fully expanded state. The airflow parameters calculated from Eqs. (6.39) to (6.41) correspond to the nozzle exit parameters. When $Ma > 1.0$in Eq. (6.41), the nozzle is in a supercritical under-expanded state. In this case, the exit Mach number is 1.0, and the static temperature T_{sout} and static pressure P_{sout} at the nozzle exit are determined by iteratively solving when the Mach number is 1.0. The effective area A_{8eff} at the nozzle exit is calculated from the mass flow rate $\dot{W}_{out}$, static temperature T_{sout}, static pressure P_{sout}, and airflow velocity V_{out} at the nozzle exit:

$$A_{8eff} = \frac{\dot{W}_{out} R T_{sout}}{P_{sout} V_{out}} \tag{6.42}$$

The relationship between the nozzle exit effective area A_{8eff} and the geometric area is:

$$A_8 = {A_{8eff}}\Big/{C_D} \tag{6.43}$$

where C_D is the nozzle discharge coefficient.

The discharge coefficient C_D is a function of the nozzle blowdown ratio π_{noz} and the nozzle contraction angle κ (as shown in Fig. 6.2). Given the nozzle blowdown

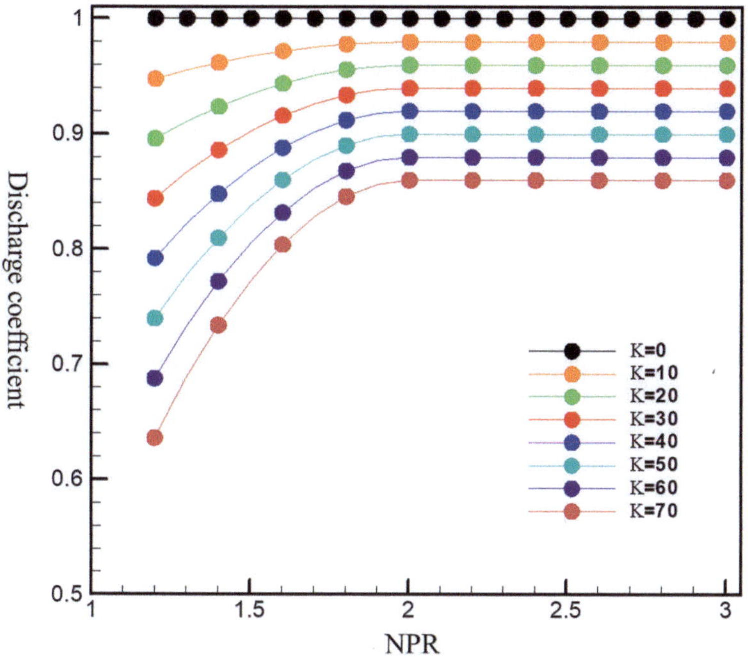

Fig. 6.2 Relationship between nozzle blowdown ratio and discharge coefficient at different contraction angles

ratio π_{noz} and the contraction angle κ, the nozzle discharge coefficient C_D can be determined through interpolation.

6.2.2 Overall Engine Matching and Performance Calculation Methods for Aero-Engines

6.2.2.1 Performance Calculation Method for Design Point

As the name implies, the engine design point refers to a condition where, under specified flight conditions, engine operating parameters are selected to meet performance requirements such as thrust and specific fuel consumption. Based on the thrust, the engine size (inlet frontal area) and component characteristic dimensions (such as nozzle throat area) are further determined. Therefore, the purpose of thermodynamic calculation at the design point is to compute the airflow parameters from the inlet of the intake duct to the outlet of the nozzle for the selected engine operating parameters such as bypass ratio, overall pressure ratio, turbine inlet total temperature, and component efficiencies or loss coefficients, in order to obtain the variation patterns of specific engine performance with respect to operating parameters. Since the component parameters are already fixed in design point calculations, no iteration or nonlinear equation solving is required. A one-pass thermodynamic calculation from engine inlet to outlet based on the given design parameters suffices, and therefore the detailed process is not elaborated here.

6.2.2.2 Overall Engine Matching and Performance Calculation Method for Non-design Points

In actual operation, changes in throttle position, flight Mach number, flight altitude, and atmospheric conditions cause the engine to deviate from its original design point, entering a non-design point. The relationships between engine performance parameters such as thrust and specific fuel consumption with variables like throttle position, flight Mach number, flight altitude, and atmospheric conditions are generally referred to as engine characteristics. For ease of study, engine characteristics are typically categorized into altitude–velocity characteristics (related to altitude and speed) and throttling characteristics (related to throttle position under given altitude and Mach number conditions).

In the calculation of the engine non-design point (i.e. off-design calculation), in addition to the influence of control variables, component efficiencies and total pressure recovery coefficients in each flow path also deviate from those at the design point. Therefore, the off-design calculation mainly solves the relevant parameters of each component through balance equations among components, and then performs thermodynamic calculations based on the solved parameters to finally obtain engine performance parameters such as thrust and specific fuel consumption. Taking the off-design thermodynamic cycle calculation of a mixed-flow engine as an example, under a single-parameter control strategy (e.g., controlling the relative corrected speed of the fan), the assumed unknowns in the cycle calculation are seven, namely the fan bypass ratio, fan pressure ratio, compressor pressure ratio, combustor outlet total temperature, high-pressure compressor relative corrected speed, high-pressure turbine blowdown ratio, and low-pressure turbine blowdown ratio (in actual calculations, the values of these variables at the design point can be used as their initial values). During multi-point off-design calculations, to accelerate the convergence, the calculated values of the above variables from the previous point can be used as initial values for the next point. These variables are determined based on the requirement that all engine components operate in coordination, meaning the engine must satisfy the following seven balance equations under off-design conditions:

(1) Mass flow continuity between the fan core outlet and compressor inlet
(2) Mass flow continuity between the combustor outlet and low-pressure turbine inlet
(3) Mass flow continuity between the low-pressure turbine outlet and high-pressure turbine inlet
(4) Static pressure balance of mixing flows in the mixer
(5) Power balance between fan and low-pressure turbine

(6) Power balance between compressor and high-pressure turbine
(7) Mass flow continuity between mixer outlet and nozzle inlet

These seven balance equations form the system of equations for off-design engine calculations, which can be written in parameter form as:

$$\left.\begin{aligned} E_1 &= f_1\left(X_1, X_2, \ldots, X_7\right) \\ E_2 &= f_2\left(X_1, X_2, \ldots, X_7\right) \\ &\vdots \\ E_7 &= f_7\left(X_1, X_2, \ldots, X_7\right) \end{aligned}\right\} \tag{6.44}$$

where X_1~X_7 are the seven unknown variables, and E_1~E_7 are their respective residual functions. The system of equations in Eq. (6.44) is typically solved using the Newton-Raphson iterative method. When the residuals of the seven variables fall below a specified tolerance, their solutions can thus be obtained.

6.3 Surrogate Model-based Fully Coupled Calculation Method for Advanced Components and Engine Systems

6.3.1 Selection of Component Calculation Points Based on Design of Numerical Experiments

The construction of surrogate models requires a sufficient number of sample points, typically obtained through high-fidelity simulations of the physical system. For complex systems, however, such simulations are often computationally expensive. It is therefore desirable to minimize the number of sample points while still ensuring the accuracy of the surrogate model. Achieving an optimal balance between computational cost and model fidelity necessitates the use of statistical Design of Experiments methods to efficiently plan the sampling strategy.

Design of Experiments (DOE), a well-established branch of statistics, is widely employed in parameter optimization, model evaluation, etc. (Sehonlau 1997; Mi et al. 2012). DOE is mainly used for identifying key experimental factors, determining optimal parameter combinations, and constructing empirical formulas and surrogate models. In this context, the primary purpose of DOE is to reasonably arrange the test scheme for building surrogate models. For clarity, the key terms used in DOE are explained as follows:

- **Experiment**: In numerical computation, this refers to a numerical experiment, i.e. a single execution of system simulation.
- **Factor**: The independent variable that influences the experimental response, corresponding to the design variables in the simulation system.

- **Level**: The number of distinct values a factor takes during the experiment; for example, two levels indicate two distinct values, and so on.

Commonly used DOE methods include full factorial design, orthogonal arrays, central composite design, and Latin hypercube design (Liu et al. 2011; He et al. 2012).

Full Factorial Design is the most fundamental approach in experimental design. It involves the complete combination of all factor levels, meaning that every possible configuration of the experimental factors is evaluated. The total number of experimental points in a full factorial design is given by m^n, where m represents the number of levels for each factor and n denotes the number of factors. The most commonly used configurations are the two-level (2^n) and three-level (3^n) designs. An example of a two-level, three-factor full factorial design is illustrated in Fig. 6.3a, where each dot "·" represents a selected experimental point. The full factorial design method is straightforward and provides high accuracy. However, the number of experimental points increases exponentially with the number of factors and their respective levels. For instance, if an experiment involves three factors, each with five levels, there would be 5^3 distinct combinations. Conducting a simulation for each combination would require 125 evaluations. In multidisciplinary design problems, the number of design variables is often large. Thus, using full factorial design to construct surrogate models becomes computationally prohibitive. Consequently, this method is only suitable for cases involving a small number of factors and levels.

Orthogonal arrays arrange experiments efficiently by employing pre-constructed orthogonal tables. It is a fast and effective Design of Experiments method. Common orthogonal tables include L_4, L_8, L_9 and L_{27}, among others. Table 6.1 presents an orthogonal table for a 4-factor, 3-level design. For additional orthogonal tables and detailed usage, readers may refer to relevant literature. The key advantages of orthogonal design lie in its balanced level combinations for each factor, uniform data point distribution, and significant reduction in the number of required experiments. Figure 6.3b illustrates the selected test points for a three-factor, two-level orthogonal arrays. As shown, compared with the full factorial design, orthogonal arrays significantly reduce the number of experiments.

Central Composite Design (CCD), also known as second-order regression rotational design, extends the design space to incorporate higher-order information. It offers advantages such as simplicity in design and good predictive capability. CCD is constructed by augmenting the 2^n full factorial design with a central point, along with two axial points added along each factor axis at a distance α from the center. These axial points are commonly referred to as "star points." A schematic of a three-factor CCD is shown in Fig. 6.3c. When $\alpha = 1$, the star points lie at the centers of each face of the design cube, a configuration referred to as Face-Centered Central Composite Design (FCCCD). In FCCCD, all design points can be represented by a central coordinate and a single radius factor, making it particularly suitable for programming and also applicable to designs with more than three factors.

The total number of experiments required in a CCD is given by $2^{n-k} + 2n + 1$, where n is the number of factors and k is the experiment control coefficient. To avoid rapid growth in the number of design points with increasing factors, the k

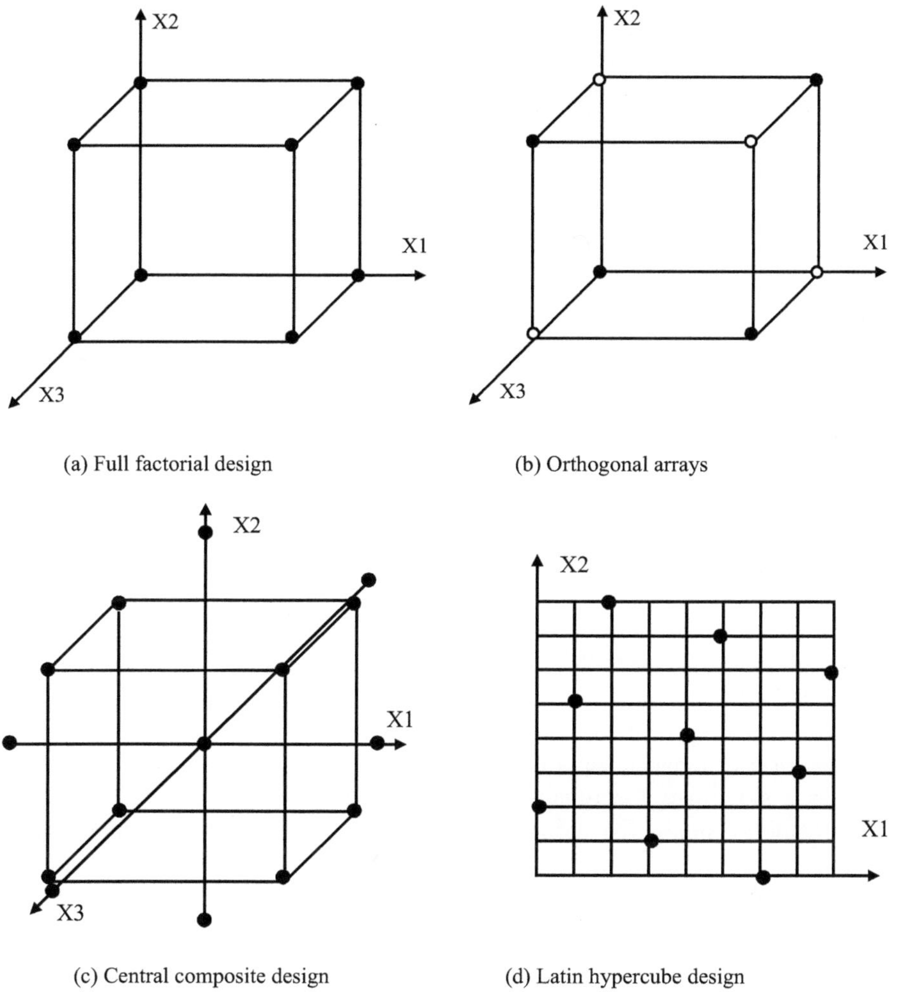

Fig. 6.3 Schematic diagrams of different design of experiments methods. (**a**) Full factorial design. (**b**) Orthogonal arrays. (**c**) Central composite design. (**d**) Latin hypercube design

value can be increased to control the number of required experiments effectively. Compared to full factorial design, CCD significantly reduces the number of experiments, which has led to its widespread application.

Latin Hypercube Design (LHD), first introduced by McKay and Beckman, is a sampling method developed specifically for computer simulation experiments. Its core principle involves dividing each factor's design range into m equal intervals (i.e., m levels) within an n-dimensional design space. Then, m points are randomly selected such that each level of each factor is used exactly once.

Due to the even partitioning of the design space for each factor and the constraint that each level is selected only once, LHD ensures that the sample points are evenly

Table 6.1 Orthogonal arrays tableL_9

No.	X1	X2	X3	X4
1	1	1	1	1
2	1	2	2	2
3	1	3	3	3
4	2	1	2	3
5	2	2	3	1
6	2	3	1	2
7	3	1	3	2
8	3	2	1	3
9	3	3	2	1

distributed throughout the design space. It captures the variability of the entire space using fewer sample points. For example, in a 2-factor, 9-level experiment, a full factorial design requires 81 (9×9) sample points, while LHD only requires 9. An illustration of LHD is shown in Fig. 6.3d.

6.3.2 Research on High-Accuracy Surrogate Modeling Methods

Surrogate modeling methods are techniques used to approximate the relationship between a set of independent variables and system responses. A surrogate model employs a simplified functional relationship to approximate and substitute the actual complex simulation model, thereby significantly reducing computational complexity and enhancing efficiency, while offering substantial value in engineering applications (Liu and Guo 2009). Surrogate modeling methods initially relied on polynomial fitting to construct surrogate models, which later evolved into various function fitting techniques. Depending on the approach used to construct the fitting functions, surrogate models can be categorized into polynomial surrogate models, Kriging surrogate models, radial basis function surrogate models, etc.

The key to constructing a surrogate model lies in how to effectively establish an approximate functional relationship between system responses and input variables, so that the surrogate model can approximate the true model response as closely as possible, and thus replace the complex real model in system analysis (Wang and Wang 2005). The basic idea is to analyze the input parameters and response values of sample points based on numerical experiment design, and then establish a surrogate model for the research object, which can be used to replace the original complex model for computational analysis or parameter optimization (Frederic and Crossley 2000; Guo et al. 2006). The general steps for constructing a response surface are as follows:

(1) Select the type of surrogate model;

(2) Construct the surrogate model based on numerical experimental sample points;
(3) Evaluate the constructed surrogate model;
(4) Apply the surrogate model that meets the performance requirements to the studied system.

The surrogate model describes the approximate relationship between independent variables and system response, which can generally be expressed as:

$$y(x) = \hat{y}(x) + \varepsilon \tag{6.45}$$

where:

$y(x)$——the true model of the response (accurate model), generally an unknown or complex function;
$\hat{y}(x)$——the surrogate model of the response;
ε——the random error between the approximate value and the actual value. Generally, ε is assumed to follow a standard normal distribution $N(0, \sigma^2)$.

The polynomial surrogate model uses polynomial regression fitting on the sample points to obtain an approximate functional relationship between the response and input variables. It is the most commonly used method for constructing surrogate models. According to the Weierstrass approximation theorem, any type of function can be approximated using a polynomial. Therefore, complex systems can always be approximated using corresponding polynomial surrogate models. In addition, since polynomial fitting is used to construct the surrogate model, the modeling process is straightforward, offering high fitting accuracy and excellent computational efficiency, making it widely adopted in practical applications. The fitting function of a polynomial surrogate model can generally be expressed as:

$$\hat{y}(x) = \sum_{i=1}^{n} b_i f_i(x) \tag{6.46}$$

where:

$f_i(x)$——polynomial basis function;
b_i——polynomial fitting coefficients;
n——number of fitting coefficients.

In polynomial fitting, a higher polynomial degree is not always better. As the degree increases, the number of undetermined coefficients in the polynomial grows exponentially, which in turn affects the model-building and analysis processes. Additionally, a higher degree requires more sample points, which increases the complexity of the design of numerical experiments. These sample points must also be obtained through precise model calculations, significantly affecting computational efficiency. Therefore, quadratic polynomials are typically used in practice. For linear problems, a first-degree polynomial may be used, while for nonlinear problems, a second-degree polynomial is preferred. However, quadratic polynomials may not adequately describe higher-order nonlinear problems, making them unsuitable for fitting high-order nonlinear models. The general form of a quadratic polynomial fitting is shown below:

$$\hat{y}(x) = b_0 + \sum_{i=1}^{n} b_i x_i + \sum_{i=1}^{n} b_{ii} x_i^2 + \sum_{1 \le i < j \le n} b_{ij} x_i x_j \tag{6.47}$$

where:

n——the number of fitting coefficients;

b_i——the polynomial fitting coefficients, also known as regression coefficients. For quadratic polynomial fitting, the number of regression coefficients can be calculated by $(n + 1)(n + 2)/2$. When the variables are two-dimensional, the number of regression coefficients is 6.

The regression coefficients can be solved using the least squares method. Once the regression coefficients are obtained, the polynomial surrogate model can be formed, followed by an evaluation of the model's accuracy.

The main evaluation methods for surrogate models are the relative root mean square error (RMSE) and the coefficient of determination (R^2). These two methods effectively reflect the fitting quality of the surrogate model, with the coefficient of determination R^2 being more widely used.

The RMSE is defined as follows:

$$\text{RMSE} = \frac{1}{n\overline{y}} \sqrt{\sum_{i=1}^{n} \left(y_i - \hat{y}_i \right)^2} \tag{6.48}$$

where:

n——the number of sample points;

y_i——actual response value;

$\hat{y}_i$——the fitted value obtained from the surrogate model;

$\overline{y}$——the mean of the actual response values.

RMSE reflects the difference between the surrogate model and the true values; the smaller the RMSE, the higher the fitting accuracy of the surrogate model.

The coefficient of determination R^2 is another very effective method for evaluating the fitting accuracy of a surrogate model. It assesses the degree of fit by defining a variable that represents the fitting quality. It is defined as:

$$R^2 = 1 - \frac{\sum_{i=1}^{n} \left(y_i - \hat{y} \right)^2}{\sum_{i=1}^{n} (y_i - \overline{y})^2} \quad \left(0 \le R^2 \le 1 \right) \tag{6.49}$$

For the coefficient of determination R^2, if R^2 is close to 1, it indicates a high degree of fit for the surrogate model. A value of R^2 between 0 and 0.25 suggests no correlation between the surrogate model and sample response values; A value between

0.25 and 0.5 indicates weak correlation; A value between 0.5 and 0.75 indicates strong correlation; A value between 0.75 and 1 suggests very good correlation. If the fitted values of all sample points are perfectly matched by the surrogate model, i.e. R^2=1, , it indicates that the surrogate model has perfectly fitted the sample response values.

6.3.3 Research on Surrogate Model-Based Engine Component and Overall Performance Model

The key to the coupling method between the engine component model and the overall engine performance model lies in utilizing the boundary conditions of the component model, derived from the overall engine performance model calculations, and then using these conditions to obtain the simulation performance results of the engine component model. These performance simulation results are then fed into the iterative solution process of the overall engine performance model until the final converged result is obtained.

Taking a mixed-flow turbofan engine as an example, the common operation process of the various components in a mixed-flow turbofan engine (without an afterburner) is shown in Fig. 6.4. The basic idea is to calculate the aerodynamic and thermodynamic parameters from the 0-th to the 9-th cross-sections, assuming initial values for any unknowns encountered. These unknowns are then solved for their precise values based on equilibrium conditions. Typically, the thermodynamic cycle calculation for a mixed-flow turbofan engine involves six unknowns: fan pressure

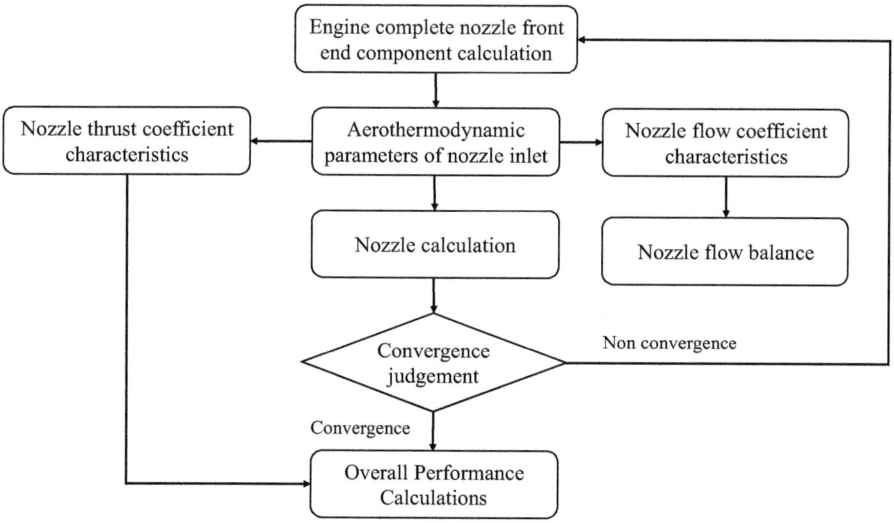

Fig. 6.4 Common operation process of the turbofan engine

ratio, compressor pressure ratio, fan relative equivalent speed, compressor relative equivalent speed, high-pressure turbine inlet equivalent flow, and low-pressure turbine inlet equivalent flow. The same number of equations as unknowns is required to solve for the unknowns. Based on the common operating characteristics of the various components of the aero-engine, the following six equilibrium equations can be established:

(1) Flow balance equation between the low-pressure turbine and the fan;
(2) Flow balance equation between the high-pressure turbine and the high-pressure compressor;
(3) Power balance equation between the low-pressure turbine and the fan;
(4) Power balance equation between the high-pressure turbine and the high-pressure compressor;
(5) Static pressure balance equation for the airflow inside and outside the mixing chamber inlet;
(6) Flow balance equation between the mixing chamber exit and the nozzle.

Solving the above balance equations will therefore conduct the solutions of the unknown variables. The 6 balance equations can be written in the following parametric form:

$$\left.\begin{aligned} E_1 &= f_1(V_1, V_2, \ldots, V_6) \\ E_2 &= f_2(V_1, V_2, \ldots, V_6) \\ &\vdots \\ E_6 &= f_6(V_1, V_2, \ldots, V_6) \end{aligned}\right\} \tag{6.50}$$

where V_1~V_6 represents the six variables, and E_1~E_6 their respective residual functions. This system of equations can be solved iteratively using the Newton-Raphson method. When the residual functions of the six variables fall below a specified tolerance, the six variable values can be obtained, thus determining the operating point of the aero-engine. Using the thermodynamic cycle parameters of the engine, the thrust, specific fuel consumption, and other parameters of the aero-engine can be calculated.

As shown in Fig. 6.4, the coupling between the nozzle model and the overall engine model mainly relies on flow balance. Therefore, the surrogate modeling process of nozzle and engine involves substituting the discharge coefficient from the nozzle surrogate model into the flow balance equation in the engine performance calculation, thus achieving flow matching between the two. In addition to flow matching, the calculation of the overall engine thrust must consider the impact of the nozzle thrust coefficient. Therefore, the nozzle thrust coefficient is substituted into the engine thrust calculation, obtaining thrust performance and fuel consumption characteristics that match both the nozzle and the overall engine model.

6.4 Coupled Calculation of Serpentine Nozzle and Turbofan Engine

6.4.1 *Construction and Evaluation of High-Precision Surrogate Model for Serpentine Nozzle*

From the design process of the serpentine nozzle, it is evident that there are numerous geometric parameters involved, as shown in Fig. 6.5. Through the study of their parametric effects, it was found that the parameters influencing the aerodynamic performance of the double serpentine nozzle include: centerline variation pattern, nozzle length-to-diameter ratio (L/D), exit width-to-height ratio (W_e/H_e), axial length ratio of the two bends (L_1/L_2), the first bend exit area (A_1), the first bend exit width (W_1), the first bend longitudinal offset (ΔY_1), and the nozzle pressure ratio (NPR). Since the centerline variation pattern is difficult to quantify, this section selects the other seven parameters as independent variables, including: L/D, L_2/L_1, $L_1/\Delta Y_1$, A_1/A_{in}, W_1/D, W_e/H_e, and NPR. For the baseline nozzle, the values are: L/D=2.8, L_2/L_1=2, $L_1/\Delta Y_1$=3, A_1/A_{in}=0.65, W_1/D = 1.1, W_e/H_e = 7, and NPR = 2.2. The centerline variation pattern for the double serpentine nozzle is taken as "sharp at the front, gradual at the rear," and the nozzle configuration considers the constraint of fully shielding the high-temperature turbine.

The aerodynamic performance parameters of the serpentine nozzle mainly include: total pressure recovery coefficient δp, discharge coefficient C_D, and thrust coefficient C_{fg}. This section compares and analyzes the aerodynamic performance of the serpentine nozzle under different operating conditions and geometric configurations to obtain the influence patterns of key parameters on the nozzle's aerodynamic performance. Table 6.2 provides the sample space for the research problem:

The Box-Behnken Design (BBD) method was selected to complete the design of experiments and establish the response surface model. Table 6.3 presents the R^2 values for the responses of the double serpentine nozzle approximation model. It

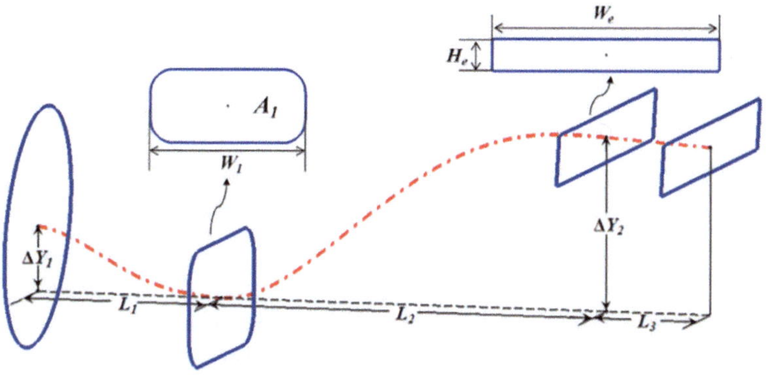

Fig. 6.5 Serpentine nozzle geometric design parameters

Table 6.2 Sample space of approximation model for double serpentine nozzle

L/D	L_2/L_1	$L_1/\Delta Y_1$	A_1/A_{in}	W_1/D	W_e/H_e	NPR
2.3	1	2	0.50	1.0	5	1.4
2.8	2	3	0.65	1.1	7	2.2
3.3	3	4	0.80	1.2	9	3.0

Table 6.3 R^2 values for the double serpentine nozzle approximation model

	C_D	C_{fg}
R^2	0.8048	0.8714

can be seen that the two responses of interest in this section both have R^2 values above 0.80, indicating a very good correlation between the approximation model and the response values. This model can thus be used to replace the precise model.

The relationships between the obtained responses and the independent variables are as follows:

Discharge coefficient C_D:

$$\begin{aligned} C_D &= 0.55319 + 0.13209\times(L/D) + 0.032201\times(L_2/L_1) + 0.026766\times(L_1/\Delta Y_1) \\ &+ 0.13828\times(A_1/A_{in}) + 0.16917\times(W_1/D) - 2.70771e-3\times(W_e/H_e) \\ &+ 0.020723\times NPR - 9.7204e-3\times(L/D)\times(L_2/L_1) \\ &- 3.31313e-3\times(L/D)\times(L_1/\Delta Y_1) - 0.042112\times(L/D)\times(W_1/D) \\ &- 0.014983\times(L_1/\Delta Y_1)\times(W_1/D) - 9.49359e-3\times(A_1/A_{in})\times(W_1/D) \\ &+ 1.58547e-3\times(W_1/D)\times(W_e/H_e) - 9.52777e-3\times(L/D)^2 \\ &- 1.11927e-3\times(L_2/L_1)^2 - 0.086584\times(A_1/A_{in})^2 - 3.8083e-3\times NPR^2 \end{aligned} \quad (6.51)$$

Thrust coefficient C_{fg}:

$$\begin{aligned} C_{fg} &= 0.43093 + 0.15569\times(L/D) + 0.034059\times(L_2/L_1) + 0.031082\times(L_1/\Delta Y_1) \\ &+ 0.11937\times(A_1/A_{in}) + 0.25342\times(W_1/D) - 4.23854e-3\times(W_e/H_e) \\ &+ 0.042696\times NPR - 0.012082\times(L/D)\times(L_2/L_1) \\ &- 4.86937e-3\times(L/D)\times(L_1/\Delta Y_1) - 0.059925\times(L/D)\times(W_1/D) \\ &+ 0.016257\times(L_1/\Delta Y_1)\times(A_1/A_{in}) - 0.023978\times(L_1/\Delta Y_1)\times(W_1/D) \\ &- 0.022829\times(A_1/A_{in})\times(W_1/D) + 2.44425e-3\times(W_1/D)\times(W_e/H_e) \\ &- 8.63651e-3\times(L/D)^2 - 0.092927\times(A_1/A_{in})^2 - 7.51538e-3\times NPR^2 \end{aligned} \quad (6.52)$$

As can be seen from the above, the relationships between C_D, C_{fg}, and the independent variables include linear and quadratic terms, as well as linear interaction term. These reflect the interaction between two factors and represent the interdependent characteristics of multiple parameters. The interaction term also plays a very important role in the model. Tables 6.4 and 6.5 show the results of the variance analysis of the discharge coefficient C_D and thrust coefficient C_{fg} for double serpentine nozzles. Table 6.4 shows that for the discharge coefficient approximation model: the NPR, $(L/D)^2$, $(A_1/A_{in})^2$, andNPR^2 in the linear and quadratic terms have highly significant effects, while L/D and W_1/D have significant effects, $L_1/\Delta Y_1$ and $(L_2/L_1)^2$have effects, and L_2/L_1, A_1/A_{in} and W_e/H_e have no effects. For the first-order interaction terms, L/D-L_2/L_1 has a highly significant effect, L/D-W_1/D has a significant effect, L/D-$L_1/\Delta Y_1$ and $L_1/\Delta Y_1$-W_1/D have effects, while A_1/A_{in}-W_1/D and W_1/D-W_e/H_e have no effects. Table 6.5 shows that for the thrust coefficient approximation model, NPR and NPR^2 in the linear and quadratic terms have highly significant effects, while L/D, $L_1/\Delta Y_1$ and W_1/D have significant effects, and L_2/L_1, A_1/A_{in} and W_e/H_e have no effects. For the first-order interaction terms, L/D-L_2/L_1 has a highly significant effect, L/D-$L_1/\Delta Y_1$, L/D-W_1/D, $L_1/\Delta Y_1$-A_1/A_{in} and $L_1/\Delta Y_1$-W_1/D have significant effects, while A_1/A_{in}-W_1/D and W_1/D-W_e/H_e have no effects.

Table 6.4 Approximation model ANOVA for the discharge coefficient C_D

Source	Sum of squares	df	Mean square	F Value	p-value Prob > F
Model	1.025e−003	17	6.027e−005	10.67	<0.0001
A-L/D	3.186e−005	1	3.186e−005	5.64	0.0220
B-L_2/L_1	6.171e−006	1	6.171e−006	1.09	0.3016
C-$L_1/\Delta Y_1$	1.625e−005	1	1.625e−005	2.88	0.0970
D-A_1/A_{in}	3.624e−009	1	3.624e−009	6.417e−004	0.9799
e−W_1/D	3.028e−005	1	3.028e−005	5.36	0.0253
F-W_e/H_e	2.571e−007	1	2.571e−007	0.046	0.8320
G-NPR	2.417e−004	1	2.417e−004	42.79	<0.0001
AB	1.890e−004	1	1.890e−004	33.46	<0.0001
AC	2.195e−005	1	2.195e−005	3.89	0.0550
AE	3.547e−005	1	3.547e−005	6.28	0.0160
CE	1.796e−005	1	1.796e−005	3.18	0.0815
DE	1.622e−007	1	1.622e−007	0.029	0.8662
EF	8.044e−007	1	8.044e−007	0.14	0.7077
A^2	8.112e−005	1	8.112e−005	14.36	0.0005
B^2	1.791e−005	1	1.791e−005	3.17	0.0819
D^2	5.426e−005	1	5.426e−005	9.61	0.0034
G^2	8.494e−005	1	8.494e−005	15.04	0.0003
Residual	2.485e−004	44	5.648e−006		
Lack of fit	2.485e−004	39	6.373e−006		
Pure error	0.000	5	0.000		
Cor total	1.273e−003	61			

Table 6.5 Approximation model ANOVA for the thrust coefficient C_{fg}

Source	Sum of squares	df	Mean square	*F* Value	*p*-value Prob > *F*
Model	3.006e−003	17	1.768e−004	17.53	<0.0001
A-L/D	6.727e−005	1	6.727e−005	6.67	0.0132
B-L_2/L_1	1.253e−006	1	1.253e−006	0.12	0.7262
C-$L_1/\Delta Y_1$	4.155e−005	1	4.155e−005	4.12	0.0485
D-A_1/A_{in}	8.689e−008	1	8.689e−008	8.616e−003	0.9265
E-W_1/D	6.114e−005	1	6.114e−005	6.06	0.0178
F-W_e/H_e	5.712e−007	1	5.712e−007	0.057	0.8130
G-NPR	1.424e−003	1	1.424e−003	141.20	<0.0001
AB	2.920e−004	1	2.920e−004	28.95	<0.0001
AC	4.742e−005	1	4.742e−005	4.70	0.0356
AE	7.182e−005	1	7.182e−005	7.12	0.0106
CD	4.757e−005	1	4.757e−005	4.72	0.0353
CE	4.600e−005	1	4.600e−005	4.56	0.0383
DE	9.381e−007	1	9.381e−007	0.093	0.7618
EF	1.912e−006	1	1.912e−006	0.19	0.6654
A^2	6.742e−005	1	6.742e−005	6.68	0.0131
D^2	6.322e−005	1	6.322e−005	6.27	0.0161
G^2	3.346e−004	1	3.346e−004	33.17	<0.0001
Residual	4.438e−004	44	1.009e−005		
Lack of fit	4.438e−004	39	1.138e−005		
Pure error	0.000	5	0.000		
Cor total	3.450e−003	61	61		

Table 6.6 Parameters of each component of a certain turbofan engine under takeoff conditions

Design point parameters	Value
Airflow/kg	285
Fan pressure ratio	2.08
Fan efficiency	0.86
Fan physical speed/%	91.5
Bypass ratio	2.33
High-pressure compressor pressure ratio	9.35
High-pressure compressor efficiency	0.85
High-pressure compressor physical speed/%	97.5
Turbine inlet temperature/K	1450
High-pressure turbine efficiency	0.89
Low-pressure turbine efficiency	0.91

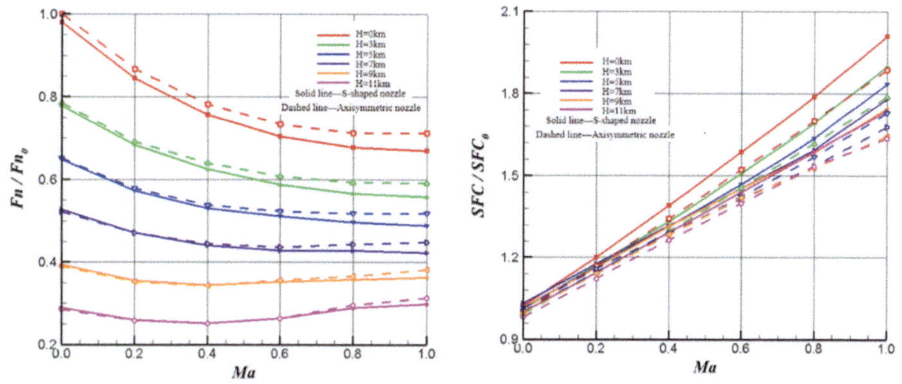

(a) Thrust

(b) Fuel consumption rate

Fig. 6.6 Altitude and velocity characteristics of aero-engine with serpentine nozzle. (**a**) Thrust. (**b**) Fuel consumption rate

6.4.2 Aerodynamic Performance Calculation and Analysis of Serpentine Nozzle in the Environment of a Turbofan Engine

This article establishes an engine simulation model based on a certain turbofan engine. The engine is a two-spool, mixe-flow, twin-shaft turbofan, while the nozzle is a subsonic, fixed-area nozzle. The takeoff condition is used as the design point, and the parameters of the design point are shown in Table 6.6.

Figures 6.6 and 6.7 show the altitude characteristics, velocity characteristics, and throttling characteristics of an aero-engine equipped with serpentine nozzle, and compare to the characteristics of an engine equipped with a conventional axisymmetric nozzle. The control rule for the engine is that the physical speed of the compressor remains constant, and the throttling characteristics are obtained under flight conditions of H = 0 km and *Ma* = 0. Since the matching between the serpentine nozzle and the engine is primarily based on flow (discharge coefficient) matching, the altitude characteristics, velocity characteristics, and throttling characteristics of the engine equipped with the serpentine nozzle are similar to those of the engine equipped with a conventional axisymmetric nozzle:

Under low-altitude conditions, engine thrust decreases as the flight Mach number *Ma* increases. Under high-altitude conditions, engine thrust first decreases and then increases as the flight Mach number *Ma* increases. Fuel consumption rate increases with *Ma*. This is because, as *Ma* increases, the total temperature at the engine inlet increases, and both the fan and compressor equivalent speeds decrease, leading to a reduction in fan pressure ratio and compressor pressure ratio. However, as *Ma* increases, the increase in the intake duct pressure ratio is greater, and the slower decrease in fan pressure ratio and compressor pressure ratio causes the total

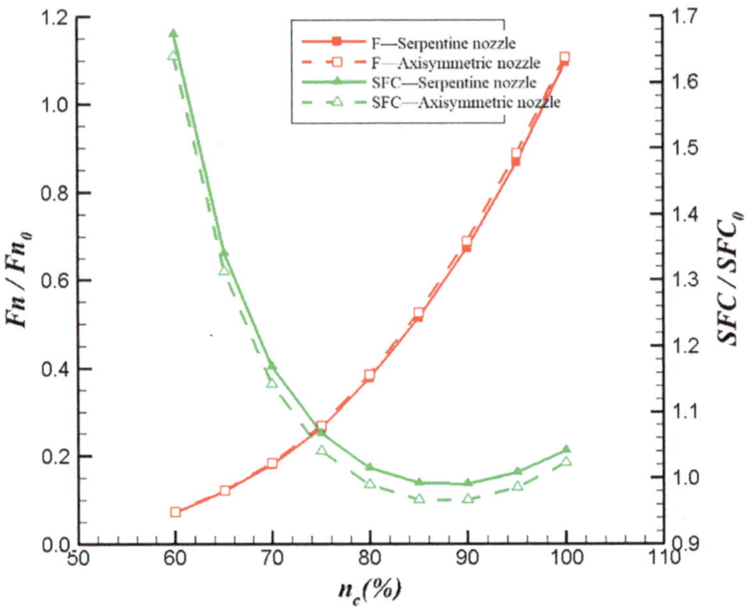

Fig. 6.7 Throttling characteristics of aero-engine with serpentine nozzle

pressure ratio to grow at a slower rate than the intake duct pressure ratio, resulting in a slower increase in exhaust velocity relative to the flight speed. Consequently, the engine's specific thrust decreases as *Ma* increases. On the other hand, the engine's flow increases as *Ma* rises. Therefore, at a given flight altitude, when the increase in flow is slower than the decrease in specific thrust, the engine thrust decreases with increasing Mach number *Ma*. However, when the increase in flow surpasses the decrease in specific thrust, the engine thrust increases. As the flight *Ma* increases, the increase in exhaust velocity is slower than the increase in flight speed, thus improving the engine's propulsion efficiency. The increase in overall pressure ratio enhances the thermal efficiency by improving heat utilization, which results in an increase in the engine's thermal efficiency. Therefore, the overall efficiency increases as *Ma* increases. From the relationship between fuel consumption rate, flight speed, and overall efficiency, it can be seen that the increase in flight speed outweighs the improvement in overall efficiency, causing the fuel consumption rate to rise as the flight *Ma* increases.

As the flight altitude increases from 0 to 11 kilometers, the engine's thrust rapidly decreases, while the fuel consumption rate decreases only slightly. At a given flight speed, as the altitude increases, the atmospheric temperature decreases, and the decrease in the engine's inlet total temperature causes the fan and compressor equivalent speeds to increase, moving the operating point upward along the working line. As a result, the pressure ratios of both the fan and the compressor increase, leading to a rise in the engine's total pressure ratio. At the same time, the air heating ratio within the engine also increases. The simultaneous increase of these two factors results in an increase in specific thrust. However, as the surrounding atmospheric pressure and density decrease with increasing altitude, the air flow decreases with altitude. Since the decrease in air flow is more significant than the increase in specific thrust, the engine thrust decreases with increasing altitude. The decrease in fuel consumption rate is due to the improvement in heat utilization as a result of the increase in the total pressure ratio.

When the engine speed decreases from the design speed, the engine thrust drops sharply, while the fuel consumption rate first slightly decreases, and then increases as the engine speed decreases and then increases again, indicating the presence of an optimal economic speed. During throttling, both the low-pressure and high-pressure speeds decrease, and the common operation point moves downward along the common operation line. As a result, the mass flow of air through the fan and compressor decreases, and the decrease in both low-pressure and high-pressure speeds causes the fan pressure ratio and compressor pressure ratio to reduce, thus lowering the engine's cycle thermal efficiency. On the other hand, the decrease in turbine inlet temperature reduces the heating ratio. The reductions in both the pressure ratio and heating ratio result in a decrease in specific thrust. Therefore, the engine's thrust sharply decreases as the engine speed drops from the design speed. During the initial stage of throttling, the decrease in turbine inlet temperature, fan pressure ratio, and compressor pressure ratio causes the exhaust velocity of the nozzle to decrease, significantly reducing the kinetic energy loss. As a result, the fuel consumption rate slightly decreases at first, but further throttling leads to a dominant effect of reduced thermal efficiency, causing the fuel consumption rate to increase.

From the comparison between the serpentine nozzle and the axisymmetric nozzle, it can be seen that under the same flight *Ma*, altitude, and speed, the thrust of the aircraft engine with the serpentine nozzle is always smaller than that of the engine with the axisymmetric nozzle, with thrust reduction ranging from 0.04% to 5.95%. The fuel consumption rate is higher for the engine with the serpentine nozzle, increasing by 1.43% to 6.48%.

Figure 6.8 shows the variations in the fan and compressor operation lines of the aero-engine with the serpentine nozzle and the axisymmetric nozzle under different altitude and speed conditions. It can be observed that on the compressor performance map, the operation lines of both engines coincide exactly under the same conditions. However, on the fan performance map, the operation line of the engine with the serpentine nozzle shifts toward the surge boundary. This is related to the change in the engine's common operation line: because the engine control rule is that the physical speed of the compressor is constant, and both the high-pressure turbine nozzle throat and the low-pressure turbine nozzle throat are at critical conditions, the high-pressure turbine expansion ratio remains unchanged. As a result, on the compressor performance map, the operation lines of the engine with the serpentine nozzle and the axisymmetric nozzle coincide completely. From the mixed-flow turbofan engine's common operation process, it can be inferred that due to the lower discharge coefficient of the serpentine nozzle compared to the axisymmetric nozzle, the introduction of the serpentine nozzle reduces both the external and internal flow

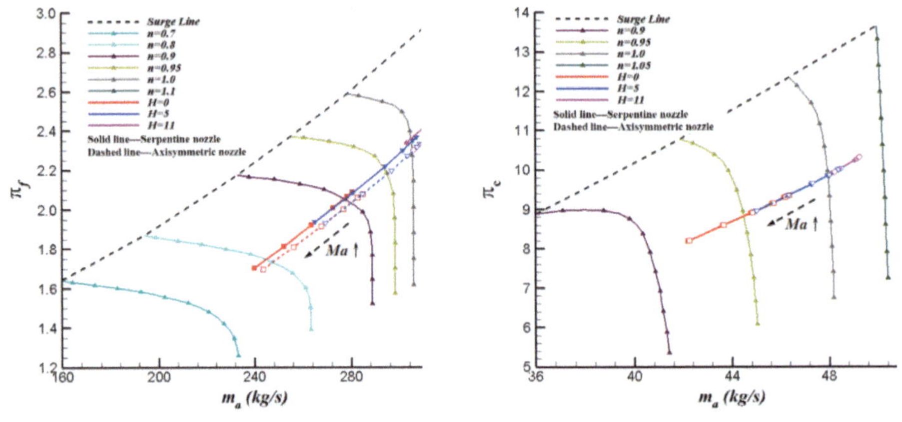

(a) Fan operation line (b) Compressor operation line

Fig. 6.8 Influence of serpentine nozzle on fan and compressor operation lines of aero-engine at different flight altitudes and speeds. (**a**) Fan operation line. (**b**) Compressor operation line

at the mixed chamber's inlet. The reduction in external flow leads to a decrease in the bypass ratio, indicating reduced fan flow capacity and causing the fan's operation line to shift toward the surge boundary. At the same time, the reduction in internal flow results in a decrease in the low-pressure turbine expansion ratio, causing the fan's operation line to shift away from the surge boundary. Ultimately, the movement of the fan's operation line depends on the rate of change of both the internal and external flow, which is related to the design bypass ratio B. When B>0.4~0.5, the external flow decreases faster, causing the fan's operation line to shift toward the surge boundary. Since the engine studied in this article has a design bypass ratio of B=2.33, the operation line of the engine with the serpentine nozzle shifts toward the surge boundary on the fan performance map.

Figure 6.9 shows the variations in fan and compressor operation lines of aero-engines with serpentine nozzles and axisymmetric nozzles at different rotational speeds. As seen, the serpentine nozzle does not affect the compressor's operation line, but it causes the fan's operation line to move closer to the surge boundary.

From the above analysis, it can be seen that, compared to aero-engines with axisymmetric nozzles, the reduced discharge coefficient of the serpentine nozzle causes the fan operation line to move closer to the surge boundary, increasing the fan pressure ratio and thus increasing the engine's thrust. However, the decrease in thrust coefficient leads to a reduction in thrust. Since the thrust coefficient of the serpentine nozzle is significantly lower than that of the axisymmetric nozzle, the reduction in thrust coefficient plays a dominant role. Therefore, the thrust of aero-engines with serpentine nozzles is smaller than that of those with axisymmetric nozzles, and the fuel consumption rate is higher.

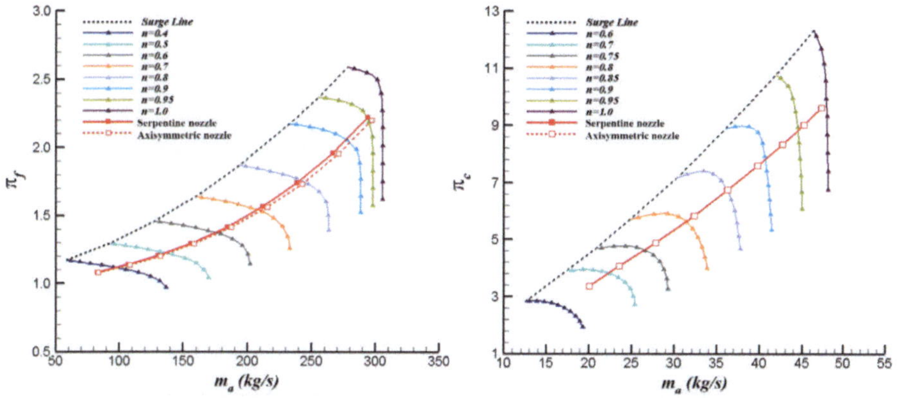

(a) Fan Operation Line (b) Compressor Operation Line

Fig. 6.9 Influence of serpentine nozzle on fan and compressor operation lines of aero-engine at different rotational speeds. (**a**) Fan operation line. (**b**) Compressor operation line

6.5 Coupled Calculation of Three-Bearing Swivel Nozzle and Turbofan Engine

6.5.1 Construction and Evaluation of High-Precision Surrogate Model for Three-Bearing Swivel Nozzle

Figure 6.10 shows the schematic of the three-bearing swivel nozzle contour. Here, d is the diameter of the first segment of the nozzle's inlet section, and l_1, l_2 and l_3 are the axial lengths of the three nozzle segments. The dimensionless lengths of the three segments are defined as $L_1=l_1/d$, $L_2=l_2/d$ and $L_3=l_3/d$. For the reference model, L_1=0.65, L_2=0.75 and L_3=0.5. The three rotatable nozzle segments are connected by bearings, and each stage of the nozzle rotates around its bearing axis, causing the axis of the third-stage nozzle to deflect by an angle β, thus achieving vector thrust (Liu et al. 2014, 2015).

For the three-bearing swivel nozzle studied in this section, the aerodynamic performance of the nozzle under various aerodynamic and geometric parameters is evaluated using the total pressure recovery coefficient σ, the discharge coefficient C_d, the thrust coefficient C_{fg}, and the thrust vector angle δ_p. The 5 main parameters affecting the aerodynamic performance of the three-bearing swivel nozzle are summarized in Table 6.7:

Table 6.8 gives the range of values for each influencing parameter.

According to the BBD design of experiment principles, when there are 5 independent variables, the design requires 46 test runs, with 40 runs for the factorial part and 6 repeated runs at the center point. The results of the total pressure recovery coefficient σ, discharge coefficient C_d, thrust coefficient C_{fg}, and thrust vector angle δ_p from numerical calculations were analyzed using quadratic analysis through the

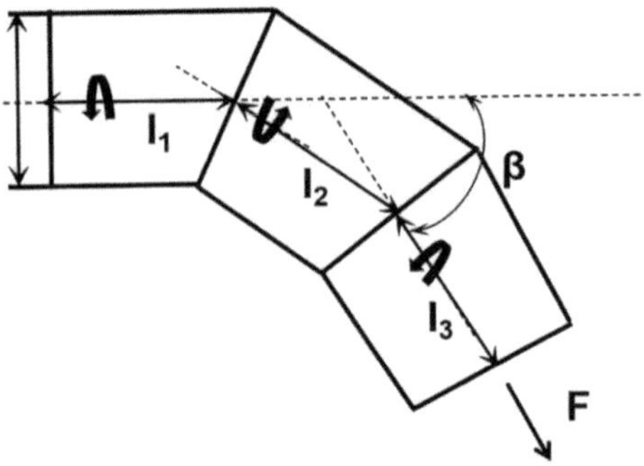

Fig. 6.10 Schematic of the three-bearing swivel nozzle contour

Design-Expert software to obtain the quadratic regression equation. This equation includes constant terms, linear terms, interaction terms, and square terms. Stepwise regression was used to perform significance analysis on each influencing parameter, and after excluding terms with negligible effects, the significance analysis of each influencing variable is shown in Table 6.9.

After the model was revised, the signal-to-noise ratio reached 30.192. The model was subjected to an R^2 test, and $R^2 = 0.9411$ for the total pressure recovery coefficient σ, which is greater than 0.8, indicating that the precision meets the required standard and the model is usable. Therefore, the relationship between the response variable, total pressure recovery coefficient σ, and the independent variables is obtained as follows:

$$\begin{aligned} \sigma = 0.95518 \quad & 2.21179e^{-5} \times \beta + 0.042034 \times L_1 \\ & +0.059620 \times L_2 - 5.66575e^{-3} \times L_3 \\ & +2.35825e^{-4} \times \beta \times L_2 - 4.47448e^{-6} \times \beta^2 \\ & -0.026114 \times L_1^{\ 2} - 0.043547 \times L_2^{\ 2} \end{aligned} \tag{6.53}$$

Next, we analyze the approximate modeling results for the discharge coefficient C_d. The Analysis of Variance (ANOVA) method for the discharge coefficient C_d's approximation model is similar to the one described above. Table 6.10 presents the significance analysis results of the influencing variables after omitting the less significant factors.

After the model was revised, the signal-to-noise ratio reached 123.427. The model was subjected to an R^2 test, and $R^2 = 0.9966$ for the discharge coefficient C_d, which is greater than 0.8, indicating that the precision meets the required standard and the model is usable. Therefore, the relationship between the response variable, discharge coefficient C_d, and the independent variables is obtained as follows:

$$\begin{aligned} C_d = 0.71054 & + 0.17632 \times NPR - 4.61302e^{-4} \times \beta \\ & +0.033673 \times L_1 + 0.040736e^{-3} \times L_3 \\ & +2.05120e^{-4} \times NPR \times \beta + 2.73432e^{-4} \times \beta \times L_2 \\ & -0.032845 \times NPR^2 - 6.19529e^{-6} \times \beta^2 \\ & -0.018210 \times L_1^{\ 2} - 0.030896 \times L_2^{\ 2} \end{aligned} \tag{6.54}$$

Table 6.7 Influencing parameters

Research parameters	Symbol
Mainstream blowdown ratio	*NPR*
Nozzle deflection angle	β
Dimensionless length of the first segment of the nozzle	L_1
Dimensionless length of the second segment of the nozzle	L_2
Dimensionless length of the third segment of the nozzle	L_3

Table 6.8 Range of values for each influencing parameter

Variable	Range	Unit
NPR	1.4–3.0	
β	0–90	(°)Degree
L_1	0.3–1.0	
L_2	0.5–1.0	
L_3	0.25–0.75	

Table 6.9 Approximation model ANOVA for total pressure recovery coefficient σ after excluding terms with negligible effects

Source	Sum of squares	df	Mean square	*F* Value	p-value Prob > *F*
Model	0.00305	8	0.000381	73.89211	<0.0001
B-β	0.001992	1	0.001992	386.0664	<0.0001
C-L1	0.000128	1	0.000128	24.8352	<0.0001
D-L2	2.41E−05	1	2.41E−05	4.675897	0.0371
E-L3	3.21E−05	1	3.21E−05	6.221502	0.0172
BD	2.82E−05	1	2.82E−05	5.456639	0.0250
B^2	0.000811	1	0.000811	157.2447	<0.0001
C^2	0.000101	1	0.000101	19.59991	<0.0001
D^2	7.32E−05	1	7.32E−05	14.18802	0.0006
Residual	0.000191	37	5.16E−06		
Lack of fit	0.000191	32	5.97E-06		
Pure error	0	5	0		
Cor total	0.003241	45			

Next, we analyze the approximate modeling results for the thrust coefficient C_{fg}. The ANOVA method for the thrust coefficient C_{fg}'s approximation model is similar to the one described above. Table 6.11 presents the significance analysis results of the influencing variables after omitting the less significant factors.

After the model was revised, the signal-to-noise ratio reached 47.829. The model was subjected to an R^2 test, and $R^2 = 0.9798$ for the thrust coefficient C_{fg}, which is greater than 0.8, indicating that the precision meets the required standard and the model is usable. Therefore, the relationship between the response variable, thrust coefficient C_{fg}, and the independent variables is obtained as follows:

Table 6.10 Approximation model ANOVA for discharge coefficient C_d after excluding terms with negligible effects

Source	Sum of squares	df	Mean square	*F* Value	p-value Prob>*F*
Model	0.027112	11	0.002465	912.7893	<0.0001
A-NPR	0.017243	1	0.017243	6385.903	<0.0001
B-β	0.004259	1	0.004259	1577.12	<0.0001
C-L1	0.000196	1	0.000196	72.60313	<0.0001
D-L2	4.48E−05	1	4.48E−05	16.60454	0.0003
E-L3	4.57E−05	1	4.57E−05	16.90703	0.0002
AB	0.000218	1	0.000218	80.77682	<0.0001
BD	3.78E−05	1	3.78E−05	14.01746	0.0007
A^2	0.004166	1	0.004166	1542.906	<0.0001
B^2	0.001484	1	0.001484	549.5734	<0.0001
C^2	4.69E−05	1	4.69E−05	17.37495	0.0002
D^2	3.52E−05	1	3.52E−05	13.02033	0.0010
Residual	9.18E−05	34	2.7E−06		
Lack of Fit	9.18E−05	29	3.17E−06		
Pure Error	0	5	0		
Cor Total	0.027204	45			

Table 6.11 Approximation model ANOVA for thrust coefficient C_{fg} after excluding terms with negligible effects

Source	Sum of squares	df	Mean square	*F* Value	p-value Prob > *F*
Model	0.002904	12	0.000242	133.0617	<0.0001
A-NPR	0.000652	1	0.000652	358.5937	<0.0001
B-β	0.001468	1	0.001468	807.2371	<0.0001
C-L1	8.15E−05	1	8.15E−05	44.79275	<0.0001
D-L2	2.54E−05	1	2.54E−05	13.96662	0.0007
E-L3	2.12E−05	1	2.12E−05	11.65646	0.0017
AB	3.71E−05	1	3.71E−05	20.39217	<0.0001
BD	2.5E−05	1	2.5E−05	13.73102	0.0008
CE	5.71E−06	1	5.71E−06	3.140026	0.0856
A^2	9.15E−05	1	9.15E−05	50.2877	<0.0001
B^2	0.000566	1	0.000566	311.072	<0.0001
C^2	4.74E−05	1	4.74E−05	26.04546	<0.0001
D^2	3.24E−05	1	3.24E−05	17.83745	0.0002
Residual	6E−05	33	1.82E−06		
Lack of Fit	6E−05	28	2.14E−06		
Pure Error	0	5	0		
Cor Total	0.002964	45			

Table 6.12 Approximation model ANOVA for thrust vector angle δ_p after excluding terms with negligible effects

Source	Sum of squares	df	Mean square	*F* Value	p-value Prob > *F*
Model	34073.68	12	2839.473	71220.28	<0.0001
A-NPR	0.596453	1	0.596453	14.96035	0.0005
B-β	34050.44	1	34050.44	854060.4	<0.0001
C-L1	0.429268	1	0.429268	10.76699	0.0024
D-L2	0.553142	1	0.553142	13.87402	0.0007
E-L3	9.247432	1	9.247432	231.9461	<0.0001
AE	0.118796	1	0.118796	2.979664	0.0937
BD	0.184968	1	0.184968	4.639413	0.0386
BE	1.999227	1	1.999227	50.14503	<0.0001
A^2	0.779643	1	0.779643	19.55517	0.0001
B^2	5.024159	1	5.024159	126.0171	<0.0001
D^2	0.152877	1	0.152877	3.834504	0.0587
E^2	2.003531	1	2.003531	50.253	<0.0001
Residual	1.315673	33	0.039869		
Lack of Fit	1.315673	28	0.046988		
Pure Error	0	5	0		
Cor Total	34075	45			

$$\begin{aligned} C_{fg} &= 0.93859 + 0.025586 \times NPR - 2.21244e^{-4} \times \beta \\ &+0.023405 \times L_1 + 0.039563 \times L_2 - 0.013480 \times L_3 \\ &+9.45806e^{-5} \times NPR \times \beta + 2.22096e^{-4} \times \beta \times L_2 \\ &+0.013655 \times L_1 \times L_3 - 4.86631e^{-3} \times NPR^2 \\ &-3.82520e^{-6} \times \beta^2 - 0.018297 \times L_1{}^2 \\ &-0.029678 \times L_2{}^2 \end{aligned} \tag{6.55}$$

Next, we analyze the approximate modeling results for the thrust vector angle δ_p. The ANOVA method for the thrust vector angle δ_p's approximation model is similar to the one described above. Table 6.12 presents the significance analysis results of the influencing variables after omitting the less significant factors.

After the model was revised, the signal-to-noise ratio reached 883.528 , indicating that the model's response became more pronounced. The model was subjected to an R^2 test, and $R^2 = 1.0000$ for the thrust vector angle δ_p, which is greater than 0.8, indicating that the precision meets the required standard and the model is usable. Therefore, the relationship between the response variable, thrust vector angle δ_p, and the independent variables is obtained as follows:

$$\begin{aligned}\delta_P &= -4.02643 + 2.64914 \times NPR + 0.97563 \times \beta \\ &-0.46799 \times L_1 - 2.93962 \times L_2 + 9.48431 \times L_3 \\ &-0.86167 \times NPR \times L_3 - 0.019115 \times \beta \times L_2 \\ &+0.062842 \times \beta \times L_3 - 0.44931 \times NPR^2 \\ &+3.60482e^{-4} \times \beta^2 + 2.03736 \times L_2^{\ 2} \\ &-7.37556 \times L_3^{\ 2}\end{aligned} \tag{6.56}$$

6.5.2 *Aerodynamic Performance Calculation and Analysis of Three-Bearing Swivel Nozzle in the Environment of Turbofan Engine*

This section establishes an engine simulation model based on a certain turbofan engine. The engine is a two-spool, mixed-flow, twin-shaft turbofan, while the nozzle is a subsonic, fixed-area nozzle. The simulation model is shown in Fig. 6.11.

The parameters at the design point are shown in Table 6.13.

The key to the matching study between the three-bearing swivel nozzle and the entire engine lies in matching the flow rate through the nozzle with the flow rate through the engine (Liu 2016). Therefore, the process of modeling the three-bearing swivel nozzle and the overall engine involves substituting the discharge coefficient C_d from the nozzle's approximation model into the engine's performance calculation flow balance equation to achieve their matching.

In addition to flow matching, the calculation of engine thrust needs to consider the influence of the nozzle's thrust coefficient. Therefore, the thrust coefficient C_{fg} from the three-bearing swivel nozzle's approximate model is incorporated into the engine's thrust calculation to obtain the thrust performance and fuel consumption characteristics of the three-bearing swivel nozzle matched with the overall engine.

Figure 6.12 shows the throttling characteristics of the turbofan engine with the three-bearing swivel nozzle at the design condition. The design point is at H = 0 km, Ma = 0, with the three-bearing nozzle at a 90° state, and a comparison is made with the characteristics of the engine equipped with a conventional nozzle. Since the matching between the three-bearing swivel nozzle and the engine is mainly based on flow (discharge coefficient) matching, the throttling characteristics of the engine

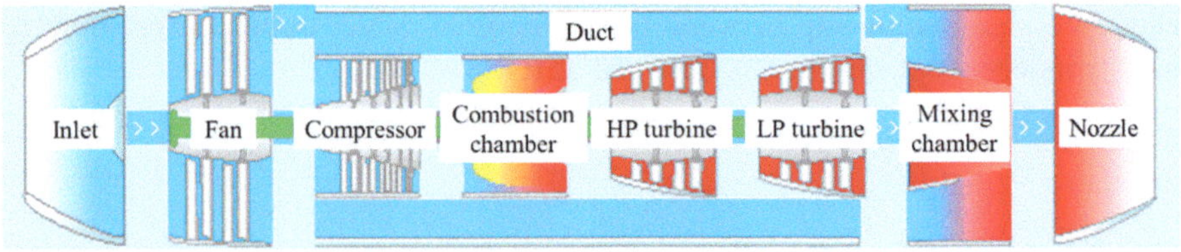

Fig. 6.11 Simulation model of the mixed-flow turbofan engine

Table 6.13 Parameters of each component of a certain turbofan engine at the design point

Design point parameters	Value
Airflow/kg	90
Fan pressure ratio	2.3
Fan efficiency	0.86
Fan physical speed/%	100
Bypass ratio	0.8
High-pressure compressor pressure ratio	9
High-pressure compressor efficiency	0.85
High-pressure compressor physical speed/%	100
Turbine inlet temperature/K	1450
High-pressure turbine efficiency	0.9
Low-pressure turbine efficiency	0.9

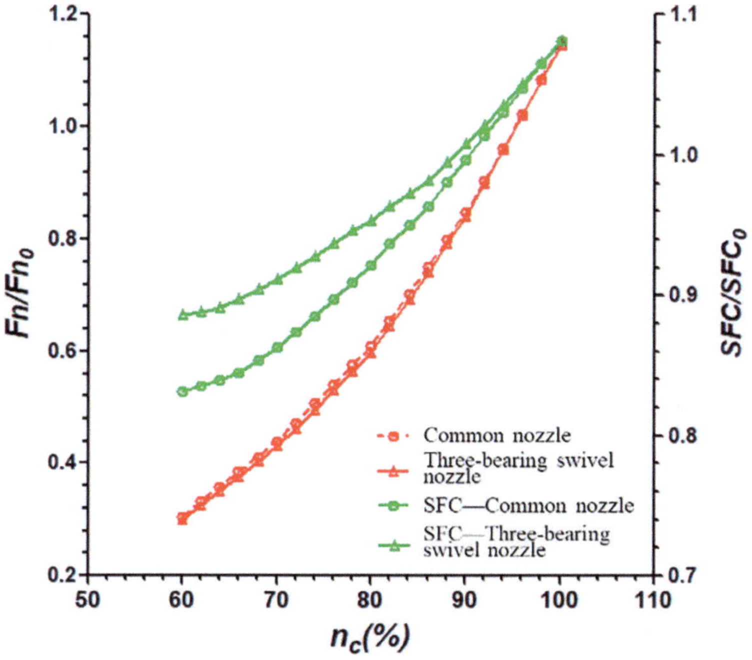

Fig. 6.12 Throttling characteristics of the turbofan engine equipped with a three-bearing swivel nozzle

with the three-bearing swivel nozzle are similar to those of the engine equipped with a conventional nozzle:

As the engine speed decreases from the design speed, both the thrust and fuel consumption of the engine drop sharply, with the rate of fuel consumption reduction gradually slowing down. During throttling, both low-pressure and high-pressure speeds decrease, and the common operation line moves downward along the operating line. Therefore, the mass flow of air passing through the fan and compressor decreases. At the same time, the reduction in low-pressure and high-pressure speeds leads to a decrease in both fan pressure ratio and compressor pressure ratio, which lowers the engine's thermal efficiency. On the other hand, the decrease in turbine inlet temperature reduces the heating ratio, and the reduction in both pressure ratio and heating ratio causes a decrease in specific thrust. As a result, the engine's thrust decreases rapidly as the engine speed drops from the design speed. As the engine speed continues to drop, the decrease in turbine inlet temperature, fan pressure ratio, and compressor pressure ratio leads to a reduction in the nozzle exhaust velocity. Consequently, kinetic energy loss significantly decreases, causing fuel consumption to decrease rapidly. Further throttling increases the impact of the reduced thermal efficiency, leading to a slower reduction in fuel consumption rate.

From the comparison between the three-bearing swivel nozzle and the conventional nozzle, it can be seen that at the same flight Mach number, flight altitude, and engine speed, the engine equipped with the three-bearing swivel nozzle produces less thrust and higher fuel consumption compared to the engine equipped with a conventional nozzle.

Figures 6.13, 6.14, 6.15 and 6.16 show the effects of the geometric parameters of the three-bearing swivel nozzle on key engine section parameters and overall performance parameters under the conditions of H = 0 km and Ma = 0.

From Fig. 6.14, it can be observed that as the nozzle deflection angle β increases, the trends in the thrust coefficient and discharge coefficient are similar. Initially, they slightly increase, then rapidly decrease. The variation in the discharge coefficient causes the fan pressure ratio, fuel flow, and engine thrust to first decrease and then increase. As β increases, the fuel consumption rate consistently decreases.

From Fig. 6.14, it can be seen that the change in L_1 has little impact on the thrust coefficient and discharge coefficient. As L_1 increases, thrust first decreases and then increases. The change in fuel consumption rate follows the opposite trend of thrust, first increasing and then decreasing.

From Fig. 6.15, it can be seen that as L_2 increases, both the thrust coefficient and discharge coefficient increase simultaneously, thrust gradually decreases, while the fuel consumption rate continuously increases.

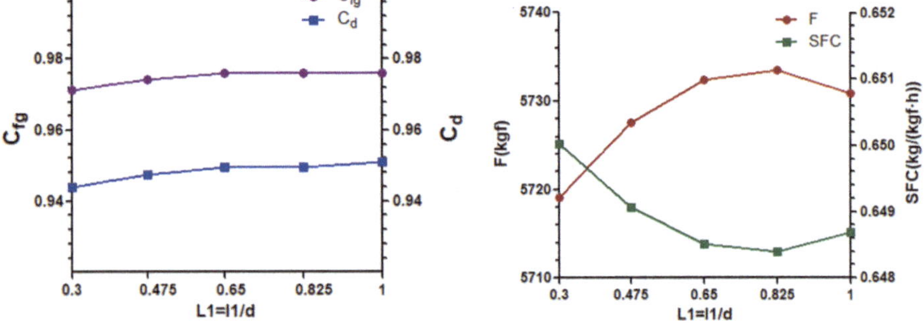

Fig. 6.14 The impact of L_1 on engine overall performance parameters

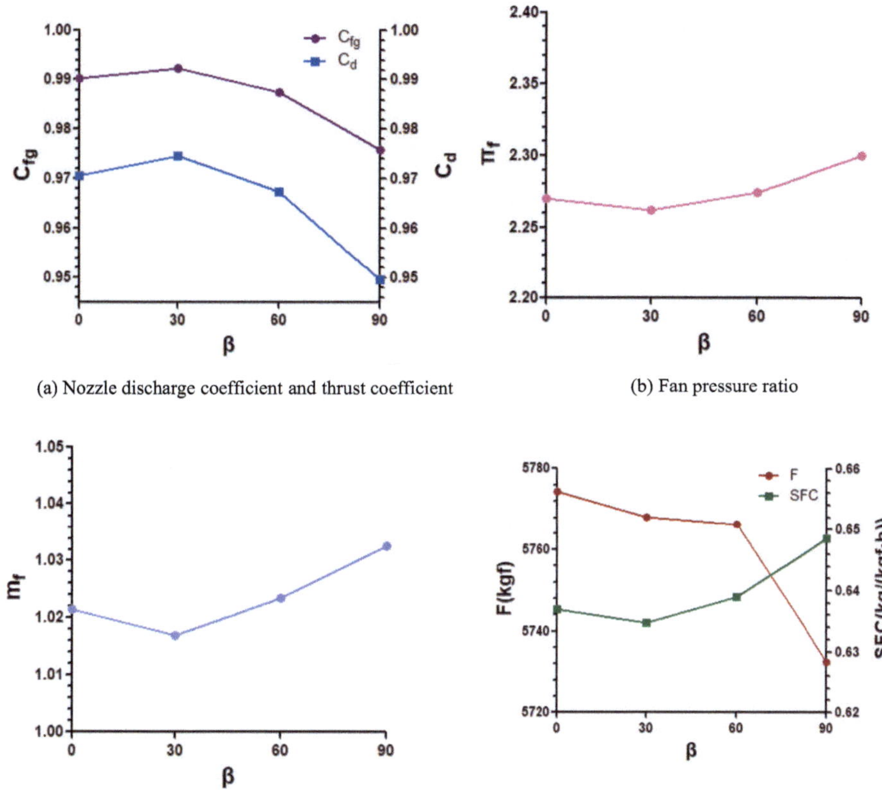

(a) Nozzle discharge coefficient and thrust coefficient

(b) Fan pressure ratio

(c) Fuel flow

(d) Thrust and fuel consumption rate

Fig. 6.13 The impact of β on key engine section parameters and overall performance parameters. (**a**) Nozzle discharge coefficient and thrust coefficient. (**b**) Fan pressure ratio. (**c**) Fuel flow. (**d**) Thrust and fuel consumption rate

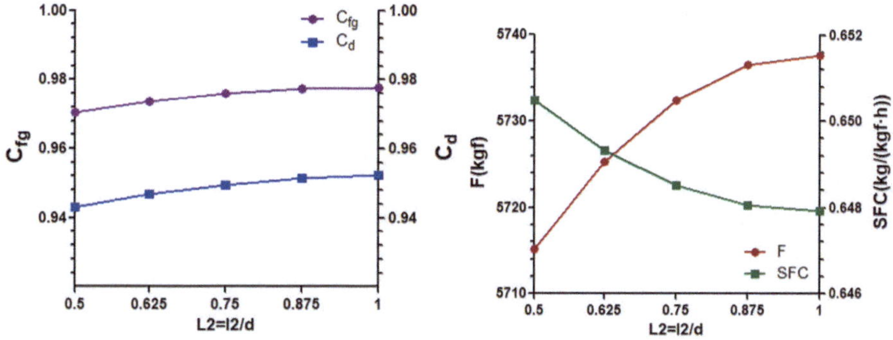

Fig. 6.15 The impact of L_2 on engine overall performance parameters

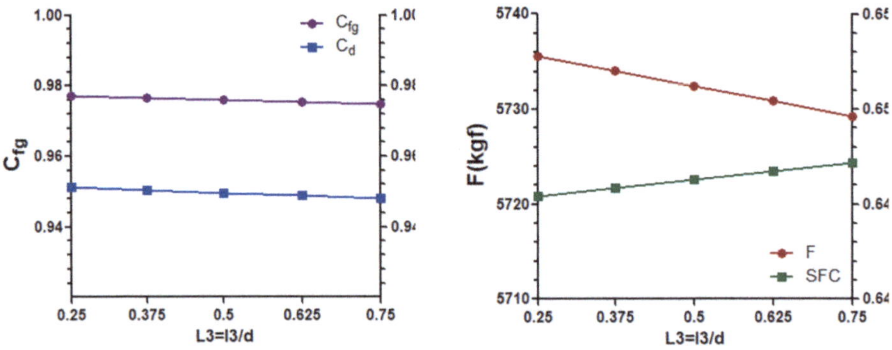

Fig. 6.16 The impact of L_3 on engine overall performance parameters

From Figure 6.16, it can be seen that the change in L_3 has little impact on the thrust coefficient and discharge coefficient. As L_3 increases, both the thrust coefficient and discharge coefficient slightly decrease, thrust increases linearly, and fuel consumption rate decreases with the same trend.

6.6 Coupled Calculation of Single Expansion Ramp Nozzle and TBCC Engine

6.6.1 Construction and Evaluation of High-Precision Surrogate Model for Single Expansion Ramp Nozzle

In order to study the aerodynamic characteristics of the single expansion ramp nozzle with respect to changes in key geometric and inflow parameters, and to obtain the performance characteristics of the nozzle, a two-dimensional model of the single expansion ramp nozzle with secondary flow injection was established using the method of characteristics (Wang et al. 2007; Liu 2007; Hao and Wang 2009), as shown in Fig. 6.17. The internal expansion ratio of the nozzle is $A_i/A_T = 1.78$, the external expansion ratio is $A_e/A_T = 6.39$, and the ratio of external expansion surface length L_r to throat height H_T is $L_r/H_T = 36.08$. The ratio of the length of the lower ramp L_c to the throat height H_T is $L_c/H_T = 1.93$, where A_T is the throat area. The opening of the secondary flow inlet is located on the lower ramp, with the distance from the exit surface being L, $L/L_c = 1/8$ (Liu and Wang 2007).

For the single expansion ramp nozzle with secondary flow studied in this section, the discharge coefficient C_d and thrust coefficient C_{fg} are used to evaluate the aerodynamic performance of the nozzle under different operating conditions. Therefore, during the approximation modeling process, the discharge coefficient and thrust coefficient are taken as response variables (Zhou et al. 2015). The main parameters influencing the aerodynamic performance of the single expansion ramp nozzle with secondary flow are five, as shown in Table 6.14:

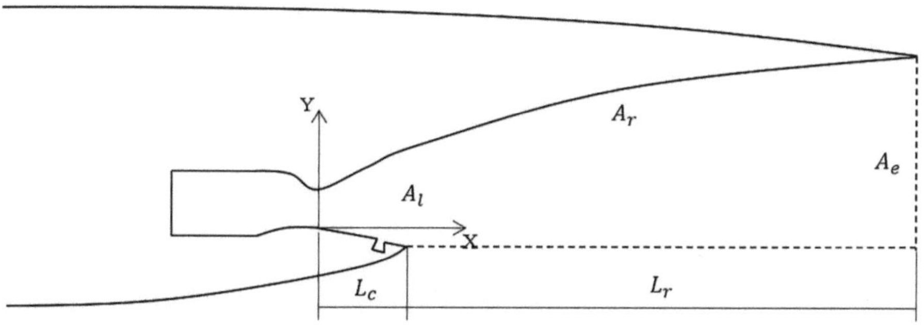

Fig. 6.17 The two-dimensional model of single expansion ramp nozzle with secondary flow injection

Table 6.14 Influencing parameters

Research parameters	Symbol
Mainstream blowdown ratio	NPR
Mainstream total temperature	T^*
Ratio of secondary flow total pressure to mainstream total pressure	π
Secondary flow injection angle	β
Ratio of secondary flow injection area to throat area	ω

Table 6.15 gives the range of values for each influencing parameter.

Since the single expansion ramp nozzle is in an over-expanded state within the given NPR range, the airflow at the throat remains at sonic conditions, meaning the discharge coefficient C_d remains nearly constant. Therefore, the test design results for the discharge coefficient are no longer analyzed. The following analysis focuses on the design of experiment results for the thrust coefficient C_{fg}.

The results of the thrust coefficient C_{fg} from numerical calculations were analyzed using quadratic analysis through the Design-Expert software to obtain the quadratic regression equation. This equation includes constant terms, linear terms, interaction terms, and square terms. Stepwise regression was used to perform significance analysis on each influencing parameter, and after excluding terms with negligible effects, the significance analysis of each influencing variable is shown in Table 6.16.

After the model was revised, the signal-to-noise ratio reached 32.843. The model was subjected to an R^2 test, and $R^2 = 0.9284$ for the thrust coefficient C_{fg}, which is greater than 0.8, indicating that the precision meets the required standard and the model is usable. Therefore, the relationship between the response variable, the thrust coefficient C_{fg}, and the independent variables is obtained as follows:

$$\begin{aligned} C_{fg} = {} & 0.83471 - 2.04759e-3 \times NPR + 0.065511 \\ & \times \pi - 1.10819e-5 \times T^* - 1.4275e-4 \times \beta - 8.5235e-4 \\ & \times \pi \times \beta + 2.23652e-4 \times NPR^2 + 3.65616e-6 \times \beta^2 \end{aligned} \tag{6.57}$$

Table 6.15 Range of values for each influencing parameter

Variable	Range	Unit
NPR	8–20	
T^*	600–1200	K
π	0.7–1.3	
β	60–120	(°)Degree
ω	0.06–0.1	

Table 6.16 Approximation model ANOVA for the thrust coefficient C_{fg} after excluding terms with negligible effects

Source	Sum of squares	df	Mean square	*F* Value	p-value Prob > *F*
Model	0.013166	7	0.001881	70.40009	<0.0001
A-NPR	0.010232	1	0.010232	382.959	<0.0001
B-π	0.000181	1	0.000181	6.761719	0.0132
C-T*	0.000177	1	0.000177	6.619055	0.0141
D-β	0.001635	1	0.001635	61.20757	<0.0001
BD	0.000235	1	0.000235	8.810181	0.0052
A^2	0.000661	1	0.000661	24.74883	<0.0001
D^2	0.00011	1	0.00011	4.133704	0.0491
Residual	0.001015	38	2.67E−05		
Lack of fit	0.001015	33	3.08E-05		
Pure error	0	5	0		
Cor total	0.014182	45			

6.6.2 Aerodynamic Performance Calculation and Analysis of Single Expansion Ramp Nozzle in the Environment of TBCC Engine

This section establishes an engine simulation model based on the TBCC engine (Fig. 6.18), with design point parameters shown in Table 6.17.

Figures 6.19 and 6.20 show the relationship between thrust and flight Mach number for the TBCC engine equipped with single expansion ramp nozzle under turbine mode and scramjet mode, respectively. When the engine operates in turbine mode, the thrust initially decreases as the Mach number increases. However, after the Mach number exceeds 1.5, the thrust increases with the Mach number. When the Mach number exceeds 2.3, the thrust decreases rapidly with the Mach number. This is because, at lower Mach numbers and flight altitudes, as the Mach number increases, the engine intake total temperature rises, the compressor equivalent speed decreases, and the compressor pressure ratio drops. The engine's flow rate increases with the Mach number. When the increase in flow rate exceeds the decrease in thrust per unit, the engine thrust gradually increases. However, as the flight altitude increases, the surrounding atmospheric pressure and density decrease, so at higher altitudes (Mach number exceeding 2.3), the reduction in airflow dominates, causing

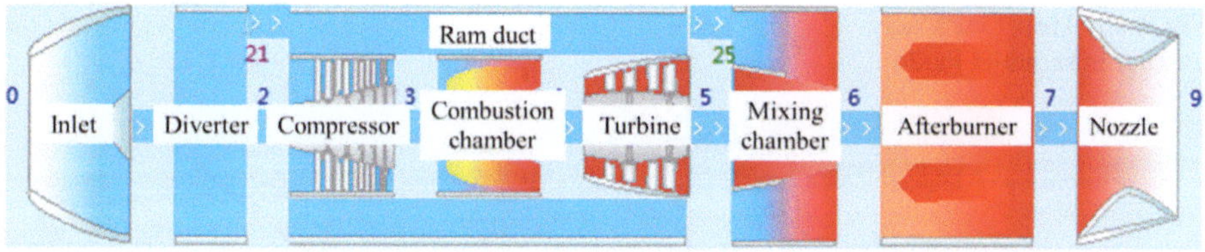

Fig. 6.18 TBCC engine

Table 6.17 TBCC engine design point parameters

Design point parameters	Value
Airflow/kg	285
Fan pressure ratio	2.08
Fan efficiency	0.86
Fan physical speed/%	91.5
Bypass ratio	2.33
High-pressure compressor pressure ratio	9.35
High-pressure compressor efficiency	0.85
High-pressure compressor physical speed/%	97.5
Turbine inlet temperature/K	1450
High-pressure turbine efficiency	0.89
Low-pressure turbine efficiency	0.91

Fig. 6.19 Thrust vs. Mach number in turbine mode

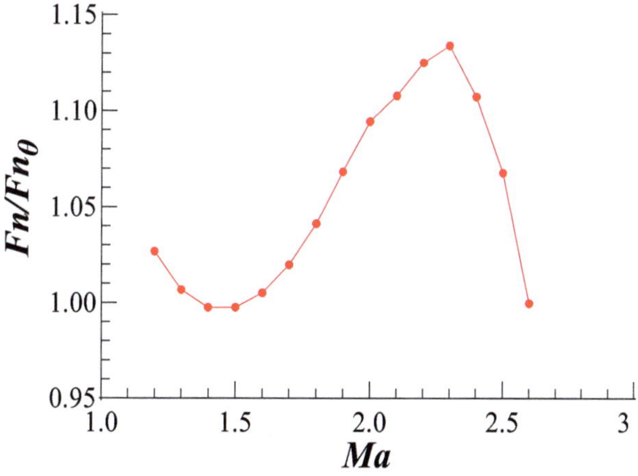

the engine thrust to decrease. When the engine operates in scramjet mode, the thrust continuously increases as the Mach number increases.

The following studies the impact of secondary flow jet angle and bleed air volume on the overall performance of the TBCC engine, mainly focusing on how the jet angle and bleed air volume affect the engine thrust, fuel consumption rate, compressor pressure ratio, and nozzle thrust coefficient under turbine and scramjet modes.

Fig. 6.20 Thrust vs. Mach number in scramjet mode

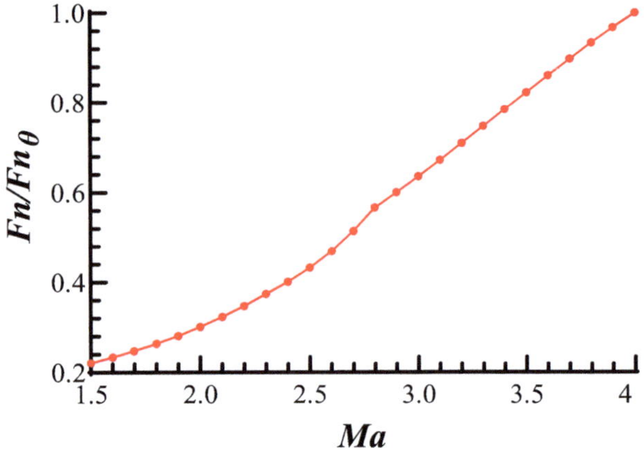

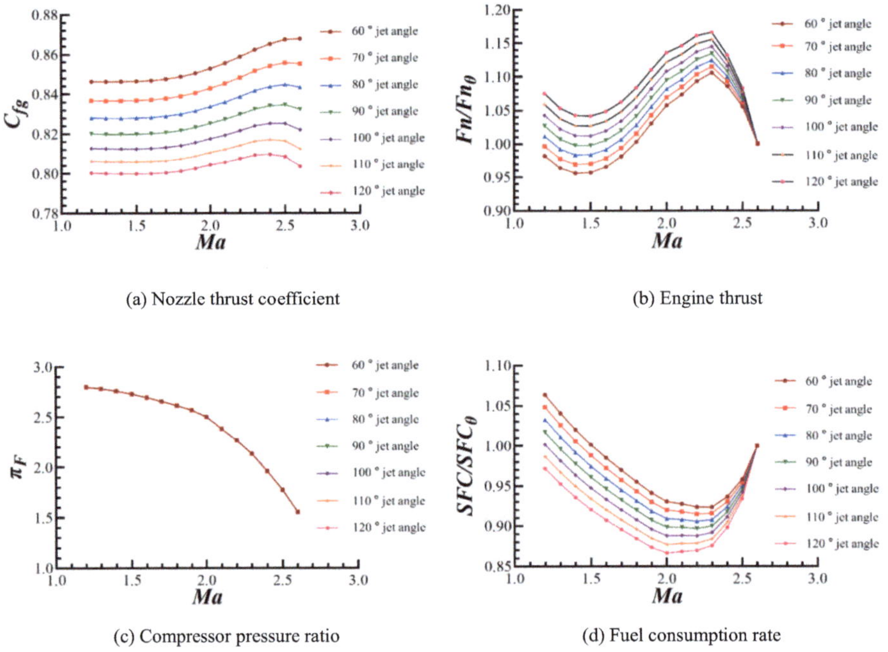

Fig. 6.21 Impact of jet angle on the overall engine performance in turbine mode. (**a**) Nozzle discharge coefficient and thrust coefficient. (**b**) Fan pressure ratio. (**c**) Fuel flow. (**d**) Thrust and fuel consumption rate

Figures 6.21 and 6.22 show the impact of jet angle on the overall performance of the TBCC engine under turbine and scramjet modes. When the engine operates in turbine mode, the nozzle thrust coefficient increases with the Mach number and then slightly decreases. Increasing the jet angle causes the nozzle thrust coefficient to continuously decrease; engine thrust shows the same trend as described earlier,

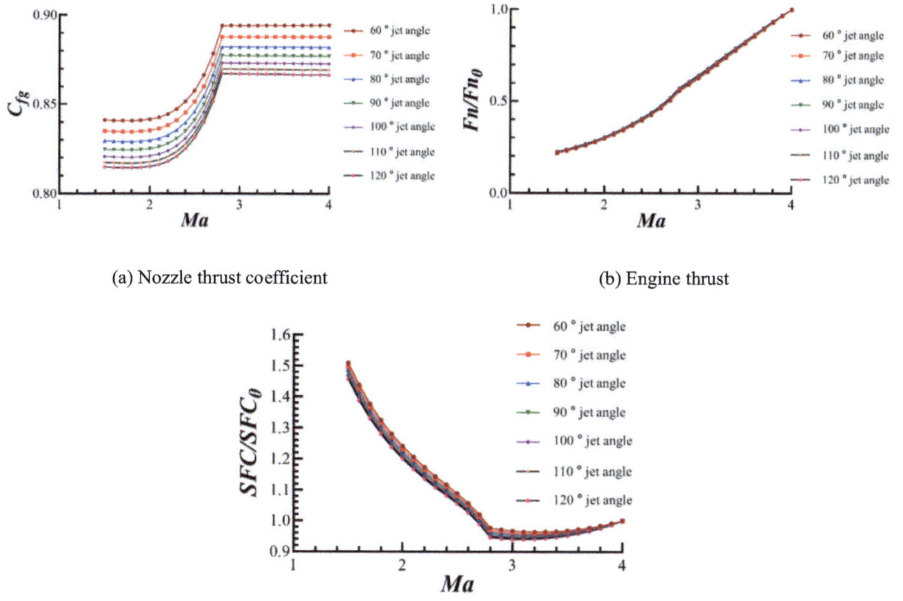

(a) Nozzle thrust coefficient

(b) Engine thrust

(c) Fuel consumption rate

Fig. 6.22 Impact of jet angle on the overall engine performance in scramjet mode. (**a**) Nozzle thrust coefficient. (**b**) Engine thrust. (**c**) Fuel consumption rate

with an increase in Mach number, and the increase in jet angle causes the thrust to gradually increase; the compressor pressure ratio decreases as the Mach number increases. Since the secondary flow injection occurs on the lower ramp surface of the single-expansion ramp nozzle, downstream of the nozzle throat, it has no effect on the compressor pressure ratio; the engine's fuel consumption rate decreases with the increase of Mach number and then increases, and the increase in jet angle causes the fuel consumption rate to continuously decrease.

When the engine operates in scramjet mode, the nozzle thrust coefficient increases with Mach number and then remains essentially unchanged. Increasing the jet angle causes the thrust coefficient to continuously decrease; engine thrust increases with Mach number, and the jet angle has little effect on the thrust; the fuel consumption rate decreases with Mach number and then slightly increases, and the jet angle has almost no effect on the fuel consumption rate.

Figures 6.23 and 6.24 show the impact of bleed air volume on the overall performance of the TBCC engine under turbine and scramjet modes. When the engine operates in turbine mode, the variation trends of nozzle thrust coefficient, engine thrust, compressor pressure ratio, and engine fuel consumption rate with Mach number are the same as analyzed above. As the secondary flow bleed air volume increases, the nozzle thrust coefficient gradually increases, engine thrust gradually

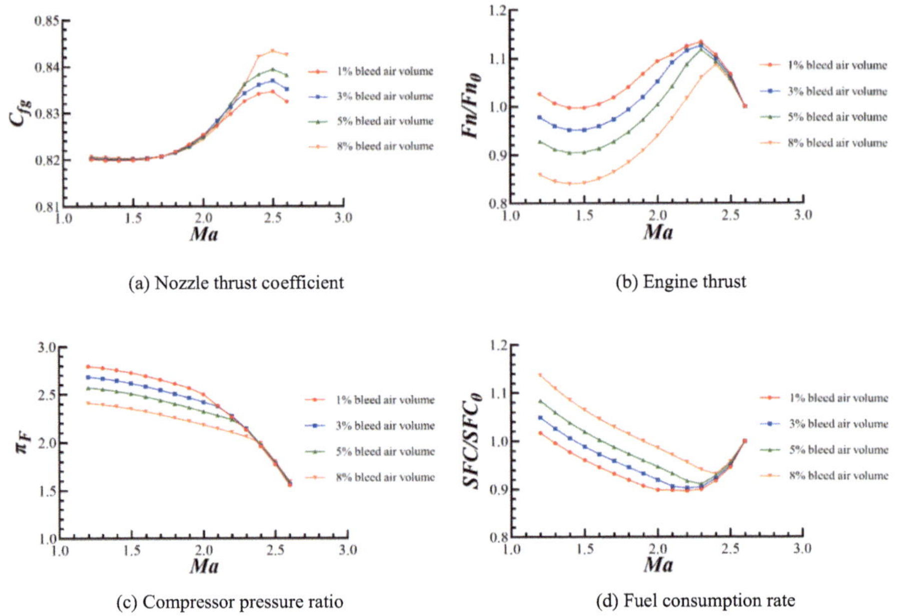

Fig. 6.23 Impact of bleed air volume on the overall engine performance in turbine mode. (**a**) Nozzle thrust coefficient. (**b**) Engine thrust. (**c**) Compressor pressure ratio. (**d**) Fuel consumption rate

decreases, compressor pressure ratio gradually decreases, and fuel consumption rate gradually increases.When the engine operates in scramjet mode, the variation trends of nozzle thrust coefficient, engine thrust, and engine fuel consumption rate are the same as analyzed above. The secondary flow bleed air volume has little effect on the nozzle thrust coefficient, engine thrust, and fuel consumption rate.

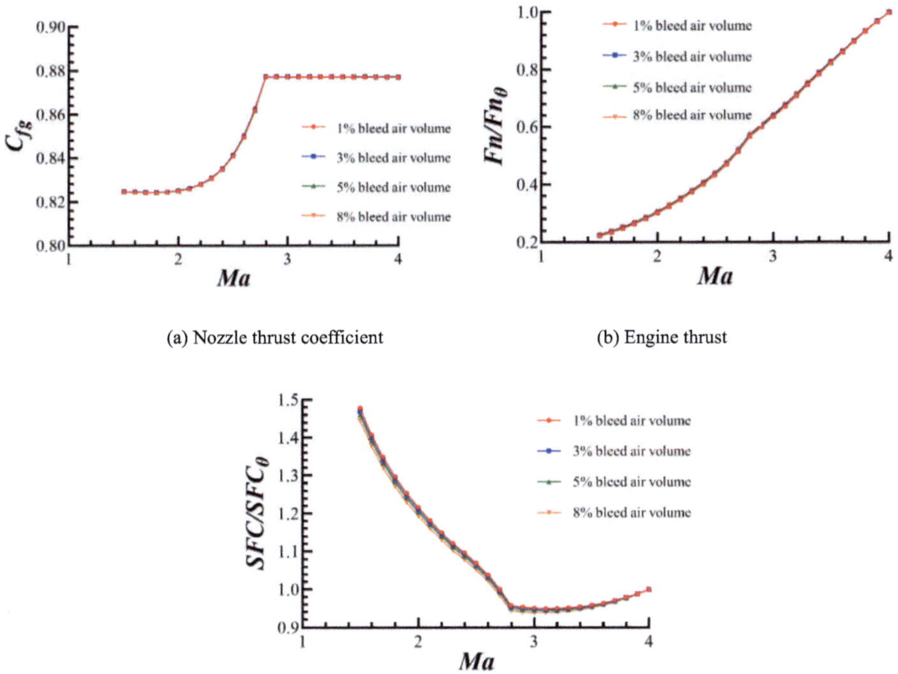

(a) Nozzle thrust coefficient (b) Engine thrust

(c) Fuel consumption rate

Fig. 6.24 Impact of bleed air volume on the overall engine performance in scramjet mode. (**a**) Nozzle thrust coefficient. (**b**) Engine thrust. (**c**) Fuel consumption rate

References

Ai J Q, Zhou L, Yang Q Z. 2017. S-shaped stealth nozzle. Beijing: National Defense Industry Press.

Alaa E K S, Abdol H, Craig A H. 2005. Numerical investigation of flow in an over-expanded nozzle with porous surfaces. AIAA 2005-4159.

Anderson S B. 1983. An overview of V/STOL aircraft development. AIAA-83-2491.

Baranwal N, Mahulikar S P. 2016. Infrared signature of aircraft engine with choked converging nozzle. Journal of Thermophysics and Heat Transfer, 30(4): 854–862.

Barata J M M. 1993. Fountain flows produced by multi-jet impingement on a ground plane. Journal of aircraft, 30(1):50-56.

Bartolotta P A, McNelis N B. 2003. High speed turbines: development of a Turbine Accelerator (RTA) for space access. AIAA 2003-6943.

Bevilaqua P M. 1990. Joint strike fighter dual-cycle propulsion system. Journal of aircraft, 27(7):577-582.

Bevilaqua P M. 2003. Future applications of the JSF variable propulsion cycle. AIAA-2003-2614.

Bevilaqua P M. 2009a. Genesis of the F-35 Joint strike fighter. Journal of Aircraft, 46(6):1825-1836.

Bevilaqua P M. 2009b. Inventing the F-35 joint strike fighter. AIAA-2009-1650.

Bradford J E. 2002. Rapid prediction of afterbody nozzle performance in SCCREAM. AIAA 2002-3605.

Brunet E, Daris T, Page A, et al. 2009. Elbow-shaped propulsion gas exhaust assembly in an aircraft. U.S. Patent: US7581400B2.

Cabrita P M, Saddington A J, Knowles K. 2002. Unsteady features of twin-jet STOVL ground effects. AIAA-2002-6014.

Capone F J, Re R J, Bare E A. 1992. Parametric investigation of single-expansion-ramp nozzles at Mach numbers from 0.60 to 1.20. NASA TP-3240.

Chali. 2002. The "metamorphosed" mysterious nighthawk - the F-117 dominating the gulf war battlefield. National Defense Science and Technology, (9):6–12.

Christopher A S, Jaime J M. 1991. The design and performance estimates for the propulsion module for the booster of a TSTO vehicle. AIAA 1991-3136.

Coates TD, Page G J. 2012. CFD based study of unconventional aeroengine exhaust systems. AIAA-2012-2775.

Crowe D S, Martin J C L. 2015. Effect of geometry on exit temperature from serpentine exhaust nozzles, AIAA-2015-1670.

Crowe D S, Martin C L. 2019. Hot streak characterization of high-performance double-serpentine exhaust nozzles at design conditions. Journal of Propulsion and Power, 35(11): 1-11.

Cui X, Xu Z H. 2018. Aero-engine exhaust nozzle and its development trend. Shandong Industrial Technology, (03): 234.

J. Shi et al., *Special Purpose Nozzles*,
https://doi.org/10.1007/978-981-95-2939-1

Deere K A, Asbury S C. 1996. An Experimental and computational and computational investigation of a translating throat single expansion ramp nozzle. AIAA 96-2540.
Dusa D J. 1987. High Mach propulsion system installation and exhaust system design considerations. AIAA-1987-2941.
Eric G, Dan H. 2004. Improving off-design nozzle performance using fluidic injection. AIAA 2004-1206.
J. I. Farah. 2007. Torque load transfer attachment hardware. US: 2007/0158527.
Frederic C S, Crossley W A. 2000. Investigating response surface approaches for drag optimization of a subsonic turbofan nacelle. AIAA 2000-4797.
Frost T P, Bishop R A. 1965. Vectored thrust engines for single and multi-engined aircraft. Journal of Aircraft, 2(1):13-19.
Gan J, Zhang J. 2016. Research status and development of stealth target detection technology. Modern Radar, 38(8): 13-16.
Gao Y X, Jin C J, Xiao Y L. 1995. Thrust vector control and thrust vectoring nozzle. Flight Dynamics, 13(2): 1-5, 12.
Gao X, Shi Y Q, Yang Q Z, et al. 2015. Influence of dielectric coating position on electromagnetic scattering characteristics of double S-duct exhaust system. Acta Physica Sinica, 64(2): 1-10.
Grarnland T, Berens T. 1995. Nozzle/Afterbody Integration for hypersonic vehicles by means of secondary air injection. AIAA 1995-6050.
Gridley M C, Walker S H. 1996. Inlet and nozzle technology for 21st century fighter aircraft. ASME Paper, 1996-GT-244.
Guo Q T, Zhang L M, Fei Q G. 2006. Research on response surface method for deterministic computational simulation and its experimental design. Acta Aeronautica et Astronautica Sinica, 27(1): 55-61.
Han J Y. 2004. Advanced mechanism theory. Beijing: China Machine Press.
Hao D X, Wang Z X. 2009. Contour design and flow field simulation for single expansion ramp nozzle with downward-sloping adjustable flap. Machinery Design & Manufacture, (12): 8-10.
He W, Xue W D, Tang B. 2012. Optimization experimental design method and data analysis. Beijing: Chemical Industry Press.
Hooker S. 1981. History of the Pegasus vectored thrust engine. Journal of Aircraft, 18(5):322-326.
Huang Q J, Liu Z C. 2013. Discussion on application of aircraft rearward infrared stealth technology. Aircraft Design, 33(1): 10–14.
Jane's Aero-Engines. 2010. Saturn RD-41. Russian Federation.
Jane's Aero-Engines. 2014. AMNTK Soyuz R79 and R179. Russian Federation.
Jiang X X. 2006. Research on shock wave induced boundary layer separation. Harbin: Master's thesis, Harbin Engineering University.
Johansson M. 2006. FOT25 2003-2005 Propulsion integration—final report.
Jürgen G, Alfred E B. 1998. Passive control for the off-design performance of a single expansion-ramp nozzle. AIAA1998-1602.
Kentfield J A C. 1967. Nozzles for jet-lift V/STOL aircraft. Journal of aircraft, 04(4):283-291.
Knight D, Yan H, Panaras A G, et al. 2003. Advances in CFD Prediction shockwave turbulent boundary interaction. Progress in Aerospace Sciences 39 (2-3): 121-184.
Laughrey J A, Drape DJ, Hiley P E. 1979. Performance evaluation of an air vehicle utilizing non-axisymmetric nozzles. AIAA-1979-1811.
Le G G, Ma D W, Ma Y Q, et al. 2003. Numerical simulation of supersonic jets of asymmetric nozzles. Journal of Experiments in Fluid Mechanics, 17(4):17–21.
Lee C C, Louis M S, Boedicker C. 1985. Subsonic diffuser design and performance for advanced fighter aircraft. AIAA-85-3073.
Lewis W D. 2000. Fundamentals of V/STOL design with joint strike fighter applications: an international cooperative effort. AIAA-2000-0805.
Li Y F, Yang Q Z, Huan X, et al. 2014. Effect of exit aspect ratio on radar cross section of serpentine two-dimensional convergent nozzle. Journal of Aerospace Power, 29(3): 645-651.
Lian X C. 2005. Principles of aero-engine. Xi'an: Northwestern Polytechnical University Press.

Liu A H. 2007. A study on the contour design and flow control of a single expansion ramp nozzle. Xi'an: PhD thesis, Northwestern Polytechnical University.
Liu D X. 2002. Aero engine design manual: Intake and exhaust devices (Vol. 7). Beijing: Aviation Industry Press.
Liu H. 2004. Development status of hypersonic vehicles. Modern Military, (12): 20-23.
Liu A H. 2007. Contour design and flow control research for single expansion ramp nozzle. Master's thesis, Northwestern Polytechnical University.
Liu Q L. 2010. Theoretical mechanics. Beijing: Science Press.
Liu S. 2016. Performance simulation of propulsion system for short take-off and vertical landing aircraft and research on design technology of three-bearing swivel nozzle. Xi'an: PhD thesis, Northwestern Polytechnical University.
Liu Y H, Chen Z Y. 2015. Effect of asymmetric lobe external expansion angle on performance of serpentine two-dimensional nozzle. Journal of Propulsion Technology, 36(11): 1616-1623.
Liu X L, Guo B. 2009. Experimental design method applicable to complex system simulation tests. Journal of National University of Defense Technology, 06: 95-99.
Liu A H, Wang Z X. 2007. Influence of secondary flow injection on nozzle flow field performance. Journal of Propulsion Technology, 28(2): 144-147.
Liu X L, Chen Y W, Jing X R, et al. 2011. Optimal Latin hypercube experimental design method and its application. Journal of National University of Defense Technology, 05: 73-77.
Liu C C, Ji H H, Huang W, et al. 2013. Numerical investigation on infrared radiation characteristics of a double S-duct two-dimensional nozzle. Journal of Aerospace Power, 28(7): 1482-1488.
Liu Z W, Li R Y, Liu S, et al. 2014. Contour design and motion law research of three-bearing swivel nozzle. Machinery Design & Manufacture, (07): 65-67.
Liu S, Wang Z X, Zhou L, et al. 2015. Vector deflection law and flow field characteristics of three-bearing swivel nozzle. Journal of Propulsion Technology, 36(5): 656-663.
Lockwood F C, Shah N G. 1981. A new radiation solution method for incorporation in general combustion prediction procedures. Symposium (International) on Combustion, 18(1): 1405–1414.
MacLean M K. 1993. Static internal performance tests of single expansion ramp nozzle concepts designed with considerations. AIAA 93-2429.
Mahulikar S P, Sonawane H R, Rao G A. 2007. Infrared signature studies of aerospace vehicles. Progress in Aerospace Sciences, 43(7): 218–245.
Mi D, Yin Z Y, Qian Z M, et al. 2012. Structural optimization design method for centrifugal impeller based on experimental design and support vector machine. Journal of Aerospace Power, 10: 2336-2341.
Miyagi H, Kimura H, Kishi K, et al. 1998.Combined cycle engine research in Japanese HYPR program. AIAA 98-3728.
Modest M F. 2013. The radiative heat transfer. New York: Academic Press.
Palmer K D. 1998 Data collection plans and metamodels for chemical process flowsheet simulators. Atlanta: Georgia Institute of Technology.
Perry J D. 2011. Navy unmanned air systems, AIAA-2011-6948.
Pomraning G C. 1973. The equations of radiation hydrodynamic. New York: Pergamon Press.
Ransone R. 2002. An overview of VSTOL aircraft and their contributions. AIAA-2002-5979.
Rao G A, Mahulikar S P. 2002. Integrated review of stealth technology and its role in airpower. Aeronautical Journal New Series, 106(1066): 629–642.
Rao G A, Mahulikar S P. 2005. Aircraft powerplant and plume infrared signature modelling and analysis. 43rd AIAA Aerospace Science Meeting and Exhibit, Reno, Nevada, AIAA 2005–221.
Saddington A J, Knowles K, Cabrita P M. 2009. Flow measurements in a short takeoff, vertical landing fountain: splayed jets. Journal of aircraft, 46(3): 874-882.
Sang J H. 2013. Aircraft stealth technology. Beijing: Aviation Industry Press.
Sang J H, Zhang Z B. 2013. Development trend of infrared stealth technology. Infrared and Laser Engineering, 42(1): 14-19.

Sang X Y, Ji H H, Wang D. 2019. Effects of length-to-diameter ratio and offset-to-diameter ratio on performance of double S-shaped two-dimensional nozzle. Infrared Technology, 41(5): 443-449.

Sehonlau M. 1997. Computer experiments and global optimization. Waterloo, Ontario, The University of Waterloo.

Siegel R, Howell J R. 2002. Thermal radiation heat transfer. New York: Taylor & Francis.

Simmons R J. 2009. Design and control of a variable geometry turbofan with an independently modulated third stream. US: The Ohio State University.

Simpson T W, Allen J K, Mistree F. 1998. Spatial correlation metamodels for global approximation in structural design optimization. Proceedings of ASME 1998 Design Engineering Technical Conferences, 13–16.

Smith M S, Chawla K, Vandalsem W R. 1991. Numerical simulation of a complete STOVL aircraft in ground effect. AIAA-91-3293.

Song X B, Lü X Y, Zhang J J. 2012. Research on aircraft infrared stealth technology. Laser & Infrared, 42(1): 3-7.

Steelant J. 2008. Achievements obtained for sustained hypersonic flight within the LAPCAT Project. AIAA 2008-2578.

Sun X L. 2018. Research on design and performance evaluation method of low observable serpentine nozzle. Xi'an: PhD thesis, Northwestern Polytechnical University.

Sun Xiao-lin, Wang Zhan-xue, Zhou Li, et al. 2016. Influences of design parameters on a double serpentine convergent nozzle. ASME Journal of Engineering for Gas Turbines and Power, 138(7): 072301.

Sun Xiao-lin, Wang Zhan-xue, Zhou Li, et al. 2018. Flow characteristics of double serpentine convergent nozzle with different inlet configuration ASME Journal of Engineering for Gas Turbines and Power, 140(8): 082602-1-082602-12.

Suo D J, Liang C H, Zhang S F, Liu J, et al. 2014. Technology research on S/VTOL fighter and its propulsion system. Aeroengine, 40(4): 7-13.

Tan H P. 2006. Numerical calculation of infrared radiation characteristics and transmission. Harbin: Harbin Institute of Technology Press.

Wang X Y. 2006. Fundamentals of gas dynamics. Xi'an: Northwestern Polytechnical University Press.

Wang Y M. 1988. A set of formulas for calculating the rotation of a rigid body about several axes. Journal of Taiyuan Institute of Machinery, 02: 99-106.

Wang Y F, Wang C G. 2005. Theory and application of response surface methodology. Journal of Minzu University of China, 14(3): 236-240.

Wang Z X, Liu A H, Cai Y H. 2007. Contour design and flow field simulation of single expansion ramp nozzle for hypersonic propulsion system. Gas Turbine Experiment and Research, (01): 8-12.

Wang Z X, Liu S, Zhou L. 2014. Development and key technology analysis of thrust vectoring nozzle for S/VTOL fighter, Aeroengine, 40(04): 1-6.

Wang D, Ji H H, Liu C C, et al. 2017. Effect of cross-section variation type on performance of double S-shaped two-dimensional exhaust system. Journal of Propulsion Technology, 38(1): 47-53.

Watanabe Y, Miyagi H, Sekido H, et al. 1993. Conceptual Design Study on Combined-Cycle Engine for Hypersonic Transport. ISABE 93-7018.

Wen Cheng, Zhanxue Wang, Li Zhou et al. 2017. Influences of shield ratio on the infrared signature of serpentine nozzle. Aerospace Science and Technology, 71: 299-311.

Wen Cheng, Zhanxue Wang, Li Zhou, et al. 2019. Investigation of Infrared Signature of Serpentine Nozzle for Turbofan. Journal of Thermophysics and Heat Transfer, 33(1): 170-178.

Wen X Q. 2001. Brief discussion on "B-2" stealth strategic bomber. Modern Defence Technology, 29(4): 1-4.

Woodward C S. 1987. Vectorable exhaust nozzle for a gas turbine engine. England: 4679732.

Xing Y L, Dong Z X. 2014-07-01. "Neuron", the rise of European UAV. China National Defense News, 13. (Note: Newspaper handled as per available info)

Xu D G, Sang J H, Luo M D. 2012. Application research of infrared stealth technology on UAV. Infrared and Laser Engineering, 41(12): 3154-3159.

Yang S X. 1985. Dynamic characteristic analysis and calculation of full-axis gimbaling flexible nozzle. Journal of Astronautics, 01: 59-77.

Yang S X. 1989. Identification of moment characteristics of flexible nozzle. Journal of Astronautics, 04: 16-22.

Yang S X. 1993. Design and analysis of full-axis gimbaling ball-and-socket nozzle. Journal of Solid Rocket Technology, 03: 23-30.

You J F, Zhang Y J, Chang X L. 2001. Contact stress and gimbal moment calculation of ball-and-socket nozzle. Journal of Solid Rocket Technology, 04: 20-24.

Yue G G, Ma D W, Ma Y Q, et al. 2003. Numerical simulation of underexpanded supersonic jet from asymmetric nozzle. Experiments and Measurements in Fluid Mechanics, 17(4). (Note: Issue number might be missing)

Yungster S, Trefny C J. 1994. Computational study of single-expansion-ramp nozzles with external burning. AIAA 1994-0024.

Zha L. 2002. "Reborn" "Mysterious Night Eagle" - F-117 dominating the Gulf War. National Defense Science & Technology, 6-12.

Zhang K S. 2005. Research on multidisciplinary optimization methods for aircraft design. Xi'an: PhD thesis, Northwestern Polytechnical University.

Zhang S L, Shan Y, Zhang Y, et al. 2012a. Numerical investigation on influence of slotted expansion surface on performance of single expansion ramp nozzle. Journal of Propulsion Technology, 33(3): 436-442.

Zhang L H, Xu J L, Mo J W, et al. 2012b. Experimental investigation of adjustable schemes for asymmetric two-dimensional nozzle. Acta Aeronautica et Astronautica Sinica, 34(3): 772-778.

Zhang W R, Ai J Q, Cui L. 2014. Analysis of infrared stealth technology for aircraft exhaust system. Aeronautical Science & Technology, 12: 5–9.

Zhou X M. 2017. Research on aerodynamic and infrared radiation characteristics of low observable double S-duct nozzle. Nanjing: PhD thesis, Nanjing University of Aeronautics and Astronautics.

Zhou L, Wang Z X, Xiao H, et al. 2015. Performance analysis of single expansion ramp nozzle with passive cavity structure. Journal of Engineering Thermophysics, 36(7): 1456-1460.

Zucrow M J, Hoffman J D. 1984. Gas dynamics. Beijing: National Defense Industry Press. (Note: Western names preserved as per original)

MIX
Papier aus verantwortungsvollen Quellen
Paper from responsible sources
FSC® C105338

If you have any concerns about our products,
you can contact us on
ProductSafety@springernature.com

In case Publisher is established outside the EU,
the EU authorized representative is:
Springer Nature Customer Service Center GmbH
Europaplatz 3, 69115 Heidelberg, Germany

Printed by Libri Plureos GmbH
in Hamburg, Germany